STAR NAMES
Their Lore and Meaning

(*formerly titled: Star-Names and Their Meanings*)

by

Richard Hinckley Allen

Dover Publications, Inc., New York

Published in Canada by General Publishing Company, Ltd., 30 Lesmill Road, Don Mills, Toronto, Ontario.

Published in the United Kingdom by Constable and Company, Ltd., 10 Orange Street, London WC 2.

This Dover edition, first published in 1963, is an unabridged and corrected republication of the work first published by G. E. Stechert in 1899, under the former title: *Star-Names and Their Meanings.*

International Standard Book Number: 0-486-21079-0
Library of Congress Catalog Card Number: 63-21808

Manufactured in the United States of America

Dover Publications Inc.
180 Varick Street
New York 14, N. Y.

CONTENTS

I GRATEFULLY DEDICATE THESE PAGES
TO THE MEMORY OF

HUBERT ANSON NEWTON

AND

WILLIAM DWIGHT WHITNEY

SENIOR PROFESSORS IN YALE UNIVERSITY

WHO FIRST ENCOURAGED ME
IN MY WORK

R. H. A.

"Wilt thou lere of sterres aught ?

. . . . . .

Elles I wolde thee have told,"
Quod he, "the sterres names, lo,
And al the hevenes signes to,
And which they ben."

Dan Geoffrey Chaucer's *Hous of Fame*.

INTRODUCTION.

This list of star-names is published in the endeavor to fill an acknowledged vacancy in our popular astronomical literature. It is not intended for the professional astronomer, who, as a rule, cares little about the old designations of the objects of his study,— alphabets, numerals, and circles being preferable, indeed needful, for his purposes of identification. Yet great scholars have thought this nomenclature not unworthy their attention, — Grotius, Scaliger, Hyde, and our own Whitney, among others, devoting much of their rare talent to its elucidation; while Ideler, of a century ago, not without authority in astronomy as in other branches of learning, wrote as to inquiry into star-names:

This is, in its very nature, coincidently a research into the constellations, and it is so much more worth while learning their history as throughout all ages the spirit of man has concerned itself with a subject that has ever had the highest interest to him,— the starry heavens.

Old Thomas Hood, of Trinity College, Cambridge, in 1590 asserted that they were "for instruction's sake . . . things cannot be taught without names"; and it is certain that knowledge of these contributes much to an intelligent pleasure when we survey the evening sky. For almost all can repeat Thomas Carlyle's lament:

Why did not somebody teach me the constellations, and make me at home in the starry heavens, which are always overhead, and which I don't half know to this day ?

Naturally these titles are chiefly from the Arabs, whose Desert life and clear skies made them very familiar with the stars, as Al Bīrūnī [1] wrote:

> He whose roof is heaven, who has no other cover, over whom the stars continually rise and set in one and the same course, makes the beginnings of his affairs and his knowledge of time depend upon them.

So that the shaykh Ilderim well told Ben Hur at the Orchard of Palms:

> Thou canst not know how much we Arabs depend upon the stars. We borrow their names in gratitude, and give them in love.

But many star-names supposed to have originated in Arabia are merely that country's translations of the Greek descriptive terms, adopted, during the rule of the Abbasids,[2] from Claudius Ptolemy's Ἡ Μεγάλη Σύνταξις τῆς Ἀστρονομίας, the Great System of Astronomy, of our second century. For it was early in this khalifate,

> in the golden prime
> Of good Haroun Alraschid

(Aaron the Just), that Ptolemy's Σύνταξις [3] was translated as *Al Kitāb al Mijisti*, the Greatest Book. This, in its various editions, substituted among the educated classes a new nomenclature; while, as revised by Al Thabit ibn Ḳurrah in the latter part of the 9th century, it eventually became, through a Latin version by Cremonaeus (Gerard of Cremona) of the 12th century, the groundwork of the first complete printed *Almagest*. This, published at Venice in 1515, so manifestly showed its composite origin that Ideler and Smyth always referred to it as the *Arabo-Latin Almagest*. The Greek text of the *Syntaxis* seems to have been practically unknown in Europe until translated into Latin from a Vatican manuscript by Trapezuntius (the monk George of Trebizond), several editions

[1] This was the celebrated Khorasmian Abū Raiḥān Muḥammad ibn Aḥmad of A. D. 1000, whose designation in literature came from his birthplace, a *bīrūn*, or suburb, of Khwārizm. His *Vestiges of Past Generations*, a chronology of ancient nations, and his *India*, are of interest and authority even now.

[2] This first organized government among the Arabs began in 749, and under "its enlightened and munificent protection Baghdad soon became what Alexandria had long ceased to be."

[3] This was subsequently designated as *Ἡ Μεγίστη* to distinguish it from his smaller astrological work in four books, the *Τετράβιβλος Σύνταξις*. Our word *Almagest* is now supposed to be composed of the principal letters of the Greek title.

of this issuing during the 16th century. From all these and kindred works have come the barbarous Graeco-Latin-Arabic words that, in a varied orthography, appear as star-names in modern lists.

But there were other purely indigenous, and so very ancient, titles from the heathen days of the Ishmaelites anterior to Mediterranean influences, perhaps even from the prehistoric "'Arab al Baidā," the Arabs of the Desert,—these titles generally pastoral in their character, as accords with such an origin. So that we find among them the nomads' words for shepherds and herdsmen with their maidens; horses, horsemen, and their trappings; cattle, camels, sheep, and goats; predatory and other animals; birds and reptiles. It should be remembered, however, that the archaic nomenclature of the Arabs—archaic properly so called, for we know nothing of its beginnings—in one respect is unique. They did not group together several stars to form a living figure, as did their Western neighbors, who subsequently became their teachers; single stars represented single creatures,—a rule that seems rarely to have been deviated from,—although the case was different in their stellar counterparts of inanimate objects. Even here they used but few stars for their geographical, anatomical, and botanical terms; their tents, nests, household articles, and ornaments; mangers and stalls; boats, biers, crosses, and thrones; wells, ponds, and rivers; fruits, grains, and nuts;—all of which they imaged in the sky.

They had, too, still another class of names peculiar to themselves, such as Al Ṣaidak, Al Simāk, Al Suhā, respectively the Trusted One, the Lofty One, the Neglected One; their Changers, Drivers, Followers, and Wardens; their Fortunate, or Unfortunate, Ones, and their Solitary Ones, etc. None of these early asterisms, however, were utilized by the scientific Arabians, but, with their titles, became merely interesting curiosities to them, as to us. These were known as "of the Arabs," while Ptolemy's figures were "of the astronomers,"—a distinction maintained in this book by the use of "Arab" or "Arabic" for the first, and "Arabian" for the last. The Persian astronomical writer, the dervish 'Abd al Raḥmān Abū al Ḥusain, now better known as Al Sufi,[1] the Mystic or Sage, made mention of this early distinction, in

[1] Al Sufi also was known as Al Razi, from his birthplace, Al Rayy, east of Teheran. A French translation of his work was published in 1874 by the late H. C. F. C. Schjellerup of Saint Petersburg.

964, in his *Description of the Fixed Stars;* Kazwini following, three centuries later, with the same expressions.

The various Arabic titles that we see applied to a single star or group, and the duplicate titles for some that are widely separated in the sky, apparently came from the various tribes, each of which had to a certain extent a nomenclature of its own.

The rest of our star-names, with but few exceptions, are directly from Greek or Latin originals,— many of these, as is the case with the Arabian, although now regarded as personal, being at first only adjectival or merely descriptive of the star's position in the constellation figure; while some are the result of misunderstanding, or of errors in translation and oft-repeated transcription. But these are now too firmly established to be discontinued or even corrected.

Vergil wrote in the *1st Georgic:*

> Navita tum stellis numeros et nomina fecit;

and Seneca, the traditional friend of Saint Paul, in his *Quaestiones Naturales:*

> Graecia stellis numeros et nomina fecit;

both of these heathen authors almost exactly following the words of the sacred psalmist, who, at least four hundred years before, had sung:

> He telleth the number of the stars;
> He giveth them all their names,

and of the prophet Isaiah:

> He calleth them all by name.

While Seneca's statement may have some foundation, and Vergil's assertion as to the sailor's influence in star-naming may be true in part, yet for most of this we should probably look to the Desert, where the stars would be as much required and relied upon for guidance as on the trackless ocean, and so necessarily objects of attentive interest and study. Indeed, Muḥammad told his followers, in the 6th *Sura* of the *Ḳur'ān:*

> God hath given you the stars to be your guides in the dark both by land and sea.

It seems safe to conclude that they were first named by herdsmen, hunters, and husbandmen, sailors and travelers, — by the common people generally, rather than by the learned and scientific; and that our modern lists are the gradual accumulation of at least three thousand years from various nations, but chiefly from the nomads, as well as the scholars, of Arabia,—

> those earthly godfathers of heaven's lights,
> That give a name to every fixed star,—

and from Greece and Rome.

It may be thought that too much attention has been paid to stellar mythology, now almost a hackneyed subject; but it serves to elucidate the literary history of the stars, and the age of its stories commands at least our interest. Indeed, we should remember that the stars were largely the source of these stories,— Eusebius, early in our 4th century, asserting in his *Praeparatio Evangelica:*

> The ancients believed that the legends about Osiris and Isis, and all other mythological fables [of a kindred sort], have reference either to the Stars, their configuration, their risings and their settings, etc.

And Proctor wrote in his *Myths and Marvels of Astronomy* that the chief charm of this study

> does not reside in the wonders revealed to us by the science, but in the lore and legends connected with its history, the strange fancies with which in old times it has been associated, the half-forgotten myths to which it has given birth.

Yet these myths, old as the present forms of some of them may be, are but modern and trivial when one goes back into the dim past to their probable fountainhead among the Himalayas and on the Ganges, or along the banks of the Euphrates, where the recent study of mythology discovers their origin in serious connection with the most ancient of earthly religions, long antedating Moses,— "attempted explanations of natural phenomena," drawn from observations on the earth and in the sky of the powers of nature and of nature's God.

The world-wide field of research that I have endeavored to traverse, containing the records of four or five millenniums, it need hardly be said demands for its exploration the best efforts, long continued, of the scientist and scholar accomplished in archaeology, astronomy, literature, and philology. None such, however, has appeared since Ideler's day, nearly a century ago; so that, with the desire of taking up again this most interesting task, and the hope of thus stimulating others more competent to carry it on, I have done what I could, although frankly confessing that I have fallen very far short of my ideal. Originality is not claimed for my book Much of it has been gathered from widely scattered sources, brought together here for the first time in readily accessible form, although doubtless with errors and certainly with much omission; for while I have sought, as did Milton's *Il Penseroso*, to

> sit and rightly spell
> Of every star that heav'n doth show,

yet in preparing my material I have seen, as Doctor Samuel Johnson wrote in the preface of his *Dictionary*,

> that one enquiry only gave occasion to another, that book referred to book, that to search was not always to find, and to find was not always to be informed.

So that, following him,

> I set limits to my work, which would in time be ended though not completed.

While to temper such criticism as may be bestowed upon my efforts, I quote again from the same source:

> Dictionaries[1] are like watches; the worst is better than none, and the best cannot be expected to go quite true.

Doctor Christian Ludwig Ideler's *Untersuchungen über den Ursprung und die Bedeutung der Sternnamen*, dated in Berlin the 2d of April, 1809, is the main critical compendium of information on stellar names — Arabic, Greek, and Latin especially. It is to him that we owe the translation of the

[1] It is greatly to be regretted that our dictionaries are, without exception, singularly unsatisfactory as to star-titles, being always deficient and too often erroneous. The recent *Century Cyclopedia of Names*, however, contains the most correct, detailed, concise, and scholarly list that we have.

original Arabic text of Kazwini's[1] *Description of the Constellations*, written in the 13th century, which forms the basis of the *Sternnamen*, with Ideler's additions and annotations from classical and other sources. From this much information in my book is derived.

The *Bedford Catalogue* in Captain (afterwards Vice-Admiral) William Henry Smyth's[2] *Cycle of Celestial Objects*, a book of exceptional value as to information on star-names and unique in its racy style, also has been drawn from.

Sir Joseph Norman Lockyer's recent *Dawn of Astronomy* — a most interesting work even if all his deductions are not accepted — has furnished many of the references to Egypt and its temple worship of various stars; this new study in orientation having been initiated by Professor Nissen of Germany, although independently so, about the same time, by Lockyer.

Professor D'Arcy Wentworth Thompson's *Glossary of Greek Birds* has been utilized as to the ornithological symbolism[3] on early coinage, sculpturing, etc.; for this, hitherto unintelligible, is now thought to be largely astronomical.

The details of star-spectra mainly are from the *Spectralanalyse der Gestirne*, of 1890, by Doctor J. Scheiner, of the Royal Astrophysical Observatory of Potsdam, translated by Professor E. B. Frost, of Dartmouth College, in 1894.

The matter connected with the astronomy of China is chiefly from Mr. John Williams' work of 1871,— the *Observations of Comets from 611 B. C. to A. D. 1640, extracted from the Chinese Annals*,— the star-names being

[1] His customary designation is from his birthplace, Ḳazwīn, in northern Persia, and has been variously given; Smyth *abbreviating* it to 'Omadu-d-dīn Abu Yahya Zakariyā Ibn-Mahmūd Ansārī al-Kazwīnī. The name is correctly written Zaḳariyā ibn Muḥammad ibn Maḥmūd al Ḳazwīnī. He was collaborator with his noted fellow-countryman Naṣr al Dīn al Ṭūsī, who, in 1270, compiled the *Ilkhanian Tables*, used in Persia perhaps to the present day.

[2] It is pleasant to us Americans to know that Smyth was a lineal descendant of Captain John Smith of Virginia fame; and of interest to all New Jersey people that his father was from the province of East Jersey, but, as a loyalist in our Revolution, was compelled to flee to England, where the son was born in 1788. He died, in 1865, after a most useful and distinguished career in the British navy and as astronomer and hydrographer.

[3] This subject originally was broached by Gorius, in 1750, in his *De Gemmis Astriferis*; and Dupuis treated of it, although in an exaggerated way, a century ago.

from that or from Mr. John Reeves' Appendix [1] to Volume I, Part 2, of the Reverend Doctor Robert Morrison's *Dictionary*, published at Macao in 1819, with Bode's star-numbers. I have also been aided by the Reverend Doctor Joseph Edkins' recent papers in the *China Review*. The translations of the names in Reeves' list are by Professor Kazutami Ukita, of the Doshisha Theological School of Kyoto, Japan; but he expresses misgivings as to the correctness of many of them in their stellar application.

Professor Richard J. H. Gottheil, of Columbia University, has very kindly supervised the transcription and translation of the Hebrew and Arabic star-names, and has added the table of the Arabic alphabet and the English equivalents of its letters. But his absence abroad while the earlier pages were going through the press will account for some errors, which, however, I have endeavored to correct in the Index. The Euphratean [2] titles are from various sources.

The star-magnitudes are from the Estimates of the *Harvard Photometry*, a list of 4260 naked-eye stars north of the 30th parallel of south declination, published in 1884 by Professor Edward C. Pickering, or from the *Uranometria Argentina* [3] of the late Doctor Benjamin A. Gould, published in 1879.

The star-maps of the northern sky to which I generally refer are those of Doctor Friedrich Wilhelm August Argelander in his *Uranometria Nova*, published at Berlin, in 1843, with 3268 stars down to the 6th magnitude; and of Doctor Eduard Heis in his *Atlas Coelestis Novus* of 1872. But

[1] The original of Reeves' list is from the 31st volume of the *Leuh Leih Yuen Yuen*, in one hundred volumes, issued in the reign of Kang Hi, with Jesuit assistance. The early native titles seem to have been arbitrarily applied to single stars or small groups, with no apparent stellar signification.

[2] The term "Euphratean" is used throughout these pages in a general way for the material lately discovered in the Euphrates Valley, the source of which — Sumerian, Akkadian, Babylonian, Chaldaean, or Assyrian — is as yet largely undetermined. The references to this material I have taken bodily from the works of Hommel, Sayce, Strassmaier and Epping, Jensen, and Robert Brown, Junior.

[3] This great work is designed to include all stars down to the 7th magnitude in that portion of the sphere within 100° of the south pole,— the favorable atmospheric conditions at Córdoba, whence the observations were made, rendering even that magnitude readily visible. It comprises, of course, all the southern constellations, with 6733 stars, and those parts of the northern, with 997 stars, that lie below the 10th degree of north declination,— 66 constellations in all, with 7730 stars.

the last-named acute observer includes those to the 6½ magnitude[1] — 5421 stars from the pole to 40° of south declination, in eight tenths of the heavens. Smyth more conservatively wrote of this oft-mooted point in observational astronomy:

The number of those seen by the naked eye at once is seldom much above a thousand; though from their scintillation, and the indistinct manner in which they are viewed, they appear to be almost infinite. Indeed, albeit the keen glances of experience might do more, the whole number that can be generally perceived by the naked eye, taking both hemispheres, is not greatly above three thousand, from the first to the sixth magnitudes, in about these proportions:

I	II	III	IV	V	VI
20	70	220	500	690	1500,—

3000 in all. Professor David P. Todd, in his *New Astronomy* of 1897, increases the number of 5th-magnitude stars to 1400, and of those of the 6th magnitude to 5000,— 7185 in all; but exceptional conditions of eyesight and atmosphere probably must exist for confirmation of this.

The star-colors generally are from Smyth's list whenever noted by him; but it should be remembered that even good authorities sometimes differ as to stellar tints, and those assigned here will not be accepted by all, and in the case of minute objects are very doubtful.

I have begun my work with brief notices of the Zodiacs,— Solar and Lunar,— that necessarily are constantly alluded to in treating of the individual Constellations; following these with three chapters on the latter,— their history among the nations, cataloguing and early treatment by authors, and their connection with astrology, art, folk-lore, literature, and religion. The detailed list of the Constellations, in alphabetical order, and of their named components follows, with the derivation, signification, and history of their titles, and some facts as to the scientific aspects of the stars. In this last feature of my book Professor Charles A. Young, of Princeton University, has afforded me much valuable assistance, for which, although very inadequately, I here return my sincere thanks. A chapter on the Galaxy ends the work.

[1] He was enabled to do this by means of special arrangements for shutting off outside light from the field of sky under view; so that the observations, although by the naked eye, were not unaided.

Where thought necessary, the accentuation of the star-titles is given in the Indices, although in some cases, from the uncertainty of origin, this may be doubtful.

In conclusion, I would acknowledge my obligations for useful suggestions to Professor Edward S. Holden, till lately the Director of the Lick Observatory; to Mr. Addison Van Name, of the Yale University Library, for access to volumes of reference and help in translations; to Messrs Theodore L. De Vinne & Co. (the De Vinne Press), for their accustomed skill in the make-up of my book; and to Mr. P. J. Cassidy, for his interest and intelligent care in its proof-reading. Lastly do I thank my young friend Miss Lucy Noble Morris, of Morristown, for long-continued aid in various ways, especially in her tasteful selection of poetical illustrations.

And now, with the hope that my work, even with its imperfections, may serve to foster a more intelligent interest in the nomenclature and "archaeology of practical astronomy," I submit it to all lovers of the stars.

RICHARD HINCKLEY ALLEN.

MEADOW VIEW,
CHATHAM, NEW JERSEY,
February 16, 1899.

HE MADE THE STARS ALSO
GENESIS i, 16

STARS INDEED FAIR CREATURES BE
HONEST GEORGE WITHER

MAKE FRIENDSHIP WITH THE STARS
MRS. SIGOURNEY

. . . a broad belt of gold of wide extent,
Wherein twelve starry animals are shown,
Marking the boundaries of Phoebus' zone.

Luiz de Camões' *Os Lusiadas.*

The Solar Zodiac.

Many theories have been propounded for the birthplace and time of formation of this; but there now seems to be general agreement of opinion that it originated, mainly as we have it, in archaic Euphratean astronomy, possibly with only the six alternate signs, Taurus, Cancer, Virgo, Scorpio, Capricornus, and Pisces, and later divided because of the annual occurrence of twelve full moons in successive parts of it. Yet Servius, about A. D. 400, said that for a long time it consisted of but eleven constellations, Scorpio and its claws being a double sign, this characteristic feature descending to Greece and Rome.

Riccioli, about 1650, cited as a "Chaldean" title **Hadronitho Demalusche,** or Circle of the Signs; but this must be taken with much allowance,[1] for in his day Babylonian study had not begun, while modern scholars think that it was known to the Akkadians as **Innum,** and as **Pidnu-sha-Shame,** the Furrow of Heaven, ploughed by the heavenly Directing Bull, our Taurus, which from about 3880 to about 1730 B. C. was first of the twelve.

Although our knowledge of that country's astronomy is as yet limited, it is certain that the Akkadian names of the months were intimately connected with the divisions of this great circle; the calendar probably being taken from the stars about 2000 B. C., according to Professor Archibald Henry Sayce, of Oxford. Thence it passed to the Jews through Assyria and Aramaea, as the identity of its titles in those countries indicates; and the eleven, or twelve, signs for a time became with that people objects of idolatrous worship, as is evident from their history detailed in the *2d Book of the Kings*, xxiii, 5.

In the Babylonian *Creation Legend*, or *Epic of Creation*, discovered by

[1] In fact, the same caution may be exercised in regard to much of the Euphratean transcription and translation throughout this work, as well as of the Chinese.

George Smith in 1872,[1] the signs were **Mizrātā,**— a very similar word appears for the Milky Way,— generally supposed to be the original of the biblical **Mazzārōth**; **Mazzālōth** being the form used in the *Targums* and later Hebrew writings. This word, although of uncertain derivation, may come from a root meaning "to watch," the constellations thus marking the watches of the night by coming successively to the meridian; but Doctor Thomas Hyde,[2] the learned translator at Oxford in 1665 of the *Zīj*, or *Tables*, of Ulug Beg, and of Al Tizini's work, derived them from Ezor, a Girdle; while the more recent Dillmann referred them to Zāhir, from Zuhrah, a Glittering Star, and so signifying something specially luminous. Still this Bible word has been variously rendered, appearing for the Greater Bear, Sirius, the planets, or even for the constellations in general; indeed it has been thought to signify the Lunar Mansions.

Another name with the Jews for the zodiac was **Galgal Hammazālōth,** the Circle of the Signs; and Bayer said that they fancifully designated it as **Opus Phrygionarum,** the Work of the Phrygians, *i.e.*, of the embroiderers in gold.

The Jewish historian Flavius Josephus, followed by Saint Clement of Alexandria, A. D. 200, surmised that the twelve stones in the breastplate of the high priest might refer to the twelve zodiacal constellations. Philo Judaeus, of about the same time, associated the latter with the stars of Joseph's dream; the modern poet Johann Christoph Friedrich Schiller, in *Die Piccolomini*, thus alluding to the ancient opinion as to its sacred character:

> Twelve! twelve signs hath the zodiac, five and seven,
> The holy numbers include themselves in twelve;

while Smyth wrote:

> The allegorical images of Jacob's blessing have been identified by several writers with the signs of the Via Solis, whence God, as bow-man, becomes Sagittarius. Hebrew antiquaries have long recognized Enoch as inventor of the Dodecatemory divisions; and both Berosus [Berōssōs as now written,— the Chaldaean historian of about 260 B. C.] and Josephus declare that Abraham was famous for his celestial observations,

and even taught the Egyptians.

As to this last people, while our twelve figures appear on the Denderah

[1] This was found on tablets of the reign of As-sur-ba-ni-pal, 600 B. C., although supposed to have been originally composed about 2350 B. C.: a supposition confirmed by Père Scheil, who recently has found a fragment of this legend on a tablet bearing the name of Am-mi-za-du-ga, King of Babylon, 2140 B. C.

[2] It was this Doctor Hyde who first described the wedge-shaped characters of the Persepolis inscriptions by the term *cuneiformes*, now a word of universal acceptation.

planisphere doubtless from Greek or Roman influence, we have little knowledge as to what was the zodiac of their native astronomy, although it perhaps represented their twelve chief divinities; and Saint Clement tells us that the White, or Sacred, Ibis, *Ibis aethiopica* or *religiosa*, was its emblem. The Jesuit Father Athanasius Kircher,[1] 1602–1680, has left to us its separate Coptic-Egyptian titles in the Greek text, with their supposed significations in Latin; but these, presumably translations from the originals, are not lexicon words. Among them, for the zodiac itself, is Ταμετοῦρο εντενίφθα, whatever that may be. But Miss Agnes M. Clerke says that when Egypt adopted the Greek figures it was with various changes that effaced its character as "a circle of living things."

In Arabia the zodiac was **Al Minṭakah al Burūj**, the Girdle of the Signs, that Bayer quoted as **Almantica** *seu* **Nitac**; and, more indefinitely, it was **Al Falaḳ**, the Expanse of the Sky.

In Greece it was *τα Δωδεκατημόρια*, the Twelve Parts, and ὁ Ζωδιακός Κύκλος; but Aristotle, the Humboldt of the 4th century before our era, called it ὁ Κύκλος τῶν Ζωδίων,[2] the Circle of Little Animals, the signs before Libra was introduced being all of living creatures. The German **Thierkreis** has the same signification. Proclus of our 5th century called it ὁ Λοξός Κύκλος, the Oblique Circle, that originally was for the ecliptic; but with Aratos, who regarded the claws as distinct from Scorpio, it was *τα* Εἴδωλα δυοκαίδεκα, the Twelve Images. As Homer and Hesiod made no allusion to it, we may consider as in some degree correct the statement that another poet, Cleostratos of Tenedos, made it known in Greece about 500 B.C., from his observations on Mount Ida.

In Rome it commonly was **Zodiacus**; the **Orbis** *qui Graecē* Ζωδιακός *dicitur* of Cicero's *De Divinatione;* and the **Orbis signiferus**, or **Circulus signifer,** of Cicero and Vitruvius, the Sign-bearing Circle, that became **Signiportant** in the *Livre de Creatures*, the 12th-century Anglo-Norman poem of Philippe de Thaun. Poetically it was **Media Via Solis** and **Orbita Solis**; the **Balteus stellatus** of Manilius, the Starry Belt; and the **varii Mutator Circulus anni** of Lucan.

Bayer's **Sigillarius** probably is a Low Latin word for the Little Images; and he quoted **Limbus textilis,** the Woven Girdle, and **Fascia,** the Band, that Ptolemy used for the Milky Way.

[1] Kircher was a distinguished mathematician and scholar to whom, as also to Roger Bacon of four centuries previously, is attributed the invention of the magic lantern. In Samuel Boteler's celebrated poem *Hudibras*, 1663–1678, he is alluded to as "the Coptic priest Kircherus." It was he who began the modern study of the Egyptian hieroglyphics.

[2] This is the first mention of the zodiac by any extant writer.

Chaucer's line in *Troilus and Criseyde* —

> and Signifer his candeles shewed brighte —

was borrowed from Claudian's *In Rufinum,* and referred to the sky; but the *Astrolabe* had

> This forseide hevenish zodiak is cleped the cercle of the signes.

Elsewhere he called the zodiac figures **Eyrish bestes** and the **Cercle of the Bestes,** for

> zodia in langage of Greek sowneth bestes in Latin tonge;

ζῶα, the original word in *The Revelation,* iv, 6, being translated "beasts" in our Authorized Version and "living creatures" in the Revised. Chaucer's terms may have been taken from Ovid's *Formasque ferarum.*

In manuscripts of the Anglo-Saxons it is **Mielan circul zodiacum,** the Great Zodiacal Circle, and **Twelf Tacna,** the Twelve Signs; but their descendants, our English ancestry of four or five centuries ago, knew it as the **Bestiary, Our Ladye's Waye,** and as the **Girdle of the Sky**; while the ecliptic was the Yoke of the Sky, or Thwart Circle, and the prime meridian, the Noonsteede, or Noonstead, Circle.

Milton, in *Paradise Lost,* thus accounts for the obliquity of the earth's axis, as if by direct interposition of the Creator:

> Some say, he bid his angels turn askance
> The poles of earth twice ten degrees or more
> From the sun's axle; they with labour push'd
> Oblique the centric globe: some say, the sun
> Was bid turn reins from th' equinoctial road
> Like distant breadth to Taurus with the seven
> Atlantic Sisters, and the Spartan Twins,
> Up to the Tropic Crab; thence down amain
> By Leo, and the Virgin, and the Scales,
> As deep as Capricorn, to bring in change
> Of seasons to each clime.

Pope, in his *Essay on Man,* called it the **Solar Walk,** and, before his day, its various divisions were the **Houses of the Sun,** and the **Monthly Abodes of Apollo.**

Dante Alighieri, 1265–1321, designated it

> The oblique circle which conveys the planets,.

and called it **Rubecchio,** the Tuscan word for a Mill-wheel whose various cogs were represented by the various signs, an image often made use of by the great poet. Longfellow translated this the Zodiac's Jagged Wheel. But many centuries, perhaps millenniums, before Dante the *Rig Veda* of India had

> The twelve-spoked wheel revolves around the heavens;
> 720 children in pairs [= 360 days + 360 nights] abide in it.

And again,

> The fellies are twelve; the wheel is one; within it are collected 360 [spokes].

A common title for it in India was **Rāsi chakra**.

In the neighboring Persia, the *Bundehesh*, or Cosmogony, in the Pahlavi dialect, of about the 8th or 9th century, a queerly mixed farrago of Persian and Semitic words, mentions our zodiacal divisions as the **Twelve Akhtārs** that lead the army of Ormuzd, while the seven Asvahtārs, or planets (including a meteor and a comet), fight for Āryamān.

But the twelve signs of that country, as also those of China and India, were gathered into four great groups marking the four quarters of the heavens, each with a Royal Star or Guardian; and the *Avesta*, or Divine Law, of Zoroaster is thought to mention a heavenly circle of figures equivalent to our zodiac.

Mr. Robert Brown, Jr., says that in China the **Kung**, or

> zodiacal signs, are the Tiger (Sagittarius); the Hare (Scorpio); the Dragon (Libra); the Serpent (Virgo); the Horse (Leo); the Ram (Cancer); the Ape (Gemini); the Cock (Taurus); the Dog (Aries); the Boar (Pisces); the Rat (Aquarius); the Ox (Capricornus). This is a *zodiac* indeed; but although the latest research [notably by the late Doctor Terrien de Lacouperie] points to a more western origin of Chinese civilization [as of about 4000 years ago], and even (a most interesting fact) to the original identity of the Chinese pictorial writing with the Akkadian Cuneiform, as both springing from one prior source, yet the Chinese Zodiac is evidently independent, and none the less so because it happens to include the Ram and the Bull, which, however, are not Aries and Taurus.

It is well shown on the Temple Money,[1] a full set of which, of uncertain age, is in my possession.

This Chinese zodiac, however, progressed in reverse order from our own, opposed to the sun's annual course in the heavens, and began with the Rat. It was known as the **Yellow Way,** the date of formation being assigned to some time between the 27th and 7th centuries before our era, and the twelve symbols utilized to mark the twelve months of the year. It was borrowed, too, by the neighboring nations ages ago, some of its features being still

[1] These are sharply minted coins, somewhat smaller than an American dime, apparently of silver and copper alloy, with a square perforation similar to that in the *tsien* or *cash*.

current among them. After the establishment in China of the Jesuits in the 16th century our zodiac was adopted, its titles being closely translated and now in current use.

In England the Venerable Bede, 673–735, substituted the eleven apostles for eleven of the early signs, as the **Corona seu Circulus sanctorum Apostolorum,** John the Baptist fitly taking the place of Aquarius to complete the circle. Sir William Drummond, in the 17th century, turned its constellations into a dozen Bible patriarchs; the Reverend G. Townsend made of them the twelve Caesars; and there have been other fanciful changes of this same character. Indeed, the Tree of Life in the *Apocalypse* has been thought a type of the zodiac, as

> bearing twelve manner of fruits, yielding its fruit every month.

Probably every nation on earth has had a solar zodiac in some form, generally one of animals. Even in Rhodesia, the aboriginal Mashona [1] Land of South Africa, there has recently been found a stone tablet thirty-eight inches in diameter, with the circle of the zodiacal signs on the edge; and early Mandaean tradition makes its figures children of their creative spirits Ur and Rūhā.

The introduction of the twelve figures into the walls or pavements of early churches, cathedrals, and public edifices, as well as, sometimes, private houses, is often to be noticed in Europe, and still more frequently in the temples of the East; [2] while all visitors to the New York State Building in the World's Columbian Exposition of Chicago in 1893 will recall the striking octagonal zodiac [3] designed by Messrs. McKim, Mead, and White, and laid in brass in the floor of the entrance hall, which, although not astronomically correct, greatly added to the interior effect of that beautiful structure.

The zodiacal constellations being of unequal extent, Hipparchos more scientifically divided the ecliptic circle into twelve equal spaces of 30° each, the twelve signs still in almanac use; but these are not now coincident with the similarly named constellations, having retrograded about 33° on the sphere since their formation.

The constellation north or south of the one of the zodiac that rose or set synchronically with it in Greece was known, in later days, as its paranatellon.

[1] This word is Anglicized from Amashuina, the Baboons, the nickname given by the Matabele to their neighbors the Makalanga, the natives of Mashona Land.

[2] Miss Clerke has much information as to this in her interesting article on the zodiac in the *Encyclopaedia Britannica*, as has Brown in the 47th volume of *Archaeologia*.

[3] This is now in the Boston Public Library.

Their number is, if you want to count them,
Twenty stars, and a number 8 after them.

An *Arabic Rhyme* quoted by Al Bīrūnī.

The Lunar Mansions

once bore an important part in observational astronomy, especially in that of Arabia, China, and India, and of Khiva — the ancient Khorasmia — and Bokhara — the ancient Sogdiana; while recent research finds them well established in the Euphrates valley, Coptic Egypt and Persia, perhaps originating in the first.

They lay for the most part along the celestial equator or in the zodiac, varying in extent, although theoretically each was supposed to represent the length of the moon's daily motion in its orbit. They sometimes were twenty-seven, but usually twenty-eight in number, the lunar month being between twenty-seven and twenty-eight days, and possibly long antedated the general constellations, or even the solar zodiac. They seem to have been among the earliest attempts at stellar science; indeed with the Khorasmians, to whom Al Bīrūnī attributed great knowledge of the stars, an astronomer was called Akhtar Wēnīk, Looking to the Lunar Stations; and they have largely been made use of in the astrology of all ages, as well as in early poetry and prose, even in Arabic doggerel.

Their astrological characters were various, eleven being considered fortunate, ten the reverse, and seven of uncertain influence; but each, at least in India, was associated with some occurrence of life. Their antiquity is proved by the fact that there, and probably elsewhere, the list began with the Pleiades, when those stars marked the vernal equinox, although this was changed about the beginning of our era, owing to precession, to stars in Aries, the 27th of the early series, and further from the fact that many of their titles occur in the most ancient books of China, and are positively claimed there as of at least 2500 B. C.

While these lunar asterisms in the main agree as to their component stars, — eighteen are coincident, — some of the Hindu and Chinese are located in our Andromeda, Aquila, Boötes, Crater, Delphinus, Hydra, Lyra, Orion, and Pegasus, outside of the moon's course. Nor are their titles similar, except in the 16th, 17th, and 28th of China and Arabia; but our great Sanskrit scholar Whitney thought that this can hardly be fortuitous, and claimed, from this and other points of resemblance, that they are "three derivative forms of the same original."

They have been much disputed about,[1] yet no substantial agreement has been reached as to the date of their formation, or their place of origin. Whitney's résumé of the discussion appears in his *Lunar Zodiac*, his conclusion being that the moon stations were adopted into India, perhaps everywhere, from Mesopotamia, their birthplace.

Biot, early in this century, said that they were of Chinese origin, and Sedillot, that they came from Arabia; but Miss Clerke considers India as their source, and that they were first published in Arabia, in Al Ferghani's *Elements of Astronomy*, under the Khalif Al Mamun, in the early part of the 9th century, when Hindu cultivation in art, literature, and the sciences was much looked up to by the Arabians. Yet in the year 1000 Al Bīrūnī wrote, in his *India*, about its astronomers:

> I never came across any one of them who knew the single stars of the lunar stations from eyesight, and was able to point them out to me with his fingers.

The Hindus knew them as **Nakshatras**, Asterisms, the **Jufūr** of Al Bīrūnī, and thought them influential in their worship, and selected from the list the names of their months; but, although in some form or other they were very ancient in India, they do not seem to have been fully recognized there until the 7th or 8th century before Christ, when they appeared in the *Brahmanas*.

Unlike their counterparts in Arabia and China, each seems to have been represented by some special figure, in no way associated with the title.

In Arabia they were **Al Nujum al Aḣdh,** the Stars of Entering, and **Al Ribāṭat,** the Roadside Inns, although better known as **Al Manāzil al Ḳamr,** the Mansions, or Resting-Places, of the Moon; *manzil*, in the singular, signifying the noonday halt of camel and rider in the desert. Readers of *Ben Hur* will recall this in connection with Balthasar, the Egyptian, at the meeting of the Magi in their search for Him "that is born King of the Jews," after they saw

> his star in the east, and are come to worship him.

They are alluded to in the 10th *Sura* of the *Ḳur'ān*, where, referring to the moon, it says that God

[1] Professors Whitney and Newton have done the most to elucidate the subject in all its details by their article of 1858 in the *Journal of the American Oriental Society* on the *Sūrya Siddhānta*, the Straight (or Standard) Book of the Sun, the most important astronomical book of India, and claimed by the Hindus to be of divine origin, although Al Bīrūnī asserted that it was composed by Lāṭa.

hath appointed her stations, that ye might know the number of years, and the computation of time;

but long before the Prophet the authors of the Chaldaean *Creation Legend* and of *Genesis* wrote similarly; while in the 104*th Psalm*, that noble nature-psalm for Whitsunday, we read:

He appointed the moon for seasons.

In China they were **Sieu,** Houses, the series commencing with Kio,—α and ζ Virginis,—at the September equinox; and some are disposed to regard them there not merely as lunar divisions, but also as determinant points in reference to the movements of the sun and planets. Differing, however, from the analogous divisions of other nations, they generally were located along the equator. In the legends of that country they were the sky representatives of twenty-eight celebrated generals.

They also were introduced into Japan at an early day, and the chronicler of Magellan's voyage in 1521 found them familiarly known in the Malay Archipelago, and their astrological influence well recognized.

These Hindu, Arabic, and Chinese lunar asterisms have long been familiar to us, but the Persian have more recently been found in the *Bundehesh*, and Brown has only lately published transcriptions and translations of the Chaldaean, Khorasmian, and Sogdian titles,—the originals of the last two from Al Bīrūnī,—as also the significations of the Coptic and Persian. Their names and locations are given in connection with their component stars throughout this work; and they have been charted in detail by Williams and by Newton.

Other divisions of the sky, somewhat analogous to these, were the **Decans** of the Chaldaeans, Egyptians, and Greeks, "belts of stars extending round the heavens, the risings of which followed each other by ten days or so," but of much greater extent north and south than the Lunar Mansions, and thirty-six in number instead of twenty-eight. Miss Clerke writes of them:

The Chaldaeans chose three stars in each sign to be the "councillor gods" of the planets. These were called by the Greeks "decans," because ten degrees of the ecliptic and ten days of the year were presided over by each. The college of the decans was conceived as moving, by their annual risings and settings, in an "eternal circuit" between the infernal and supernal regions.

They are mentioned by Manilius as **Decania,** by others as **Decanica, Decane, Decanon, Degane, Deganae,** and **Decima**; while the lords of the decans were known as Decani and their titles have been preserved to us by Maternus Julius Firmicus, the prose writer of Constantine's reign. They appear in representations of ancient zodiacs on temple walls and astrological monuments in Egypt, as probably elsewhere.

The sky domed above us with its heavenly frescoes painted by the thought of the Great Artist.
Allen Throckmorton's *Sketches.*

The Constellations,

now designated by arbitrary lines outside and entirely independent of the figures, in ancient times were confined within the outlines of the forms that they were supposed to represent, although any resemblance was only occasionally noticeable. All stars adjacent to but beyond these were called by the Greeks *αμόρφωτοι*, unformed, and *σποράδες*, scattered, which Latin authors followed in their *extra*, *informes*, *dispersae*, *disseminatae*, and *sparsiles;* and the Arabians in their *Al H'ārij min Al Ṣūrah*, Outside of the Image.

In our day, however, every star is within the limits of some one of the constellations, although the boundaries of these are not in all cases agreed upon by astronomers. Still those adopted by Argelander are generally accepted for the northern figures, as those of Gould are for the southern; Gould's boundaries largely agreeing with the suggestions of Sir John Herschel, *i. e.*, formed by arcs of meridians and parallels of declination for a given epoch.

The figures were variously known by the Greeks as *Σήματα* and *Τείρεα*, Signs; *Σώματα*, Bodies; *Ζώδια*, Animals; and as *Μετέορα*, Things in Heaven, our word Meteors. Hipparchos said *'Αστερισμόι*, as did Ptolemy, but also alluded to them as *Μορφώσεις*, Semblances, and *Σχήματα*, Figures.

Pliny and other Latins called them **Astra, Sidera,** and **Signa,** while later on **Constellatio** appeared, that in the 1515 *Almagest* is **Stellatio;** and the Arabians knew them as **Al Ṣuwar,** Figures.

Aratos, in the Φαινόμενα of 270 B. C., mentioned forty-five, but many of these probably had been formed millenniums previously by the Chaldaeans, or even by their predecessors; in fact, he is not supposed to have invented any that he described. Eratosthenes, nearly a century after Aratos, reduced the number to forty-two in the Καταστερισμοί that were attributed to his authorship until Bernhardy's time; as did Gaius Julius Hyginus Historia, about the beginning of our era, in his reputed work, the *Poeticon Astronomicon*, and Decimus Magnus Ausonius, the Christian poet of nearly four centuries later.

The *Catalogue* of Hipparchos, now lost except as preserved by Ptolemy, is said to have contained forty-nine constellations with 1080 stars; but his *Commentary on Eudoxos and Aratos*, that we still have, mentions only forty-six. It was of this great astronomer that Pliny wrote in the year 78, as translated by Philemon Holland, in 1634, in his *Historie of the Worlde:*

> The same man went so farre that he attempted (a thing even hard for God to perform) to deliver to posteritie the just number of starres;

and asserted that this was induced by the appearance, in 134 B. C., of the bright *nova*, or temporary star, in Scorpio. The observations of Hipparchos seem to have been made between 162 and 127 B. C.

Pliny, although but a poor cosmographer, devoted two chapters to astronomy in the *Historia Naturalis*, and, according to the usual rendering, mentioned seventy-two[1] asterisms with 1600 stars; but this, if the original be correctly understood, could have been only by separately counting parts of the old figures, for nowhere does he allude to any that are new, unless it be his Thronos Caesaris, probably the Southern Cross.

Ptolemy scientifically followed with those now known as the ancient forty-eight, in the 7th and 8th books of the *Syntaxis*, twelve of the zodiac with twenty-one northern and fifteen southern, made up by 1028 stars, including 102 ἀμόρφωτοι, all probably from Hipparchos, although with some acknowledged alterations by himself; for in the 5th chapter of his 7th book he wrote:

> we employ not the same Figures of the Constellations that those before us did, as neither did they of those before them, but frequently make Use of others that more truly represent the Forms for which they are drawn.

His catalogue was supposed to comprise all the stars above the 54th de-

[1] In Chilmead's *Treatise* is an attempted explanation of this, from Scaliger's *Commentaries on Manilius:* "that he might untie this knot, reads those words of Pliny thus . . . *discreta in duo de L. signa, &c.*, where for seventy two, hee would have it to be wanting two: which is 48. the just number reckoned by Ptolemy."

gree of south declination, his earliest recorded observations being in A. D. 127 and the last in 151; and we find with him the first comparative list of star magnitudes.

In the year 1252 Europe resumed its old position in astronomical work by the compilation of *Los Libros del Saber de Astronomia*, the celebrated *Alfonsine Tables*, by Arabian or Moorish astronomers, at Toledo, under the patronage of the Infante, afterward King Alfonso X, El Sabio, the Wise, and the Astronomer, of Leon and Castile, who "abandoned the crown for the astrolabe and forgot the earth for the sky."

These *Tables* and their Latin translations are strongly Arabicized, as plainly appears in our modern star-titles drawn from them; while the whole work is in the main only copied from Ptolemy with some necessary corrections. But it probably fairly represents the science of the Middle Ages, and was in use until at least the 16th century; for Eden,[1] in 1555, quoted from Gemma Phrysius' *On the Maner of Fyndynge the Longitude:* "Then eyther by the Ephemerides or by the tables of Alphonsus . . ." Various editions have been printed: the first in 1483, two hundred years after Alfonso's death; again, in 1492 and 1521, all at Venice and in Latin; in 1545 at Paris; in 1641 at Madrid; and, lastly, splendidly reproduced there in 1863–1867, in the earliest accessible Spanish text, with illustrations, supposed copies of the original.

It was this Alfonso who has so often been condemned for his remark:

> Had I been present at the Creation, I would have given some useful hints for the better ordering of the universe;

but as he was speaking of the absurd Ptolemaic system, it does not seem so irreverent now as it did before Copernicus' day. Carlyle quoted it in his *History of Friedrich II of Prussia*,—

> that it seemed a crank machine; that it was pity the Creator had not taken advice!

and said that this, and this only, of his many wise sayings is still remembered by mankind.

From Ptolemy's time, with the exception of the *Alfonsine Tables*, no advance was made in astronomical science for 1300 years, and the *Syntaxis* continued to be the standard of the world's astronomy, "a sort of astronomical Bible, from which nothing was taken, and to which nothing material in principle was added."

[1] Rycharde Eden was one of the principal authors of the reign of Mary Tudor, and the translator of the writings of Peter Martyr on the early navigators Vespucci, Corsali, Pigafetta, and others. His *Decades of the newe worlde or west India* was the third English book on America, or Armenica as he called it, published in London in 1555.

In the 15th century, however, it was corrected and copied under the auspices of the celebrated Ulug Beg, grandson of the great Tatar conqueror, Timur i Leng, Timur the Lame, our Tamerlane, and, as his *Tables*, was published at Samarkhand, with the date of the 5th of July, 1437. The constellation descriptions in these are from Al Sufi's translation of five centuries previously, the titles of a few groups being changed; and the intrinsic excellence of the work, as well as the deservedly great reputation of its author as an astronomer, supported by many able assistants, made it a standard authority for nearly two centuries. Following Ulug Beg, but from Europe, came in 1548–51 the globes of Gerardus Mercator (Gerhard Kramer), on which were located fifty-one asterisms with 934 stars, besides numerous *informes*. About this time Copernicus' great work laid the foundations of modern astronomy, and was soon followed by Tycho Brahē's posthumous catalogue of 1602, with forty-six constellations, but only 777 stars, the mystic number, and so perhaps by design, for the author, although the first real observer of modern days, was still under the influence of astrology.

In the succeeding year appeared the *Uranometria* of Johann Bayer, the great Protestant lawyer of Augsburg, a work also much tinctured with the occult science, in which the author probably followed Tycho. This contained spirited drawings, after Dürer, of the ancient forty-eight figures, with a list of 1709 stars and twelve new southern asterisms. These last were its noticeable feature, with the fact that in the plates of the ancient constellations for the first time formally appeared Greek and Roman letters to indicate the individual stars, and so conveniently taking the place of the cumbersome descriptions till then in vogue.[1] Although this lettering did not come into general use until the succeeding century, Bayer had been anticipated in it fifty years before by Piccolomini of Siena, and even the Persians and Hebrews are said to have had something similar. Dr. Robert Wittie, of London, in his 'Ουρανοσκοπία of 1681, wrote of this last people:

> Aben Ezra tells that they first divided the Stars into Constellations, and expressed them all by the Hebrew Letters, which when they had gone through, they added a second Letter to express the shape, and oft-times a third to set forth the Nature of the Constellation.

After Bayer new constellations were published in the *Planisphaerium Stellatum* of 1624 by Jakob Bartsch (Bartschius); in the *Rudolphine Tables* of 1627, Kepler's edition of Tycho's catalogue; in Augustin Royer's work of 1679; and in the *Catalogue of Southern Stars* of the same year, by Doctor Edmund Halley, from his observations at Saint Helena. The *Prodromus Astronomiae* of 1690, by Johann Hewel, or Hoevelke (Hevelius), and its

1 No lettering, however, was applied by Bayer to stars of the twelve new southern figures.

appendix with plates, the *Firmamentum Sobiescianum*, also gave new figures, as did the *Historia Coelestis Britannica* of the Reverend Doctor John Flamsteed, completed in 1729 by Crosthwait and Sharp after Flamsteed's death in 1719. This comprised fifty-four constellations, the stars being consecutively numbered in the order of their right ascension; the companion *Atlas* following in 1753, and again in 1781. The Abbé Nicolas Louis de La Caille, "the true Columbus of the southern sky,"[1] in his *Mémoires* of 1752 and his *Coelum Stelliferum* of 1763, introduced fourteen new groups, "to which he assigned the names of the principal implements of the sciences and fine arts"; while a few others were formed by Pierre Charles Le Monnier from 1741 to 1755, and by Joseph Jerome Le Français (*dit* de La Lande) from 1776 to 1792, the 3d edition of La Lande's *Astronomie* containing a total of eighty-eight constellations. Lastly, in 1800, Johann Ellert Bode published nine new figures in his *Uranographia*, although some of these were by La Lande; a 2d edition, entitled *Die Gestirne*, being issued in 1805. But none of these inventions of the last three authors are now recognized.

The greater part of the new constellations were of course in the south, a quarter of the heavens which, although alluded to by a writer of the time of Pharaoh Neku, who sent a Phoenician fleet to circumnavigate Africa about 600 B. C., practically was unknown till the discovery of the New World stimulated the efforts of the early voyagers at the beginning of the 16th century. Some of these have left records of their stellar observations — among them the Italians Corsali, Pigafetta, and Vespucci, and the Dutch Pieter Theodor of Embden (Embdanus), *alias* Pieter Dircksz Keyser, and Friedrich Houtmann. But the results did not formally appear till a century later in the works of Bayer and Kepler, although they were mentioned in the *Decades* of Peter Martyr[2] and in Eden's translations of it and similar works; and some of the figures were inserted on the now almost unknown globes of 'Emeric Mollineux, Jodocus Hondius, and Jansenius Caesius (Willem Jansson Blaeu), of 1592 and the years following.

The hitherto unfigured space around the south pole, the object of these observations, was an eccentric one as to the pole, although in itself circular, reaching from Argo, Ara, and Centaurus, now within 20° of that point on

1 It is interesting to know that La Caille's observations were made with a half-inch glass.

2 Peter Martyr — not the great reformer Vermigli — was Pietro Martire d'Anghiera, Angleria, or Angliera, from his supposed birthplace near Milan. His work *De Rebus Oceanicis et Orbe Novo*, issued from 1511 to 1521, is a most interesting source of information on the early voyages to our country, largely derived from Columbus.

one side, to Cetus and Piscis Australis, within 60° on the other; while its centre, near γ Hydri and the Nubecula Minor, was the pole of 2000 to 2400 B. C., when α Draconis corresponded to it on the north. From this fact came Proctor's ingenious argument that such was the date of formation of the latest of the ancient constellations.

It is perhaps worthy of notice that the *Ductor in linguas*, or *Guide into Tongues*, the polyglot dictionary of 1617–27, by John Minsheu (Minshaeus), at the word *Asterisme* in the later editions alluded to

Eighty-four in all besides a few found out of late by the Discoverers of the South Pole;

but he gave no detailed list, and doubtless erred in his statement.

In our day there is discrepancy in the number of constellations accepted by astronomers, few of whom entirely agree in recognition of the modern formations. For, although Ideler described 106, with allusions to others entirely obsolete, or of which nearly all traces had been lost, Argelander catalogued only eighty-six, Vela, Puppis, and Carina being included under Argo; and the *British Association Catalogue* of 1845 only eighty-four. Professor Young recognizes sixty-seven as in ordinary use, although he catalogues eighty-four, Argo being divided into Carina, Puppis, and Vela; Upton's *Star Atlas*, of 1896, eighty-five; and the *Standard Dictionary* eighty-nine, but the latter's list of 188 star-names is disappointing. Nor should I forget to mention a very popular book in its day, the *Geography of the Heavens*, with its *Atlas* by Elijah H. Burritt, published in various editions from 1833 to 1856. This described fifty well-recognized constellations visible from the latitude of Hartford, Connecticut, 41° 46′; although his table of those in the entire heavens included ninety-six, most of which appeared in the accompanying maps, the figures being taken from Wollaston's drawings. Although not an original work of great scientific value, and erroneous as well as deficient in its stellar nomenclature, it had a sale of over a quarter of a million copies, and much influence in the dissemination of astronomical knowledge in the generation now passing away. I am glad to pay here my own tribute to the memory of the author, in acknowledgment of the service rendered me in stimulating a boyhood interest in the skies.

From eighty to ninety constellations may be considered as now more or less acknowledged; while probably a million stars are laid down on the various modern maps, and this is soon to be increased perhaps to three millions upon the completion of the present photographic work for this object by the international association of eighteen observatories engaged upon it in different parts of the world. The first instalment in print of these observations may be expected in a few years; the whole perhaps in twenty-five or thirty years.

It has been the fashion with astronomers to decry this multiplicity of sky figures, and with good reason; for, as Miss Clerke writes in her monograph on *The Herschels and Modern Astronomy:*

> Celestial maps had become "a system of derangement and confusion," of confusion "worse confounded." New asterisms, carved out of old, existed precariously, recognized by some, ignored by others; waste places in the sky had been annexed by encroaching astronomers as standing-ground for their glorified telescopes, quadrants, sextants, clocks; a chemical apparatus had been set up by the shore of the river Eridanus, itself a meandering and uncomfortable figure; while serpents and dragons trailed their perplexing convolutions through hour after hour of right ascension;

with more to the same effect. This condition of things led the Royal Astronomical Society, in 1841, to depute to Sir John Herschel and Mr. Francis Baily the task of attempting a reform. But although improvement was made by the discarding of several figures and the subdivision of others, their changes were too sweeping and were not successful, so that as the constellations stood then, in the main do they stand to-day, and so will they probably remain, at least with the people.

The change from the old system of star-designations, however, has been much more thorough, and, except in the popular mind, has been practically accomplished; but now in turn is there confusion in their substitutes, the various catalogue numbers and letters, even among the astronomers, and certainly with us unscientific star-gazers. As to this Miss Clerke graphically continues:

> palpable blunders, unsettled discrepancies, anomalies of all imaginable kinds, survive in an inextricable web of arbitrary appellations, until it has come to pass that a star has often as many aliases as an accomplished swindler.

II.

What were the dates of formation and places of origin of the earliest of the present sky figures are questions that have often been asked, but till recently impossible to be answered, and now only in part, and that tentatively. Greece and Rome, Egypt and Chaldaea, China, India, Aethiopia, and Phoenicia, and perhaps other countries, all lay claim to the honor, while history, theory, and tradition are all cited in proof; but we may safely agree with La Place that their forms and names have not been given them by chance.

Aratos,[1] the first Greek poetical writer on astronomy now extant, described them as from the most ancient times, and wrote in the *Phainomena:*

> Some man of yore
> A nomenclature thought of and devised,
> And forms sufficient found.
>
> . . . .
>
> So thought he good to make the stellar groups,
> That each by other lying orderly,
> They might display their forms. And thus the stars
> At once took names and rise familiar now.

His sphere, probably identical with that of Eudoxos of a century previous, accurately represented the heavens of about 2000 to 2200 B. C., a fact which has induced many to think it a reproduction from Babylonia; and the disagreement in the poet's description with the sky of his day led Hipparchos, the first commentator on the *Phainomena*, to much needless although in some cases well-founded criticism; for Aratos was, as Cicero said, *hominem ignarum astronomiae.* Still his poem is now apparently our sole source of knowledge as to the arrangement of the early constellations, and has been closely followed in all star-maps as an indispensable guide. It seems to have been a versification of its now lost prose namesake by Eudoxos, somewhat influenced by the writings of Theophrastus, and had a great run in its day. Landseer[2] wrote in his *Sabaean Researches* of 1823:

> When the poem entitled the Phenomena of Aratus was introduced at Rome by Cicero and other leading characters, we read that it became the polite amusement of the Roman ladies to work the celestial forms in gold and silver on the most costly hangings; and this had previously been done at Athens, where concave ceilings were also emblazoned with the heavenly figures, under the auspices of Antigonus Gonatas,

King of Macedonia and patron of Aratos. It has always been much translated, versified, commented upon, and quoted from; and we know of thirty-five Greek commentaries on this work. "It continued to be used as a practical manual of sidereal astronomy as late as the 6th century of our era." Cicero translated it in his youth, seventy years before the appearance of Vergil's *Aeneid;* Germanicus Cæsar did the same about A. D. 15; and Rufus Festus Avienus versified it in our 4th century: all commented on by Hugo Grotius in his *Syntagma Arateorum* of 1600. Of several English

[1] Aratos is supposed to have been the *quis alter* who, with Conon, was shown on the "beechen bowls, the carved work of the divine Alcimedon," that Menalcas wagers with Damoetas in the 3d Eclogue of the *Bucolica*.

[2] John Landseer, engraver and writer on art, was the father of Thomas and Sir Edwin Landseer.

translations the most literal and useful is that of Mr. Robert Brown, Jr., in 1885.

Saint Paul's supposed quotation from it in *The Acts of the Apostles*, xvii, 28, perhaps made it popular with the Christians of his and subsequent times, for apparent references to it occur in the writings of the early fathers.

It may be assumed that, with the exception of Ursa Minor, Equuleus, and Libra in its present shape, the sources of the old forty-eight have been lost in their great antiquity. Yet Pliny asserted that Aries and Sagittarius were formed by Cleostratos at some time between 548 and 432 B. C.; and the rest, with equal improbability, have been ascribed by Aristotle's pupil Eudemos to the Pythagorean Oinopides of Chios as of about 500 B. C., but from Egyptian dictation.

Whatever may be the facts as to all this, we know that a long line of notable Greeks, from Homer and Hesiod to Ptolemy, were interested in, and have preserved to us, their constellated heavens. Of these the first astronomers were Thales, 640–546 B. C., who gave us Ursa Minor; Eudoxos, who, according to common story, brought the constellations from Egypt, and, about 366 B. C., was the first to publish them in the original prose *Phainomena*, Cicero calling him the greatest astronomer that ever lived; while Hipparchos,[1] of whom Pliny said *nunquam satis laudatus*, is the acknowledged founder of our modern science. His works, however, are now lost, except his *Commentary* and the star-catalogue reproduced by Ptolemy. All these are mentioned with respect even by the astronomers of to-day; and it is certain that we find in their country the immediate source of most of the constellations as they now appear on our maps, and of the stories connected therewith. Yet these unquestionably are in many cases variations of long antecedent, perhaps prehistoric, legends and observations from the Euphrates, Ganges, and Nile; indeed the Greek astronomers always acknowledged their indebtedness to Chaldaea and Egypt, but gave most of the credit to the latter.

While we have few individual star-titles from Greece, the characters of the Argonautic Expedition are largely represented in the heavens; and Saint Clement, followed by many,— even by the great Sir Isaac Newton,— attributed the invention of the constellations to Chiron, the reputed preceptor of Jason, for the latter's use on that celebrated voyage, fixing its date as about 1420 B. C. And, coincidently as to the time of their formation, that good authority Seneca said that they were from the Greeks of about 1500 B. C.,

[1] The Abarchis and Abrachys of the Arabians.

which may be true to the extent that they then adopted them from some earlier nation. But the mythologists ascribed them to Atlas, the Endurer, the father of the Hyades and Pleiades, so skilled in knowledge of the skies that he was shown as their supporter; and they had a story fitted to every heavenly figure.

But much of this is more than unreliable, even childish, and we are only sure that Greece originated our scientific astronomy and gave great attention to it from the times of Thales and Anaximander; this culminating in the work of the Alexandrian School,[1] Egyptian in location, but entirely Greek in character.

To the Romans we owe but little in the way of astronomy,— indeed they always were ready to acknowledge the superiority of Greece in this respect, — although we find much of stellar mythology and meteorology in their poetry and prose. No real astronomer, however, appeared among them; and when Julius Caesar needed such for his reform of the calendar, albeit himself somewhat skilled in the science, as his *De Astris* shows, he was compelled to call Sosigenes to his aid. The architect Vitruvius (Marcus Vitruvius Pollio), just before the beginning of our era, apparently was the most scientific among them, and in the 9th book of his *De Architectura* tells us much of their star-lore in connection with the proper location of sun-dials; while Columella, of our 1st century, in his *De Re Rustica* made many allusions to stars and constellations in their supposed connection with the weather and crops.

Many have maintained that Egypt was the first to give shapes and names to the star-groups; Dupuis, perhaps inspired by Macrobius of our 5th century, tracing the present solar zodiac to that country, and placing its date 13,000 years anterior to our era, when the flow of the Nile with its consequent harvests, and the seasons, coincided with the positions of the separate figures and the characters assigned to them. In this he has been followed by others even to our day.

The little that we know of Egypt's early constellations indicates that they apparently were of native origin, and in no respect like those of Greece, which, if adopted at all, were so at a very late time in that history, and from the influence of the dominant Greeks, perhaps aided by recollections

[1] This great school was begun by such men as the two Arystilli and Timochares, under Ptolemy Soter, 300 B. C., the first really scientific astronomers who initiated the observations that are generally supposed to have led Hipparchos to his discovery of the phenomenon of precession; and it was carried on by Aristarchos, Eratosthenes (the inventor of the armillary sphere), Euclid (the geometrician), Conon, Sosigenes, and lastly Ptolemy, who ended the famous list in A. D. 151, although the school was nominally maintained till the final destruction of the great Alexandrian Library in the 7th century.

of Chaldaea. Diodorus the Sicilian, of the 1st century before Christ, and Lucian, of three centuries later, distinctly assert this.

The following are among the native stellar groups of Egypt so far as at present can be thought assured: **Sahu,** identified with Orion, although by some limited to the head of that figure; **Sept, Set, Sothis,** etc., with Sirius; the **Hippopotamus,** a part of our Draco; the **Thigh,** our Ursa Major; the **Deer,** our Cassiopeia, although some place the **Leg** here. The doubtful ones were **Mena,** or **Menat,** an immense figure if Renouf[1] be correct in his statement that it included Antares and Arcturus; the **Many Stars,** our Coma Berenices; **Arit,** that Renouf thought may have been marked by β Andromedae; the **Fleece,** indicated by some stars of Aries; the **Goose,** by α Arietis; **Chu,** or **Chow,** the Pleiades; the **Cynocephalus,** claimed by La Lande for Ara's stars; the **Servant,** that Brugsch says was our Pegasus, although the Denderah planisphere shows a Jackal here; the **Two Stars,** that we may guess were Castor and Pollux; and the **Lute-Bearer,** or **Repā,** the Lord, perhaps our Spica.

Those so far unidentified were the **Stars of the Water; Mena's Herald; Mena's Followers; Necht,** in the vicinity of our Draco; the **Lion,** but not our Leo; and the **Hare,** with some others that La Lande indefinitely alluded to as lying on the borders of Ophiuchus and Scorpio and in Aquarius.

A reference is made in Egypt's veritable history to the vernal equinox, then in our Taurus, 3285 B. C.; yet the astronomy of that country was not scientific, and we know little of it except as connected with religion, the worship in the north, about 5200 B. C., of the northern stars being associated with the god An, Annu, Ant, or On, under the supposed government of Set, or Typhon, the god of darkness, recognized under many synonyms. That of the east and west stars was indicated by the Ghizeh temples and pyramids, about 4000 B. C.; while in southern Egypt the worship of the southern stars, as early as 6400 B. C., perhaps much earlier, was presided over by Horus, a southern sun-god, although later he occasionally appeared as a northern divinity. The rising stars represented the youthful goddesses; those setting, the dying gods; while a figure of three stars together symbolized divinity.

Assertions as to India being the first home of astronomy, and the birthplace of the constellation figures, have been made by many—notably, a century ago, by Sir William Jones and Messrs. Colebrooke, Davis, and Von Schlegel; but modern research finds little in Sanskrit literature to confirm this belief, while it seems to be generally acknowledged that the Hindus

[1] The eminent Egyptologist Sir Peter Le Page Renouf, who died in 1897.

borrowed much from Greece, perhaps beginning with Pythagoras, who is said to have traveled there and even listened to Zoroaster's teachings. Indeed, Aryabhata, of our 5th or 6th century, reckoned by the same signs as Hipparchos; and their most noted later astronomer, Varāha Mihira,[1] of 504, in writing of the constellations, used the Grecian titles, changed, however, to suit his native tongue. But Arabia also probably exercised influence over them, as over the rest of Asia.

Professor Whitney's opinion as to this is summed up thus:

> We regard the Hindu science as an offshoot from the Greek, planted not far from the commencement of the Christian era, and attaining its fully developed form in the course of the fifth and sixth centuries;

but unfavorably criticizes it, as did Al Bīrūnī. The annals of China, a country never backward in claiming the invention of almost everything, new or old, on earth or in the sky, ascribe the formation of constellations to Tajao, the prime minister of Hwang Ti, 2637 B. C., and make much of an observation of the Pleiades, 2537 B. C., from an observatory said to have been erected 2608 B. C. But real stellar work in that country seems to have begun only about ten or twelve centuries before our era, and then almost solely in the interests of astrology.

The attainment of the Chinese in the science, probably very highly overrated, however, is thought to be largely due to Chaldaea, and later on the Arabians, in the times of the khalifs, apparently exercised influence over them; while all their recent advance is due to the Jesuit missionaries who settled among them in the 16th century, during the early years of the present Tsing dynasty, and introduced the knowledge of our Western figures. These were thenceforward to a great extent adopted, and our own star-titles in the translations which the Chinese called Sze Kwo Ming, the Western Nation Names, became common, especially in the case of the constellations visible only from south of the parallel of Peking, 40°. The indigenous titles were Chung Kwo Ming, the Middle Nation Names, Edkins saying as to these that there were two great periods of star-naming: the first about 2300 B. C. by the people, and the second from 1120 to 220 B. C., during the Chow dynasty, that plainly shows an imperial origin. And it was during this period, about 600 B. C., that a chart was drawn with 1460 stars correctly laid down. This is now in the Royal Library of Paris.

In all its history in China astronomy has been under the special care of the state, and the regulator of all affairs of life, public and private.

[1] Al Bīrūnī mentioned this author as an excellent astronomer, and quoted much from his work the *Brihatsamhitā*, or Collection.

The early Chinese included the twenty-eight *sieu* and the twelve *kung*, or zodiac figures, in four larger equal spaces,— **Tsing Lung,** the Azure Dragon; **Heung Woo,** the Dark Warrior; **Choo Neaou,** the Red Bird, Phoenix, Pheasant, or Quail; and **Pih Hoo,** the White Tiger. And they marked off, in their general constellations, three large *yuen*, or inclosures,— **Tsze Wei,** the circumpolar stars; **Tien She** and **Tai Wei,** containing the rest that were visible to them.

Williams' *Observations of Comets* is accompanied by a full set of maps of 351 early asterisms traced over Flamsteed's figures; but, large as is this number, M. Gustave Schlegel, in his *Uranographie Chinoise* of 1875, cited 670 that he asserted could be traced back to 17000 B. C.!

In the-neighboring Japan some, even of its wise men, thought that the stars were made to guide navigators of foreign peoples, with their tribute, to the land of the Mikados.

Aethiopia's claim to the invention of the constellations probably can be entertained only by considering that country as the Kush of southwestern Asia, — Homer's eastern Aethiopia, — stretching along the Arabian and Persian gulfs, whence early migrations across the Red Sea at the Strait of Babd al Mandab may have carried astronomical knowledge directly to the Nile, or, by a roundabout way, to Meroë in western Aethiopia, the modern Nubia, and thence northward into Egypt.

Of Phoenician stellar science little is known, and assertions as to its antiquity rest largely upon the fact that this people was the great maritime nation of ancient times, and hence some knowledge of the heavenly bodies was a necessity with them. Yet Thales, the father of astronomy and a teacher of the Greeks in the science, — indeed one of their Seven Sages, — probably was Phoenician by birth; and Samuel Bochart, the Oriental scholar of the 17th century, as well as other authorities, thought that many of our older groups in the sky are merely reproductions of the figureheads on the Carthaginian, Sidonian, and Tyrian ships. This, if correct, might account for the incompleteness of such as Argo, Pegasus, and Taurus, as well as for the marine character of many of them. But the general opinion is that the Phoenicians drew from Chaldaea such astronomy as they may have had.

Ideler, in his *Sternkunde der Chaldaer* of 1815, asserted that the constellations originated on the Euphrates, — "reduplications of simpler ideas connected with natural phenomena," — and conviction as to the truth of this seems to be growing with students of stellar archaeology. Indeed recent discoveries make it apparently safe to say that those of the zodiac at least were first formed in the Akkad country, probably in almost prehistoric times, and that there, as among all the earliest nations, "their order and harmony is contrasted with and opposed to the supposed disorderly motion of the planets." It is also probable that many of the extra-zodiacal groups, in somewhat the same form and location as we have them now, came from the Valley of the Great River, as well as the myths associated with them, originally introduced by Northern invaders; for Bailly said that the science current in Chaldaea, as well as in India and Persia, belonged to a latitude higher than that of Babylon, Benares, and Persepolis.

With the Babylonians the chief stars represented their chief gods, and they connected the several constellations with particular nations over whose destiny they were thought to dominate. Cuneiform characters arranged in stellar form were the ideograph of Ilu, Divinity; while, combining business and religion, their *Ku-dur-ru*, or Division Stones, recently unearthed, that marked the metes and bounds of city lots and farm lands, are often inscribed with some constellation figure, probably the one representing the tutelar god of the owner. But whatever may be our conclusions as to the beginning of astronomy in the Euphrates valley, it can be considered settled that astrology in the present sense of the word had its origin there, and that the modern astrological characters of the sun, moon, and planets are those current on that river and in all ages since.

The prophet Isaiah, 700 B. C., in pronouncing the Almighty's judgment on Babylon, contemptuously referred to

> the astrologers, the stargazers, the monthly prognosticators;

Daniel, a century later, knew his captors as accomplished in the art, although himself and his companions were "ten times better"; while the terms "Babylonians" and "Chaldaeans" have come down almost to our own time as synonymous with observers of, and diviners from, the stars, whatever their individual nationality.

But the art became widely spread elsewhere, and especially in vogue in Rome, where its devotees, known as Babylonii, Chaldaei, Astronomi, Astrologi, Genethliaci, Mathematici, and Planetarii, seem to have flourished notwithstanding the efforts made to suppress them and the ridicule cast on them by Cicero, Juvenal, and others of the time. Indeed they were driven out of the city by law in 139 B. C., and frequently afterward, but as often returned. In Greece, Eudoxos and Aristarchos of Samos felt it needful to urge their countrymen against it, although Berōssōs taught it there soon after them; and its influence everywhere up to two hundred years ago is well known. Dante's belief in it is frequently shown throughout the *Divina Commedia*, while in Shakespeare's day — indeed for a century after him —

reliance upon it was well-nigh universal, and much was made of it in all drama and poetry, Kent, in *King Lear*, only expressing prevalent opinion when he said:

> It is the stars,
> The stars above us that govern our conditions.

Cecil, Baron of Burghley, calculated the nativity of Queen Elizabeth; Lilly was consulted by King Charles I, in 1647, as to his escape from Carisbrooke Castle; Flamsteed drew a horoscope of the heavens at the moment of laying the foundation of the Royal Observatory, on the 10th of August, 1675, although he added to it *Risum teneatis amici;* and about the same time astrologers were called into the councils of Parliament. The art still obtained even among the educated classes of the succeeding century; for astrological evidence was received in a court of justice as late as 1758, and Sir Walter Scott made Guy Mannering cast a horoscope for the young laird of Ellangowan that the latter preserved till of mature age.

It is not unlikely that the decadence of astrology in England was hastened by the publication of Boteler's *Hudibras*, in which the practice and its great exponent William Lilly, under the title Sidrophel, were so successfully and popularly satirized. Among its passages we read of its devotees:

> in one case they tell more lies,
> In figures and nativities,
> Than th' old Chaldean conjurors
> In so many hundred thousand years.

Dean Swift followed in the same vein in his *Predictions for the year* 1708 *by Isaac Bickerstaff, Esq.*

On the Continent astrology had been still more prevalent, and even men of science were seriously interested in it. Gassendi began his distinguished career in its practice; Tycho predicted from the comet of 1577, and, as it happened, successfully, the achievements and time of death of Gustavus Adolphus; the still greater Kepler prophesied from the stars a coming hard winter, and so it proved. Miss Maria Mitchell wrote of these two astronomers:

> Both of these philosophers leaned to the astrological opinions of their times; and Kepler was certainly a believer in them. He calculated nativities when pressed for money, and published astrological almanacs, though he admitted that such procedures were little better than begging, and his work but "worthless conjectures";

and he plaintively said:

> The scanty rewards of an astronomer would not provide me with bread, if men did not entertain hopes of reading the future in the heavens.

The horoscope of Wallenstein by one or the other of these great men is still preserved in the library of the Poulkowa Observatory. Napoleon's belief in his guiding star is well known. But as an occult science astrology practically died out in England with the astronomers of the 17th century. It still flourishes, however, in the East, especially among the Chinese and Parsis. The recent advent of a little son to the Chinese consul-general in New York was the occasion of much telegraphing to the chief astrologers of the Celestial Kingdom who were to predict his future; and the horoscope of the Parsi even now is carefully preserved during life, burned at his death, and its ashes scattered over the Sacred River. In a measure it lingers among the people everywhere, for its almanacs and periodicals are still published; its advertisements and signs are daily to be seen in our large cities; a society for its study, called the Zodiac, was established in New York City in 1897; and even now there are many districts in Germany where the child's horoscope is regularly kept with the baptismal certificate in the family chest.

It should not be forgotten that astrology, Kepler's "foolish daughter of a wise mother," originally included astronomy, Seneca being the first in classical times to make distinction between the meanings of the two words; and he was followed in this by Saint Isidore of Seville (Isidorus Hispalensis), the Egregius Doctor of the 7th century, and author of the *Origines et Etymologiae;* although even as late as the 17th century we see confusion in their use, for Minsheu mentioned the "astrologers" as having formed the "asterismes," and the diarist John Evelyn wrote of "Mr. Flamsteed the learned astrologer."

Contrariwise, and not long previously, the word "astronomer" was applied to those whom we would now call astrologers. Shakespeare devoted his 14*th Sonnet* to the subject, beginning thus:

> Not from the stars do I my judgment pluck,
> And yet methinks I have astronomy;

and in *Troilus and Cressida* we read

> When he performs astronomers foretell it.

But this is a long digression from my subject.

Arabia's part in early astronomy was slight, for although the tribes before Muḥammād's day doubtless paid much attention to the heavenly bodies, this was entirely unscientific, merely observational and superstitious; and only in their subsequent days of peace and power, after the Prophet had solidified them into an active nation, did their more cultured class seriously take up the study of the sky. Even this was solely along the lines laid down by Ptolemy, and they originated little. Still we owe them and their Jewish assistants much of gratitude for their preservation of the beginnings of modern astronomy during the thousand years of the Dark and Mediaeval Ages; while, as we have seen, our star-names are largely due to them.

The heathen Arabs were star-worshipers,— Sabaeans,— as still are the Parsis of our own special star, the sun; indeed this worship was very general in antiquity. It was universal in earliest India, and constantly alluded to in their sacred books; Egyptian priests showed to Plutarch stars that had been Isis and Osiris; in Greece Aristophanes made special mention of it in his *Pax*, 419 B. C., and Aristotle wrote to Alexander:

> Heaven is full of the gods to whom we give the name of stars.

In Plato's *Timaeus* we read of his supreme divinity:

> And after having thus framed the universe, he allotted to it souls equal in number to the stars, inserting each in each. . . . And he declared also, that after living well for the time appointed to him, each one should once more return to the habitation of his associate star, and spend a blessed and suitable existence;

Dante adopting this in the *Paradiso*:

> Parer tornarsi l'anime alle stelle,
> Secondo la sentenza di Platone;

while Vergil wrote in the *Georgics*:

> viva volare
> Sideris in numerum, atque alto succedere coelo;

Milton, in *Paradise Lost*:

> Those argent fields more likely habitants,
> Translated saints, or middle spirits hold,
> Betwixt the angelical and human kind;

and Wordsworth, almost of our own day, in his *Poems of the Imagination*:

> The stars are mansions built by nature's hand,
> And, haply, there the spirits of the blest
> Dwell clothed in radiance, their immortal vest.

Indeed this thought has been current in all history and tradition, in civilized as in savage life, on every continent, and in the isles of the sea.

The Christian father Origen, following the supposed authority of the *Book of Job*, xxv, 5, and perhaps influenced by the 43d verse of chapter xiii of the *Gospel of Saint Matthew*, said that the stars themselves were living beings; and Dionysius Exiguus, the chronologist of our 6th century, established in the constellations the hierarchies of the genii, assigning to the cherubim the domain of the fixed stars. Shakespeare has many allusions to this stellar attribute. In *King Henry VI*, Bedford, invoking the ghost of Henry V, said

> a far more glorious star thy soul will make
> Than Julius Caesar;

and in *Pericles* we see

> Heavens make a star of him.

Even now, according to Mr. Andrew Lang, German folk-lore asserts that when a child dies God makes a new star—a superstition also found in New England fifty years or more ago. The German peasant tells his children that the stars are angels' eyes; and the English cottager impresses it on the youthful mind that it is wicked to point at the stars, though why he cannot tell.

In much the same way Al Bīrūnī cited from Varāha Mihira:

> Comets are such beings as have been on account of their merits raised to heaven, whose period of dwelling in heaven has elapsed and who are then redescending to the earth.

Cicero, in *De Natura Deorum*, asserted that the constellations were looked upon as divine; and Statius, that the sea nymphs were the constellations of the sea, the divine inhabitants of the waters, as the others were of the heavens. Yet this same author elsewhere represented Aurora as driving the stars out of heaven with a scourge like so many beasts; and Manilius called them a flock going on like sheep; while Shelley, in his *Prometheus Unbound*, writing of the astronomer's work, said:

> Heaven's utmost deep
> Gives up her stars, and like a flock of sheep
> They pass before his eye, are number'd, and roll on.

In Upper India even now women teach their children that the stars are kine, and the moon their keeper.

Following the opinion of Josephus, Origen said that the constellations were known long before the days of the patriarchs by Noah, Enoch, Seth, and Adam — indeed were mentioned in the *Book of Enoch* as "already named and divided"; and he claimed that ancient longevity was a blessing specially bestowed to give opportunity for a long-continued period of observation and comparison of the heavenly bodies.

In early Christian art a star became the peculiar emblem of sanctity, and often appeared over the heads or on the breasts of representations of the saints.

III.

Some allusion should be made to what Smyth called the Biblical School and the Mosaicists, who at various times have sought to alter the sky figures to others drawn from sacred history and its interpretation. Beginning with the Venerable Bede, this school has come to our time, but their efforts, fortunately, have been in vain; for, although their motives may have been praiseworthy, our scheme of the heavenly groups is of too much historical value and too useful and interesting a source of popular instruction for us to wish it discarded.

Among the number of these stellar iconoclasts was the unfortunate Giordano Bruno of the 16th century, who, in his *Spaccio della Bestia Trionfante*, sought to substitute for the ancient figures the moral virtues, Law, Mercy, Prudence, Truth, Universal Judgment, Wisdom, etc.; and others, most numerous in the 17th century, were Caesius, Jeremias Drexelius, Novidius, Postellus, Bartsch, Schickard, Harsdörffer, and Julius Schiller of Augsburg; while in our day the Reverend Doctor John Lamb, the versifier of Aratos, and Proctor wrote in somewhat the same vein. The recent efforts of Miss Frances Rolleston and the Reverend Doctor Joseph A. Seiss are especially remarkable. Proctor made other changes in constellation titles, although he followed the old lines; but his changes have not been adopted, and, Chambers says, "were far more barbarous than the originals which he condemned"; indeed in his later works he abandoned the effort as impracticable.

The following remarks by Professor Holden on the history of the delineation of our stellar figures are interesting:

> The contribution of Albrecht Dürer to astronomy is . . . unknown, I believe, to all his biographers.

But this statement he subsequently modified by a reference to Thausing's *Life of Dürer*, in which this artist's map-work is mentioned:

> Hipparchus (B. C. 127) and Ptolemy (A. D. 136) fixed the positions of stars by celestial latitudes and longitudes, and named the stars so fixed by describing their situation in some constellation figure. The celestial globes of that day have all disappeared, and we have only a few Arabian copies of them, not more ancient than the XIIIth century, so that we

may say that the original constellation figures are entirely lost. The situations of the principal stars in each one of the forty-eight classic constellations are verbally described by Ptolemy. In La Lande's *Bibliographie Astronomique* we find that in A. D. 1515 Albrecht Dürer published two star-maps, one of each hemisphere, engraved on wood, in which the stars of Ptolemy were laid down by Heinfogel, a mathematician of Nuremberg. The stars themselves were connected by constellation figures drawn by Dürer. These constellation figures of Dürer, with but few changes, have been copied by Bayer in his *Uranometria* (A. D. 1603); by Flamsteed in the *Atlas Coelestis* (1729); by Argelander in the *Uranometria Nova* (1843); and by Heis in the *Atlas Coelestis Novus* (1872), and have thus become classic.

It is a matter of congratulation that designs which are destined to be so permanent should have come down to us from the hands of so consummate a master.

I would add to this that Ptolemy's catalogue of stars was published at Cologne in 1537, in folio, with the forty-eight drawings by Dürer.

It seems singular that of the world's artists few, save he and Raphael, have done anything for this most ancient, exalted, and interesting of the sciences; others, famous or forgotten, introduce the subject into their compositions with generally sad result.[1] One instance especially absurd, although not strictly astronomical, is worthy repetition. Mrs. Jameson, in her *Sacred and Legendary Art*, describes, from an old French print,

St. Denis at Heliopolis, seated on the summit of a tower or observatory, contemplating, through a telescope, the crucifixion of our Saviour seen in the far distance.

And much the same may be said of most of our authors. Pope thus mistranslated Homer's allusion to Sirius:

> rises to the sight
> Through the thick gloom of some tempestuous night;

Henry Kirk White, in *Time*, had

> Orion in his Arctic tower;

Shelley, in the *Witch of Atlas*, wrote of the minor planets as

> those mysterious stars
> Which hide themselves between the earth and Mars;

and in *Prince Athanase* thus ignored the apparent motion of the stars:

> far o'er southern waves immovably
> Belted Orion hangs;

Dickens, in *Hard Times*, doing the same in his description of Stephen Blackpool's death, comforted the sufferer by a star shining brightly for

[1] This is especially the case with the moon, which is rarely correctly located or drawn.

hours down to the bottom of the Old Hell Shaft. In the poor man's own words:

Often as I coom to myseln, and found it shinin' on me down there in my trouble, I thowt it were the star as guided to Our Saviour's home.

Carlyle, who at one time aspired to the position of astronomer at the Edinburgh University, thus alluded, in his *French Revolution*, to the scenes in Paris on the night of the 9th of August, 1792:

the night . . . "is beautiful and calm"; Orion and the Pleiades glitter down quite serene,

although the former did not rise till daybreak; and again, still more blunderingly:

Overhead, as always, the Great Bear is turning so quiet round Boötes;

while Dickens, in *Our Mutual Friend*, made perhaps the worst mistake of all when, in describing the voyage that "brought a baby Bella home," a revolution of the earth around the sun marks a month instead of a year. Wallace, in *Ben Hur*, makes the shaykh Ilderim give impossible star-names to the parents of his great team — Sirius, from the hated Roman tongue instead of the beautiful Al Shirā of the Desert; and Mira, unknown to him, or indeed to any one, till nearly sixteen centuries thereafter; while the unlikely Greek Antares was given to one of the victorious four.

Errors as to the moon and planets are notoriously frequent, Venus and the new moon often being made to rise at sunset. Shakespeare, although contemporary with Galileo and Kepler, has many such; yet he seems to have known the action of the moon, his "governess of the floods," on the tides,[1] for we find in *Hamlet*

the moist star
Upon whose influence Neptune's empire stands;

and in *King Henry IV*,

being governed as the sea is by the moon.

Marryat, sea-captain though he was, wrote of a waning crescent moon seen in the early evening; and H. Rider Haggard has something similar in *King Solomon's Mines* — a book, by the way, that was once ordered for the library of a school of mineralogy! Charles Wolfe, in his *Burial of Sir John Moore* after the battle of Corunna, January 16, 1809, said that it was

By the struggling moonbeams' misty light,

[1] Dante showed similar knowledge in *Paradiso*, xvi, 82, 83.

whereas the moon did not shine that night, whether misty or clear; and Coleridge, in the *Rime of the Ancient Mariner*, had

> The horned moon with one bright star
> Within the nether tip.

The astronomy of the modern newspaper is notorious — ridiculous were not the fact of such prevalent ignorance lamentable.

Classical writers abounded in stellar allusions far more than do authors of our day; in fact, Quintilian, of our 1st century, in his *Institutio Oratoria*, insisted that a knowledge of astronomy was absolutely necessary to a proper understanding of the poets. And these allusions generally were correct, at least for their day.

The same may be said of Dante, whose thorough acquaintance with the stellar science of the 14th century appears everywhere in his works — in fact, the *Paradiso* may be called a poetical frame for the Ptolemaic system; and it has been well written of Milton, "the poetical historian of the astronomy of his day," that in astronomy the accuracy of his facts fairly divides the honors with the beauty of his language; but he slipped when he located Ophiuchus "in th' Arctic sky," and it is not till late in his works that we see the abandonment of Ptolemy's theories.

Tennyson makes many beautiful allusions to stars and planets, and is always accurate, unless we except his "moonless Mars," which, however, was before Asaph Hall's discovery; while our Longfellow and Lowell knew the stars well, and well showed this in their works.

★

> Andromeda! Sweet woman! why delaying
> So timidly among the stars: come hither!
> Join this bright throng, and nimbly follow whither
> They all are going.
>
> John Keats' *Endymion.*

Andromeda, the Woman Chained,

the 'Ανδρομέδη of Aratos and 'Ανδρομέδα of Eratosthenes, Hipparchos, and Ptolemy, represents in the sky the daughter of Cepheus and Cassiopeia, king and queen of Aethiopia, chained in exposure to the sea monster as punishment of her mother's boast of beauty superior to that of the Nereids. Sappho, of the 7th century before Christ, is supposed to mention her, while Euripides and Sophocles, of the 5th, wrote dramas in which she was a character; but she seems to go far back of classical times, and we probably must look to the Euphrates for her origin, with that of her family and Cetus. Sayce claims that she appeared in the great Babylonian *Epic of Creation*, of more than two millenniums before our era, in connection with the story of Bēl Mardūk and the dragon Tiāmat, that doubtless is the foundation of the story of Perseus and Andromeda. She was noted, too, in Phoenicia, where Chaldaean influence was early felt.

As a constellation these stars have always borne our title, frequently with the added **Mulier Catenata,** the Woman Chained, and many of the classical Latins alluded to her as familiar and a great favorite. Caesar Germanicus called her **Virgo Devota**; a scholiast, **Persea,** as the bride of Perseus; while Manilius, and Germanicus again, had **Cepheis,** from her father.

In some editions of the *Alfonsine Tables* and *Almagest* she is **Alamac,** taken from the title of her star γ; and **Andromada,** described as *Mulier qui non vidit maritum*, evidently from Al Bīrūnī, this reappearing in Bayer's *Carens Omnino viro*. Ali Aben Reduan (Haly), the Latin translator of the Arabian commentary on the *Tetrabiblos,* had **Asnade,** which in the *Berlin Codex* reads **Ansnade** *et est mulier quae non habet vivum maritum;* these changed by manifold transcription from **Alarmalah,** the Widow, applied by the Arabians to Andromeda; but the philologist Buttmann said from **Anroneda,** another erroneous form of our word. The **Antamarda** of the Hindus is their variation of the classical name.

The original figure probably was, as Dürer drew it, that of a young and beautiful woman bound to the rocks, Strabo said at Iope, the biblical Joppa; and Josephus wrote that in his day the marks of her chains and the bones of her monster foe were still shown on that sea-shore. But this author, "who did not receive the Greek mythology, observes that these marks attest not the truth but the antiquity of the legend."

Others, who very naturally thought her too far from home at that spot, located Iope in Aethiopia and made her a negress; Ovid expressing this in his *patriae fusca colore suae*, although he followed Herodotus in referring her to India. Manilius,[1] on the contrary, in his version of the story described her as *nivea cervice;* but the Aethiopia of this legend probably was along the Red Sea in southwestern Arabia.

1 Manilius, author of the *Poeticon Astronomicon*, frequently quoted throughout these pages, flourished under Augustus and Tiberius, and probably was the first Latin author to write at length on astronomy and astrology; but he adhered closely to Aratos' scheme of the constellations, making no mention of Berenice's Hair, Equuleus, or the Southern Crown. The text, as we have it, is from a manuscript exhumed in the 15th century from an old German library by Poggius, the celebrated Gian Francesco Poggio Bracciolini, who rescued so much of our classic literature from the dust of ages.

Arabian astronomers knew these stars as **Al Mar'ah al Musalsalah,** their equivalent of the classical descriptive title,— Chilmead's **Almara Almasulsala,**— for Western mythological names had no place in their science, although they were familiar with the ideas. But they represented a **Sea Calf,** or **Seal,** *Vitulus marinus catenatus*, as Bayer Latinized it, with a chain around its neck that united it to one of the Fishes; their religious scruples deterring them from figuring the human form. Such images were prohibited by the *Ḳur'ān;* and in the oral utterances attributed by tradition to the Prophet is this anathema:

Woe unto him who paints the likeness of a living thing: on the Day of Judgment those whom he has depicted will rise up out of the grave and ask him for their souls. Then, verily, unable to make the work of his hands live, will he be consumed in everlasting flames.

This still is the belief of the Muslīm, for William Holman Hunt was warned of it, while painting his *Scape Goat in the Wilderness*, by the shaykh under whose protection he was at the time.

The Spanish edition of the *Alfonsine Tables* pictures Andromeda with an unfastened chain around her body, and two fishes, one on her bosom, the other at her feet, showing an early connection with Pisces; the *Hyginus*, printed at Venice *anno salutifere incarnationis*, 7th of June, 1488, by *Thomas de blauis de alexandria*, with some most remarkable illustrations, has her standing between two trees, to which she is bound at the outstretched wrists; in the *Leyden Manuscript*[1] she is partly clothed on the sea beach, chained to rocks on either side.

Caesius[2] said that she represented the biblical **Abigail** of *The Books of Samuel;* and Julius Schiller, in 1627, made of her stars **Sepulchrum Christi,**[3] the "new Sepulchre wherein was never man yet laid."

[1] The figures in this old manuscript are spirited, many of them beautiful, and all studded with stars, but with no attempt at orderly arrangement; and, although in perfect preservation, high antiquity has been claimed for them as of ancient Roman times. Hugo Grotius reproduced them in his *Syntagma Arateorum*, and the *Manuscript* is still preserved in the University Library at Leyden.

[2] The work of Caesius (Philip Zesen), the *Coelum Astronomico-Poeticum*, published by Joannes Blaeu at Amsterdam in 1662, is much quoted by La Lande, and is a most interesting source of information as to star-names and the mythology of the constellations, with many extracts from Greek and Roman authors. He mentions sixty-four figures, but some of his star-titles, as also perhaps those of other astronomical writers, would seem merely to be synonyms for the human originals erroneously assumed as for their sky namesakes.

[3] This appeared in the *Coelum Stellatum Christianum*, which, according to its title-page, was the joint production of Schiller and Bayer, an enlarged reprint of the *Uranométria* of 1603; and Gould says that it was in reality the 2d edition of Bayer's work, almost ready for the press at the latter's death in 1625, but appropriated by Schiller to embody his own absurd constellation changes.

The apparently universal impulse of star-gazers to find earthly objects in the heavens is shown in the Cross which is claimed for some of Andromeda's stars; β, γ, and δ marking the upright, α and κ the transverse. But a much more noticeable group, an immense Dipper, is readily seen in following up its γ and β to the Square of Pegasus, far surpassing, in extent at least, the better-known pair of Dippers around the pole.

Andromeda is bounded on the north by Cassiopeia and Perseus; on the east by Perseus; on the south by Pisces and Triangulum; and on the west by Lacerta and Pegasus.

Milton's passage in *Paradise Lost*, where Satan surveys our world

> from eastern point
> Of Libra to the fleecy star that bears
> Andromeda far off Atlantic seas
> Beyond th' Horizon,

seems to have puzzled many; but the poet was only seeking to show the comprehensive view had by the arch-fiend east and west through the six signs of the zodiac from the Scales to the Ram with the golden fleece; Andromeda, above the latter, apparently being borne on by him to the westward, and so, to an observer from England, over the Atlantic.

Kingsley's *Andromeda* well describes her place:

> I set thee
> High for a star in the heavens, a sign and a hope for the seamen,
> Spreading thy long white arms all night in the heights of the aether,
> Hard by thy sire and the hero, thy spouse, while near thee thy mother
> Sits in her ivory chair, as she plaits ambrosial tresses;
> All night long thou wilt shine;

these members of the royal family, Andromeda, Cassiopeia, Cepheus, and Perseus, lying contiguous to each other, wholly or partly in the Milky Way.

The stars that mark her right arm may be seen stretching from σ to ι and κ; ζ marking the left arm with the end of the chain towards Lacerta; but in early days she was somewhat differently located, and even till recently there has been confusion here; for Smyth wrote:

> Flamsteed's Nos. 51 and 54 Andromedae are ψ and υ Persei, though placed exactly where Ptolemy wished them to be — on the lady's foot: so also α in this asterism has been lettered δ Pegasi by Bayer, and β has been the lucida of the Northern Fish.

Argelander has 83 stars here, and Heis 138.

La Lande and Dupuis asserted that the Phoenician sphere had a broad **Threshing-floor** in this spot, with stars of Cassiopeia as one of the **Gleaners** in the large **Wheat-field** that occupied so much of that people's sky; its exact boundaries, however, being unknown to us.

α, Double, magnitudes, 2.2 and 11, white and purplish.

Alpheratz, Alpherat, and **Sirrah** are from the Arabians' **Al Surrat al Faras,** the Horse's Navel, as this star formerly was associated with Pegasus, whence it was transferred to the Woman's hair; and some one has strangely called it **Umbilicus Andromedae.** But in all late Arabian astronomy taken from Ptolemy it was described as **Al Rās al Mar'ah al Musalsalah,** the Head of the Woman in Chains.

Aratos designated it as ξῦνός ἀστήρ, *i. e.*, common to both constellations, and it is still retained in Pegasus as the δ of that figure, although not in general use by astronomers.

In England, two centuries ago, it was familiarly known as **Andromeda's Head.**

With β Cassiopeiae and γ Pegasi, as the **Three Guides,** it marks the equinoctial colure, the prime meridian of the heavens; and, with γ Pegasi, the eastern side of the Great Square of Pegasus.

In the Hindu lunar zodiac this star, with α, β, and γ Pegasi,— the Great Square,— constituted the double *nakshatra,*— the 24th and 25th,— **Pūrva** and **Uttara Bhādrapadās,** the Former and the Latter Beautiful, or Auspicious, Feet; also given as **Proshthapadās,** Footstool Feet; while Professor Weber of Berlin says that it was **Praṭishthana,** a Stand or Support, which the four bright stars may represent.

With γ Pegasi, the determinant star, it formed the 25th *sieu* **Pi,** or **Peih,** a Wall or Partition, anciently **Lek,** and the *manzil* **Al Fargu,** from Al Farigh al Mu'ah·h·ar, the Hindmost Loiterer; or, perhaps more correctly, the Hind Spout of the Water-jar, for Kazwini called it **Al Farigh al Thānī,** the Second Spout; a Well-mouth and its accompaniments being imagined here by the early Arabs.

The Persian title for this lunar station, **Miyan**; the Sogdian, **Bar Farshat**; the Khorasmian, **Wabir**; and the Coptic, **Artulosia,** all have somewhat similar meanings.

In astrology α portended honor and riches to all born under its influence. It comes to the meridian — culminates — at nine o'clock [1] in the evening of the 10th of November.

[1] All culminations mentioned in this work are for this hour.

β, 2.3, yellow.

Mirach was described in the *Alfonsine Tables* of 1521 as *super mirat*, from which has been derived its present title, as well as the occasional forms **Mirac, Merach, Mirar, Mirath, Mirax,** etc.; *mirat* probably coming from the 1515 *Almagest's super mizar*, the Arabic mi'zar, a girdle or waist-cloth. Scaliger, the great critical scholar of the 15th century, adopted this **Mizar** as a title, and Riccioli followed him in its use, thus confounding the star with ζ Ursae Majoris. The **Mirae** of Smyth doubtless is a typographical error, although Miraë had appeared in Chilmead's *Treatise*[1] of 1639 for the same word applied to β Ursae Majoris.

Hipparchos seems to refer to it in his *ζώνη*; and, synonymously, some have termed it **Cingulum**; others, **Ventrale,** from its former position in the figure, although now it is on the left hip. In later Arabian astronomy it marked the right side of Andromeda, and so was known as **Al Janb al Musalsalah,** the Side of the Chained Woman. β appeared in very early drawings as the *lucida* of the northern of the two Fishes, and marked the 26th *manzil* **Al Baṭn al Ḥūt,** the Belly of the Fish, or **Al Ḳalb al Ḥūt,** the Heart of the Fish; and the corresponding *sieu* **Goei,** or **Kwei,** the Man Striding, or the Striding Legs, anciently **Kwet.** In this location it was **Al Rishā,** the Band, Cord, Ribbon, or Thread, as being on the line uniting the Fishes; but this title now belongs to α Piscium.

Brown includes it, with υ, φ, and χ Piscium, in the Coptic lunar station **Kuton,** the Thread; and Renouf, in **Arit,** an asterism indigenous to Egypt. It lies midway between α and γ, about 15° distant from each; and in astrology was a fortunate star, portending renown and good luck in matrimony.

γ, Binary,—and perhaps ternary, 2.3, 5.5, and 6.5, orange, emerald, and blue.

This is **Alamac** in the *Alfonsine Tables* and 1515 *Almagest;* Riccioli's **Alamak**; Flamsteed's **Alamech**; now **Almach, Almak, Almaack,** and **Almaac** or **Almaak**; all from **Al 'Anāḳ al 'Arḍ,** a small predatory animal of Arabia, similar to a badger, and popularly known there as Al Barīd. Scaliger's conjecture that it is from Al Mauk, the Buskin, although likely enough for a star marking the left foot of Andromeda, is not accepted; for

[1] This book, a *Learned Treatise on Globes*, was a translation by Master John Chilmead, of Oxford, of two early Latin works by Robert Hues and Io. Isa. Pontanus. It is an interestingly quaint description of the celestial globes of that and the preceding century, with their stellar nomenclature.

Ulug Beg, a century and a half previously, as well as Al Tizini[1] and the Arabic globes before him, gave it the animal's title in full. But the propriety of such a designation here is not obvious in connection with Andromeda, and would indicate that it belonged to very early Arab astronomy.

Bayer said of it, *perperam* **Alhames,** an erroneous form of some of the foregoing. Riccioli[2] also mentioned this name, but only to repudiate it.

Muḥammād al Achsasi[3] al Muwakkit designated γ as **Al Ḥāmis al Na'āmāt,** his editor translating this *Quinta Struthionum*, the 5th one of the Ostriches; but I have not elsewhere seen the association of these birds with this constellation.

Hyde gives another Arabian designation for γ as **Al Rijl al Musalsalah,** the Woman's Foot.

In the astronomy of China this star, with others in Andromeda and in Triangulum, was **Tien Ta Tseang,** Heaven's Great General. Astrologically it was honorable and eminent.

Its duplicity was discovered by Johann Tobias Mayer of Göttingen in 1778; and Wilhelm Struve,[4] in October, 1842, found that its companion was closely double, less than 1″ apart at a position angle of 100°, and probably binary. The two larger components are 10″.4 apart with a position

[1] The catalogue of this author, Muḥammād abu Bekr al Tizini al Muwakkit, was published at Damascus in 1533 with 302 stars, and from its long list of purely Arabic star-names was regarded as worthy of translation and republication by Hyde, in 1665, with the original text. The *muwakkit* of his title indicates that he was shaykh of the grand mosque.

[2] This last author, to whom I shall make frequent reference, was Joanne Baptista Riccioli, of the Society of Jesus, whose *Almagestum Novum* of 1651 and *Astronomia Reformata* of 1665 were famous in their day, and are interesting in ours, as preserving to us much of the queer mediaeval stellar nomenclature, as well as of the general astronomical knowledge of the times. In the 2d volume of this last work is a long list of titles, curiosities in philology, with this heading: *Nomina Stellarum Peregrinum & Plerumque Arabica;* while the comment thereon, *ne mirere Lector, si eidem Stellae diversa nomina videbis adscripta, pro diversitate Dialectorum aut codicum fortasse corruptorum,* might well have served as a motto for this book. He is noted, too, as having drawn for his *Almagest* the 2d map of the moon,—Hevelius preceding him in this by four years,—and as having given the various names to its various features, more than two hundred of these being still in use, while all but six of those given by his justly more celebrated contemporary have been discarded. His lunar titles naturally were Jesuitical; nor was he overmodest, for his own name appears first in the list, and that of his colleague Grimaldi immediately succeeding.

[3] The Arabic manuscript of this author, with its star-list of about the year 1650, has been reviewed by Mr. E. B. Knobel in the *Monthly Notices* of the Royal Astronomical Society for June, 1895. It contains 112 stars, perhaps taken from Al Tizini's catalogue of the preceding century. The Achsasi of his title was from the village of similar name in the Fayūm, doubtless his birthplace; and, like Tizini, he was shaykh of the grand mosque in Cairo, where his work was written.

[4] Struve was the first director of the Russian National Observatory at Poulkowa, where he was succeeded by his son Otto; and two of the grandsons bear names already celebrated in astronomy.

angle of 63°.3. The contrast in their colors is extraordinarily fine. Sir William Herschel wrote of it in 1804:

> This double Star is one of the most beautiful Objects in the Heavens. The striking difference in the colour of the two Stars suggests the idea of a Sun and its Planet, to which the contrast of their unequal size contributes not a little; but Webb thought them stationary.

It is readily resolved by a 2¼-inch glass with a power of forty diameters, and it seems singular that its double character was not sooner discovered.

From its vicinity radiate the **Andromedes II**, the **Bielid meteors** of November, so wonderfully displayed on the 27th of that month in 1872 and 1885, and on the 23d in 1892, and identified by Secchi and others with the celebrated comet discovered by Biela in 1826, which, on its return in 1832, almost created a panic in France. The stream completes three revolutions in about twenty years, although subject to great perturbations from Jupiter, and doubtless was that noticed on the 7th of December, 1798, and in 1838. These objects move in the same direction as the earth, and so with apparent slowness,— about ten miles a second,— leaving small trains of reddish-yellow sparks. The radiant, lying northeast from γ, is remarkable for its extent, being from 7 to 10 degrees in diameter. The Mazapil iron meteorite which fell in northern Mexico on the 27th of November, 1886, has been claimed "as being really a piece of Biela's comet itself."

δ, Double, 3 and 12.5, orange and dusky.

Burritt added to the letter for this the title **Delta,** perhaps from its forming a triangle with ε and a small adjacent star.

It marks the radiant point of the **Andromedes I** of the 21st of July.

The components are 27″.9 apart, at a position angle of 299°.3.

θ, a 4.7-magnitude star, with ρ and σ, was the Chinese **Tien Ke,**[1] the Heavenly Stable.

ξ, 4.9,

is **Adhil**, first appearing in the *Almagest* of 1515, and again in the *Alfonsine Tables* of 1521, from **Al Dhail,** the Train of a Garment, the Arabic equivalent of Ptolemy's σύρμα; but Baily thought the title better applied to the slightly fainter A, which is more nearly in that part of the lady's dress; and

[1] The star-names of China that appear in this work are few in comparison with the total in the great number of that country's constellations. I occasionally cite them merely to indicate the general character of Chinese stellar nomenclature.

Bayer erroneously gave it to the 6th-magnitude *b*, claiming—for he was somewhat of an astrologer, although the *Os Protestantium* of his day—that, with the surrounding stars, it partook of the nature of Venus.

φ, Binary, 4.9 and 6.5, yellow and green, and χ, 5,

in Chinese astronomy, were **Keun Nan Mun,** the Camp's South Gate; they lie in the train near the star *σύρμα*. The components of ϕ were observed by Burnham in 1879, 0″.3 apart, at a position angle of 272°.4.

N. G. C.[1] 224, or 31 M.,[2]

the **Great Nebula,** the **Queen of the Nebulae,** just northwest of the star ν, is said to have been known as far back as A. D. 905; was described by Al Sufi as the **Little Cloud** before 986; and appeared on a Dutch star-map of 1500. But otherwise there seems to be no record of it till the time of Simon Marius (Mayer of Gunzenhausen), who, in his rare work *De Mundo Joviali*, tells us that he first examined it with a telescope on the 15th of December, 1612. He did not, however, claim it as a new discovery, as he is reported to have fraudulently done of the four satellites of Jupiter,[3] when he gave them their present but rarely used names, **Io, Europa, Ganymede,** and **Kallisto,** that are now known as I, II, III, and IIII, in the order of their distances from the planet. Halley, however, did so claim it in 1661 in favor of Bullialdus (Ismail Bouillaud), who, although he doubtless again brought it into notice as the *nebulosa in cingulo Andromedae*, expressly mentioned that it had been observed 150 years previously by some anonymous but expert astronomer.

Hevelius catalogued it in his *Prodromus*, and Flamsteed inserted it in his *Historia* as *nebulosa supra cingulum* and *nebulosa cinguli;* but Hipparchos, Ptolemy, Ulug Beg, Tycho Brahē, and Bayer did not allude to it, from which some have inferred an increase, or variability, in its light; but there is no positive evidence as to this, and it does not seem probable.

Marius said that it resembled the diluted light from the flame of a candle seen through horn,[4] while others of our early astronomers described it differently; discordances probably owing to the different means employed. Its true character seems as yet undetermined, although astro-photography

[1] This is the *New General Catalogue* of Doctor J. L. E. Dreyer, published in 1887.

[2] Messier's *Catalogue*.

[3] This planet was known to the Greeks as Ζεύς, and as Φαέθων, the Shining One.

[4] This reminds us of Dante's beautiful simile in the *Paradiso*, although of a different object:

So that fire seemed it behind alabaster.

"has proved it to be a vast Saturniform body, a great, comparatively condensed nucleus, surrounded by a series of rings, elliptical as they appear to us, but probably only so from the angle under which they are presented to our view"; "masses of nebulous matter partially condensed into the solid form"— a new and enormous solar system in formation.

Its length, or diameter, about 3½°, is estimated at more than thirty thousand times the distance from the earth to the sun.

Its attendant companion, visible as a nebula in the same field if a low-power be used, is the star-cluster N. G. C. 221, 32 M., discovered in 1749 by Le Gentil. It is nearly circular in form, and apparently ⅛ the size of the Great Nebula. Sir William Huggins and others have suggested that the small nebulae near the latter may be planets in process of formation.

S Andromedae, the *nova* of 1885 that excited so much interest, was first seen about the middle of August, 16″ of arc to the southeast of the nucleus, and, for a brief period, of the 6th to the 7th magnitude; but it soon disappeared to ordinary glasses, and Hall last saw it with the 26-inch refractor at Washington on the 1st of February, 1886, as of the 16th magnitude.

★

In dreams it seemed to me I saw suspended
An eagle in the sky, with plumes of gold,
With wings wide open, and intent to stoop,
And this, it seemed to me, was where had been
By Ganymede his kith and kin abandoned,
When to the high consistory he was rapt.

Longfellow's translation of Dante's *Purgatorio.*

Antinoüs

lies in the Milky Way, directly south from the star Altair; the head of the figure at η and σ, the rest of the outline being marked by θ, ι, κ, λ, ν, and δ, all now in Aquila. Flamsteed omitted σ and ν from his catalogue, but added *i*.

The constellation is said to have been introduced into the sky, in the year 132, by the Emperor Hadrian, in honor of his young Bithynian favorite, whose soul his courtiers had shown him shining in its *lucida* after the youth's self-sacrifice by drowning in the Nile from his belief that his master's life might thus be prolonged. This was because the oracle at Beza had asserted that only by the death of the object which the emperor most loved could great danger to the latter be averted. The new asterism, however, was little known among early astronomers; and although Ptolemy alluded to it, he did so but slightingly in calling half a dozen of the ἀμόρφωτοι of Aquila ἐφ ὧν ὁ ἀντίνοος.

After his day it seems unnoticed till Mercator put it on his celestial globe of 1551 with six components; Bayer following him in illustrating it with Aquila, although with no distinct list of its stars. Tycho also utilized it; but it first separately appeared in print on a plate in Kepler's *Stella Nova* of 1606, and in his *Rudolphine Tables.* Longomontanus (Christian Longberg of Denmark) had it in his *Astronomica Danica* of 1640; Hevelius included it in the *Prodromus*, but added a Bow and Arrow, the ancient Sagitta; Flamsteed mentioned it in the *Historia Coelestis* as **Aquila Antinous, Aquila vel Antinous,** and **Aquila cum Antinoo**; and the Hungarian Jesuit Abbé Maximilian Hell had it in constant use in his *Ephemerides Astronomicae* of 1769 and 1770. Bode also distinctly catalogued and illustrated it; but Argelander omitted its title from his *Uranometria Nova* of 1843, although he showed it as a part of Aquila. It is now hardly recognized, its stars being included with those of the latter constellation.

Bayer substituted **Ganymedes** for Antinoüs, and others have used both names indiscriminately; Tennyson describing the youth as

> Flush'd Ganymede, his rosy thigh
> Half buried in the Eagle's down.

This same name occasionally has appeared for Aquarius, but is given by La Lande, with many other titles, for our Antinoüs; among these are **Puer Adrianaeus, Bithynicus, Phrygius,** and **Troicus**; **Novus Aegypti Deus**; **Puer Aquilae**; **Pincerna** and **Pocillator,** the Cup-bearer.

Caesius saw in it the **Son of the Shunammite** raised to life by the prophet Elisha; and La Lande said that some had identified it with the bold Ithacan, one of Penelope's suitors slain by Ulixes.

Two of the Arabic globes bear the stars δ, θ, κ, and λ Aquilae, which mark the distinguishing rhombus of Antinoüs, as **Al Ṭhalīmain,** the Two Ostriches; but Ideler assigned this title to ι and λ; giving δ, η, and θ as **Al Mīzān,** the Scale-beam. Simone Assemani said that they were **Alkhalimain,** that more correctly is **Al Ḣalīlain,** the Two Friends, or **Al Ḥalimatain,** the Two Papillae; but his assertions as to star-names are often unfavorably criticized by Ideler as "a confused medley, raked together without criticism."

These globes are so frequently referred to as indicative of the character and progress of the astronomy of Arabia, that I may be pardoned a brief digression as to them.

One, of the year 1225, now rests in the museum established by the Cardinal Borgia at Villetri; another, of 1289, is in the Mathematical Salon at Dresden; Mr. A. V. Newton claims the early date of the 11th century for one lettered in Arabo-Cufic characters, now in the Bibliothèque Nationale of Paris, as does Signor F. Meucci for one in Florence; another, of bronze, from Arabian times, the stars lettered in silver, but not figured, is in the rooms of the Royal Astronomical Society of London; and the Emperor Frederick II of Italy, in the 13th century, is said to have had one of gold, the stars being shown by inlaid pearls. All these seem to have been of comparatively small dimensions, five to eight inches in diameter, a great contrast to the six-foot globe of Tycho Brahē, now in the castle at Prague. Those of Mercator were about sixteen inches.

But celestial globes were known long anterior to these. One that is considered very correct as to the location of the early constellations, although it does not show the individual stars, is in the Farnese collection of antiquities, surmounting the statue of Atlas. This globe, supposed to be a copy of the sphere of Eudoxos, and perhaps antedating Ptolemy, although somewhat defaced, has preserved to us more than forty of the sky figures of its day; while another, of brass, said to have been constructed by Ptolemy himself,— doubtless an apocryphal statement,— was found in 1043 in an old public library in Kahira, the modern Cairo. Ptolemy described the globe of Hipparchos that is illustrated in Halma's edition of the *Syntaxis*, published with a French translation in Paris in 1813–16; Eudoxos is said to have constructed one 366 B. C., as did Anaximander of Miletus 584 B. C.

The actual invention of celestial globes has been credited to Thales, as the mythical was to Atlas; but Flammarion nearly rivals this last when he seriously tells us of Chiron's sphere — "the most ancient sphere known, constructed about the epoch of the Trojan War, 1300 B. C."; and Sir Isaac Newton, induced by an incorrect translation from Diogenes Laertius, asserted that Musaeus, one of the Argo's crew, was the first to make a celestial sphere, on which he located the ship and many others of the Greek constellations derived from the story and characters of the Argonauts.

★

Antlia Pneumatica, the Air Pump,

is La Caille's **Machine Pneumatique,** at first Latinized as **Machina Pneumatica** (which occurs in Burritt, and is the Italian name); but astronomers know it as simple **Antlia.** In Germany it is the **Luft Pumpe.**

The constellation lies just south of Crater and Hydra, bordering on the Vela of Argo along the branches of the Milky Way, and culminates on the 6th of April; Gould assigning to it eighty-five naked-eye stars.

He thinks that *α*, the red *lucida,* may be a variable, as his observers had variously noted it as of from the 4th to the 5th magnitude, and Argelander entered both of these.

La Caille's *β* lies within the present limits of Hydra.

Although inconspicuous, and without any named star, Antlia is of special interest to astronomers from containing the noted variable[1] S, discovered in 1888 by Paul of Washington, and confirmed by Sawyer. Chandler gives its maximum as 6.7 and its minimum as 7.3, the period being 7 hours, 46 minutes, 48 seconds,— the shortest known until it was supplanted by U Pegasi with a period of 5½ hours.

★

And all the stars that shine in southern skies
Had been admired by none but savage eyes.
John Dryden's *Ode to Doctor Charleton.*

Apus, the Bird of Paradise,

or **Apous,** as Caesius wrote it from the Greek, lies immediately below the Southern Triangle, about 13° from the pole. It is the French **Oiseau de Paradis**; the German **Paradies Vogel**; and the Italian **Uccello Paradiso.**

Its avian original is found only in the Papuan Islands, and the title is from Ἄπους, Without Feet, for the ancient Greek swallow, but well applied to this bird that has been thus fabled, as witness Keats' "legless birds of paradise," in his *Eve of Saint Mark.*

Bayer strangely had it **Apis Indica** on his planisphere of the new southern figures, where the typical bird is shown, as also in the corresponding page of text; but the universal consent as to the name **Apus,** or **Avis,** and its appearance as **Apus Indica** and **Indianischer Vogel** in the abridged German edition of Bayer's work issued in 1720, with the fact that he had another, and correct, Apis, would indicate a typographical and engraver's error in the original; but I have not seen this alluded to till now. The drawing always has been of the typical bird of our title, which Caesius adopted in his **Paradisaeus Ales**; but it sometimes is **Avis Indica,** the Indian Bird.

[1] Chandler's *Third Catalogue of Variable Stars*, 8th July, 1896, describes 393, to which have been added 36 to the 19th of August, 1898,— a total of 429, not including those still awaiting notation, and those found in star-clusters by the Harvard observers.

The planisphere in Gore's English edition of Flammarion's *Astronomie Populaire* has the constellation as the **House Swallow,** probably taken from early ornithological lists or the lexicons; for our *Andrews-Freund* translates *Apus* as the Black Martin, the English synonym of the *Hirundo apus* of Linnaeus,— the *Cypselus apus* of William Yarrell,— not a swallow, however, but a well-known swift of the Old World, with perfectly formed, although small, legs and feet, yet appropriate enough to its mode of life; and the stellar bird appears in Willis' *Scholar of Thebet Ben Khorat* as

> Hirundo with its little company;
> And white-brow'd Vesta lamping on her path
> Lonely and planet-calm;

with this explanatory note:

> An Arabic constellation placed instead of the Piscis Australis, because the swallow arrives in Arabia about the time of the heliacal rising of the Fishes.

I have not met with these hirundine star-titles except in these two instances, and think them both incorrect. Mr. Willis' idea may have come from the Χελιδόνιας of the zodiacal pair, but he errs in ascribing the figure to Arabia and in considering it a substitute for the Southern Fish, as well as in confusing it with the older Pisces.

But all this poem is beautiful in stellar allusions. Here is another bit:

> Where has the Pleiad gone?
> Where have all missing stars found light and home?
> Who bids the Stella Mira go and come?
> Why sits the Pole-star lone?
> And why, like banded sisters, through the air
> Go in bright troops the constellations fair?

Apus similarly appears in China as **E Cho,** the Curious Sparrow; and as the **Little Wonder Bird.** Schiller included it with the Chamaeleon and the Southern Fly in his biblical **Eve.** Gould details sixty-seven naked-eye stars in Apus, its *lucida*, γ, being 3.9. It culminates about the middle of July, but of course is invisible from northern latitudes.

This is one of the twelve new southern constellations with which Bayer's name generally is associated, although he only adopted them and, Gould says, took them from one of the globes of Jacob, or Arnold, Florent van Langren; but Bayer distinctly attributed their formation to "Americus Vespucius, Andreas Corsalius, Petrus Medinensis and Petrus Theodorus," navigators of the early part of the 16th century, giving to the last most of the credit of their publication; and Smyth ascribed their invention to "Peter Theodore," and their publication to another sailor, Andrea Corsali, in 1516. In Chilmead's *Treatise* they are indefinitely ascribed to "*the* Portugals, Hollanders, *and* English *sea-faring men*."

Willem Jansson Blaeu, the celebrated globe-maker of Amsterdam, Chilmead's contemporary, credited their introduction to Friedrich Houtmann, who observed from the island of Sumatra; while the latter, Semler asserted, took his ideas from the Chinese. Although Ideler denied this, yet he acknowledged that the latter nation knew Phoenix, Indus, and Apus as the Fire Bird, the Persian and the Little Wonder Bird, almost exact translations of the Western titles; and summed up his account of them with the opinion that their origin "is involved in an obscurity that it is scarcely possible to penetrate."

★

> The sun his locks beneath Aquarius tempers,
> And now the nights draw near to half the day,
> What time the hoar frost copies on the ground
> The outward semblance of her sister white,
> But little lasts the temper of her pen.
>
> Longfellow's translation of Dante's *Inferno*.

Aquarius, the Waterman,

il Aquario in Italy, **le Verseau** in France, **der Wassermann** in Germany, has universally borne this or kindred titles; Ideler assigning as a reason the fact that the sun passed through it during the rainy season. In connection with this the proximity of other analogous stellar forms is worthy of note: Capricornus, Cetus, Delphinus, Eridanus, Hydra, Pisces, and Piscis Australis, all the watery shapes in the early heavens, with Argo and Crater, are in this neighborhood; some of whose stars Aratos said "are called the Water"; indeed in Euphratean astronomy this region of the sky was the **Sea,** and thought to be under the control of Aquarius.

The constellation immemorially has been represented, even on very early Babylonian stones, as a man, or boy, pouring water from a bucket or urn, with an appropriate towel in the left hand, the human figure sometimes being omitted; while the Arabians, who knew of the latter but did not dare to show it, depicted a mule carrying two water-barrels; and again simply a water-bucket. This last was Ulug Beg's idea of it, his original word being rendered by Hyde **Situla,** the Roman Well-bucket; but Al Bīrūnī had it in his astrological charts as **Amphora,** a Two-handled Wine-jar, that he may have adopted from Ausonius the poet of our 4th century. Even Vercingetorix, Caesar's foe in Gaul, 52 B. C., is said to have put the similar figure on his *stateres* with the title **Diota,** a Two-eared Jar.

On a Roman zodiac it was a **Peacock,** the symbol of Juno, the Greek Herē, in whose month Gamelion—January–February—the sun was in the sign; and at times it has been shown as a **Goose,** another bird sacred to that goddess.

New Testament Christians of the 16th and 17th centuries likened it appropriately enough to **John the Baptist,** and to **Judas Thaddaeus** the Apostle, although some went back to **Naaman** in the waters of Jordan, and even to **Moses** taken out of the water.

Its nomenclature has been extensive but consistent. In Greek literature it was Ὑδροχόος, the epic Ὑδροχοεύς, or Water-pourer, transliterated by Catullus as **Hydrochoüs,** and by Germanicus as **Hydrochoös**; although the latter also called it **Aquitenens** and **Fundens latices,** saying that it personified Deucalion of the Greek Deluge, 1500 B. C. Ausonius had **Urnam qui tenet**; Manilius, **Aequoreus Juvenis,** or simply **Juvenis,** and **Ganymedes,** the beautiful Phrygian boy, son of Tros and cup-bearer of Jove, of whom Statius wrote in his *Thebais:*

> Then from the chase Jove's towering eagle bears,
> On golden wings, the Phrygian to the stars.

This title also appeared with Cicero, Hyginus, and Vergil; and with Ovid, in the *Fasti*, as **Ganymede Juvenis, Puer Idaeus,** and **Iliacus,** from his birthplace, and **Juvenis gerens aquam**; while in a larger sense it was said to represent the creator Jove, the pourer forth of water upon the earth.

We find it, too, as **Aristaeus,** their Elijah, who brought rain to the inhabitants of Ceos, and **Cecrops,** from the cicada nourished by the dew, whose eggs were hatched by the showers; while Appian, the historian of our 2d century, called it **Hydridurus,** which reappeared in the 1515 *Almagest* as **Idrudurus** and **Hauritor aquae.** The great Grecian lyric poet Pindar asserted that it symbolized the genius of the fountains of the Nile, the life-giving waters of the earth. Horace added to its modern title **Tyrannus aquae,** writing of it as "saddening the inverted year," which James Thomson, 1700–1748, followed in the Winter of his *Seasons:*

> fierce Aquarius stains th' inverted year;

and Vergil, calling it *frigidus*, similarly said that when coincident with the sun it closed the year with moisture:

> Extremoque inrorat Aquarius anno.

In Babylonia it was associated with the 11th month Shabatu, the Curse of Rain, January–February; and the *Epic of Creation* has an account of the Deluge in its 11th book, corresponding to this the 11th constellation; each of its other books numerically coinciding with the other zodiacal signs. In that country its Urn seems to have been known as **Gu,** a Water-jar overflowing, the Akkadian **Ku-ur-ku,** the Seat of the Flowing Waters; and it also was **Rammān** or **Rammānu,** the God of the Storm, the still earlier **Imma,** shown pouring water from a vase, the god, however, frequently being omitted. Some assert that **Lord of Canals** is the signification of the Akkadian word for Aquarius, given to it 15,000 years ago (!), when the sun entered it and the Nile flood was at its height. And while this statement carries the beginnings of astronomy very much farther back than has generally been supposed, or will now be acknowledged, yet for many years we have seen Egyptian and Euphratean history continuously extended into the hitherto dim past; and this theory would easily solve the much discussed question of the origin of the zodiac figures if we are to regard either of those countries as their source, and the seasons and agricultural operations as giving them names.

Aben Ezra called it the Egyptians' **Monius,** from their *muau*, or Mῶ, Water; Kircher said that it was their Υπευθέριαν, *Brachium beneficum*, the Place of Good Fortune; which Brown, however, limits to its stars α, γ, ζ, and η as a Coptic lunar station; and our Serviss writes that "the ancient Egyptians imagined that the setting of Aquarius caused the rising of the Nile, as he sank his huge urn in the river to fill it."

With the Arabians it was **Al Dalw,** the Well-bucket; and Kazwini's **Al Sākib al Mā',** the Water-pourer; from the first of which came the **Edeleu** of Bayer, and the **Eldelis** of Chilmead. The Persians knew it as **Dol** or **Dūl**; the Hebrews, as **Deli** (Riccioli's **Delle**); the Syrians, as **Daulo,** like the Latin *Dolium;* and the Turks, as **Kugha,**—all meaning a Water-bucket. In the Persian *Bundehesh* it is **Vahik.**

In China, with Capricornus, Pisces, and a part of Sagittarius, it constituted the early Serpent, or Turtle, **Tien Yuen**; and later was known as **Hiuen Ying,** the Dark Warrior and Hero, or Darkly Flourishing One, the **Hiuen Wu,** or **Hiuen Heaou,** of the Han dynasty, which Dupuis gave as **Hiven Mao.** It was a symbol of the emperor Tchoun Hin, in whose reign was a great deluge; but after the Jesuits came in it became **Paou Ping,** the Precious Vase. It contained three of the *sieu*, and headed the list of zodiac signs as the **Rat,** which in the far East was the ideograph for "water," and still so remains in the almanacs of Central Asia, Cochin China, and Japan.

Some of the minor stars of Aquarius,—ι, λ, σ, and ϕ,—with others of

Capricornus and Pisces, formed the asterism **Luy Peih Chin,** the Camp with Intrenched Walls.

On the Ganges, as in China, it began the circle of the zodiacal signs; and Al Bīrūnī said that at one time in India it was **Khumba,** or **Kumbaba,** which recalls the Elamite divinity of that name, the Κόμβη, or Storm God, of Hesychios. This, too, was the Tamil title for it; La Lande writing it **Coumbum.** Varāha Mihira, under the influence of Greek astronomy, called it **Hridroga** and **Udruvaga,** in which we can see Ὑδροχόος.

With the Magi and Druids it represented the whole science of astronomy.

The Anglo-Saxons called it **se Waeter-gyt,** the Water-pourer; while not long after them John of Trevisa, the English translator, in 1398 thus quaintly recalled the classical form:

> The Sygne Aquarius is the butlere of the goddes and yevyth them a water-potte.

English books immediately succeeding had **Aquary, Aquarye,** and, still later, the queer title **Skinker.** This last, which has puzzled more than one commentator, is found in the rare book of 1703, *Meteorologiae by Mr. Cock, Philomathemat.:*

> Jupiter in the Skinker opposed by Saturn in the Lion did raise mighty South-west winds.

But the passage affords its own explanation that ought not to have been delayed till now; for we know our sign to be the opposite of Leo, while the dictionaries tell us that this archaic or provincial word signifies a Tapster, or Pourer-out of liquor, which Aquarius and Ganymede have notably been in all ages of astronomy.

Although early authors varied in their ascription of the twelve zodiacal constellations to the twelve tribes of Israel, yet they generally were in accord in assigning this to Reuben, "unstable as water." But the fountainheads of all this Jewish banner story, Jacob's death-bed address to his sons in Egypt, and Moses' dying song on Mount Nebo, are not clear enough to justify much positiveness as to the proper assignment of any of the tribal symbols, if indeed the Israelites had any at all. The little that we have on the subject is from Josephus and the *Chaldee Paraphrase.*

Dante, in the 19th canto of *Il Purgatorio,* wrote that here

> geomancers their Fortuna Major
> See in the Orient before the dawn
> Rise by a path that long remains not dim;

which Longfellow explains in his notes on the passage:

> Geomancy is divination by points in the ground, or pebbles arranged in certain figures, which have peculiar names. Among these is the figure called the Fortuna Major, which

is thus drawn, ⁑ ⁑ and, by an effort of the imagination, can also be formed out of some of the last stars *⁎* in Aquarius and some of the first in Pisces.

In astrology it was the **Airy Trigon,** Gemini and Libra being included, and

a sign of no small note, since there was no disputing that its stars possessed influence, virtue, and efficacy, whereby they altered the air and seasons "in a wonderful, strange, and secret manner";

and an illuminated manuscript almanac of 1386, perhaps the earliest in our language that has been printed, says of the sign: "It is gode to byg castellis, and to wed, and lat blode." With Capricorn it was the **House of Saturn,** governing the legs and ankles; and when on the horizon with the sun the weather was always rainy. When Saturn was here, he had man completely in his clutches—*caput et collum;* while Jupiter, when here, had *humeros, pectus et pedes.*

As **Junonis astrum** it was a diurnal sign, Juno and Jove being its guardians, and bore rule over Cilicia and Tyre; later, over Arabia, Tatary, Denmark, Russia, Lower Sweden, Westphalia, Bremen, and Hamburg.

Proctor's *Myths and Marvels of Astronomy* has a list of the astrological colors of the zodiac signs attributing to Aquarius an aqueous blue; while Lucius Ampelius, of our 2d century, assigning in his *Liber Memorialis* the care of the various winds to the various signs, intrusts to this the guardianship of Eurus and Notus, which blew from the east, or southeast, and from the south.

The astronomers' symbol for the sign, ♒, showing undulating lines of waves, is said to have been the hieroglyph for Water, the title of Aquarius in the Nile country, where a measuring-rod may have been associated with it; indeed Burritt drew such in the hand of the figure as **Norma Nilotica,** a suggestion of the ancient Nilometer.

Brown, in the 47th volume of *Archaeologia*, has these interesting remarks on the symbols of the signs:

Respecting these Mr. C. W. King observes: "Although the planets are often expressed by their emblems, yet neither they nor the signs are ever to be seen represented on antique works by those symbols so familiar to the eye in our almanacs. Wherever such occur upon a stone it may be pronounced without any hesitation a production of the cinque-cento, or the following century. . . . As for the source of these hieroglyphics, I have never been able to trace it. They are to be found exactly as we see them in very old medieval MSS."; and Mr. King is inclined, in default of any other origin, "to suspect they were devised by Arab sages"—an opinion which I do not follow. The subject is certainly shrouded in great obscurity; and even Professor Sayce recently informed me that he had been unable to trace the history of the zodiacal symbols up to their first appearance in Western literature.

While Miss Clerke writes that they are found in manuscripts of about the 10th century, but in carvings not until the 15th or 16th. Their origin is unknown; but some, if not all, of them have antique associations.

Hargrave's *Rosicrucians* has an illustration of an object showing an Egyptian cross and disk with our present symbols of Leo and Virgo, or Scorpio, purporting to be from the breast of a mummy in the museum of the London University. If this statement be correct, a much earlier origin can be claimed for these symbols[1] than has hitherto been supposed.

From his researches into the archaic astronomical symbolism on classic coins, monuments, etc., Thompson concludes that the great bas-relief of the Asiatic Cybele, now in the Hermitage Museum at Saint Petersburg, was designed to represent the ancient tropics of Aquarius and Leo; and that Aquarius, Aquila,—or more probably the other *Vultur*, our Lyra,—Leo, and Taurus appear in the familiar imagery of *Ezekiel* i, 10, and x, 14, and of *The Revelation* iv, 7.

Aquarius is not conspicuous, being chiefly marked by the stars γ, ζ, η, and π,—the Urn, the familiar Y,—called by the Greeks Κάλπη, Κάλπις, Κάλπεις, and **Situla**, or **Urna**, by the Latins, Pliny making a distinct constellation of the latter; and by the line of fainter stars, λ, φ, χ, ψ, ω, and others indicating the water running down into the mouth of the Southern Fish, or, as it is occasionally drawn, uniting with the river Eridanus. Spence, commenting on this figure on the Farnese globe and its description by Manilius, *Ad juvenem, aeternas fundentem Piscibus undas*, and *Fundentis semper Aquarii*, wrote:

Ganymedes, the cup-bearer of Jupiter. He holds the cup or little urn in his hand, inclined downwards; and is always pouring out of it: as indeed he ought to be, to be able from so small a source to form that river, which you see running from his feet, and making so large a tour over all this part of the globe.

Manilius ended his lines on Aquarius with *Sic profluit urna*, which Spence translated "And so the urn flows on"; adding:

which seems to have been a proverbial expression among the antients, taken from the ceaseless flowing of this urn; and which might be not inapplicable now, when certain ladies are telling a story; or certain lawyers are pleading.

Geminos, in his Ἐισαγωγή, about 77 B. C., made a separate constellation of this stream as Χύσις ὕδατος, the Pouring Forth of Water; but Aratos also had called it this as well as the **Water,** although in the latter he included β Ceti and the star Fomalhaut. Cicero gave it as **Aqua**; and the

[1] An interesting article on the symbols appears in Bailly's *Histoire de l'Astronomie Ancienne*, Paris, 1775.

scholiast on Germanicus, as **Effusio aquae**; while **Effusor** and **Fusor aquae** were common titles. The modern Burritt has **Fluvius Aquarii** and **Cascade.**

The stars marking the ribs of the figure in this constellation are, in some maps, mingled with *c* and others in Capricorn.

Although of astronomical importance chiefly from its zodiacal position and from its richness in doubles, clusters, and nebulae, it also is interesting from the fact that one of its three stars *ψ* was occulted by the planet Mars on the 1st of October, 1672. This occultation was predicted by Flamsteed, and, on his suggestion, observed and verified in France and by Richer at Cayenne; and the several independently accordant results are considered reliable, although made more than two centuries ago. These have enabled our modern astronomers, especially Leverrier, accurately to ascertain the mean motion of Mars, and materially aid them in calculating the mass of the earth and our distance from the sun.

Aquarius lies between Capricornus and Pisces, the sun entering it on the 14th of February, and leaving it on the 14th of March.

Argelander catalogues here 97 naked-eye stars; Heis, 146.

La Lande, citing Firmicus and the Egyptian sphere of Petosiris,[1] wrote in *l'Astronomie:*

> *Aquarius se lève, avec un autre constellation qu'il nomme Aquarius Minor avec la Faulx, le Loup, le Lièvre & l'Autel;*

but elsewhere I find no allusion to this **Lesser Waterman,** and the statement is incorrect as to the other constellations; indeed the **Faulx** is entirely unknown to us moderns.

α, 3.2, pale yellow.

Sadalmelik is from the Arabic **Al Sa'd al Malik,** the Lucky One of the King, sometimes given as **Al Sa'd al Mulk,** the Lucky One of the Kingdom, under which last title Kazwini and Ulug Beg combined it with *o*. It similarly was **Sidus Faustum Regis** with the astrologers. Burritt called it **El Melik** and **Phard,** but this last seems unintelligible.

The Rucbah of the *Century Cyclopedia* is erroneous for this star — indeed was intended for *a* Sagittarii.

Sadalmelik lies on the right shoulder of the figure, 1° south of the celestial equator, and has a distant 11th-magnitude gray companion.

With *ε* and *θ* Pegasi it made up the 23d *sieu* **Goei,** or **Wei,** Steep, or Danger, anciently **Gui**; but Brown says that the word signifies Foundation. *a* was the determinant star of this lunar station.

[1] Petosiris, the philosopher of Necepsos, the astronomical King of Saïs, was an almost mythical character to the Greeks; for Ptolemy termed him ἀρχαῖος, although he is generally assigned to the period of 900–700 B.C.

Gould called it red, and of 2.7 magnitude. It culminates on the 9th of October. From between α and η radiate the **Eta Aquarids,** the meteors visible from April 29th to May 2d.

β, 3.1, pale yellow.

Sadalsuud — not **Sund** nor **Saud,** as frequently written — is from **Al Sa'd al Su'ud,** liberally translated the Luckiest of the Lucky, from its rising with the sun when the winter had passed and the season of gentle, continuous rain had begun. This title also belongs to the 22d *manzil*, which included the star with ξ of Aquarius and c of Capricornus.

β and ξ also constituted the Persian lunar station **Bunda** and the similar Coptic **Upuineuti,** the Foundation; but β alone marked the *sieu* **Heu, Hiu,** or **Hü,** Void, anciently **Ko,** the central one of the seven *sieu* which, taken together, were known as **Heung Wu,** the Black Warrior, in the northern quarter of the sky. It is found in Hindu lists as **Kalpeny,** of unknown signification. On the Euphrates it was **Kakkab Nammaχ,** the Star of Mighty Destiny, that may have given origin to the title of the *manzil*, as well as to the astrologers' name for it — **Fortuna Fortunarum.**

Al Firuzabadi of Khorasan, editor of *Al Ḳāmūs*, the great Arabic dictionary of the 14th century, called some of the smaller stars below this **Al Au'ā,** the plural of Nau', a Star, but without explanation, and they certainly are inconspicuous.

γ, 4.1, greenish,

on the right arm at the inner edge of the Urn, and the westernmost star in the Y, is **Sadachbia,** from **Al Sa'd al Aḣbiyah,** which has been interpreted the Lucky Star of Hidden Things or Hiding-places, because when it emerged from the sun's rays all hidden worms and reptiles, buried during the preceding cold, creep out of their holes! But as this word Aḣbiyah is merely the plural of Ḣibā', a Tent, a more reasonable explanation is that the star was so called from its rising in the spring twilight, when, after the winter's want and suffering, the nomads' tents were raised on the freshening pastures, and the pleasant weather set in. This idea renders Professor Whitney's "Felicity of Tents" a happy translation of the original. ζ, η, and π are included with γ under this designation by Ulug Beg — ζ, in the centre, marking the top of the tent; Kazwini, however, considered this central star as **Al Sa'd,** and the three surrounding ones his tents.

All these stars, with α, formed the 23d *manzil*, bearing the foregoing title.

γ, ζ, η, π, and τ were the Chinese **Fun Mo,** the Tomb.

It was near γ that the Capuchin friar of Cologne, Schyraelus de Rheita,[1] in 1643, thought that he had found five new satellites attendant upon Jupiter, which he named Stellae Urbani Octavi in compliment to the reigning pontiff; and a treatise, *De novem Stellae circa Jovem*, was written by Lobkowitz upon this wonderful discovery. "The planet, however, soon deserted his companions, and the stars proved to be the little group in front of the Urn."

δ, 3.4,

the **Scheat** of Tycho, and **Scheat Edeleu** of Riccioli, is **Skat** in modern lists, and variously derived: either from Al Shi'at, a Wish, said to be found for it on Arabic globes; or from Al Ṣāk, the Shin-bone, near which it is located in the figure. But Hyde, probably following Grotius, said that it was from Al Saʿd of the preceding stars.

On the Euphrates it seems to have been associated with Hasisadra or Xasisadra, the 10th antediluvian king and hero of the Deluge; while, with β, κ, and others adjacent, it was the lunar station **Apin,** the Channel, and individually the Star of the Foundation. The corresponding stations, **Khatsar** in Persia, **Shawshat** in Sogdiana, and **Mashtawand** in Khorasmia, were also determined by this star.

The Chinese knew it, with τ, χ, the three stars ψ, and some in Pisces, as **Yu lin Keun,** the Imperial Guard.

From near δ issues a meteor stream, the **Delta Aquarids,** from the 27th to the 29th of July, and not far away Mayer noted as a fixed star, on the 25th of September, 1756, the object that nearly twenty-five years later Sir William Herschel observed as a comet, but afterwards ascertained to be a new planet, our Uranus.

ϵ, 3.4,

was **Al Bali,** the brightest one of the 21st *manzil*, **Al Saʿd al Bulaʿ,** the Good Fortune of the Swallower, which included μ and ν; these last also known as **Al Bulāān** in the dual. Kazwini said that this strange title came from the fact that the two outside stars were more open than α and β of Capricorn,

1 De Rheita is more deservedly famous as a supposed inventor, in 1650, of the planetarium, an honor also claimed for Archimedes of the 3d century before Christ, for Posidonius the Stoic, mentioned by Cicero in *De Natura Deorum*, and for Boëtius about the year A. D. 510. This instrument is the orrery of modern days, named by Sir Richard Steele after Charles Boyle, Earl of Orrery, for whom one was made in 1715 by Rowley, from designs by the clockmaker George Graham. Professor Roger Long constructed one eighteen feet in diameter, in 1758, for Pembroke Hall, Cambridge, where it probably still remains; and Doctor William Kitchiner mentioned one by Arnold, annually exhibited in London about the year 1825, that was 130 feet in circumference.

so that they seemed to swallow, or absorb, the light of the other! The corresponding *sieu*, **Mo, Mu, Niu, Nü,** or **Woo Neu,** a Woman, anciently written **Nok,** was composed of these stars with the addition of another, unidentified, ϵ being the determinant; and the same three were the Euphratean lunar asterism **Munaχa,** the Goat-fish, and the Coptic **Upeuritos,** the Discoverer.

Bayer mentioned for it **Mantellum** and **Mantile,** marking the Napkin or Towel held in the youth's hand; but in some early drawings this was shown as a **Bunch of Grain Stalks.**

Grotius had **Ancha** and **Pyxis,** but neither appropriate; while in our day the former is applied only to θ, and the latter is never seen as a stellar title except in La Caille's Pyxis Nautica in Argo.

Eastward from ϵ, near ν, is the **Saturn Nebula,** N. G. C. 7009, that the largest telescopes show somewhat like the planet.

ζ, Binary, 4 and 4.1, very white and white.

Although unnamed, this is an interesting star at the centre of the Y of the Urn, and almost exactly on the celestial equator.

Mayer discovered its duplicity in 1777, and its binary character, first noted by Herschel in 1804, was confirmed by his son in 1821; but the period is not yet determined, although it is very long.

The components are $3''.3$ apart, and the position angle 322°.

θ, 4.3,

is **Ancha,** the Hip, although on most modern atlases the star lies in the belt on the front of the figure. The word is from the Latin of the Middle Ages, and still appears in the French *hanche*, our haunch.

Reeves says that in China it was **Lei,** a Tear.

κ, 5.5.

Situla is applied to this, from the classical Latin term for a Water-jar or -bucket, the later Arabian word being the somewhat similar **Saṭl,** and the earlier **Al Dalw.**

Gassendi, however, derived it from *sitis*, thirst, the Waterman's Urn having been figured by some as an Oven!

Theon the Younger, father of the celebrated Hypatia of our 5th century, termed this star Ὀινοχοεία, the Outpouring of Wine, as if by Ganymede; and others, Κάλπη, and **Urna,** the southern edge of which, near the outflow, it marks.

Keats, in *Endymion*, very fancifully wrote of this Urn:

> Crystalline brother of the belt of heaven,
> Aquarius! to whom King Jove has given
> Two liquid pulse streams 'stead of feather'd wings,
> Two fan-like fountains,— thine illuminings
> For Dian play.

In China κ was **Heu Leang,** the Empty Bridge.

λ, 3.8, red,

is the most prominent of the first stars in the Stream.

Proclus followed Aratos in calling it Ὕδωρ, the Water; and others, Ἔκχυσις, the Outpouring; Aratos describing it,

> Like a slight flow of water here and there
> Scattered around, bright stars revolve but small;

although these titles, appropriated by Bayer for λ, originally were for the whole group set apart as the Stream.

λ, with about 100 stars surrounding it, was the 23d *nakshatra* **Çatabhishaj,** the Hundred Physician, whose regent was Varuna, the goddess of the waters and chief of the Adityas, the various early divinities of Hindu mythology, and all children of Aditi, the Sky and the Heavens.

With ι, σ, and φ, it was the Chinese asterism **Luy Peih Chin,** the Camp with Intrenched Walls; but this included stars in Capricornus and Pisces.

ο, 4.7, a little to the southwest of α, was associated with it under the title **Al Sa'd al Mulk.** In China it was **Kae Uh,** the Roof.

π, 4.8, was called **Seat** by Grotius, as one of the group Al Sa'd al Ahbiyah.

Sundry other four or five small stars in Aquarius were given by Reeves as **Foo Yue,** the Headsman's Ax.

★

> si quaeritis astra
> Tunc oritur magni praepes adunca Jovis.
>
> — Ovid's *Fasti.*

> Jove for the prince of birds decreed,
> And carrier of his thunder, too,
> The bird whom golden Ganymede
> Too well for trusty agent knew.
>
> — Gladstone's translation of Horace's *Odes.*

Aquila, the Eagle,

the French **Aigle,** the German **Adler,** and the Italian **Aquila,** next to and westward from the Dolphin, is shown flying toward the east and across the Milky Way; its southern stars constituting the now discarded Antinoüs. Early representations added an arrow held in the Eagle's talons; and Hevelius included a bow and arrow in his description; but on the Heis map the Youth is held by Aquila, for the Germans still continue this association in their combined title **der Adler mit dem Antinoüs.**

Our constellation is supposed to be represented by the bird figured on a Euphratean uranographic stone of about 1200 B. C., and known on the tablets as **Idχu Zamama,** the Eagle, the Living Eye.

It always was known as Aquila by the Latins, and by their poets as **Jovis Ales** and **Jovis Nutrix,** the Bird, and the Nurse, of Jove; **Jovis Armiger** and **Armiger Ales,** the Armor-bearing Bird of Jove in this god's conflict with the giants; while **Ganymedes Raptrix** and **Servans Antinoüm** are from the old stories that the Eagle carried Ganymede to the heavens and stood in attendance on Jove. Ovid made it **Merops,** King of Cos, turned into the Eagle of the sky; but others thought it some Aethiopian king like Cepheus, and with the same heavenly reward.

As the eagles often were confounded with the vultures in Greek and Roman ornithology, at least in nomenclature, Aquila also was **Vultur volans,** the stars β and γ, on either side of a, marking the outstretched wings; this title appearing even as late as Flamsteed's day, and its translation, the **Flying Grype,** becoming the Old English name, especially with the astrologers, who ascribed to it mighty virtue.

'Αετός, the Eagle, in a much varied orthography, was used for our constellation by all the Greeks; while poetically it was *Διός Ὄρνις*, the Bird of Zeus; and Pindar had *'Οιωνῶν Βασιλεῦς*, the King of Birds, which, ornithologically, has come to our day. Later on it was *Βάσανος*, *Βασανισμός*, and *Βασανιστήριον*, all kindred titles signifying Torture, referred by Hyde to the story of the eagle which preyed on the liver of Prometheus. Similarly we find **Aquila Promethei** and **Tortor Promethei**; but Ideler said that this idea came from a confounding by Scaliger of the Arabic 'Ikāb, Torture, and 'Oḳāb, Eagle.

Dupuis fancifully thought that its name was given when it was near the summer solstice, and that the bird of highest flight was chosen to express the greatest elevation of the sun; and he asserted that the famous three Stymphalian Birds of mythology were represented by Aquila, Cygnus, and Vultur cadens, our Lyra, still located together in the sky; the argument being that these are all paranatellons of Sagittarius, which is the fifth in the line of zodiacal constellations beginning with Leo, the Nemean lion, the object of Hercules' first labor, while the slaying of the birds was the fifth. Appropriately enough, like so much other stellar material, these creatures came from Arabia, migrating thence either to the Insula Martis, or to Lake Stymphalis, where Hercules encountered them.

Thompson thinks that the fable, in Greek ornithology, of the eagle attacking the swan, but defeated by it, is symbolical of "Aquila, which rises in the East, immediately after Cygnus, but, setting in the West, goes down a little while before that more northern constellation."

A similar thought was in the ancient mind as to the eagle in opposition to the dolphin and the serpent; their stellar counterparts, Aquila, Delphinus, and Serpens, also being thus relatively situated.

In connection with the story of Ganymede, the eagle appeared on coins of Chalcis, Dardanos, and Ilia; and generally on those of Mallos in Cilicia and of Camarina; while it is shown perched on the Dolphin on coins of Sinope and other towns, chiefly along the Black Sea and Hellespont. One, bearing the prominent stars, was struck in Rome, 94 B. C., by Manius Aquilius Nepos,[1] the design being evidently inspired by his name; and a coin of Agrigentum bears Aquila, with Cancer on the reverse,—the one setting as the other rises.

To the Arabians the classical figure became **Al 'Okāb,** probably their Black Eagle, Chilmead citing this as **Alhhakhab**; while their **Al Naṣr al Ṭāir,** the Flying Eagle, was confined to α, β, and γ; although this was contrary to their custom of using only one star for a sky figure. Grotius called the whole **Altair** and **Alcair**; Bayer said **Alcar** and **Atair.** Al Achsasi, however, mentioned it as **Al Ghurāb,** the Crow, or Raven, probably a late Arabian name, and the only instance that I have seen of its application to the stars of our Aquila.

Persian titles were **Alub, Gherges,** and **Shahin tara zed,** the Star-striking Falcon of Al Naṣr al Dīn, but now divided for β and γ. In the *Ilkhanian Tables*, as perhaps elsewhere, it was Γύψ πετόμενος, the Flying Vulture; the Turks call it **Taushaugjil,** their Hunting Eagle;—all these for the three bright stars.

The Hebrews knew it as **Neshr,** an Eagle, Falcon, or Vulture; and the *Chaldee Paraphrase* asserted that it was figured on the banners of Dan; but as these tribal symbols properly were for the zodiac, Scorpio usually was ascribed to Dan. This confusion may have originated from the fact, asserted by Sir William Drummond, that in Abraham's day Scorpio was figured as an Eagle. Caesius said that Aquila represented the **Eagle of military Rome,** or the **Eagle of Saint John**; but Julius Schiller had already made it **Saint Catherine the Martyr**; and Erhard Weigel, a

[1] This was the consul defeated and captured by Mithridates, who put him to death by pouring molten gold down his throat in punishment for his rapacity.

professor at Jena in the 17th century, started a new set of constellations, based on the heraldry then so much in vogue, among which was the **Brandenburg Eagle,** made up from Aquila, Antinoüs, and the Dolphin. Hevelius said that the stellar Eagle was a fitting representation of that bird on the Polish and Teutonic coats of arms.

The Chinese have here the **Draught Oxen,** mentioned in the book of odes entitled *She King*, compiled 500 years before Christ by K'ung fu tsu, Kung the Philosopher (Confucius), — the passage being rendered by the Reverend Doctor James Legge:

> Brilliant show the Draught Oxen,
> But they do not serve to draw our carts;

and the three bright stars are their **Cowherd,** for whom the Magpies' Bridge gives access to the **Spinning Damsel,** our Lyra, across the River of the Sky, the Milky Way. This story appears in various forms; stars in the Swan sometimes being substituted for those in the Eagle, Lyra becoming the **Weaving Sisters.**

The Korean version, with more detail, turns the Cowherd into a Prince, and the Spinster into his Bride, both banished to different parts of the sky by the irate father-in-law, but with the privilege of an annual meeting if they can cross the River. This they accomplish through the friendly aid of the good-natured magpies, who congregate from all parts of the kingdom during the 7th moon, and on its 7th night form the fluttering bridge across which the couple meet, lovers still, although married. When the day is over they return for another year to their respective places of exile, and the bridge breaks up; the birds scattering to their various homes with bare heads, the feathers having been worn off by the trampling feet of the Prince and his retinue. But as all this happens during the birds' moulting-time, the bare heads are not to be wondered at; nor, as it is the rainy season, the attendant showers which, if occurring in the morning, the story-tellers attribute to the tears of the couple in the joy of meeting; or if in the evening, to those of sorrow at parting. Should a magpie anywhere be found loitering around home at this time, it is pursued by the children with well-merited ill-treatment for its selfish indifference to its duty. Nor must I forget to mention that the trouble in the royal household originated from the Prince's unfortunate investment of the paternal *sapekes* in a very promising scheme to tap the Milky Way and divert the fluid to nourish distant stars.

Another version is given by the Reverend Doctor William Elliot Griffis in his *Japanese Fairy World*, where the Spinning Damsel is the industrious princess Shokujo, separated by the Heavenly River from her herd-boy lover,

Kinjin. But here the narrator makes Capricorn and the star Wega represent the lovers.

The native Australians knew the whole of Aquila as **Totyarguil,** one of their mythical personages, who, while bathing, was killed by a kelpie; their stellar Eagle being Sirius.

It was in the stars of our constellation, to the northwest of Altair, that Professor Edward E. Barnard discovered a comet from its trail on a photograph taken at the Lick Observatory on the 12th of October, 1892—the first ever found by the camera.

Argelander catalogued 82 naked-eye stars in Aquila, including those of Antinoüs; Heis gives 123.

α, 1.3, pale yellow.

Altair is from a part of the Arabic name for the constellation; but occasionally is written **Althair, Athair, Attair,** and **Atair;** this last readers of *Ben Hur* will remember as the name of one of the shaykh Ilderim's horses in the chariot race at Antioch. And the word has been altered to **Alcair, Alchayr,** and **Alcar.**

In the *Syntaxis* it was 'Αετός, one of Ptolemy's few stellar titles, probably first applied to α, and after the formation of the figure transferred to the latter, as in other instances in the early days of astronomy. Even six or seven centuries before Ptolemy it was referred to as 'Αιετός where the chorus in the 'Ρῆσος, until recently attributed to Euripides, says:

> What is the star now passing?

the answer being:

> The Pleiades show themselves in the east,
> The Eagle soars in the summit of heaven.

It is supposed that long antecedent to this it was the Euphratean **Idχu,** the Eagle, or **Erigu,** the Powerful Bird, inscriptions to this effect being quoted by Brown, who thinks that it also was the Persian **Muru,** the Bird; the Sogdian **Shad Mashir,** and the Khorasmian **Sadmasij,** the Noble Falcon.

In Mr. J. F. Hewitt's *Essays on the Ruling Races of Prehistoric Times* it is asserted that later Zend mythology knew Altair as **Vanant,** the Western Quarter of the heavens, which earlier had been marked by our Corvus.

With β and γ it constituted the twenty-first *nakshatra* **Çravana,** the Ear, and probably was at first so drawn, although also known as **Çrona,** Lame, or as **Açvattha,** the Sacred Fig Tree, Vishnu being regent of the asterism; these stars representing the Three Footsteps with which that god strode through the heavens, a Trident being the symbol.

In China α, β, and γ were **Ho Koo,** a River Drum.

In astrology Altair was a mischief-maker, and portended danger from reptiles.

Ptolemy, who designated the degrees of star brilliancy by Greek letters, applied β to this as being of the 2d magnitude, whence some think that it has increased in light since his day. It is now the standard 1st magnitude according to the Pogson, or "absolute," photometric scale generally adopted by workers in stellar photometry, and is largely used in determining lunar distances at sea; while Flamsteed made it the fundamental reference star in his observations on the sun and in the construction of his catalogue.

Its parallax,[1] $0''.214$, considered by Elkin as nearly or quite exact, indicates a distance of about $15\frac{1}{5}$ light years.

Its spectrum is of Pickering's class Xb of Secchi's first type, but peculiar, with very hazy solar lines between the broad hydrogen lines.

Altair has the large proper motion of $0''.65$ annually; and Gould thought it slightly variable.

It marks the junction of the right wing with the body, and rises at sunset about the 15th of June, culminating on the 1st of September.

Near it appeared, in A. D. 389, an object, whether a temporary star or a comet is not now known, said by Cuspinianus to have equaled Venus in brilliancy, which vanished after three weeks' visibility; and there is record of another, of sixty years previous, in this constellation.

5° to the eastward of Altair, according to Denning, lies the radiant point of the **Aquilids,** the meteor stream visible from the 7th of June to the 12th of August.

β, 3.9, pale orange.

Alshain is from Shahin, a portion of the Persian name for the constellation; but Al Achsasi termed it **Al Unuḳ al Ghurāb,** the Raven's Neck.

It is the southern of the two stars flanking Altair; yet, although it bears the second letter, is not as bright as γ or δ.

γ, 3, pale orange.

Tarazed, or **Tarazad,** from the same Persian title, lies north of Altair.

These three stars constitute the **Family of Aquila,** the line joining them being 5° in length.

[1] A parallax of $1''$ represents a distance from the earth of 3.26 light years; a light year, the astronomers' unit in measuring stellar distances,—light traveling 186,327 miles in a second of time,—being about 63,000 times the distance of the earth from the sun. But no star thus far investigated has so large a parallax; that of the nearest, α Centauri, being only $0''.75$.

Just north of γ is π, the only pretty and fairly easy double in the constellation. The components, of 6 and 6.8 magnitudes, $1''.5$ apart, are at a position angle of $120^{\circ}.7$.

δ, η, and θ, of 3d to 4th magnitudes, in Antinoüs, were **Al Mīzān,** the Scale-beam, of early Arabia, from their similar direction and nearly equal distances apart.

ϵ, 4.3, and ζ, 3.3, green.

Each of these is known as **Deneb,** from **Al Dhanab al 'Okāb,** the Eagle's Tail, which they mark.

In China they were **Woo** and **Yuë,** names of old feudal states.

η, in Antinoüs, is a noteworthy short-period variable of the 2d type, discovered by Pigott in 1784, yellow in tint, and fluctuating in brilliancy from 3.5 to 4.7 in a period of about seven days and four hours, and thus a convenient and interesting object of observation for midsummer evenings.

Its spectrum is similar to that of our sun, and Lockyer and Belopolsky think it a spectroscopic binary.

θ was the Chinese **Tseen Foo,** the Heavenly Raft.

ι, 4.3, and λ, 3.6,

were **Al Thalīmain,** the Two Ostriches, by some confusion with the not far distant stars of like designation in Sagittarius; but the Grynaeus *Syntaxis* of 1538 gave λ, with some others unlettered, as belonging to the Dolphin.

ι, with δ, η, and κ, was **Yew Ke** in China, the Right Flag; ρ being **Tso Ke,** the Left Flag.

λ, with h, g, and some stars in Scutum, was **Tseen Peen,** the Heavenly Casque.

★

And this you note but little time aloft;
For opposite Bear-watcher doth it rise.
And whilst his course is high in air,
It quickly speeds beneath the western sea.

Robert Brown, Junior's, translation of the *Phainomena* of Aratos.

Ara, the Altar,

is **Altar** in Germany, **Altare** in Italy, **Autel** and **Encensoir** in France.

It is located as Aratos described it —

'neath the glowing sting of that huge sign
The *Scorpion*, near the south, the *Altar* hangs;

and in classical times was intimately associated with Centaurus and Lupus, which it joined on the west before Norma was formed.

The Latins knew it under our title, often designated as **Ara Centauri, Ara Thymiamatis,** and as **Thymele,** the altar of Dionysus; and occasionally in the diminutive **Arula.** It also was **Altare, Apta Altaria, Altarium**; **Sacrarium** and **Sacris**; **Acerra,** the small altar on which perfumes were burned before the dead; **Batillus,** an Incense Pan; **Prunarum Conceptaculum,** a Brazier; **Focus, Lar,** and **Ignitabulum,** all meaning a Hearth; and 'Εστία, or **Vesta,** the goddess of the hearth.

Thuribulum and **Turribulum,** a Censer, more correctly **Turibulum,** were customary titles down to the 18th century.

Pharus also appears, altars often being placed upon the summits of temple towers and thus serving the ancients as lighthouses, of which the Alexandrian Pharos was the great example.

The *Alfonsine Tables* added to some of these titles **Puteus,** a Pit; **Sacrarius,** and **Templum,** a Sacred Place; but represented it as a typical altar. The *Leyden Manuscript* made it a tripod censer with incense burning; the illustrated editions of *Hyginus* of 1488 and 1535, an altar from which flames ascend, with demons on either side; and an illustrated German manuscript of the 15th century showed the Pit with big demons thrusting little ones into the abyss. This recalls the story of Jove's punishment of the defeated giants after he had, as Manilius wrote,

> Rais'd this Altar, and the Form appears
> With Incense loaded, and adorn'd with Stars;

the occasion being the war with the Titans, when the gods needed an altar in heaven for their mutual vows. That poet also described it as

> ara ferens turris, stellis imitantibus ignem,

which would show that the flame was conceived of as rising northwards through the Milky Way, or that the latter itself was the smoke and flame; and it was so thought of and represented by the ancients, and down to the times of Arabic globes and Middle Age manuscripts. But from Bayer's day to ours it has been shown in an inverted position, which for a southern constellation is appropriate.

Aratos called it Θυτήριον; others, Θυσιαστήριον, both signifying an Altar; Proclus and Ptolemy, Θυμιατήριον, a Censer; and Bayer cited 'Εχάρα that should be 'Εσχάρα, a Brazier; Πυράμνη, not a lexicon word; and Λιβανωτίς, by which he doubtless intended the Λιβανωτρίς, or Censer, where the votive plant was burned. Eratosthenes had Νέκταρ ἡ Θῦτήριον, which Ideler and Schaubach, however, did not understand, and thought a corrupted reading.

Its varied classical names show disagreement as to its form, yet great familiarity with its stars, on the part of early observers, with whom it was of importance as portending changes in the winds and weather; Aratos devoting twenty-eight lines — a large proportionate space — of the *Phainomena* to this character of Ara.

In Arabia it was **Al Mijmarah,** a Censer, which, being its only title in that country, implies that it was unformed there before the introduction of Greek astronomy. Derivations from this word are found in the **Almegramith** of Riccioli and the **Almugamra** of Caesius.

This last author said that Ara represented **one of the altars raised by Moses,** or the permanent **golden one in the Temple** at Jerusalem; but others of the biblical school considered it the **Altar of Noah** erected after the Deluge. Euphratean research seems to show a stellar Altar differently located, which Brown says probably was the lost zodiacal sign subsequently represented by the Claws and afterwards by the Balance; and identifies it with the 7th Akkadian month and sign **Tul-Ku,** the Holy Altar, or the Illustrious Mound, perhaps a reference to the mound-altar of the Tower of Babel. When these changes were accomplished this early zodiacal Altar was removed to its present position, and its diversified altar-censer form retained from the Euphratean figuring. This recollection of the first Altar will perhaps account for the otherwise strange prominence given in classical times to our visually unimportant Ara, when Manilius called it **Mundi Templum;** this last word also having another stellar signification, for Varro used it to indicate a division of the sky.

Other details of this early Euphratean Altar are noted under Libra.

Ara is not wholly visible now north of the 23d degree of latitude; and its brief period above the horizon — only about four hours — explains Aratos' allusion in our motto; his horizon being about the same as that of the city of New York.

Gould catalogues in it eighty-five stars, from 2.8 to the 7th magnitude; but none seem to be named except in China. There α, 2.9 magnitude, was **Choo,** a Club or Staff; and with β, γ, and ι, **Low,** Trailing.

With θ it marks the top of the Altar's frame, culminating, on the 24th of July, just above the horizon in the latitude of New York,—40° 42′ 43″ at the City Hall.

Bayer's map carries the latter star several degrees too far to the southwest; similar errors being found in others of his constellation figures of the southern heavens.

β, a 2.8-magnitude, γ, δ, ε, and ζ mark the flame rising toward the south.

In China δ, 3.7, with ζ, was **Tseen Yin,** the Dark Sky; ε, a 4th-magnitude, was **Tso Kang,** the Left Watch; and *e* 602 of Reeves was **Tseen O,** Heaven's Ridge.

La Lande stated that a constellation was supposed to exist here, containing Ara's stars, that was represented on the Egyptian sphere of Petosiris as a **Cynocephalus.**

★

So when the first bold vessel dar'd the seas,
High on the stern the Thracian rais'd his strain
While Argo saw her kindred trees
Descend from Pelion to the main.
Transported demi-gods stood round.

Pope's *Ode on St. Cecilia's Day.*

Argo Navis, the Ship Argo,

generally plain **Argo,**— erroneously **Argus,** from confusion with its genitive case,— and **Navis,** is the German **Schiff,** the French **Navire Argo,** and the Italian **Nave Argo.**

It lies entirely in the southern hemisphere, east of Canis Major, south of Monoceros and Hydra, largely in the Milky Way, showing above the horizon of New York city only a few of its unimportant stars; but it covers a great extent of sky, nearly seventy-five degrees in length,— Manilius calling it *nobilis Argo,*— and contains 829 naked-eye components. The centre culminates on the 1st of March.

La Caille used for it nearly 180 letters, many of them of course duplicated, so that although this notation was adopted in the *British Association Catalogue*, recent astronomers have subdivided the figure for convenience in reference, and now know its three divisions as **Carina,** the Keel, with 268 stars, **Puppis,** the Stern, with 313, and **Vela,** the Sail, with 248. This last is the German **Segel.**

La Caille, moreover, formed from stars in the early subordinate division **Malus,** the Mast, **Pyxis Nautica,** the Nautical Box or Mariner's Compass, the German **See Compass,** the French **Boussole** or **Compas de Mer,** and the Italian **Bussola**; and this is still recognized by some good astronomers as **Pyxis.**

From other stars Bode formed **Lochium Funis,** his **Logleine,** our **Log and Line,** now entirely fallen into disuse.

The Ship appears to have no bow, thus presenting the same sectional character noticeable in Equuleus, Pegasus, and Taurus, and generally is so shown on the maps. It was in reference to this that Aratos wrote:

> Sternforward *Argō* by the *Great Dog's* tail
> Is drawn; for hers is not a usual course,
> But backward turned she comes, as vessels do
> When sailors have transposed the crooked stern
> On entering harbour; all the ship reverse,
> And gliding backward on the beach it grounds.
> Sternforward thus is Jason's *Argō* drawn.

This loss of its bow is said to have occurred

> when Argo pass'd
> Through Bosporus betwixt the justling rocks —

the Symplegades, the Cyanean (azure), or the Planctae Rocks at the mouth of the Euxine Sea. Yet Aratos may have thought it complete, for he wrote:

> All Argo stands aloft in sky,

and

> Part moves dim and starless from the prow
> Up to the mast, but all the rest is bright;

and it has often been so illustrated and described by artists and authors. The *Alfonsine Tables* show it as a complete double-masted vessel with oars, and Lubienitzki, in the *Theatrum Cometicum* of 1667, as a three-masted argosy with a tier of ports and all sails set full to the wind.

Mythology insisted that it was built by Glaucus, or by Argos, for Jason, leader of the fifty Argonauts, whose number equaled that of the oars of the ship, aided by Pallas Athene, who herself set in the prow a piece from the speaking oak of Dodona; the Argo being "thus endowed with the power of warning and guiding the chieftains who form its crew." She carried the famous expedition from Iolchis in Thessaly to Aea in Colchis,[1] in search of the golden fleece, and when the voyage was over Athene placed the boat in the sky.

Another Greek tradition, according to Eratosthenes, asserted that our constellation represented the first ship to sail the ocean, which long before

[1] Colchis was the district along the eastern shore of the Euxine Sea, now Mingrelia.

Jason's time carried Danaos with his fifty daughters from Egypt to Rhodes and Argos, and, as Dante wrote,

Startled Neptune with the aid of Argo.

Egyptian story said that it was the ark that bore Isis and Osiris over the Deluge; while the Hindus thought that it performed the same office for their equivalent Isi and Iswara. And their prehistoric tradition made it the ship Argha for their wandering sun, steered by Agastya, the star Canopus. In this Sanskrit *argha* we perhaps may see our title; but Lindsay derives Argo from *arek*, a Semitic word, used by the Phoenicians, signifying "long," this vessel having been the first large one launched.

Sir Isaac Newton devoted much attention to the famous craft, fixing the date of its building about 936 B. C., forty-two years after King Solomon.

With the Romans it always was Argo and Navis, Vitruvius writing *Navis quae nominatur Argo*; but Cicero called it **Argolica Navis** and **Argolica Puppis**; Germanicus, **Argoa Puppis**; Propertius, the elegiac poet of the 1st century before our era, **Iasonia Carina**; Ovid, **Pagasaea Carina** and **Pagasaea Puppis,** from the Thessalian seaport where it was built; Manilius, **Ratis Heroum,** the Heroes' Raft,

which now midst Stars doth sail;

and others, **Navis Jasonis,** or **Osiridis, Celox Jasonis, Carina Argoa, Argo Ratis,** and **Navigium Praedatorium,** the Pirate Ship. While somewhat similar are **Currus Maris,** the Sea Chariot, the **Currus Volitans** of Catullus, who said that in Egypt it had been the **Vehiculum Lunae.**

It also was **Equus Neptunius**; indeed Ptolemy asserted that it was known as a **Horse** by the inhabitants of Azania, the modern Ajan, on the north-eastern coast of Africa, south of Cape Gardafui.

The Arabians called it **Al Safīnah,** a Ship, and **Markab,** something to ride upon, that two or three centuries ago in Europe were transcribed **Alsephina** and **Merkeb.**

Grotius mentioned **Cautel** as a title for Puppis, "from the Tables," but he added *Hoc quid sit nescio.*

The biblical school of course called it **Noah's Ark,** the **Arca Noachi,** or **Archa Noae** as Bayer wrote it; Jacob Bryant, the English mythologist of the last century, making its story another form of that of Noah. Indeed in the 17th century the **Ark** seems to have been its popular title.

In Hewitt's *Essays* we find a reference to "the four stars which marked the four quarters of the heavens in the Zendavesta, the four Loka-pālas, or nourishers of the world," of the Hindus; and that author claims these for

Sirius in the east, the seven stars of the Greater Bear in the north, Corvus in the west, and Argo in the south. He gives the latter's title as **Sata Vaēsa,** the One Hundred Creators; all these imagined as forming a great cross in the sky. The differing Persian conception of this appears in the remarks on Regulus,— *a* Leonis.

The Chinese asterism **Tien Meaou** probably was formed from some components of Argo.

The constellation is noticeable in lower latitudes not only from its great extent and the splendor of Canopus, but also from possessing the remarkable variable *η* and its inclosing nebula.

Near the star *z'* Carinae appeared, between March 5 and April 8, 1895, a *nova* with a spectrum similar to those of the recent *novae* in Auriga and Norma.

> . . . like a meadow which no scythe has shaven,
> Which rain could never bend or whirl-blast shake,
> With the Antarctic constellations paven,
> Canopus and his crew, lay the Austral lake.
>
> Percy Bysshe Shelley's *The Witch of Atlas.*

α Carinae, —0.4, white.

Κάνωβος, in the early orthography of the Greeks, apparently was first given to this star by Eratosthenes, but Κάνωπος later on by Hipparchos. Ptolemy used the former word, among his few star-names, which Halley and Flamsteed transcribed into **Canobus**; but now it universally is **Canopus,** Al Sufi's translator having **Kanupus** as an Arabian adaptation of the Greek.

Aratos, Eudoxos, and Hipparchos also, designated it as Πηδάλιον, the Rudder, Cicero's **Gubernaculum,** Aratos writing:

> The slackened rudder has been placed beneath
> The hind-feet of the Dog.

Ancient ships had a rudder on each side of the stern, in one of which our star generally was figured, thus differing from the modern maps that locate it in the bank of oars.

Strabo, the geographer of the century preceding our era, said that its title was "but of yesterday," which may have been true of the word that we now know it by; but an Egyptian priestly poet of the time of Thothmes III — 1500 years before Strabo — wrote of it as **Karbana,**

> the star
> Which pours his light in a glance of fire,
> When he disperses the morning dew;

and this still was seen a millennium later in the **Kabarnit** of As-sur-ba-ni-pal's time.

Our name for it is that of the chief pilot of the fleet of Menelaos, who, on his return from the destruction of Troy, 1183 B. C., touched at Egypt, where, twelve miles to the northeastward from Alexandria, Canopus died and was honored, according to Scylax, by a monument raised by his grateful master, giving his name to the city [1] and to this splendid star, which at that time rose about 7½° above that horizon.

The foregoing derivation of the word Canopus is an early and popular one; but another, perhaps as old, and more probable, being on the authority of Aristides, is from the Coptic, or Egyptian, **Kahi Nub,** Golden Earth. Ideler, coinciding in this, claimed these words as also the source of other titles for Canopus, the Arabic **Wazn,** Weight, and **Ḥaḍar,** Ground; and of the occasional later **Ponderosus** and **Terrestris.** Although I find no reason assigned for the appropriateness of these names, it is easy to infer that they may come from the magnitude of the star and its nearness to the horizon; this last certainly made it the *περίγειος* of Eratosthenes.

Similarly the universal Arabic title was **Suhail,** written by Western nations **Suhel, Suhil, Suhilon, Sohayl, Sohel, Sohil,** and **Soheil, Sahil, Sihel,** and **Sihil**; all taken, according to Buttmann, from Al Sahl, the Plain.

This word also was a personal title in Arabia, and, Delitzsch says, the symbol of what is brilliant, glorious, and beautiful, and even now among the nomads is thus applied to a handsome person. Our word Canopus itself apparently had a somewhat similar use among early writers; for Eden translated from Vespucci's account of his third voyage and *Of the Pole Antartike and the Starres abowt the Same*:

Amonge other, I sawe three starres cauled Canopi, wherof two were exceadynge cleare, and the thyrde sumwhat darke;

and again, after describing the "foure starres abowte the pole":

When these are hydden, there is scene on the lefte syde a bryght Canopus of three starres of notable greatnesse, which beinge in the myddest of heaven representeth this figure ⁂ ;

with more to the same effect in connection with the Nubeculae; for it is to

[1] Ancient Canopus is now in ruins, but its site is occupied by the village of Al Bekūr, or Aboukir, famous from Lord Nelson's Battle of the Nile, August 1, 1798, and from Napoleon's victory over the Turks a year afterwards; and it is interesting to remember that it was here, from the terraced walls of the Serapeum, the temple of Serapis, that Ptolemy made his observations.

Serapis was the title of the great Osiris of Egypt as god of the lower world; his incarnation as god of the upper world being in the bull Apis.

these Clouds that the Canopus of Vespucci would seem to refer in much of his description. But I have never seen any explanation of this title as used by him, and Vespucci's fame certainly does not rest upon his knowledge of the skies. The great *New English Dictionary* erroneously quotes some of the foregoing as being references to our *a* Carinae, strangely ignoring this different use of the star's title.

Among the Persians Suhail is a synonym of wisdom, seen in the well-known Al Anwār i Suhaili, the Lights of Canopus.

A note to Humboldt's *Cosmos* tells us that this name was given to other stars in Argo, and Hyde asserted the same as to its use for stars in neighboring constellations. Thus he found Suhel Alfard, Suhel Aldabaran, and Suhel Sirius; in fact this last star, Karsten Niebuhr[1] said, was commonly known thus in Arabia a century and more ago.

The *Alfonsine Tables* had **Suhel ponderosus,** that appeared in a contemporary chronicle as **Sihil ponderosa,** a translation of **Al Suhail al Wazn.** In the 1515 *Almagest* it was **Subhel;** and in the *Graeco-Persian Tables* of Chrysococca (the 14th-century Greek astronomer, author, and physician resident in Persia), edited by Bullialdus in his *Astronomia Philolaica*, it was Σοαὶλ Ἰαμανῆ. This was from the Arabs' **Al Suhail al Yamaniyyah,** the Suhail of the South, or perhaps an allusion to the old story, told in connection with our Procyon, that Suhail, formerly located near Orion's stars, the feminine Al Jauzah, had to flee to the south after his marriage to her, where he still remains. Others said that Suhail only went a-wooing of Al Jauzah, who not only refused him, but very unceremoniously kicked him to the southern heavens.

Another occasional early title was **Al Fahl,** the Camel Stallion. Allusions to it in every age indicate that everywhere it was an important star, especially on the Desert. There it was a great favorite, giving rise to many of the proverbs of the Arabs, their stories and superstitions, and supposed to impart the much prized color to their precious stones, and immunity from disease. Its heliacal rising, even now used in computing their year, ripened their fruits, ended the hot term of the summer, and set the time for the weaning of their young camels, thus alluded to by Thomas Moore in his *Evenings in Greece :*

> A camel slept — young as if wean'd
> When last the star Canopus rose.

And in a general way it served them as a southern pole-star.

[1] This Niebuhr was the noted Danish traveler in the East between 1761 and 1767, and subsequently the father of the great historian. His discoveries at Persepolis gave the clue to the decipherment of cuneiform inscriptions.

It was worshiped by the tribe of Tai, as it probably still is by the wilder of the Badāwiyy; and in this connection Carlyle wrote of it in his *Heroes and Hero Worship:*

Canopus shining-down over the desert, with its blue diamond brightness (that wild, blue, spirit-like brightness far brighter than we ever witness here), would pierce into the heart of the wild Ishmaelitish man, whom it was guiding through the solitary waste there. To his wild heart, with all feelings in it, with no *speech* for any feeling, it might seem a little eye, that Canopus, glancing-out on him from the great, deep Eternity; revealing the inner splendour to him.

Cannot we understand how these men *worshipped* Canopus; became what we call Sabeans, worshipping the stars? . . .

To us also, through every star, through every blade of grass, is not a God made visible, if we will open our minds and eyes?

We do not worship in that way now: but is it not reckoned still a merit, proof of what we call a "poetic nature," that we recognize how every object has a divine beauty in it; how every object still verily is "a window through which we may look into Infinitude itself"?

Moore wrote of it in *Lalla Rookh:*

The Star of Egypt, whose proud light,
Never hath beam'd on those who rest
In the White Islands of the West;

again alluding to it, in the same poem, as the cause of the unfailing cheerfulness of the Zingians.[1] And, as the constellation was associated on the Nile with the great god Osiris, so its great star became the **Star of Osiris**; but, later on, Capella and the scholiast on Germanicus called it **Ptolemaeon** and **Ptolemaeus,** in honor of Egypt's great king Ptolemy Lagos; and at times it has been **Subilon,** but the appropriateness of this I have been unable to verify. The Σάμπιλος, cited by Hyde as from Kircher, and so presumably Coptic, is equally unintelligible.

While all this knowledge of Canopus is ancient, it seems "but of yesterday" when we consider the star's history in worship on the Nile. Lockyer tells us of a series of temples at Edfū, Philae, Amada, and Semneh, so oriented at their erection, 6400 B. C., as to show Canopus heralding the sunrise at the autumnal equinox, when it was known as the symbol of Khons, or Khonsu, the first southern star-god; and of other similar temples later. At least two of the great structures at Karnak, of 2100 and 1700 B. C., respectively, pointed to its setting; as did another at Naga, and the temple of Khons at Thebes, built by Rameses III about 1300 B. C., afterwards restored and en-

[1] The inhabitants of Zinge, a large village forty miles northeast of Mosul, in Kurdistan, and not far from Kazwin.

larged under the Ptolemies. It thus probably was the prominent object in the religion of Southern Egypt, where it represented the god of the waters.

Some of the Rabbis have asserted — and Delitzsch in modern times — that this star, and not Orion, was the **H·asīl** of the *Bible*, arguing from the similarity in sound of that word to the Suhail of Arabia, and from other reasons fully explained, although not accepted, by Ideler; while, coincidently, there are able commentators who have thought that the Kesīlīm of *Isaiah* xiii, 10, now translated "Constellations," means the brightest stars, which often are those now referred to in the use of the word Suhail. Delitzsch, in his commentary on the *Book of Job*, quotes much, from Wetzstein and others, of this identity of Canopus with H·asīl, illustrating it with stellar stories and proverbs of the present-day Arabs of the Haurān, the patriarch's traditional home.

The Hindus called it **Agastya,** one of their Rishis, or inspired sages,— and helmsman of their Argha,— a son of Varuna, the goddess of the waters; and Sanskrit literature has many allusions to its heliacal rising in connection with certain religious ceremonies. In the *Avesta* it is mentioned as "pushing the waters forward"— governing the tides (?).

The late George Bertin identified it with **Sugi,** the Euphratean Chariot Yoke; but others claim that title for some stars in the zodiac as yet perhaps unascertained, but probably the *lucidae* of Libra.

In China it was **Laou Jin**, the Old Man, and an object of worship down to at least 100 B. C.

Since the 6th century it has been the **Star of Saint Catharine,** appearing to the Greek and Russian pilgrim devotees as they approached her convent and shrine at Sinai, on their way from Gaza, their landing-place.

In early German astronomical books it was the **Schif-stern**, or Ship-star.

With Achernar and Fomalhaut, corresponding stars in Eridanus and Piscis Australis, it made up the **Tre Facelle** of Dante's *Purgatorio*, symbolizing Faith, Hope, and Charity,—

> those three torches,
> With which this hither pole is all on fire.

Hipparchos was wont to observe it from Rhodes in latitude 36° 30′; and, even before him, Posidonius[1] of Alexandria, about the middle of the 3d century before Christ, utilized it in his attempt to measure a degree on the earth's surface on the line between that city and Rhodes, making his ob-

[1] This Posidonius should not be confounded with the Stoic philosopher contemporary with Cicero, although the Stoic himself was somewhat of an astronomer, and, it has been said, the inventor of the planetarium.

servations from the old watch-tower of Eudoxos at Cnidos in the Asian Caria,—possibly the earliest attempt at geodetic measurement, as this observatory was the first one mentioned in classical days. Manilius poetically followed in his path by using it, with the Bear, to prove the sphericity of the earth.

The confusion in the titles of Canopus and Coma Berenices is noted under that constellation.

Lying 52° 38′ south of the celestial equator, about 35° below Sirius, this star is invisible to observers north of the 37th parallel; but there it is just above the horizon at nine o'clock in the evening of the 6th of February, and conspicuous from Georgia, Florida, and our Gulf States. Sirius follows it in culmination by about twenty minutes.

Canopus is so brilliant that observers in Chile, in 1861, considered it brighter than Sirius; and Tennyson, in his *Dream of Fair Women*, made it a simile of intensest light,—in Cleopatra's words,—

> lamps which outburn'd Canopus.

Yet Elkin obtained a parallax of only 0″.03,— practically *nil*,— indicating a distance from our system at least twelve times that of its apparently greater neighbor. Its spectrum is similar to that of the latter.

See discovered, in 1897, a 15th-magnitude bluish companion 30″ away, at a position angle of 160°.

β, 2.

Miaplacidus is thus written in Burritt's *Geography* of 1856, but is **Maiaplacidus** in his *Atlas* of 1835, the meaning and derivation of which I cannot learn, unless it be in part, as Higgins asserts in his brief work on star-names, from Miyah, the plural of the Arabic Mā, Water. The original, however, is better transcribed Mi'ah.

β lies in the Carina subdivision and is the α of Halley's Robur Carolinum, 25° east of Canopus, and 61° south of Alphard of the Hydra; but Baily said that he could find no star corresponding to this as Bayer laid it down on his map of Argo.

γ, Triple, 2, 6, and 8, white, greenish white, and purple,

was the Arabs' **Al Suhail al Muḥlif,** the Suhail of the Oath, as with ζ and λ it formed one of the several groups **Al Muḥlifaïn, Muḥtalifaïn,** or **Muḥnithaïn,** by which reference was made to the statement that at their rising some mistook them for Suhail, and the consequent arguments were the occasion of much profanity among the disputatious Arabs. As, however, it would seem impossible that Canopus could be mistaken for any neighboring star, this derivation is as absurd as the proper location of the Muḥlīfaïn was doubtful, for they have been assigned not only to the foregoing, but also to stars in Canis Major, Centaurus, and Columba.

γ lies in the Vela subdivision, and is visible from all points south of 42° of north latitude. Like β, it seems to have been incorrectly laid down on the *Uranometria*, for Baily wrote that he could not find Bayer's γ in the sky.

This is the only conspicuous star that shows the Wolf-Rayet type of a continuous spectrum crossed with bright lines; and its superb beauty is the admiration of the spectroscopic observer. Eddie calls it the **Spectral Gem** of the southern skies.

δ, 2.2, and ω, with stars in Canis Major, were the Chinese **Koo She,** the Bow and Arrow.

ζ, 2.5, at the southeastern extremity of the Egyptian X, is the **Suhail Ḥaḍar** of Al Sufi, and the **Naos,** or Ship, of Burritt's *Atlas;* while, with γ and λ, it was one of the Muhlīfaïn.

Its south declination in 1880 was 39° 40′, and so it is plainly visible from the latitude of the State of Maine, coming to the meridian on the 3d of March.

η, Irregularly variable, $>$ 1 to 7.4, reddish,

lies in the Carina subdivision, but is invisible from north of the 30th parallel.

This is one of the most noted objects in the heavens, perhaps even so in almost prehistoric times, for Babylonian inscriptions seem to refer to a star, noticeable from occasional faintness in its light, that Jensen thinks was η. And he claims it as one of the temple stars associated with Ea, or Ia, of Eridhu,[1] the Lord of the Waves, otherwise known as Oannes,[2] the mysterious human fish and greatest god of the kingdom.

In China η was **Tseen She,** Heaven's Altars.

1 Eridhu, or Eri-duga, the Holy City, Nunki, or Nunpe, one of the oldest cities in the world, even in ancient Babylonia, was that kingdom's flourishing port on the Persian Gulf, but, by the encroachments of the delta, its site is now one hundred miles inland. In its vicinity the Babylonians located their sacred Tree of Life.

2 Berōssōs described Oannes as the teacher of early man in all knowledge; and in mythology he was even the creator of man and the father of Tammuz and Ishtar, themselves associated with other stars and sky figures. Jensen thinks Oannes connected with the stars of Capricorn; Lockyer finds his counterpart in the god Chnemu of Southern Egypt; and some have regarded him as the prototype of Noah.

The variations in its light are as remarkable in their irregularity as in their degree. The first recorded observation, said to have been by Halley in 1677, although it is not in his *Southern Catalogue*, made η a 4th-magnitude, but since that it has often varied either way, at longer or shorter intervals, from absolute invisibility by the naked eye to a brilliancy almost the equal of Sirius. Sir John Herschel saw it thus in December, 1837, as others did in 1843; but, gradually declining since then, it touched its lowest recorded magnitude of 7.6 in March, 1886. It is now, however, on the increase; for on the 13th of May, 1896, it was 5.1, or about a half-magnitude higher than its maximum of the preceding year.

The nebula, N. G. C. 3372, surrounding this star has been called the **Keyhole** from its characteristic features; but the most brilliant portion, as drawn by Sir John Herschel, seems to have disappeared at some time between 1837 and 1871. That great observer saw 1203 stars scattered over its surface.

Near η is a vacant space of irregular shape that Abbott has called the **Crooked Billet**; and there are two remarkable coarse clusters in its immediate vicinity.

ι, 2.9, pale yellow.

This was the Latins' **Scutulum**, or Little Shield, the Arabians' **Turais**, probably referring to the ornamental Aplustre at the stern of the Ship in the subdivision Carina; but Hyde, quoting it as **Turyeish** from Tizini, said that the original was *verbum ignotum*, and suggested that some one else should make a guess at it and its meaning. Smyth wrote of it as "corresponding to the 'Ασπιδίσκε of Ptolemy"; but the latter described it as being in the 'Ακροστόλιον, Gunwale, and located κ, ξ, ο, π, ρ, σ, and τ in the 'Ασπιδίσκε, or Aplustre, where they are shown to-day. The *Century Atlas* follows Smyth in calling ι **Aspidiske.** It is visible from the latitude of New York City.

κ, 3.9, is **Markab** and **Markeb,** probably from the *Alfonsine Tables* of 1521, where this last word is found plainly applied to it as a proper name. This also is visible from the latitude of New York, culminating on the 25th of March.

λ, 2.5, in Vela is Al Sufi's **Al Suhail al Wazn,** Suhail of the Weight; and, with γ and ζ, one of the Muḥlīfaïn.

ξ, 3.4, has been called **Asmidiske** by an incorrect transliteration of the 'Ασπιδίσκε where it is located with the star ι.

ψ, 3.7, in Vela is given by Reeves as **Tseen Ke,** Heaven's Record; a star that he letters A, as **Hae Shih,** the Sea Stone; and one numbered 1971, as **Tseen Kow,** the Heavenly Dog.

Grotius mentioned **Alphart** as the title of some star in Navis, although without locating it, and very correctly added *sed hoc ad lucidam Hydrae pertinet;* but as the top of the Mast is in some maps very close to this *lucida,* Alphard, the explanation would seem obvious.

Baily said that Flamsteed's star 13 Argūs, strangely placed 20° from Argo across Monoceros, should be Fl. 15 Canis Minoris.

From stars in Argo, behind the back of the Greater Dog, was formed by Bartsch the small asterism **Gallus, the Cock,** but it has long since been forgotten.

*

. . . the fleecy star that bears
Andromeda far off Atlantic seas
Beyond th' Horizon.

Milton's *Paradise Lost.*

Aries, the Ram,

is **Ariete** in Italy, **Bélier** in France, and **Widder** in Germany — Bayer's **Wider;** in the Anglo-Saxon tongue it is **Ramm,** and in the Anglo-Norman of the 12th century, **Multuns.** The constellation is marked by the noticeable triangle to the west of the Pleiades, 6° north of the ecliptic, 20° north of the celestial equator, and 20° due south from γ Andromedae.

With the Greeks it was Κριός, and sometimes 'Αιγόκερως, although this last was more usual for Capricorn.

It always was **Aries** with the Romans; but Ovid called it **Phrixea Ovis;** and Columella, **Pecus Athamantidos Helles, Phrixus,** and **Portitor Phrixi;** others, **Phrixeum Pecus** and **Phrixi Vector,** Phrixus being the hero-son of Athamas, who fled on the back of this Ram with his sister Helle to Colchis to escape the wrath of his stepmother Ino. It will be remembered that on the way Helle fell off into the sea, which thereafter became the Hellespont, as Manilius wrote:

First Golden Aries shines (who whilst he swam
Lost part of's Freight, and gave the Sea a Name);

and Longfellow, in his translation from Ovid's *Tristia:*

The Ram that bore unsafely the burden of Helle.

On reaching his journey's end, Phrixus sacrificed the creature and hung its fleece in the Grove of Ares, where it was turned to gold and became the object of the Argonauts' quest. From this came others of Aries' titles : **Ovis aurea** and **auratus, Chrysomallus,** and the Low Latin **Chrysovellus.**

The **Athamas** used by Columella was a classical reproduction of the Euphratean Tammuz Dum-uzi, the Only Son of Life, whom Aries at one time represented in the heavens, as did Orion at a previous date, perhaps when it marked the vernal equinox 4500 B. C.

Cicero and Ovid styled the constellation **Cornus**; elsewhere it was **Corniger** and **Laniger; Vervex,** the Wether; **Dux opulenti gregis; Caput arietinum**; and, in allusion to its position, **Aequinoctialis. Vernus Portitor,** the Spring-bringer, is cited by Caesius, who also mentioned **Arcanus,** that may refer to the secret rites in the worship of the divinities whom Aries represented.

From about the year 1730 before our era he was the **Princeps signorum coelestium, Princeps zodiaci,** and the **Ductor exercitus zodiaci,** continuing so through Hipparchos' time; Manilius writing of this:

> The *Ram* having pass'd the *Sea* serenely shines,
> And leads the Year, the Prince of all the Signs.

But about A. D. 420 his office was transferred to Pisces.

Brown writes as to the origin of the title Aries, without any supposition of resemblance of the group to the animal:

> The stars were regarded by a pastoral population as flocks; each asterism had its special leader, and the star, and subsequently the constellation, that led the heavens through the year was the Ram.

Elsewhere he tells us that when Aries became chief of the zodiac signs it took the Akkadian titles **Ku, I-ku,** and **I-ku-u,** from its *lucida* Hamal, all equivalents of the Assyrian **Rubū,** Prince, and very appropriate to the leading stellar group of that date, although not one of the first formations.

He also finds, from an inscription on the *Tablet of the Thirty Stars*, that the Euphratean astronomers had a constellation **Gam,** the Scimetar, stretching from Okda of the Fishes to Hamal of Aries, the curved blade being formed by the latter's three brightest components. This was the weapon protecting the kingdom against the Seven Evil Spirits, or Tempest Powers.

Jensen thinks that Aries may have been first adopted into the zodiac by the Babylonians when its stars began to mark the vernal equinox; and that the insertion of it between Taurus and Pegasus compelled the cutting off a part of each of those figures,— a novel suggestion that would save much theorizing as to their sectional character.

The Jewish Nīsān, our March-April, was associated with Aries, for Josephus said that it was when the sun was here in this month that his people were released from the bondage of Egypt; and so was the same month Nisanu of Assyria, where Aries represented the Altar and the Sacrifice, a ram usually being the victim. Hence the prominence given to this sign in antiquity even before its stars became the leaders of the rest; although Berōssōs and Macrobius attributed this to the ancient belief that the earth was created when the sun was within its boundaries; and Albumasar,[1] of the 9th century, in his *Revolution of Years* wrote of the Creation as having taken place when "the seven planets"— the Sun, Moon, Mercury, Venus, Mars, Jupiter, and Saturn — were in conjunction here, and foretold the destruction of the world when they should be in the same position in the last degree of Pisces.

Dante, who called the constellation **Montone,** followed with a similar thought in the *Inferno:*

> The sun was mounting with those stars
> That with him were, what time the Love Divine
> At first in motion set those beauteous things.

To come, however, to a more precise date, Pliny said that Cleostratos of Tenedos first formed Aries, and, at the same time, Sagittarius; but their origin probably was many centuries, even millenniums, antecedent to this, and the statement is only correct in so far as that he may have been the first to write of them.

Many think that our figure was designed to represent the Egyptian King of Gods shown at Thebes with ram's horns, or veiled and crowned with feathers, and variously known as Amon, Ammon, Hammon, Amen, or Amun, and worshiped with great ceremony at his temple in the oasis Ammonium, now Siwah, 5° west of Cairo on the northern limit of the Libyan desert. Kircher gave Aries' title there as *Ταμετοῦρο Αμοῦν*, Regum Ammonis. But there is doubt whether the Egyptian stellar Ram coincided with ours, although Miss Clerke says that the latter's stars were called the Fleece.

[1] This author, known also as 'Abū Ma'shar and Ja'phar, was from Balh' in Turkestan, celebrated as an astrologer and quoted by Al Bīrūnī, but with the caution that he was a very incorrect astronomer. The Lenox Library of New York has a copy of his *Opus introductorii in astronomiā Albumazaris abalachi, Idus Februarii,* 1489, published at Venice with illustrations. Its similarity to the *Hyginus* of the preceding year would indicate that they issued from the same press.

As the god Amen was identified with Ζεύς and Jupiter of the Greeks and Romans, so also was Aries, although this popularly was attributed to the story that the classical divinity assumed the Ram's form when all the inhabitants of Olympus fled into Egypt from the giants led by Typhon. From this came the constellation's titles **Jupiter Ammon**; **Jovis Sidus**; **Minervae Sidus**, the goddess being Jove's daughter; the **Jupiter Libycus** of Propertius, **Deus Libycus** of Dionysius, and **Ammon Libycus** of Nonnus.

The Hebrews knew it as **Telī**, and inscribed it on the banners of Gad or Naphtali; the Syrians, as **Amru** or **Emru**; the Persians, as **Bara**, **Bere**, or **Berre**; the Turks, as **Kuzi**; and in the Parsi *Bundehesh* it was **Varak**: all these being synonymous with Aries. The unexplained **Arabib**, or **Aribib**, also is seen for it. The early Hindus called it **Aja** and **Mesha**, the Tamil **Mesham**; but the later followed the Greeks in **Kriya**.

An Arabian commentator on Ulug Beg called the constellation **Al Kabsh al ʽAlīf**, the Tame Ram; but that people generally knew it as **Al Ḥamal**, the Sheep, — **Hammel** with Riccioli, **Alchamalo** with Schickard, and **Alhamel** with Chilmead.

As one of the zodiacal twelve of China it was the **Dog**, early known as **Heang Low**, or **Kiang Leu**; and later, under Jesuit influence, as **Pih Yang**, the White Sheep; while with Taurus and Gemini it constituted the **White Tiger**, the western one of the four great zodiac groups of China; also known as the **Lake of Fullness**, the **Five Reservoirs of Heaven**, and the **House of the Five Emperors**.

Chaucer and other English writers of the 14th, 15th, and 16th centuries Anglicized the title as **Ariete**, which also appeared in the Low Latin of the 17th century. It was about this time, when it was sought to reconstruct the constellations on Bible lines, that Aries was said to represent **Abraham's Ram** caught in the thicket; as also **Saint Peter**, the bishop of the early church, with Triangulum as his Mitre. Caesius considered it the **Lamb** sacrificed on Calvary for all sinful humanity.

Aries generally has been figured as reclining with reverted head admiring his own golden fleece, or looking with astonishment at the Bull rising backward; but in the *Albumasar* of 1489 he is standing erect, and some early artists showed him running towards the west, with what is probably designed for the zodiac-belt around his body. A coin of Domitian bears a representation of him as the **Princeps juventutis**, and he appeared on those of Antiochus of Syria with head towards the Moon and Mars — an appropriate figuring; for, astrologically, Aries was the lunar house of that planet. In common with all the other signs, he is shown on the zodiacal rupees generally attributed to the great Mogul prince Jehangir Shah, but really struck by Nūr Mahal Mumtaza, his favorite wife, between 1616 and 1624, each figure being surrounded by sun-rays with an inscription on the reverse.

Its equinoctial position gave force to Aratos' description of its "rapid transits," but he is strangely inexact in his

> faint and starless to behold
> As stars by moonlight —

a blunder for which Hipparchos seems to have taken him to task. Aratos however, was a more successful versifier than astronomer.

Among astrologers Aries was a dreaded sign indicating passionate temper and bodily hurt, and thus it fitly formed the **House of Mars**, although some attributed guardianship over it to Pallas Minerva, daughter of Jove whom Aries represented. It was supposed to hold sway over the head and face; in fact the Egyptians called it **Arnum**, the Lord of the Head; while, geographically, it ruled Denmark, England, France, Germany, Lesser Poland and Switzerland, Syria, Capua, Naples and Verona, with white and red as its colors. In the time of Manilius it was naturally thought of as ruling the Hellespont and Propontis, Egypt and the Nile, Persia and Syria; and, with Leo and Sagittarius, was the **Fiery Trigon.**

Ampelius said that it was in charge of the Roman Africus, the Southwest Wind, the Italians' Affrico, or Gherbino; but the Archer and Scorpion also shared this duty. Pliny wrote that the appearance of a comet within its borders portended great wars and wide-spread mortality, abasement of the great and elevation of the small, with fearful drought in the regions over which the sign predominated; while 17th-century almanacs attributed many troubles to men, and declared that "many shall die of the rope" when the sun was in the sign; but they ascribed to its influence "an abundance of herbs."

Its symbol, ♈, probably represents the head and horns of the animal.

The eastern portion is inconspicuous, and astronomers have mapped others of its stars somewhat irregularly, carrying a horn into Pisces and a leg into Cetus.

Argelander assigns to it 50 naked-eye components; Heis, 80.

The sun now passes through it from the 16th of April to the 13th of May.

A *nova* is reported to have appeared here in May, 1012, described by Epidamnus, the monk of Saint Gall, as *oculos verberans.*

α, 2.3, yellow.

Hamal, from the constellation title, was formerly written **Hamel, Hemal, Hamul,** and **Hammel**; Riccioli having **Ras Hammel** from **Al Rās al Ḥamal,** the Head of the Sheep.

Burritt's **El Nāṭh,** from **Al Nāṭiḥ,** the Horn of the Butting One, is appropriate enough for this star, but in our day is given to β Tauri; still Burritt had authority for it, as Kazwini, Al Tizini, Ulug Beg, and the Arabic globes all used the word here; and Chaucer wrote, in 1374:

He knew ful wel how fer Alnath was shove ffro the heed of thilke fixe Aries above.

The title of the whole figure also is seen in **Arietis,** another designation for this star, as was often the case with many of the *lucidae* of the constellations.

In Ptolemy's and Ulug Beg's descriptions it was "over the head"; but both of these mentioned Hipparchos as having located it over the muzzle, and near to that feature it was restored by Tycho, in the forehead, as we now have it.

Renouf identified it with the head of the **Goose** supposed to be one of the early zodiacal constellations of Egypt.

Strassmaier and Epping, in their *Astronomisches aus Babylon*, say that there its stars formed the third of the twenty-eight ecliptic constellations,—**Arku-sha-rishu-ku,** literally the Back of the Head of Ku,—which had been established along that great circle millenniums before our era; and Lenormant quotes, as an individual title from cuneiform inscriptions, **Dil-kar,** the Proclaimer of the Dawn, that Jensen reads **As-kar**, and others **Dil-gan,** the Messenger of Light. George Smith inferred from the tablets that it might be the **Star of the Flocks**; while other Euphratean names have been **Lu-lim,** or **Lu-nit,** the Ram's Eye; and **Si-mal** or **Si-mul,** the Horn Star, which came down even to late astrology as the **Ram's Horn.** It also was **Anuv,** and had its constellation's titles **I-ku** and **I-ku-u,**—by abbreviation **Ku,**—the Prince, or the Leading One, the Ram that led the heavenly flock, some of its titles at a different date being applied to Capella of Auriga.

Brown associates it with Aloros, the first of the ten mythical kings of Akkad anterior to the Deluge, the duration of whose reigns proportionately coincided with the distances apart of the ten chief ecliptic stars beginning with Hamal, and he deduces from this kingly title the Assyrian **Ailuv,** and the Hebrew **Ayil**; the other stars corresponding to the other mythical kings being Alcyone, Aldebaran, Pollux, Regulus, Spica, Antares, Algedi, Deneb Algedi, and Scheat.

The interesting researches of Mr. F. C. Penrose on orientation in Greece have shown that many of its temples were pointed to the rising or setting of various prominent stars, as we have seen to be the case in Egypt; this feature in their architecture having doubtless been taken by the receptive, as well as "somewhat superstitious," Greeks from the Egyptians, many of whose structures are thought to have been so oriented six or seven millenniums before the Christian era, although our star Hamal was not among those thus observed on the Nile, for precession had not yet brought it into importance. Of the Grecian temples at least eight, at various places and of dates ranging from 1580 to 360 B. C., were oriented to this star; those of Zeus and his daughter Athene being especially thus favored, as Aries was this god's symbol in the sky.

It was perhaps this prevalence of temple orientation, in addition to their many divinities and especially ὁ Ἄγνωστος Θεός, the Unknown God, which furnished an appropriate text for Saint Paul's great sermon on the Areopagus to the "men of Athens," when, in order to prove our source of being from Him, he quoted, as in *Acts* xvii, 28, from the celebrated fifth verse of the *Phainomena:*

τοῦ γάρ καί γένος ἐσμίν [1]
(For we are also his offspring).

To this work this quotation generally is ascribed, and naturally so, for the poet and apostle were fellow-countrymen from Cilicia; but the same words are found in the *Hymn to Jupiter* by Cleanthes the Stoic, 265 B. C. As Saint Paul, however, used the plural τίνες in his reference, "certain even of your own poets," he may have had both of these authors in mind.

Hamal lies but little north of the ecliptic, and is much used in navigation in connection with lunar observations. It culminates on the 11th of December.

Vogel finds it to be in approach to our system at the rate of about nine miles a second. Its spectrum is similar to that of the sun.

β, 2.9, pearly white.

Sharatan and **Sheratan** are from **Al Sharaṭain,** the dual form of Al Sharaṭ, a Sign, referring to this and γ, the third star in the head, as a sign of the opening year; β having marked the vernal equinox in the days of

[1] The Christian fathers Eusebius and Clement of Alexandria made this same quotation; while frequent references to Aratos' poem appear in the writings of Saints Chrysostom and Jerome, and of Oecumenius. The heathen Manilius similarly wrote,

. . . nostrumque parentem
Stirps sua,

to prove the immortality of the soul.

Hipparchos, about the time when these stars were named. Bayer's **Sartai** is from this dual word.

These were the 1st *manzil* in Al Bīrūnī's list, the earlier 27th, but some added α to the combination, calling it **Al Ashrāṭ** in the plural; Hyde saying that λ also was included. **Al Nāṭiḥ** was another name for this lunar station, as the chief components are near the horns of Aries.

β and γ constituted the 27th *nakshatra* **Açvini,** the Ashwins, or Horsemen, the earlier dual **Açvinău** and **Açvayujău,** the Two Horsemen, corresponding to the Gemini of Rome, but figured as a Horse's Head. α sometimes was added to this lunar station, but β always was the junction star with the adjoining Bharani. About 400 years before our era this superseded Krittikā as leader of the *nakshatras.* They were the Persian **Padevar,** the Protecting Pair; the Sogdian **Bashish,** the Protector; and the equivalent Coptic **Pikutorion;** while in Babylonia, according to Epping, they marked the second ecliptic constellation **Mahrū-sha-rishu-ku,** the Front of the Head of Ku.

α, β, and γ were the corresponding *sieu* **Leu,** or **Low,** the Train of a garment, β being the determinant.

♈, Double, 4.5 and 5, bright white and gray,

has been called the **First Star in Aries,** as at one time nearest to the equinoctial point.

Its present title, **Mesarthim,** or **Mesartim,** has been connected with the Hebrew Mᵉshārᵉtīm, Ministers, but the connection is not apparent; and Ideler considered the word an erroneous deduction by Bayer from the name of the lunar station of which this and β were members. In Smyth's index it is **Mesartun**; and Caesius had **Scartai** from Sharaṭain. α, β, and γ may have been the Jewish **Shalisha,**— more correctly **Shālīsh,**— some musical instrument of triangular shape, a title also of Triangulum. And they formed one of the several **Athāfiyy,** Trivets or Tripods; this Arabic word indicating the rude arrangement of three stones on which the nomad placed his kettle, or pot, in his open-air kitchen; others being in our Draco, Orion, Musca, and Lyra.

Gamma's duplicity was discovered by Doctor Robert Hooke while following the comet of 1664, when he said of it, "a like instance to which I have not else met in all the heaven";[1] but it was an easy discovery, for the components are 8″.8 apart, readily resolved by a low-power.

The position angle has been about 0° for fifty years.

[1] Huygens is said to have seen three stars in ϑ^1 Orionis in 1656, and Riccioli two in ζ Ursae Majoris in 1650.

δ, 4.6.

Botein is from **Al Buṭain,** the dual of Al Baṭn, the Belly, probably from some early figuring, for in modern maps the star lies on the tail.

With ζ it was **Tsin Yin** in China.

δ, ε, and ρ^3 generally were considered the 28th *manzil,* **Al Buṭain,** but Al Bīrūnī substituted π for ρ^3, and others, ζ; while still others located this station in our Musca, the faint little triangle above the figure of the Ram.

ε marks the base of the tail, and is the radiant point of the **Arietids,** the meteors of the 11th to the 24th of October. It is a double star of 5th and 6.5 magnitudes, 0″.5 apart, and probably binary. Its present position angle is about 200°. Gould thinks it variable.

Williams mentions *b*, *e*, *o*, and *z* as the Chinese **Teen Ho.**

★

Thou hast loosened the necks of thine horses, and goaded their flanks with affright,
To the race of a course that we know not on ways that are hid from our sight.
As a wind through the darkness the wheels of their chariot are whirled,
And the light of its passage is night on the face of the world.

Algernon Charles Swinburne's *Erechtheus.*

Auriga, the Charioteer or Wagoner,

in early days the **Wainman,** is the French **Cocher,** the Italian **Cocchiere,** and the German **Fuhrmann.**

It is a large constellation stretching northward across the Milky Way from its star γ, which also marks one of the Bull's horns, to the feet of Camelopardalis, about 30° in extent north and south and 40° east and west; and is shown as a young man with whip in the right hand, but without a chariot, the Goat being supported against the left shoulder and the Kids on the wrist. This, with some variations, has been the drawing from the earliest days, when, as now, it was important, chiefly from the beauty of Capella and its attendant stars so prominent in the northwest in the spring twilight, and in the northeast in early autumn. But the *Hyginus* of 1488 has a most absurd Driver in a ridiculously inadequate four-wheeled car, with the Goat and Kids in their usual position, the reins being held over four animals abreast — a yoke of oxen, a horse, and a zebra (!); while the *Hyginus* of Micyllus, in 1535, has the Driver in a two-wheeled cart with a pair of horses and a yoke of oxen all abreast. A Turkish planisphere shows these stars depicted as a **Mule,** and they were so regarded by the early Arabs, who did not know — at all events did not picture — the Driver, Goat, or Kids. In this form Bayer Latinized it as the **Mulus clitellatus,** the Mule with Panniers.

Ideler thinks that the original figure was made up of the five stars *α*, *β*, *ε*, *ζ*, and *η*; the Driver, represented by *α*, standing on an antique sloping Chariot marked by *β*; the other stars showing the reins. But later on the Chariot was abandoned and the reins transferred to their present position, the Goat being added by a misunderstanding, the word Ἄιξ, analogous to Ἀιγίς, simply meaning a Storm Wind that, apparently, in all former times the stars *α*, *η*, and *ζ* have portended at their heliacal rising, or by their disappearance in the mists. Still later to *α* as the Goat were added the near-by *η* and *ζ* as her Kids, the Ἔριφοι,— an addition that Hyginus said was made by Cleostratos.

But the results of modern research now give us reason to think that the constellation originated on the Euphrates in much the same form as we have it, and that it certainly was a well-established sky figure there millenniums ago. A sculpture from Nimroud is an almost exact representation of Auriga with the Goat carried on the left arm; while in Graeco-Babylonian times the constellation **Rukubi,** the Chariot, lay here nearly coincident with our Charioteer, perhaps running over into Taurus.

Ἑνίοχος, the Rein-holder, was transcribed **Heniochus** by Latin authors, and personified by Germanicus and others as **Erechtheus,** or more properly **Erichthonius,** son of Vulcan and Minerva, who, having inherited his father's lameness, found necessary some means of easy locomotion. This was secured by his invention of the four-horse chariot which not only well became his regal position as the 4th of the early kings of Athens, but secured for him a place in the sky. Manilius thus told the story:

Near the bent Bull a Seat the Driver claims,
Whose skill conferr'd his Honour and his Names.
His Art great Jove admir'd, when first he drove
His rattling Carr, and fix't the Youth above.

Vergil had something similar in his 3d *Georgic.*

These names appear as late as the 17th century with Bullialdus and Longomontanus, Riccioli writing **Erichtonius.**

Others saw here **Myrtilus,** the charioteer of Oenomaus, who betrayed his master to Pelops; or **Cillas,** the latter's driver; **Pelethronius,** a Thessalian; and **Trethon;** while Euripides and Pausanias identified him with the unfortunate **Hippolytus,** the Hebrew Joseph of classical literature. Additional titles in Greece were Ἀρμελάτης, Διφρηλάτης, Ἱππηλάτης, and Ἐλάσιππος, all signifying a Charioteer; while La Lande's **Bellerophon** and **Phaëthon** are appropriate enough, and his **Trochilus** may be, if the word be degenerated from τροχαλός, running; but his **Absyrthe,** correctly Ἄψυρτος, the young brother of Medea, is unintelligible.

Although Auriga was the usual name with the Latins, their poets called it **Aurigator**; **Agitator currus retinens habenas**; **Habenifer** and **Tenens habenas,** the Charioteer and the Rein-holder; some of these titles descending to the *Tables* and *Almagests* down to the 16th century. **Arator,** the Ploughman, appeared with Nigidius and Varro for this, or for Boötes; in fact the same idea still holds with some of the Teutonic peasantry, among whom Capella and the Kids are known as the **Ploughman with his Oxen.** Grimm mentions for the group **Voluyara,** as stars that ploughmen know. The **Acator** occasionally seen may be an erroneous printing of Arator.

From the Goat and Kids came **Custos caprarum, Habens capellas, Habens haedos,** and **Habens hircum. Habens oleniam capram** and **Oleniae sidus pluviale Capellae** of Ovid's *Metamorphoses* are from the Ὠλενίνην of Aratos, thought to be derived from ὠλένη, the wrist, on which the Kids are resting. Some, however, with more probability have referred the word to Olenus, the father and birthplace of the nymph Amalthea in ancient Aetolia.

Isidorus of Hispalis[1] — Saint Isidore — called it **Mavors,** the poetical term for Mars, the father of Romulus and so the god of the shepherds; Nonius, the Portuguese Pedro Nuñez of the 16th century, similarly said that it was **Mafurtius**; and Bayer found for it **Maforte**: but his **Ophiultus,** probably a Low Latin word also applied to *a*, seems to be without explanation.

Some have thought that Auriga was **Horus** with the Egyptians; but Scaliger said that the **Hora** of the translation of Ptolemy's Τετράβιβλος should be **Roha,** Bayer's **Roh,** a Wagoner; Beigel, however, considered it a misprint for **Lora,** the Reins.

The barbarous **Alhaior, Alhaiot, Althaiot, Alhaiset, Alhatod, Alhajot, Alhajoth, Alhojet, Alanac, Alanat,** and **Alioc,**— even these perhaps do not exhaust the list,— used for both constellation and *lucida,* are probably degenerate forms of the Arabs' **Al 'Anz** and **Al 'Ayyūḳ,** specially applied to Capella as the Goat, which they figured as the desert Ibex, their *Bădan;* and Ideler thinks that this may have been the earliest Arabic designation for the star.

The 1515 *Almagest* says, "*et nominatur latine antarii . . . id est collarium,*"— this **Collarium** perhaps referring to the collar in the Charioteer's har-

[1] This early Hispalis, the modern Seville, was the site of the first European observatory of our era, erected by the Moor Geber in 1196.

ness; but the **Antarii** has puzzled all, unless it be Professor Young, who suggests that it may be the reins diverging from the Driver's hand like guy-ropes, which the original means as used by Vitruvius in his description of a builder's derrick.

The Arabians translated the classic titles for the Rein-holder into **Al Dhu al 'Inān, Al Māsik al 'Inān,** and **Al Mumsik al 'Inān,**— Chilmead's **Mumassich Alhanam;** but the Rabbi Aben Ezra[1] mixed things up by calling the figure **Pastor** *in cujus manu est frenum.*

Some have illustrated it as **Saint Jerome,** but Caesius likened it to **Jacob** deceiving his father with the flesh of his kids; and Seiss says that it represents the **Good Shepherd** who laid down his life for the sheep. A Chariot and Goat are shown on coins of consular Rome, and a Goat alone on those of Paros, that may have referred to this constellation.

Argelander counts 70 naked-eye stars here, and Heis 144.

Capella's course admiring landsmen trace,
But sailors hate her inauspicious face.

Lamb's *Aratos.*

α, 0.3, white.

This has been known as **Capella,** the Little She-goat, since at least the times of Manilius, Ovid, and Pliny, all of whom followed the Κινῆσαι Χειμῶνας of Aratos in terming it a *Signum pluviale* like its companions the Haedi, thus confirming its stormy character throughout classical days. Holland translated Pliny's words the **rainy Goat-starre;** Pliny and Manilius treated it as a constellation by itself, also calling it **Capra, Caper, Hircus,** and by other hircine titles.

Our word is the diminutive of **Capra,** sometimes turned into **Crepa,** and more definitely given as **Olenia, Olenie, Capra Olenie,** and the **Olenium Astrum** of Ovid's *Heroides.* In the present day it is **Cabrilla** with the Spaniards, and **Chevre** with the French.

Amalthea came from the name of the Cretan goat, the nurse of Jupiter and mother of the Haedi, which she put aside to accommodate her foster-child, and for which Manilius wrote:

The Nursing Goat's repaid with Heaven.

From this came the occasional **Jovis Nutrix.**

[1] This celebrated man, often cited in bygone days as Abenare, Avenore, Evenare, was Abraham ben Meir ben Ezra of Toledo, the great Hebrew commentator of the 12th century, an astronomer, mathematician, philologist, poet, and scholar, and the first noted biblical critic.

But, according to an earlier version, the nurse was the nymph Amalthea, who, with her sister Melissa, fed the infant god with goat's milk and honey on Mount Ida, the nymph Aigē being sometimes substituted for one or both of the foregoing; or Adrasta, with her sister Ida, all daughters of the Cretan king Melisseus. Others said that the star represented the Goat's horn broken off in play by the infant Jove and transferred to the heavens as **Cornu copiae,** the Horn of Plenty, a title recalled by the modern Lithuanian **Food-bearer.** In this connection, it was 'Αμαλθείας κέρας, also brought absurdly enough into the *Septuagint* as a translation of the words Keren-happuch, the Paint-horn, or the Horn of Antimony, of the *Book of Job* xlii, 14,—the Cornus tibii of the *Vulgate.* Ptolemy's Αἴξ probably became the Arabo-Greek 'Αιοὺκ of the Graeco-Persian Chrysococca's book, and the **Ayyūḳ, Alhajoc, Alhajoth, Alathod, Alkatod, Alatudo, Atud,** etc., which it shared with the constellation; but Ideler thought 'Ayyuḳ an indigenous term of the Arabs for this star. Assemani's **Alcahela** may have come from Capella. The Tyrians called it **'Iyūthā,** applied also to Aldebaran and perhaps to other stars; but the Rabbis adopted the Arabic 'Ayyūḳ as a title for their heavenly Goat, although they greatly disagreed as to its location, placing it variously in Auriga, Taurus, Aries, and Orion. The "armborne she goat," however, of Aratos, derived from the priests of Zeus, would seem to fix it positively where we now recognize it. Hyde devoted three pages of learned criticism to this important (!) subject, but insisted that the Arabic and Hebrew word **'Āsh** designated this star.

With ζ and η, the Kids, it formed the group that Kazwini knew as **Al 'Ināz,** the Goats, but others as **Al 'Anz,** in the singular.

The early Arabs called it **Al Rākib,** the Driver; for, lying far to the north, it was prominent in the evening sky before other stars became visible, and so apparently watching over them; and the synonymous **Al Hādī** of the Pleiades, as, on the parallel of Arabia, it rose with that cluster. Wetzstein, the biblical critic often quoted by Delitzsch, explains this last term as "the singer riding before the procession, who cheers the camels by the sound of the *hadwa,* and thereby urges them on," the Pleiades here being regarded as a troop of camels. An early Arab poet alluded to this Hādī as overseer of the *Meisir* game, sitting behind the players, the other stars.

Bayer's **Ophiultus** now seems unintelligible.

Capella's place on the Denderah zodiac is occupied by a mummied cat in the outstretched hand of a male figure crowned with feathers; while, always an important star in the temple worship of the great Egyptian god **Ptah,** the Opener, it is supposed to have borne the name of that divinity and probably was observed at its setting 1700 B. C. from his temple, the noted edifice at Karnak near Thebes, the No Amon of the books of the prophets *Jeremiah* and *Nahum*. Another recently discovered sanctuary of Ptah at Memphis also was oriented to it about 5200 B. C. Lockyer thinks that at least five temples were oriented to its setting.

It served, too, the same purpose for worship in Greece, where it may have been the orientation point of a temple at Eleusis to the goddess Diana Propyla; and of another at Athens.

In India it also was sacred as **Brahma Ridaya,** the Heart of Brahma; and Hewitt considers Capella, or Arcturus, the **Āryaman,** or **Airyaman,** of the *Rig Veda*.

The Chinese had an asterism here, formed by Capella with β, θ, κ, and γ, which they called **Woo Chay,** the Five Chariots — a singular resemblance in title to our Charioteer; although Edkins says that this should be the Chariots of the Five Emperors.

The Akkadian **Dil-gan I-ku,** the Messenger of Light, or **Dil-gan Babili,** the Patron star of Babylon, is thought to have been Capella, known in Assyria as **I-ku,** the Leader, *i. e.* of the year; for, according to Sayce, in Akkadian times the commencement of the year was determined by the position of this star in relation to the moon at the vernal equinox. This was previous to 1730 B. C., when, during the preceding 2150 years, spring began when the sun entered the constellation Taurus; in this connection the star was known as the **Star of Mardūk,** but subsequent to that date some of these titles were apparently applied to Hamal, Wega, and others whose positions as to that initial point had changed by reason of precession. One cuneiform inscription, supposed to refer to our Capella, is rendered by Jensen **Askar,** the Tempest God; and the *Tablet of the Thirty Stars* bears the synonymous **Ma-a-tu**; all this well accounting for its subsequent character in classical times, and one of the many evidences adduced as to the origin of Greek constellational astronomy in the Euphrates valley.

The ancient Peruvians, the Quichuas, whose language is still spoken by their descendants, appear to have devoted much attention to the stars; and José de Acosta, the Spanish Jesuit and naturalist of the 16th century, said that every bird and beast on earth had its namesake in their sky. He cited several of their stellar titles, identifying this star with **Colca,** singularly prominent with their shepherds, as Capella was with the same class on the Mediterranean in ancient days; indeed in later also, for the **Shepherd's Star** has been applied to it by our English poets, although more commonly to the planet Venus.

In astrology Capella portended civic and military honors and wealth.

Tennyson, in some fine lines in his *Maud*, mentions it as "a glorious crown."

As to its color astronomers are not agreed; Smyth calling it bright white; Professor Young yellow; and others say blue or red, which last it was asserted to be by Ptolemy, Al Ferghani, and Riccioli; while those whose eyes are specially sensitive to that tint still find it such.

Capella perhaps has increased in lustre during the present century; but, brilliant as it is, its parallax of 0″.095, obtained from Elkin's observations, indicates a distance from our system of 34¼ light years; and, if this be correct, the star emits 250 times as much light as our sun.

Its spectrum resembles that of the latter; indeed spectroscopists say that Capella is virtually identical with the sun in physical constitution, and furnishes the model spectrum of the Solar type,[1] yellow in tinge and ruled throughout with innumerable fine dark lines.

Vogel thinks it receding from our system at the rate of 15¼ miles a second. It is the most northern of all the 1st-magnitude stars, rising in the latitude of New York City at sunset about the middle of October, and culminating at nine o'clock in the evening of the 19th of January. Thus it is visible at some hour of every clear night throughout the year.

β, 2.1, lucid yellow.

Menkalinan, Menkalinam, and **Menkalina** are from **Al Mankib dhi'l 'Inān,** the Shoulder of the Rein-holder, which it marks, the solstitial colure passing it 2° to the east; the star itself being about 10° east of Capella. It is supposed to be a very close binary, receding from us about 17½ miles a second; the two practically equal stars that compose the pair being only 7½ millions of miles apart, and revolving in a period of about four days, with a relative velocity of fully 150 miles a second. This discovery was made by Pickering from spectroscopic observations in 1889. The lines in the spectrum double and undouble every two days.

γ, 2.1, brilliant white,

was **Al Ḳa'b dhi'l 'Inān,** the Heel of the Rein-holder, of Arabian astronomy, so showing its location in the figure of Auriga. From the earliest days of descriptive astronomy it has been identical with the star **Al Nath,** the *β* of Taurus at the extremity of the right horn, and Aratos so mentioned it. Vitruvius, however, said that it was **Aurigae Manus,** because the Charioteer was supposed to hold it in his hand, which would imply a very different drawing from that of Rome, Greece, and our own; and Father Hell, in 1769,

1 This is the 2d of the classification of Father Angelo Secchi, the modern Roman astronomer.

correctly had this expression for the star θ. The later Arabian astronomers also considered it in Taurus by designating it as **Al Ḳarn al Thaur al Shamālīyyah,** the Northern Horn of the Bull; but Kazwini adhered to Auriga by giving "the two in the ankles" as **Al Tawābi' al 'Ayyūḳ,** the Goat's Attendants, Ideler identifying these with γ and ι.

δ, 4.1, yellow,

is on the head of the Charioteer. It is unnamed with us, but, inconspicuous as it is, the Hindus called it **Praja-pāti,** the Lord of Created Beings, a title also and far more appropriately given to Orion and to Corvus. The *Sūrya Siddhānta* devotes considerable space to it; but "why so faint and inconspicuous a star should be found among the few of which Hindu astronomers have taken particular notice is not easy to discover."

The Chinese include it, with ξ, *h*, *k*, *i*, and others near Cassiopeia, in their asterism **Pa Kuh,** the Eight Cereals.

ϵ, variable, 3 to 4.5.

Hyde cited Arabic authority for this, being at one time **Al Ma'az,** the He Goat, and later on it so appeared in one of the commentaries on Ulug Beg; but Kazwini knew it by the general title **Al 'Anz,** although it was not in his **Al 'Ināz,** the group of Goats,—α, ζ, and η. Some modern lists include it with the Kids.

Its variability, in an irregular period, was suspected by Fritsch in 1821, confirmed by Schmidt in 1843, and independently discovered by Heis in 1847. ζ and η are about 5° southwest of Capella.

ζ, 4, orange,

is the western one of the Ἔριφοι, or Kids, of Hipparchos and Ptolemy, the **Haedi** of the Latins. Pliny made of them a separate constellation.

The poet Callimachus, 240 B. C., wrote in an epigram of the *Anthologia:*

Tempt not the winds forewarned of dangers nigh,
When the Kids glitter in the western sky;

Vergil, commending in the *Georgics* their observation to his farmer neighbors, made special allusion to the *dies Haedorum*, and with Horace and Manilius called them *pluviales*, the latter author's

Stormy Haedi . . . which shut the Main
And stop the Sailers hot pursuit of gain.

Horace similarly knew them as *horrida et insana sidera* and *insana Caprae sidera;* and Ovid as *nimbosi*, rainy. They thus shared the bad repute in which Capella was held by mariners, and were so much dreaded, as presaging the stormy season on the Mediterranean, that their rising early in October evenings was the signal for the closing of navigation. All classical authors who mention the stars alluded to this direful influence, and a festival, the *Natalis navigationis*, was held when the days of that influence were past. Propertius wrote of them, in the singular, as **Haedus**; Albumasar, as **Agni**, the Lambs; the Arabians knew them as **Al Jadyain**, the Two Young He Goats; and Bayer, in the plural, as **Capellae.**

ζ appeared in the original edition of the *Alfonsine Tables* as **Sadatoni**; but in the later, and in the *Almagest* of 1515, as **Saclateni**: both strangely changed, either from **Al Dhat al 'Inān**, the Rein-holder, or more probably from **Al Saiḍ al Thani**, the Second Arm, by some confusion with the star β that is thus located; or because itself was in that part of an earlier conception of the figure.

η is a half-magnitude brighter than ζ, but not individually named.

ι, 3.1,

was Al Tizini's **Al Ḳa'b dhi'l 'Inān**, which other authors gave to γ; and Kazwini included it with the latter in his **Al Tawābi' al 'Ayyūḳ.**

λ, Double, 5 and 9½, pale yellow and plum color; μ, 5.1; and σ, 5.3,

in the centre of the figure, were Kazwini's **Al Ḥibā'**, the Tent; but he had other such in Aquarius, the Southern Crown, and Corvus, for this naturally was a favorite simile with the Arabs.

It is this star that may be the one lettered Al Ḥurr, the Fawn, on the Borgian globe.

The 5th-magnitudes μ, ρ, and σ were **Tseen Hwang**, the Heavenly Pool; and ν, τ, υ, ϕ, χ, with another unidentified star, **Choo,** a Pillar.

2° south from χ, on the 24th of January, 1892, an amateur observer, the Reverend Doctor Thomas D. Anderson of Edinburgh, discovered with an opera-glass a 5th-magnitude yellowish *nova*, now known as T Aurigae, which has excited so much interest in the astronomical world by the character of its spectrum. Subsequent to the optical discovery it was identified on a photographic plate taken on the 10th of December previously, but not on one taken on the 8th, thus indicating its appearance in the sky between those two dates. Other photographs show that its maximum, 4.4, occurred about the 20th. Its conflagration, however, is supposed to have occurred at least a hundred, perhaps many hundred, years ago, so great is its distance from our system. It became invisible towards the end of April, 1892, but was rediscovered from Mount Hamilton on the 19th of August as a planetary nebula, the second instance in astronomical history of such a change of character, the *nova* Cygni of 1877 having been the first. It was still visible in 1895, its spectrum continuing distinctly nebular in its character; and it is worthy of notice that two others of the new stars discovered since the application of the spectroscope to this class of investigations have had nearly identical histories. Scheiner, who gives a detailed account of this phenomenon in his *Spectralanalyse*, alludes to the velocity of the two constituent bodies as being 400 miles or more a second; if indeed — which some doubt — the peculiar separation of the bright and dark lines of hydrogen noted in its spectrum is to be accounted for by the relative motion of gaseous masses involved in the phenomenon.

ψ^1 to ψ^{10}, 5th-magnitude stars, were the Βουλήγες, or Goads, the Latin **Dolones**, called **Stimulus** by Tibullus. Bayer said of them: *Decem stellulae flagellum constituentes.* As figured by Dürer they are the several lashes of the whip in the Charioteer's hands.

★

Boötes' golden wain.
Pope's *Statius His Thebais.*

Boötes only seem'd to roll
His Arctic charge around the Pole.
Byron's 3d Ode in *Hours of Idleness.*

Boötes,

the Italians' **Boöte** and the French **Bouvier,** is transliterated from Βοώτης, which appeared in the *Odyssey*, so that our title has been in use for nearly 3000 years, perhaps for much longer; although doubtless at first applied only to its prominent star Arcturus. Degenerate forms of the word have been **Bootis** and **Bootres.**

It has been variously derived: some say from Βοῦς, Ox, and ὠθεῖν, to drive, and so the **Wagoner,** or **Driver, of the Wain;** Claudian writing:

Boötes with the wain the north unfolds;

or the Ploughman of the Triones that, as **Arator,** occurs with Nigidius and Varro of the century before our era. But in recent times the figure has been imagined the Driver of Asterion and Chara in their pursuit of the Bear around the pole, thus alluded to by Carlyle in *Sartor Resartus:*

> What thinks Boötes of them, as he leads his Hunting Dogs over the zenith in their leash of sidereal fire?

Others, and perhaps more correctly, thought the word Βοητής, Clamorous, transcribed as **Boetes,** from the shouts of the Driver to his Oxen,—the Triones,—or of the Hunter in pursuit of the Bear; Hevelius suggesting that the shouting was in encouragement of the Hounds. In translations of the *Syntaxis* this idea of a Shouter was shown by **Vociferator, Vociferans, Clamans, Clamator, Plorans,** the Loud Weeper, and even, perhaps, by **Canis latrans,** the Barking Dog, that Aben Ezra applied to its stars in the Hebrew words **Kelebh hannabāh.**

The Arabians rendered their similar conception of the figure by **Al 'Awwā',**— Chilmead's **Alhava.**

The not infrequent title **Herdsman,** from the French Bouvier, also is appropriate, for not only was he associated with the Oxen of the Wain, but in Arab days the near-by circumpolar stars were regarded as a **Fold** with its inmates and enemies.

Other names were 'Αρκτοφύλαξ and 'Αρκτοῦρος, the **Bear-watcher** and the **Bear-guard,** the latter first found in the Ἔργα καὶ Ἡμέραι, the *Works and Days*, "a Boeotian shepherd's calendar," by Hesiod, eight centuries before our era. But, although these words were often interchanged, the former generally was used for the constellation and the latter for its *lucida*, as in the *Phainomena* and by Geminos and Ptolemy. Still the poets did not always discriminate in this, the versifiers of Aratos confounding the titles notwithstanding the exactness of the original; although Cicero in one place definitely wrote:

> Arctophylax, vulgo qui dicitur esse Bootes.

Transliterated thus,—or **Artophilaxe,**—and as **Arcturus,** both names are seen for the constellation with writers and astronomers even to the 18th century; Chaucer having "ye sterres of Arctour." The scientific Isidorus knew it as **Arcturus Minor,** his *Major* being the Greater Bear. Smyth derived this word from Ἄρκτου οὐρά, the Bear's Tail, as Boötes is near that part of Ursa Major; but this is not generally accepted—indeed is expressly condemned by the critic Buttmann.

Statius also called it **Portitor Ursae;** Vitruvius had **Custos** and **Custos Arcti,** the Bear-keeper; Ovid, **Custos Erymanthidos Ursae;** the *Alfonsine*

Tables, **Arcturi Custôs**; while the **Bear-driver** is often seen with early English writers.

Although Manilius knew it in connection with the Bear, he changed the simile when he wrote:

> whose order'd Beams
> Present a Figure driving of his Teams;

and Aratos long before had united the two thoughts and titles:

> Behind and seeming to urge on the Bear,
> Arctophylax, on earth Boötes named,
> Sheds o'er the Arctic car his silver light.

Plaustri Custos, the Keeper of the Wain, was another name for it that altered the character of Boötes' duties; Ovid following in this with:

> interque Triones
> Flexerat obliquo plaustrum temone Bootes.

It has been **Lycaon,** the father, or grandfather, of Kallisto, when that nymph was identified with Ursa Major; as well as **Arcas,** her son; Ovid distinctly asserting in the 2d of the *Fasti* that Arctophylax in the skies was the earthly Arcas, although it is often wrongly supposed that the latter is represented by Ursa Minor; it was **Septentrio,** from its nearness to the north, so taking one of the Bear's titles; and **Atlas,** because, near to the pole, it sustained the world.

Hesychios, of about A. D. 370, called it **Orion,** but this seems unintelligible unless originating from a misunderstanding of Homer's lines, translated by Lord Derby:

> Arctos call'd the Wain, who wheels on high
> His circling course, and on Orion waits,

as if they were in close proximity. Or the title may come from some confusion with the **Orus,** or **Horus,** of the Egyptians, that was associated with both Orion and Boötes. La Lande alluded to this when he wrote:

> Arctouros ou l'Orus voisin de l'Ourse, pour le distinguer de la constellation méridionale d'Orion;

and, in considering this very different derivation of our word Arcturus, it should be remembered that Κάνδαος and Κανδάων were the titles also applied to Boötes, as the latter Greek word was to Orion by the Boeotians. It would be interesting to know more of this connection.

Philomelus is another designation, as if he were the son of the neighboring Virgo Ceres; and the early title **Venator Ursae,** the Hunter of the Bear, again appears as **Nimrod,** the Mighty Hunter before the Lord, with the biblical school of two or three centuries ago; although this was more usual for Orion.

Pastor, the Shepherd, presumably is from the Arabic idea of a Fold around the pole, or from the near-by flock in the Pasture towards the southeast, in our Hercules and Ophiuchus; or perhaps by some confusion with Cepheus, who also was a Shepherd with his Dog.

Pastinator is Hyde's rendering of a supposed Arabic title signifying a Digger or Trencher in a vineyard. A commentator on Aratos called it Τρυγετής, the Vintager, as its rising in the morning twilight coincided with the autumnal equinox and the time of the grape harvest; Cicero repeating this in his **Protrygeter;** but both of these names better belonged to the star Vindemiatrix, our ε Virginis.

Still its risings and settings were frequently observed and made much of in all classical days, and even beyond the Augustan age, although many, perhaps most, of these allusions were to its bright star. As a calendar sign it was first mentioned by Hesiod, thus translated by Thomas Cooke:

> When in the rosy morn Arcturus shines,
> Then pluck the clusters from the parent vines;

and again, but for a different season of the year:

> When from the Tropic, or the winter's sun,
> Thrice twenty days and nights their course have run;
> And when Arcturus leaves the main, to rise
> A star bright shining in the evening skies;
> Then prune the vine.

Columella, Palladius, Pliny, Vergil, and others have similar references to Boötes, or to Arcturus, as indicating the proper seasons for various farm-work, as in the 1st *Georgic:*

> Setting Boötes will afford the signs not obscure.

Icarus, or **Icarius,** also was a title for our constellation, from the unfortunate Athenian who brought so much trouble into the world by his practical expounding of Bacchus' ideas as to the proper use of the grape, and who was so unworthily exalted to the sky, with his daughter Erigone as Virgo, and their faithful hound Maera as Procyon or Sirius. From this story came the **Icarii boves** applied to the Triones by Propertius, and in the Andrews-Freund *Lexicon* to Boötes himself.

Ceginus, Seginus, and **Chegninus,** as well as the **Cheguius** of the *Arabo-Latin Almagest,* may have wandered here in strangely changed form from the neighboring Cepheus; although Buttmann asserted that they probably came, by long-repeated transcription and consequent errors, from **Kheturus,** the Arabian orthography for Arcturus. Bayer had **Thegius,** as usual without explanation; still I find in Riccioli's *Almagestum Novum : Arabicē* **Theguius,** *quasi plorans aut vociferans;* but Arabic scholars do not confirm this.

La Lande cited **Custos Boum,** the Keeper of the Oxen, and **Bubulus,** or **Bubulcus,** the Peasant Ox-driver, although Ideler denied that the latter ever was used for Boötes. Juvenal, however, had it, and Minsheu defined Boötes as **Bubulcus coelestis.** Landseer, following La Lande, said that the **Herdsman** was the national sign of ancient Egypt, the myth of the dismemberment of Osiris originating in the successive settings of its stars; and that there it was called **Osiris, Bacchus,** or **Sabazius,** the ancient name for Bacchus and Noah; and that Kircher's planisphere showed a **Vine** instead of the customary figure, thus recalling incidents in the histories of those worthies, as well as of Icarius.

Homer characterized the constellation as *ὀψὲ δύων,* late in setting, a thought and expression now become hackneyed by frequent repetition. Aratos had it:

> he, when tired of day,
> At even lingers more than half the night;

Manilius somewhat varying this by

> Slow Boötes drives his ling'ring Teams;

Claudian, Juvenal, and Ovid, by *tardus*, slow, and *piger*, sluggish, which their later countryman Ariosto, of the 16th century, repeated in his *pigro* **Arturo**; and Minsheu, in the 17th century, wrote of it as

> Bootes, *or the* **Carman,** a slow mooving starre, *seated* in the North Pole *neere* to Charles Waine, *which it followes.*

And all this because, as the figure sets in a perpendicular position, eight hours are consumed in its downward progress, and even then the hand of Boötes never disappears below the horizon — a fact more noticeable in early days than now. The reverse, however, takes place at its rising in a horizontal position; hence the *ἀθρόος*, all at once, of Aratos.

Some say that these expressions of sluggishness are from its setting late in the season when the daylight is curtailed, or a reference to the natural gait of the Triones that Boötes is driving around the pole; while still others, more astronomically inclined, attributed them to his comparative nearness to

> that point where slowest are the stars,
> Even as a wheel the nearest to its axle,

that Dante wrote of in the *Purgatorio.*

Boötes' association with the Mons Maenalus, on which he is sometimes shown, is unexplained unless by the suggestion found under that constellation heading. This association was current even in early days, if Landseer be correct where he says:

> Eusebius, quoting an ancient oracle which has apparent reference to this constellation as formerly represented, writes —
>
> A mystic goad the mountain herdsman bears.

Brown says that it was known in Assyria as **Riu-but-same,** "that reappears in Greek as Boötes"; and thus

> the idea of the ox-driving Ploughman or Herdsman, as applied to the constellation, is Euphratean in character.

Among its Arabian derivatives are **Nekkar,** often considered as **Al Naḳḳār,** the Digger, or Tearer, analogous to the classic Trencher in the vineyard; but Ideler showed this to be an erroneous form of **Al Baḳḳār,** the Herdsman, found with Ibn Yunus (or Yunis).

Alkalurops, which appeared for Boötes in the *Alfonsine Tables* as **Incalurus,** is from Κἄλαῦροψ, a herdsman's Crook or Staff, with the Arabic article prefixed; this now is our title for the star μ. The staff, ultimately figured as a Lance, gave rise to the name **Al Rāmiḥ,** which came into general use among the Arabians, but subsequently degenerated in early European astronomical works into **Aramech, Ariamech,** and like words for the constellation as well as for its great star.

The same figure is seen in **Al Ḥāmil Luzz,** the Spear-bearer, or, as Caesius had it, **Al Kameluz,** Riccioli's **Kolanza,** and the **Azimeth Colanza** of Reduan's translator, which Ideler compared to the Latin *cum lancea* and the Italian *colla lancia.* Similarly, Bayer said that on a Turkish map it was 'Οϊστοφόρος, the Arrow-bearer; and elsewhere **Sagittifer** and **Lanceator.**

Al Ḥāris al Samā' of Arabic literature originally was for Arcturus, although eventually applied to the constellation. But long before these ideas were current in Arabia, that people are supposed to have had an enormous Lion, their early **Asad,** extending over a third of the heavens, of which the stars Arcturus and Spica were the shin-bones; Regulus, the forehead; the heads of Gemini, one of the fore paws; Canis Minor, the other; and Corvus, the hind quarters. Yet there seems to be doubt as to all this, as is more fully explained under α Geminorum.

In Poland Boötes forms the **Ogka,** or Thills, of that country's much-extended **Woz Niebeski,** the Heavenly Wain; and in the Old Bohemian tongue it was **Przyczck,** as unintelligible as it is unpronounceable.

The early Catholics knew it as **Saint Sylvester;** Caesius said that it might represent the prophet **Amos,** the Herdsman, or Shepherd Fig-dresser, of Tekoa; but Weigel turned it into the **Three Swedish Crowns.**

Proctor asserted that Boötes, when first formed, perhaps included even the Crown, as we know that it did the Hunting Dogs; and that, so constituted,

> it exhibits better than most constellations the character assigned to it. One can readily picture to one's self the figure of a Herdsman with upraised arm driving the Greater Bear before him.

The drawing by Heis, after Dürer, is of a mature man, with herdsman's staff, holding the leash of the Hounds; but earlier representations are of a much younger figure: in all cases, however, well equipped with weapons of the chase, or implements of husbandry; the earliest form of these probably having been the winnowing fan of Bacchus.

The Venetian *Hyginus* of 1488 shows the Wheat Sheaf, Coma Berenices, at his feet; Argelander's *Uranometria Nova* has different figures on its two plates — one of the ancient form, the other of the modern holding the leash of the Hounds in full pursuit of the Bear.

This constellation and the Bear, Orion, the Hyades, Pleiades, and Dog were the only starry figures mentioned by Homer and Hesiod; the latter's versifier, Thomas Cooke, giving as a reason therefor — "the names of which naturally run into an hexameter verse"; but the general assumption that these great poets knew no other constellations does not seem reasonable, although it will be noticed that all those alluded to are identical with each author.

Boötes is a constellation of large extent, stretching from Draco to Virgo, nearly 50° in declination, and 30° in right ascension, and contains 85 naked-eye stars according to Argelander, 140 according to Heis.

> Poises Arcturus aloft morning and evening his spear.
>
> Emerson's translation of Hafiz' *To the Shah.*

α, 0.3, golden yellow.

Arcturus has been an object of the highest interest and admiration to all observant mankind from the earliest times, and doubtless was one of the first stars to be named; for from Hesiod's day to the present it thus appears throughout all literature, although often confounded with the Greater Bear. Indeed Hesiod's use of the word probably was for that constellation, except in two cases, already quoted, where he unquestionably referred to this star, mentioning its rising fifty days after the winter solstice, the first allusion that we have to that celestial point. And it is popularly supposed that our Arcturus is that of the *Book of Job*, xxxviii, 32; but there it merely is one of the early titles of Ursa Major, the Revised Version correctly rendering it "the Bear." Still, even now, the *Standard Dictionary* quotes for the star the Authorized Version's

Canst thou guide Arcturus with his sons?

But, like other prominent stars, it shared its name with its constellation — in fact, probably at first, and as late as Pliny's day, was a constellation by itself. Homer's Βοώτης doubtless was this, with, possibly, a few of its larger companions; and Bayer cited **Bootes** for the star; but in recent times the latter has monopolized the present title.

It was famous with the seamen of early days, even from the traditional period of the Arcadian Evander, and regulated their annual festival by its movements in relation to the sun. But its influence always was dreaded, as is seen in Aratos' δεινοῦ 'Αρκτούροιο and Pliny's *horridum sidus;* while Demosthenes, in his action against Lacritus 341 B. C., tells us of a bottomry bond, made in Athens on a vessel going to the river Borysthenes — the modern Dnieper — and to the Tauric Chersonese — the Crimea — and back, that stipulated for a rate of 22½ per cent. interest if she arrived within the Bosporus "before Arcturus," *i. e.* before its heliacal[1] rising about mid-September; after which it was to be 30 per cent. Its acronycal[2] rising fixed the date of the husbandmen's Lustratio frugum; and Vergil twice made allusion in his 1st *Georgic* to its character as unfavorably affecting the farmers' work. Other contemporaneous authors confirmed this stormy reputation, while all classical calendars[3] gave the dates of its risings and settings.

Hippocrates, 460 B. C., made much of the influence of Arcturus on the human body, in one instance claiming that a dry season, after its rising,

agrees best with those who are naturally phlegmatic, with those who are of a humid temperament, and with women; but it is most inimical to the bilious;

and that

diseases are especially apt to prove critical in these days.

1 This was its first perceptible appearance in the dawn after emergence from the sun, then about 10° or 12° away.

2 The latest rising visible at sunset.

3 Copies of these calendars, called Παράπηγματα, engraved on stone or brass, were conspicuously exposed in the market-places, and two are supposed to have come down to us, — that of Geminos, 77 B. C., and of Ptolemy, A. D. 140. While these probably in the main were accurate, the allusions to their subjects by the poets and authors generally seem to be as often wrong as right, being based upon observations taken on trust from earlier writers, or from tradition, although by various causes, and especially by the effect of precession, they had become incorrect. Hesiod's statement, in the *Works and Days*, of the heliacal rising of Arcturus is regarded as fixing his own date in history at about 800 B. C.

The Prologue of the *Rudens* of Plautus, delivered by Arcturus in person, and "one of the early opinions of the presence of invisible agents amongst mankind," declares of himself that he is considered a stormy sign at the times of his rising and setting,— as the original has it:

> Arcturus signum, sum omnium quam acerrimum.
> Vehemens sum, cum exorior, cum occido vehementior.

And the passage from Horace's *Odes* —

> Nec saevus Arcturi cadentis
> Impetus aut orientis Haedi —

is familiar to all. This same idea came down to modern days, for Pope repeated it in his verse,

> When moist Arcturus clouds the sky.

Astrologically, however, the star brought riches and honor to those born under it.

An Egyptian astronomical calendar of the 15th century before Christ, deciphered by Renouf, associates it with the star Antares in the immense sky figure **Menat**; and Lockyer claims it as one of the objects of worship in Nile temples, as it was in the temple of Venus at Ancona in Italy.

In India it was the 13th *nakshatra*, **Svati**, the Good Goer, or perhaps Sword, but figured as a Coral Bead, Gem, or Pearl; and known there also as **Nishṭya**, Outcast, possibly from its remote northern situation far outside of the zodiac, whence, from its brilliancy, it was arbitrarily taken to complete the series of Hindu asterisms. Hewitt thinks that it, or Capella, was the **Āryamān** of the *Rig Veda;* and Edkins that it was the **Tistar** usually assigned to Sirius.

The Chinese called it **Ta Kiō**, the Great Horn, four small stars near by being **Kang Che**, the Drought Lake; Edkins further writing of it:

> Arcturus is the palace of the emperor. The two groups of three small stars on its right [η, τ, υ] and left [ζ, o, π] are called **She ti**, the Leaders, because they assign a fixed direction to the tail of the Bear, which, as it revolves, points out the twelve hours of the horizon.

The Arabs knew Arcturus as **Al Simāk**[1] **al Rāmiḥ**, sometimes translated the Leg of the Lance-bearer, and again, perhaps more correctly, the Lofty

[1] This word Simāk is of disputed signification, and was a fruitful subject of discussion a century ago. It is from a root meaning "to raise on high," and is thought to have been employed by the Arabs when they wished to indicate any prominent object high up in the heavens, but with special reference to this star and to the other Simāk, Spica of the Virgin.

Lance-bearer. From the Arabic title came various degenerate forms: **Al Ramec, Aramec, Aremeah, Ascimec, Azimech,** and **Azimeth,** found in those queer compendiums of stellar nomenclature the *Alfonsine Tables* and the *Almagest* of 1515; **Somech haramach** of Chilmead's *Treatise;* and **Aramākh,** which Karsten Niebuhr heard from the Arabs 136 years ago. The **Kheturus** of their predecessors, already alluded to under Boötes, also was used for this.

The idea of a weapon again manifested itself in the Κονταράτος, Javelin-bearer, of the *Graeco-Persian Tables;* while Bayer had **Gladius, Kolanza,** and **Pugio,** all applied to Arcturus, which probably marked in some early drawing the Sword, Lance, or Dagger in the Hunter's hand. Similarly it took the title **Alkameluz** of the whole constellation.

Al Ḥāris al Samā, the Keeper of Heaven, perhaps came from the star's early visibility in the twilight owing to its great northern declination, as though on the lookout for the safety and proper deportment of his lesser stellar companions, and so "Patriarch Mentor of the Train." This subsequently became **Al Ḥāris al Simāk,** the Keeper of Simāk, probably referring to Spica, the Unarmed One.

Al Bīrūnī mentioned Arcturus as the **Second Calf of the Lion,** the early Asad; Spica being the First Calf.

It has been identified with the Chaldaeans' **Papsukal,** the Guardian Messenger, the divinity of their 10th month Tibitu; while Smith and Sayce have said that on the Euphrates it was the **Shepherd of the Heavenly Flock,** or the **Shepherd of the Life of Heaven,** undoubtedly the **Sib-zi-anna** of the inscriptions; the star η being often included in this, and thus making one of the several pairs of Euphratean Twin Stars.

The 1515 *Almagest* and the *Alfonsine Tables* of 1521 add to their list of strange titles *et nominatur* **Audiens,** which seems unintelligible unless the word be a misprint for *Audens*, the Bold One.

John de Wiclif, in his translation of *Amos* v, 8, in 1383, had it **Arture,** which he took from the *Vulgate's* Arcturus for Ursa Major; but John of Trevisa in 1398 more correctly wrote:

> **Arthurus** is a signe made of VII starres, . . . but properly Arthurus is a sterre sette behynde the tayle of the synge that hyght Vrsa maior.

With others it was **Arturis** and **Ariture,** or the **Carlwaynesterre** from the early confusion in applying the title Arcturus to Charles' Wain as well as to Boötes and its *lucida*.

Prominent as this star always has been, and one of the few to which Ptolemy assigned a name, yet its position has greatly varied in the drawings; indeed in the earliest it was located outside of the figure and so described in the *Syntaxis*. It has been put on the breast; in the girdle, whence, perhaps, came Bayer's **Arctuzona**; on the leg; between the knees, — Robert Recorde, the first English writer on astronomy, in 1556 mentioning in the *Castle of Knowledge* the "very bryghte starre called Arcturus, which standeth between Boötes his legges"; and, as some of its titles denote, on the weapon in the hand. But since Dürer's time it has usually marked the fringe of the tunic.

Smyth asserted that this is the first star on record as having been observed in the daytime with the telescope, as it was in 1635 by Morin, and subsequently, in July, 1669, by Gautier and the Abbé Picard, the sun having an elevation of 17°. Schmidt has seen it with the naked eye twenty-four minutes before sunset. While these instances serve to show its brilliancy, yet this was still more evinced when, enveloped in the Donati comet of 1858, and on the 5th of October, only 20′ from the nucleus, "it flashed out so vividly its superiority," visible for many hours. And it is somewhat remarkable that this same thing was seen 240 years before in the case of the comet of 1618; at least such is the record of John Bainbridge, "Doctor of Physicke," who wrote:

> The 27th of November, in the morning, the comet's hair was spread over the faire starre Arcturus, betwixt the thighs of Arctophylax, or Bootes.

It is interesting to know that the first photograph of a comet was of Donati's, near this star, on the 28th of September, 1858.

Ptolemy specified its color as ὑπόκιῤῥος, rendered *rutilus*, "golden red," in the 1551 *Almagest;* but Schmidt observed, on the 21st of March, 1852, that the star had lost its usual tinge, which it did not regain for several years. This phenomenon was confirmed by Argelander and by Kaiser of Leyden; but generally it has "figured immemorially in the short list of visibly fiery objects." Its rich color, in contrast with the white of Spica, the deeper red of Antares, and the sapphire of Wega, is very noticeable when all can be taken in together, at almost a single glance, on a midsummer evening.

The Germans know it as **Arctur**; the Italians and Spanish, as **Arturo.** Schiller wrote in the *Death of Wallenstein:*

> Not every one doth it become to question
> The far off high Arcturus;

but Elkin did so in 1892, his observations resulting in a parallax of 0″.016, *i. e.* insensible, the probable error being much greater than the measured parallax itself.

The star has a large proper motion,[1] given as 2″.3 annually, which probably has shifted its position southwestward on the face of the sky by somewhat more than 1° since the time of Ptolemy; and great velocity in the line of sight was assigned to it by the earlier spectroscopists, even as high as seventy miles a second; but the later and accordant determinations, at Potsdam by Vogel and at the Lick Observatory by Keeler, reduce this to between 4 and 4¾ miles.

Its spectrum is Solar, of Secchi's second type, but with a remarkable mass of dark lines in the violet.

Arcturus culminates on the 8th of June.

β, 3.6, golden yellow.

Nakkar and **Nekkar** are from the Arabic name for the whole constellation.

The Chinese knew it as **Chaou Yaou,** or **Teaou,** words meaning "to beckon, excite, or move."

With γ, δ, and μ, it constituted the trapezium **Al Dhi'bah,** the Female Wolves, or, perhaps, Hyaenas, an early asterism of the Arabs before they adopted the Greek constellation; these animals, with others similar shown by stars in Draco and near it, lying in wait for the occupants of the ancient Fold around the pole.

β marks the head of the modern figure.

γ, 3.1.

Seginus appears on Burritt's *Atlas* from the Ceginus of the constellation.

Manilius termed it *prona Lycaonia*, "sloping towards, or in front of, Lycaon," referring to the Greater Bear, as the star marks the left shoulder of Boötes near to that constellation; and Euripides similarly wrote in his Ἴων of about 420 B. C.:

> Above, Arcturus to the golden pole inclines.

Flammarion gives to it the **Alkalurops** that is better recognized for μ. The Chinese called it **Heuen Ko,** the Heavenly Spear.

It is interesting to know that the variable ν is in the telescopic field with γ.

[1] This proper motion of some of the stars, *i.e.* the angular motion across the line of sight, was first detected by Halley, in 1718, from examination of modern observations, especially those of Tycho, on Arcturus, Aldebaran, and Sirius, in comparison with the ancient records.

δ, 3.5, pale yellow.

This star does not appear to be named, but in China was part of **Tseih Kung,** the Seven Princes; the other components being μ, ν, ϕ, ψ, χ^1, and χ^2, or b, in the right hand and on the Club, 20° northeast of Arcturus.

ϵ, Binary, 3 and 6, pale orange and bluish green,

lying 10° northeast of Arcturus, bore these titles in Arabia: **Al Minṭakah al 'Awwā',** the Belt of the Shouter; **Izār,** the Girdle; and **Mi'zar,** the Waist-cloth,— all references to its place in the figure. This last word was turned by early European astronomical writers into **Micar, Mirar, Merer, Meirer, Mezen, Mezer, Merak,** and **Mirak,** similar to the title of β Andromedae, and all appropriate. The analogous **Perizoma** was used for it in the *Alfonsine Tables.*

Why it was so favored in nomenclature is not known, for with us it is noticeable only from its exquisite beauty in the telescope, whence it is fast monopolizing the name **Pulcherrima,** given to it by the elder Struve.

The components can be seen with a 2¼-inch glass, about 3″ apart, at a position angle of 325°. The period of their revolution is as yet undetermined, but they are thought to be approaching us at the rate of ten miles a second.

This pair was the chief object of Sir William Herschel's investigations for stellar parallax about 1782, in which, of course, he was unsuccessful, although he did not know the cause of his failure till years thereafter, when he recognized its binary character.

ζ, ξ, o, and π were **Tso She Ti,** an Officer, in China, on the left hand of the emperor.

η, 2.8, pale yellow.

Muphrid, Mufrid, and **Mufride,** of the Palermo and other catalogues, is from Ulug Beg's **Al Mufrid al Rāmiḥ,** the Solitary Star of the Lancer, and inexplicable unless on the supposition that it formerly was regarded as outside of the figure lines. Kazwini called it **Al Rumḥ;** and Al Tizini, with Al Naṣr al Dīn, more definitely, **Al Rumḥ al Rāmiḥ,** the Lance of the Lancebearer, although inappropriately, for they designated its position as on Al Sāḳ, the Shin-bone, and it thus appears as **Saak** in some lists; but as the figure is now drawn η lies above the left knee.

It seems to have been included with Arcturus in the Euphratean **Sib-zi-anna.**

With υ and τ in the feet, it was **Yew She Ti** in China, the Officer standing on the right hand of the emperor.

θ, 4.1; ι, Triple, 4.4, 4.5, and 8; and κ, Double, 4.5 and 6.6.

Bayer called these **Asellus,**—*primus*, *secundus*, and *tertius* respectively,—although without explanation; but the title is well known for each of the two stars in Cancer flanking Praesaepe. They mark the finger-tips of the upraised left hand just eastward from Alkaid, the last star in the Greater Bear's tail.

In China they were **Tseen Tsang,** the Heavenly Lance.

The members of the larger component of ι are 0″.8 apart; the smaller is 38″ away.

κ is pale white, and the two stars are about 12″ apart, making it an easy object in a small telescope.

All of these, with the 4th-magnitude λ on the lower part of the left arm, were **Al Aulād al Dhi'bah,** the Whelps of the Hyaenas, shown by β, γ, δ, and μ, and so given on the earliest Arabic maps and globes.

μ^1, Ternary, 4.2, 8, and 8.5, flushed white, the last two greenish white,

the small companion μ^2 being a close double.

Alkalurops was the Arabian adaptation of Κᾱλᾱυροψ, used by Hesychios for the Herdsman's Club, Crook, or Staff, analogous to the Ρόπαλον of Hyginus and the **Clava** of the Latins.

Inkalunis appears in some of the *Alfonsine Tables*; **Icalurus** in those of 1521, and **Incalurus** in the 1515 *Almagest*, all long supposed to be bungled renderings of Ptolemy's Κολλορόβος, itself probably a word of his own coining to designate the position of the star in the club; Riccioli writing it **Colorrhobus.** But Ideler, rejecting this, thought Schickard more correct in deriving these words from ἐν κολόυρω, "in the colure," a statement that was nearly right as to Arcturus 2000 years ago; the name since then having, in some way, been transferred to this star, as also to the constellation. The editor of the 1515 *Almagest* added to his title for μ *et est hastile habens canes*, which, Ideler said,—and Homer is for once caught nodding,—"is with reference to the surrounding hyaenas." This most erroneous explanation is corrected by the late Professor C. H. F. Peters of the Hamilton Observatory, whose private copy of this rare edition is now in my possession, in his autographic annotation that the original Arabic should have been rendered *ferrum curvatum* instead of *canes.* Some Latin writers have called this star **Venabulum,** a Hunting-spear.

ρ and σ, 4th- and 5th-magnitude stars, were **Kang Ho,** a river in China; and ψ, according to Assemani, with another in the right arm that may have been ε, constituted the Arabs' **Al Aulād al Nadhlāt,** which he rendered *Filii altercationis;* but the original signifies the Low, or Mean, Little Ones.

h, or Fl. 38, a 5½-magnitude hardly visible to the naked eye, is **Merga,** and marks the Reaping-hook held in the left hand of the figure. This word is from *Marra*, a Hoe, or Rake, used by Columella and Juvenal, and still is sometimes seen as **Marrha** for the star. The latter was well known to Pliny as **Falx Italica.**

★

Caelum, or Scalptorium, the Burin or Graving-tool,

sometimes incorrectly written **Cela sculptoria,** is the French **Burin,** the Italian **Bulino,** and the German **Grabstichel.**

It was formed by La Caille from stars between Columba and Eridanus, directly south of the Sceptrum Brandenburgicum; Gould now assigns to it twenty-eight components, of magnitudes from four to seven.

Burritt, in the early editions of his book, arbitrarily changed the name to **Praxiteles,** perhaps thinking thereby to avoid possible confusion with the constellation Sculptor.

Caelum comes to the meridian with the star Aldebaran on the 10th of January, and is entirely visible from the 40th parallel.

★

Camelopardalis, or Camelopardus, the Giraffe,

the French **Girafe** and Italian **Giraffa,** is long, faint, and straggling like its namesake. It stretches from the pole-star to Perseus, Auriga, and the Lynx, the hind quarters within the Milky Way.

It was formed by Bartschius, who published it, in outline only, in 1614, and wrote that it represented to him the Camel that brought Rebecca to Isaac. Was it from this that Proctor attempted to change its title to **Camelus?** — an alteration that seems to have been adopted only by Mr. J. Ellard Gore in his translation, in 1894, of Flammarion's *Astronomie Populaire.* Weigel used it with Auriga to form his heraldic figure, **the French Lilies.**

The Chinese located seven asterisms within its boundaries: **Hwa Kae,** the State Umbrella, extending beyond Camelopardalis; **Luh Kea,** a term in anatomy; **Shang Ching,** the Higher Minister; **Shàng Wei,** the Higher Guard; **Shaou Wei,** the Minor Guard; **Sze Foo,** the Four Official Supporters of the Throne; and **Yin Tih,** Unostentatious Virtue.

Argelander enumerates 84 naked-eye stars, and Heis 138; these culminating in the middle of January.

The 4th-magnitude *lucida* is 20° north of Capella, below the left hock of the animal; and two others of the same brilliancy, 1° apart, are in front of the fore quarters.

★

. . . and there a crab
Puts coldly out its gradual shadow-claws,
Like a slow blot that spreads,— till all the ground,
Crawled over by it, seems to crawl itself.

Mrs. Browning's *Drama of Exile.*

Cancer, the Crab,

der Krebs of the Germans,— **die Krippe** of Bayer; **le Cancre,** or **l'Écrevisse,** of the French; and **il Cancro** or **Granchio** of the Italians, lies next to Gemini on the east, and is popularly recognized by its distinguishing feature, the Beehive, ancient Praesaepe. Aratos called it Καρκίνος, which Hipparchos and Ptolemy followed; the **Carcinus** of the *Alfonsine Tables* being the Latinized form of the Greek word. Eratosthenes extended this as Καρκίνος, Ὄνοι, καί Φάτνη, the Crab, Asses, and Crib; and other Greeks have said Ὀπισθοβάμων and Ὀκτάπους, the **Octipes** of Ovid and Propertius. **Litoreus,** Shore-inhabiting, is from Manilius and Ovid; **Astacus** and **Cammarus** appear with various classic writers; and **Nepa** is from Cicero's *De Finibus* and the works of Columella, Manilius, Plautus, and Varro,— all signifying Crab, or Lobster, although more usual, and perhaps more correct, for Scorpio. Festus, the grammarian of the 3d century, said that this was an African word equivalent to *Sidus*, a Constellation or Star.

It is the most inconspicuous figure in the zodiac, and mythology apologizes for its being there by the story that when the Crab was crushed by Hercules, for pinching his toes during his contest with the Hydra in the marsh of Lerna, Juno exalted it to the sky; whence Columella called it **Lernaeus.** Yet few heavenly signs have been subjects of more attention in early days, and few better determined; for, according to Chaldaean and Platonist philosophy, it was the supposed **Gate of Men** through which souls descended from heaven into human bodies.

In astrology, with Scorpio and Pisces, it was the **Watery Trigon;** and has been the **House of the Moon,** from the early belief that this luminary was located here at the creation; and the **Horoscope of the World,** as being, of all the signs, nearest to the zenith. It was one of the unfortunate signs, governing the human breast and stomach; and reigned over Scotland, Holland, Zealand, Burgundy, Africa (especially over Algiers, Tripoli, and Tunis), and the cities of Constantinople and New York. In the times of Manilius it ruled India and Aethiopia, but he termed it a fruitful sign. Its colors were green and russet; and early fable attributed its guardianship to the god Mercury, whence its title **Mercurii Sidus.** When the sun was within its boundaries every thunder-storm would cause commotions, famine, and locusts; and Berōssōs asserted that the earth was to be submerged when all the planets met in Cancer, and consumed by fire when they met in Capricorn. But this was a reversal of the astrologers' rule; for, as Pascal wrote:

> They only assign good fortune with rare conjunctions of the stars, and this is how their predictions rarely fail.

It is said to have been the Akkadian **Sun of the South,** perhaps from its position at the winter solstice in very remote antiquity; but afterwards it was associated with the fourth month Duzu, our June–July, and was known as the **Northern Gate of the Sun,** whence that luminary commences its retrograde movement. **Nan-garu** is Strassmaier's transliteration of the cuneiform title; others being **Puluk-ku** and **χas,** Division, possibly referring to the solstitial colure as a dividing line. Brown has recently claimed for it the title **Nagar-asagga,** the Workman of the Waterway.

The early Sanskrit name was **Karka** and **Karkata,** the Tamil **Karkatan,** and the Cingalese **Kathaca;** but the later Hindus knew it as **Kulira,** from Κόλουρος, the term originated by Proclus for our colure.

The Persians had it **Chercjengh** and **Kalakang;** the Turks, **Lenkutch;** the Syrians, and perhaps the later Chaldaeans, **Sartono;** the Hebrews, **Sarṭān;** and the Arabians, **Al Saraṭān,** all words equivalent to Cancer. Al Bīrūnī added **Al Lihā'**, the Soft Palate, but this was an early title of the Arabs in connection with their *manzil* **Al Nathrah.**

Kircher said that in Coptic Egypt it was Κλαρία, the *Bestia seu Statio Typhonis*, the Power of Darkness; La Lande identifying this with Anubis, one of the divinities of the Nile country commonly associated with Sirius. But the Jews assigned it to the tribe of Issachar, whom Jacob likened to the "strong ass" that each of the Aselli represents; Dupuis asserting that these last titles were derived from this Jewish association.

A Saxon chronicle of about the year 1000 had "Cancer that is **Crabba**";

Chaucer had **Cancre,** probably a relic of Anglo-Norman days, for in his time it generally was **Canser**; and Milton called it the **Tropic Crab** from its having marked one of these great circles.

Showing but few stars, and its *lucida* being less than a 4th-magnitude, it was the **Dark Sign,** quaintly described as black and without eyes. Dante, alluding to this faintness and high position in the heavens, wrote in the *Paradiso:*

> Thereafterward a light among them brightened,
> So that, if Cancer one such crystal had,
> Winter would have a month of one sole day.

Jensen makes it the **Tortoise** of Babylonia, and it was so figured there and in Egypt 4000 B. C.; although in the Egyptian records of about 2000 B. C. it was described as a **Scarabaeus,** sacred, as its specific name *sacer* signifies, and an emblem of immortality. This was the Greek *κάραβος*, with its nest-ball of earth in its claws, an idea which occurs again even as late as the 12th century, when an illuminated astronomical manuscript shows a **Water-beetle.** In the *Albumasar* of 1489 it is a large **Crayfish**; Bartschius and Lubienitzki, in the 17th century, made it into a **Lobster,** and the latter added toward Gemini a small shrimp-like object which he called **Cancer minor.**

Caesius likened it to the **Breastplate of Righteousness** in *Ephesians* vi, 14; while Praesaepe and the Aselli were the **Manger** of the infant Jesus, with the Ass and Ox presumed to be standing by. Julius Schiller said that the whole represented **Saint John** the Evangelist.

Our figure appears on the round zodiac of Denderah, but in the location of Leo Minor.

This planisphere[1] is a comparatively late sculpturing, supposed to be about 34 B. C., in the time of Tiberius and Cleopatra, possibly later; but it shows, at least in part, the heavens of many centuries previous, the exact date fixed by Biot being 700 B. C., although some scholars, notably Brugsch, carry it back a thousand years earlier and assert that it was largely copied from similar works of Sargon's time. It was discovered by the French general Desaix de Voygoux in 1799, and removed in 1820 to the Bibliothèque Imperiale in Paris, where it has since remained. Its appearance is that of a very large antique sandstone medallion, 4 feet 9 inches in diameter, contained in a square of 7 feet 9 inches. With some manifest errors, it is, nevertheless, a most interesting and much-quoted object, although not of the importance once attributed to it. Of the many en-

[1] The temple which contained this was dedicated to Isis, and is the smaller of the two most celebrated at Denderah, the Tentyris of the Greeks and Tentore of the Copts, names derived from the Tan-ta-rer of ancient Egypt, signifying the Land of the Hippopotamus. It is on a site sacred long before the present edifice, of which we now have the ruins, was erected.

gravings of this, the best is found in Flammarion's journal *L'Astronomie* for September, 1888.

Cancer appears on the Farnese globe underneath a quadrangular figure, in the location of our Lynx, of which I can find no explanation.

In this constellation, with some slight variations as to boundaries at different times in Hindu astronomy,—γ and δ always being included and occasionally η, θ, and Praesaepe,— was located the 6th *nakshatra* **Pushya,** Flower, or **Tishiya,** Auspicious, with Brihaspati, the priest and teacher of the gods, as presiding divinity. It was sometimes figured as a Crescent, and again as the head of an Arrow; but Amara Sinha, the Sanskrit author of about 56 B. C., called it **Sidhaya,** Prosperous.

The *manzil* **Al Nathrah,** the Gap in the hair under the muzzle of the supposed immense ancient Lion, was chiefly formed by Praesaepe; but later on γ and δ were sometimes included, when it was **Al Ḥimārain,** the Two Asses, a title adopted from the Greeks. The Arabs also knew it as **Al Fum al Asad** and as **Al Anf al Asad,** the Mouth, and the Muzzle, of the Lion, both referring to the early figure.

The *sieu* **Kwei,** Spectre, anciently **Kut,** the Cloud-like, was made up from Praesaepe with η and θ, the latter most strangely selected, as it is now hardly distinguishable by the naked eye, and yet was the determining star,— perhaps a case of variation in brightness. This asterism, with Tsing in our Gemini, formed **Shun Show,** one of the twelve zodiacal *Kung*, which Williams translates as the Quail's Head, giving the modern title as **Keu Hea,** the Crab; this Quail being otherwise known as the Phoenix, Pheasant, or the Red Bird that, with the stars of Leo and Virgo, marked the residence of the Red, or Southern, Emperor.

Like Gemini and Taurus, it was shown rising backward, to which some of the ancients fancifully ascribed the slower motion of the sun in passing through these constellations, as well as its influence in producing the summer's heat; even Doctor Johnson, in *Rasselas*, alluded to "the fervours of the crab." Very differently, however, Ampelius associated it with the cold Septentrio, or North Wind.

Coins of Cos in the Aegean Sea bore the figure of a Crab that may have been for this constellation.

The symbol of the sign, ♋, probably is "the remains of the representation of some such creature"; but it is also referred to the two Asses that took part in the conflict of the gods with the giants on the peninsula of the Macedonian Pallene, the early Phlegra, afterwards rewarded by a resting-place in the sky on either side of the Manger.

The sun is in Cancer from the 18th of July to the 7th of August; but the solstice, which was formerly here and gave name to the tropic, is now about 33° to the westward, near η Geminorum.

The celebrated Halley comet first appeared here in 1531; and in June, 1895, all the planets, except Neptune, were in this quarter of the heavens, an unusual and most interesting occurrence. Argelander catalogues 47 stars in the constellation in addition to Praesaepe; and Heis, 91.

α, Double, 4.4 and 11, white and red.

Acubens, from the *Chelae quas Acubenae Chaldai vocant* of the *Alfonsine Tables*, is not Chaldaean, but from the Arabic **Al Zubanāh,** the Claws, on the southern one of which this star lies, near the head of Hydra. Bayer repeated this in his **Acubene** and **Azubene,** adding Pliny's names for it—**Acetabula,** the Arm Sockets of a crab, and **Cirros,**—properly **Cirrus,**—the Arms themselves, equivalent to Ovid's **Flagella,** which Bayer wrongly translated Scourge; others similarly saying **Branchiae** and **Ungulae.** Bayer also cited the "Barbarians'" **Grivenescos,** unintelligible unless it be their form of Γραψαῖος, a Crab. **Sartan** and **Sertan** are from the Arabic word for the whole figure. The star ι, marking the other claw, shares in many of these titles.

Some assign **Al Hamarein** to α,—an undoubted error, as Al Ḥimārain was the common Arabian term for the Aselli, γ and δ, that the Arabic signifies.

Acubens culminates on the 18th of March. The companion is 11''.4 distant, at a position angle of 325°.5.

β, a 4th-magnitude, is **Al Tarf,** the End, *i. e.* of the southern foot on which it lies.

Sunt in signo Cancri duae stellae parvae, aselli appellati.
Pliny's *Historia Naturalis.*

γ, 4.6, and δ, 4.3, straw color.

Asellus borealis and **Asellus australis,** the Northern and the Southern Ass Colt, were the Ὄνοι, or Asses, of Ptolemy and the Greeks; the **Aselli,** or **Asini,** of the Latins, distinguished by their position as here given, even to the present day, and now popularly .known as the **Donkeys.** The Basel *Latin Almagest* of 1551 says **Asinus** for γ only, but the *Alfonsine Tables* and the *Almagest* of 1515 have **Duo Asini**; and the Arabians similarly knew them as **Al Ḥimārain,** the Two Asses. Bailey, in his *Mystic* of 1858, calls them the **Aselline Starlets.**

Manilius is supposed to allude to these outstretched stars as the **Jugulae,** taken indirectly from *Jugum*, a Yoke, which became *Jugulum*, the Collar-bone,— in the plural *Jugula* and *Jugulae;* but Ideler asserted that this originated from an erroneous statement of Firmicus, and that reference was really made by the poet to the well-known Belt of Orion.

Riccioli's strange title, **Elnatret,** doubtless was from that of the lunar mansion **Al Nathrah,** which the Aselli and Praesaepe constituted.

In astrology they were portents of violent death to such as came under their influence; while to the weather-wise their dimness was an infallible precursor of rain, on which Pliny thus enlarges:

> If fog conceals the Asellus to the northeast high winds from the south may be expected, but if the southern star is concealed the wind will be from the northeast.

Our modern Weather Bureau would probably tell us that if one of these stars were thus concealed, the other also would be. Pliny mentioned them with Praesaepe as forming a constellation by themselves; but he was given to multiplying the stellar groups.

Inconspicuous though it be, the Babylonians used δ to mark their 13th ecliptic constellation **Arkū-sha-nangaru-sha-shūtu,** the Southeast Star in the Crab; and Brown says that the Aselli, with η, θ, and Praesaepe, were the Akkadian **Gu-shir-kes-da,** the Yoke of the Enclosure. They also marked the junction of the *nakshatras* **Pushya** and **Āçleshā.**

The following passage from Hind's *Solar System* in regard to δ will be found interesting:

> The most ancient observation of Jupiter[1] which we are acquainted with is that reported by Ptolemy in Book X, chap. iii, of the *Almagest*, and considered by him free from all doubt. It is dated in the 83d year after the death of Alexander the Great, on the 18th of the Egyptian month *Epiphi*, in the morning, when the planet eclipsed the star now known as δ Cancri. This observation was made on September 3, B. C. 240, about 18^h on the meridian of Alexandria.

ε

was applied by Bayer to the coarse extended cluster, N. G. C. 2632, 44 M., on the head of the Crab, composed of about 150 stars of magnitudes from 6½ to 10, with two noticeable triangles among them.

With us it is the well-known **Beehive,** but its history as such I have not been able to learn, although it undoubtedly is a recent designation, for nowhere is it *Apiarium*.

Scientifically it was the Νεφέλιον, or Little Cloud, of Hipparchos; the Ἀχλύς, or Little Mist, of Aratos; the Νεφελοειδής, Cloudy One, Συστροφή, Whirling Cloud, and **Nubilum,** literally a Cloudy Sky, of Bayer;

[1] This planet was known to the Greeks as Ζεύς, and as Φαέθων, the Shining One.

but the *Almagests* and astronomers generally of the 16th and 17th centuries referred to it as the *Nebula*, and *Nebulosa*, *in pectore Cancri*, for before the invention of the telescope this was the only universally recognized nebula, its components not being separately distinguishable by ordinary vision. But it seems to have been strangely regarded as three nebulous objects. Galileo, of course, was the first to resolve it, and wrote in the *Nuncius Sidereus:*[1]

> The nebula called Praesepe, which is not one star, only, but a mass of more than forty small stars. I have noticed thirty stars, besides the Aselli.

Popularly it also is the **Manger**, or **Crib**, the Φάτνη of Aratos and Eratosthenes; the Φάτνης of Ptolemy; and with the Latins, **Praesaepe, Praesaepes, Praesaepis, Praesaepia, Praesaepium**, the Alfonsine **Presepe** and Bayer's **Pesebre**,— also the modern Spanish,— flanked by the Aselli, for whose accommodation it perhaps was invented. Bayer cited for it **Melleff**, which Chilmead followed with **Mellef**, and Riccioli with **Meeleph**; these from the Arabians' **Al Ma'laf**, the Stall; and this, in turn, derived from the Greek astronomy, for their indigenous Ma'laf was in Crater. Schickard had this as **Mallephon.**

Brown includes ε with γ, δ, η, and θ in the Persian lunar station **Avra-k**, the Cloud, and the Coptic **Ermelia**, Nurturing.

Tyrtaeus Theophrastus, the first botanist-author, about 300 B. C., and Aratos, described its dimness and disappearance in the progressive condensation of the atmosphere as a sure token of approaching rain; Pliny said,

> If Praesaepe is not visible in a clear sky it is a presage of a violent storm;

and Aratos in the Διοσημεῖα (the *Prognostica*):

> A murky Manger with both stars
> Shining unaltered is a sign of rain.
> If while the northern Ass is dimmed
> By vaporous shroud, he of the south gleam radiant,
> Expect a south wind: the vaporous shroud and radiance
> Exchanging stars harbinger Boreas.

Weigel used it in the 17th century, in his set of heraldic signs, as the **Manger**, a fancied coat of arms for the farmers.

In astrology, like all clusters, it threatened mischief and blindness.

In China it was known by the unsavory title **Tseih She Ke**, Exhalation of Piled-up Corpses; and within 1° of it Mercury was observed from that

[1] This *Nuncius Sidereus*, published at Venice by Galileo in 1610, first gave to the world the results of his telescopic observations.

country, on the 9th of June, A. D. 118, one of the early records of that planet.

ζ, Ternary, 5.6, 6.3, and 6, yellow, orange, yellowish,—changing.

This lies on the rear edge of the Crab's shell, and is known as **Tegmine,** In the Covering; but, if the word be allowable at all, it should be **Tegmen,** as Avienus is supposed to have had it. Ideler, however, said that Avienus was referring to the covering shell of the marine object, and not to the stellar.

This is a system of great interest to astronomers from the singular changes in color, the probable existence of a fourth and invisible component, and for the short period of orbital revolution — sixty years — of the two closer stars. The maximum of interval between these is but 1″, the minimum 0″.2; yet they never close up as one star. The third member is 5″ away, and its orbital period must be at least 500 years.

ζ and θ, according to Peters' investigations, probably are the objects announced by Watson as two intra-Mercurial planets, discovered (?) during the total eclipse of the sun on the 29th of July, 1878.

λ, of the 6th magnitude, with adjacent stars, was in China **Kwan Wei,** the Bright Fire.

μ, a 5½-magnitude, with χ Geminorum, was **Tsih Tsin,** a Heap of Fuel.

ξ, another 5½-magnitude, with λ Leonis, formed the seventh *manzil* **Al Tarf,** the End, or, as some translate it, the Glance, *i. e.* of the Lion's Eye, the ancient Asad, which occupied so large a portion of the sky in this neighborhood. They also were the Persian **Nahn,** the Nose, and the Coptic **Piautos,** the Eye, both lunar asterisms.

ξ, with κ and stars in Leo, was the Chinese **Tsu Ke,** one of the flags of that country.

★

Boötes hath unleash'd his fiery hounds.
Owen Meredith's *Clytemnestra.*

Canes Venatici, the Hunting Dogs,

are the French **Chiens du Chasse,** or **Levriers**; the German **Jagdhunde,** and the Italian **Levrieri,** lying between Boötes and Ursa Major. Ptolemy entered their stars among the ἀμόρφωτοι of the latter constellation, and the modern forms first appear in the *Prodromus* of their inventor Hevelius. The more northern one is **Asterion,** Starry, from the little stars marking the body; and the other, which contains the two brightest stars, is **Chara,** as Dear to the heart of her master. Flamsteed followed in the use of these names, and the **Hounds** are now well established in the recognition of astronomers, as is the case with most of the stellar creations of Hevelius, which were generally placed where needed.

Proctor, in his attempt to simplify constellation nomenclature, called them **Catuli,** the Puppies; but the usual illustration is of two Greyhounds held by a leash in the hand of Boötes, ready for pursuit of the Bear around the pole; their inventor thus reviving the idea that Boötes was a hunter.

Hevelius counted 23 stars here; Argelander, 54; and Heis, 88.

The Chinese designated three stars in or near the head of Asterion as **San Kung,** the Three Honorary Guardians of the Heir Apparent.

Assemani alluded to a quadrate figure on the Borgian globe, below the tail of the Greater Bear, as **Al Karb al Ibl,** the Camel's Burden, that can be no other than stars in the heads of the Hunting Dogs.

Bartschius drew on his map of this part of the sky the **River Jordan,** his **Jordanis** and **Jordanus,** not now recognized, indeed hardly remembered. Its course was from Cor Caroli, under the Bears and above Leo, Cancer, and Gemini, through the stars from which Hevelius afterwards formed Leo Minor and the Lynx, ending at Camelopardalis. But the outlines of his stream were left somewhat undetermined, much like those of Central African waters when guessed at by map-makers thirty years or more ago. This river, however, had already existed before his day on French star-maps and -globes.

α, Double, 3.2 and 5.7, flushed white and pale lilac.

This star, the 12 of Flamsteed's list of the Hounds, stands alone, marking Chara's collar; but was set apart in 1725 by Halley, when Astronomer Royal, as the distinct figure **Cor Caroli,** not Cor Caroli II as many have it, in honor of Charles II. This was done at the suggestion of the court physician, Sir Charles Scarborough, who said that it had shone with special brilliancy on the eve of the king's return to London on the 29th of May, 1660. It has occasionally been seen on maps as the centre of a Heart-shaped figure surmounted by a crown, and its name occurs in popular lists; but Flamsteed did not insert it on his plate of the Hounds, although he distinctly wrote of it in his manuscript under this title; and the Heart perhaps is shown in the tail-piece to the preface of the *Atlas Coelestis.*

It is the French **Coeur de Charles**; the Italian **Cuor di Carlo**; and the German **Herz Karls.**

With Ulug Beg it was **Al Kabd al Asad,** the Liver of the Lion,—here a technical term indicating the highest position of any star within the compass of a figure reckoned from the equator.

In China it was **Chang Chen,** a Seat.

This is a favorite object with amateur observers, the components being about 20″ apart. Espin says, in Webb's *Celestial Objects* of 1893, that they have been relatively fixed for seventy-three years, yet show considerable proper motion, and probably are unequal stars at nearly equal distances from us; and he gives various opinions of observers as to their colors. Miss Clerke calls them pale yellow and fawn. Their present position angle is about 230°, but is slowly changing.

Cor Caroli culminates on the 20th of May.

On the line from Cor Caroli to Arcturus, and somewhat nearer the latter, in a triangle of small stars, is a beautiful globular cluster concentrated into a central blaze. This is N. G. C. 5272, 3 M., long a well-known object, but recently rendered specially noticeable by Bailey's discovery in 1895, on photographs taken by Harvard astronomers at Arequipa, Peru, of no less than ninety-six variable stars within its boundaries,—nearly ten per cent. of the whole number in the cluster distinctly photographed: the usual proportion of variables among the naked-eye stars is not quite one per cent. The stars near the centre run together and cannot be counted, but the total number in the cluster probably is many thousands.

β, 4.3, is **Chara,** the 8 of Flamsteed, and, after Cor Caroli, the brightest member of the Southern Hound.

152 Schjellerup, 5.5, brilliant red.

La Superba was so named by Father Secchi from the superbly flashing brilliancy of its prismatic rays. It is the brightest of its class of stars with spectra of the 4th type, of which only about 120 are known from our latitude, and but seven or eight of these visible to the naked eye. Variability in its light is also suspected.

It lies about 7° north and 2½° west of Cor Caroli.

A misty spot in this constellation can be seen with a low-power 3° southwest from Al Kaid (η Ursae Majoris). This is the **Spiral Nebula** of Lord Rosse, or the **Whirlpool Nebula,** N. G. C. 5194, 51 M., our long-established ideas of which have recently been somewhat modified by a photograph taken by Mr. Isaac Roberts after four hours' exposure. It now appears to be composed of a pair of curving arms issuing from opposite extremities of an oval central body, one of the arms joining itself to a second nucleus,—a new system in process of formation.

★

> Fierce on her front the blasting Dog-star glowed.
>
> Samuel Taylor Coleridge's *On the French Revolution.*

> One blazes through the brief bright summer's length,
> Lavishing life-heat from a flaming car.
>
> Christina G. Rossetti's *Later Life.*

Canis Major, the Greater Dog,

of the southern heavens, and thus **Canis Australior,** lies immediately to the southeast of Orion, cut through its centre by the Tropic of Capricorn, and with its eastern edge on the Milky Way.

It is **Cane Maggiore** in Italy; **Cães** in Portugal; **Grand Chien** in France; and **Grosse Hund** in Germany.

In early classical days it was simple **Canis,** representing Laelaps, the hound of Actaeon, or that of Diana's nymph Procris, or the one given to Cephalus by Aurora and famed for the speed that so gratified Jove as to cause its transfer to the sky. But from the earliest times it also has been the **Dog of Orion** to which Aratos alluded in the *Prognostica,* and thus wrote of in the *Phainomena* in connection with the Hare:

> The constant Scorcher comes as in pursuit,
> . . . and rises with it and its setting spies.

Homer made much of it as Κύων, but his **Dog** doubtless was limited to the star Sirius, as among the ancients generally till, at some unknown date, the constellation was formed as we have it,—indeed till long afterwards, for we find many allusions to the Dog in which we are uncertain whether the constellation or its *lucida* is referred to. Hesiod and Aratos gave this title, both also saying Σείριος, and the latter μέγας; but by this adjective he designed only to characterize the brilliancy of the star, and not to distinguish it from the Lesser Dog. The Greeks did not know the two Dogs thus, nor did the comparison appear till the days of the Roman Vitruvius.

Ptolemy and his countrymen knew it by Homer's title, and often as Ἀστροκύων, although it seems singular that the former never used the word Σείριος.

The Latins adopted their Canis from the Greeks, and it has since always borne this name, sometimes even **Canicula** in the diminutive (with the adjectival *candens*, shining), **Erigonaeus,** and **Icarius;** the last two being from the fable of the dog Maera,— which itself means Shining,— transported here; her mistress Erigone having been transformed into Virgo, and her master Icarius into Boötes. Ovid alluded to this in his *Icarii stella proterva canis;* and Statius mentioned the **Icarium astrum,** although Hyginus had ascribed this to the Lesser Dog.

Sirion and **Syrius** occasionally appeared with the best Latin authors; and the *Alfonsine Tables* of 1521 had **Canis Syrius.**

Vergil brought it into the 1st *Georgic* as a calendar sign,—

> adverso cedens Canis occidit astro,—

instructing the farmer to sow his beans, lucerne, and millet at its heliacal setting on the 1st of May; the *adverso* here generally being referred to the well-known reversed position of the figure of Taurus, but may have been intended to indicate the hostility of the Bull to the Giant's Dog that was attacking him.

Custos Europae is in allusion to the story of the Bull who, notwithstanding the Dog's watchfulness, carried off that maiden; and **Janitor Lethaeus,** the Keeper of Hell, makes him a southern Cerberus, the watch-dog of the lower heavens, which in early mythology were regarded as the abode of demons: a title more appropriate here than for the so-named modern group in the northern, or upper, sky.

Bayer erroneously quoted as proper names **Dexter, Magnus,** and **Secundus,** while others had **Alter** and **Sequens;** but these originally were designed only to indicate the Dog's position, size, and order of rising with regard to his lesser companion.

The *aestifer* of Cicero and Vergil referred to its bright Sirius as the cause of the summer's heat, which also induced Horace's *invidum agricolis;* and Bayer's Ὑδροφοβία was from the absurd notion, prevalent then as now, of the occurrence of canine madness solely during the heat from the Dog-star: an idea first seen with Asclepiades of the 3d century before Christ. Or it may have come from being confounded by Bayer, none too careful a compiler, with the Ὑδραγωγόν, which Plutarch applied to Sirius in his *De Isidoro*, signifying the Water-bringer, *i. e.* the cause of the Nile flood.

Aratos termed the constellation ποικίλος, as of varying brightness in its different parts; or mottled—the Dog, lying in as well as out of the Milky Way, being thus diversified in light.

In early Arabia, as indeed everywhere, it took titles from its *lucida*, although strangely corrupted from the original **Al Shi'rā al 'Abūr al Yamaniyyah,** the Brightly Shining Star of Passage of Yemen, in the direction of which province it set. Among these we see, in the *Latin Almagest* of 1515, "**canis**: *et est* **asehere, alahabor aliemenia**"; in the edition of 1551, **Elscheere**; in Bayer's *Uranometria*, **Elseiri** (which Grotius derived from σείριος), **Elsere, Sceara, Scera, Scheereliemini**; in Chilmead's *Treatise*, **Alsahare aliemalija**; and **Elchabar,** which La Lande, in his *l'Astronomie*, not unreasonably derived from Al Kabir, the Great.

The Arabian astronomers called it **Al Kalb al Akbar,** the Greater Dog, so following the Latins, Chilmead writing it **Alcheleb Alachbar**; and Al Bīrūnī quoted their **Al Kalb al Jabbār,** the Dog of the Giant, directly from the Greek conception of the figure. Similarly it was the Persians' **Kelbo Gavoro.**

It was, of course, important in Euphratean astronomy, and is shown on remains from the temples and mounds, variously pictured, but often just as Aratos described it and as drawn on maps of the present day,—standing on the hind feet, watching or springing after the Hare. Professor Young describes the figure as one "who sits up watching his master Orion, but with an eye out for Lepus."

Bayer and Flamsteed alone among its illustrators showed it as a typical bulldog.

A Dog, presumably this with another adjacent, is represented on an ivory disc found by Schliemann on his supposed site of Troy; and an Etruscan mirror of unknown age bears it with Orion, Lepus, the crescent moon, and correctly located neighboring stars. While both of the Dogs, the Dragon, Fishes, Swan, Perseus, the Twins, Orion, and the Hare are described as on the Shield of Hercules in the old poem of that title generally attributed to Hesiod. The Hindus knew it as **Mrigavyādha,** the Deerslayer, and as **Lubdhaka,** the Hunter, who shot the arrow, our Belt of Orion, into the infamous Praja-pāti, where it even now is seen sticking in his body; and, much earlier still, with their prehistoric predecessors it was **Saramā,** one of the Twin Watch-dogs of the Milky Way.

Among northern nations it was **Greip,** the dog in the myth of Sigurd. All of these doubtless referred solely to Sirius.

Novidius, who imagined biblical significance in every starry group, said that this was the **Dog of Tobias** in the *Book of Tobit*, v, 16, which Moxon confirmed "because he hath a tayle," and for that reason only; but Julius Schiller, another of the same school, saw here the royal **Saint David**.

Gould catalogued 178 stars down to the 7th magnitude.

> Hail, mighty Sirius, monarch of the suns!
> May we in this poor planet speak with thee?
>
> Mrs. Sigourney's *The Stars.*

α, Binary, —1.43 and 8.5, brilliant white and yellow.

Sirius, the **Dog-star,** often written **Syrius** even as late as Flamsteed's and Father Hell's day, has generally been derived from *σείριος*, sparkling or scorching, which first appeared with Hesiod as a title for this star, although also applied to the sun, and by Abychos to all the stars. Various early Greek authors used it for our Sirius, perhaps generally as an adjective, for we read in Eratosthenes:

> Such stars astronomers call *σειρίους* on account of the tremulous motion of their light;

so that it would seem that the word, in its forms *σείρ*, *σεῖρος*, and *σείριος*,— Suidas used all three for both sun and star,—originally was employed to indicate any bright and sparkling heavenly object, but in the course of time became a proper name for this brightest of all the stars. Lamb, however, thought it of Phoenician origin, signifying the Chief One, and originally in that country a title for the sun; Jacob Bryant, the mythologist, said that it was from the Egyptians' **Cahen Sihor**; but Brown considers it a transcription from their well-known **Hesiri,** the Greek **Osiris**; while Dupuis distinctly asserted that it was from the Celtic **Syr.**

Plutarch called it Προόπτης, the Leader, which well agrees with its character and is an almost exact translation of its Euphratean, Persian, Phoenician, and Vedic titles; but *Κύων*, *Κύων σείριος*, *Κύων ἀστήρ*, *Σείριος ἀστήρ*, *Σείριον ἄστρον*, or simply *το ἄστρον*, were its names in early Greek astrônomy and poetry. *Προκύων*, better known for the Lesser Dog and its *lucida*, also was applied to Sirius by Galen as preceding the other stars in the constellation.

Homer alluded to it in the *Iliad* as Ὀπωρινός, the **Star of Autumn**;[1] but the season intended was the last days of July, all of August, and part of September — the latter part of summer. Lord Derby translated this celebrated passage:

> A fiery light
> There flash'd, like autumn's star, that brightest shines
> When newly risen from his ocean bath;

[1] The Greeks had no word exactly equivalent to our "autumn" until the 5th century before Christ, when it appeared in writings ascribed to Hippocrates.

while later on in the poem Homer compares Achilles, when viewed by Priam, to

> th' autumnal star, whose brilliant ray
> Shines eminent amid the depth of night,
> Whom men the dog-star of Orion call.

The Roman farmers sacrificed to it a fawn-colored dog at their three festivals when, in May, the sun began to approach Sirius. These, instituted 238 B. C., were the Robigalia, to secure the propitious influence of their goddess Robigo in averting rust and mildew from their fields; and the Floralia and Vinalia, to ensure the maturity of their blooming flowers, fruits and grapes.

Among the Latins it naturally shared the constellation's titles, probably originated them; and occasionally was even **Canicula**; indeed, as late as 1420 the *Palladium of Husbandry* urged certain farm-work to be done "Er the caniculere, the hounde ascende"; and, more than a century later, Eden, in the *Historie of the Vyage to Moscovie and Cathay*, wrote: "**Serius** is otherwise cauled Canicula, this is the dogge, of whom the canicular dayes have theyr name."

It has been asserted that Ovid and Vergil referred to Sirius in their **Latrator Anubis**, representing a jackal- or dog-headed Egyptian divinity, guardian of the visible horizon and of the solstices, transferred to Rome as goddess of the chase; but it is very doubtful whether they had in mind either star or constellation.

Its well-known name, **Al Shi'rā**, or **Al Si'rā**, extended as **al Abūr al Yamaniyyah**, much resembles the Egyptian, Persian, Phoenician, Greek, and Roman equivalents, and, Ideler thought, may have had common origin with them from some one ancient source: possibly the Sanskrit Sūrya, the Shining One,—the Sun. The ʽAbur, or Passage, refers to the myth of Canopus' flight to the South; and the adjective to the same, or perhaps to the southerly position of the star towards Yemen, in distinction from that of Al Ghumaiṣā' in the Lesser Dog, seen towards Shām,—Syria,—in the North. From these geographical names originated the Arabic adjectives Yamaniyyah and Shamāliyyah, Southern and Northern; although the former literally signifies On the Right-hand Side, *i. e.* to an observer facing eastward towards Mecca.

In Chrysococca's *Tables* the title is *Σιαὴρ* Ιαμανὴ; and Doctor C. Edward Sachau's translation of Al Bīrūnī's *Chronology* renders it **Sirius Jemenicus.** Riccioli had **Halabor,** which the 1515 *Almagest* applied to the constellation; and Chilmead, **Gabbar, Ecber,** and **Habor**; while **Shaari lobur,** another queerly corrupted form, is found in Eber's *Egyptian Princess*. In the *Alfonsine Tables* the original is changed to **Asceher** and **Aschere Aliemini;** while Bayer gives plain **Aschere** and **Elscheere** for the star, with others similar for both star and constellation. **Scera** is cited by Grotius for the star, and **Sceara** for the whole, derived from an old lexicon; and **Alsere;** but he traced all to Σείριος.

In modern Arabia it is **Suhail,** the general designation for bright stars.

The late Finnish poet Zakris Topelius accounted for the exceptional magnitude of Sirius by the fact that the lovers Zulamith the Bold and Salami the Fair, after a thousand years of separation and toil while building their bridge, the Milky Way, upon meeting at its completion,

> Straight rushed into each other's arms
> And melted into one;
> So they became the brightest star
> In heaven's high arch that dwelt —
> Great Sirius, the mighty Sun
> Beneath Orion's belt.

The native Australians knew it as their **Eagle,** a constellation by itself; while the Hervey Islanders, calling it **Mere,** associated it in their folk-lore with Aldebaran and the Pleiades.

Sharing the Sanskrit titles for the whole, it was the **Deer-slayer** and the **Hunter,** while the *Vedas* also have for it **Tishiya** or **Tishiga, Tistrija, Tishtrya,** the **Tistar,** or **Chieftain's, Star**. And this we find too in Persia; as also **Sira.** The later Persian and Pahlavi have **Tir,** the Arrow. Edkins, however, considers Sirius, or Procyon, to be **Vanand,** and Arcturus, Tistar.

Hewitt sees in Sirius the **Sivānam,** or Dog, of the *Rig Veda* awakening the Ribhus, the gods of mid-air, who "thus calls them to their office of rain sending," a very different office from that assigned to this star in Rome. Yet these gods, philologically, had a Roman connection, for Professor Friedrich Maximilian Mueller, writing the word Arbhu, associates it with the Latin Orpheus. Hewitt also says that in the earliest Hindu mythology Sirius was **Sukra,** the Rain-god, before Indra was thus known; and that in the *Avesta* it marked one of the Four Quarters of the Heavens.

Although the identification of Euphratean stellar titles is by no means settled, especially and singularly so as to this great star, yet various authorities have found for it names more or less probable.

Bertin and Brown think it conclusively proved that it was **Kak-shisha,** the Dog that Leads, and "a Star of the South"; while **Kak-shidi** is Sayce's transliteration of the original signifying the Creator of Prosperity, a character which the Persians also assigned to it; and it may have been the Akkadian **Du-shisha,** the Director — in Assyrian **Mes-ri-e.** Epping and Strassmaier have **Kak-ban** as a late Chaldaean title, which Brown renders **Kal-bu,** the Dog, " exactly the name for Sirius we should expect to find"; Jensen has **Kakkab lik-ku**, the Star of the Dog, revived in Homer's *κύων*; and it perhaps was the Assyrian **Kal-bu Sa-mas,** the Dog of the Sun; and the Akkadian **Mul-lik-ud,** the Star Dog of the Sun. Jensen also gives **Kakkab kasti,** the Bow Star, although this may be doubtful; and Brown has, from the Assyrian, **Su-ku-du,** the Restless, Impetuous, Blazing, well characterizing the marked scintillation and color changes in its light. Hewitt cites an Akkadian title **Tis-khu.**

Its risings and settings were regularly tabulated in Chaldaea about 300 B. C., and Oppert is reported to have recently said that the Babylonian astronomers could not have known certain astronomical periods, which as a matter of fact they did know, if they had not observed Sirius from the island of Zylos in the Persian Gulf on Thursday, the 29th of April, 11542 B. C.!

It is the only star known to us with absolute certitude in the Egyptian records — its hieroglyph, a dog, often appearing on the monuments and temple walls throughout the Nile country. Its worship, chiefly in the north, perhaps, did not commence till about 3285 B. C., when its heliacal rising at the summer solstice marked Egypt's New Year and the beginning of the inundation, although precession has now carried this rising to the 10th of August. At that early date, according to Lockyer, Sirius had replaced γ Draconis as an orientation point, especially at Thebes, and notably in the great temple of Queen Hatshepsu, known to-day as Al Dēr al Bahārī, the Arabs' translation of the modern Copts' Convent of the North. Here it was symbolized, under the title of **Isis Hathor,** by the form of a cow with disc and horns appearing from behind the western hills. With the same title, and styled **Her Majesty of Denderah,** it is seen in the small temple of Isis, erected 700 B. C., which was oriented toward it; as well as on the walls of the great Memnonium, the Ramesseum, of Al Ḳurneh at Thebes, probably erected about the same time that this star's worship began. Lockyer thinks that he has found seven temples oriented to the rising of Sirius. It is also represented on the walls of the recently discovered step-temple of Saḳḳara, dating from about 2700 B. C., and supposed to have been erected in its honor.

Great prominence is given to it on the square zodiac of Denderah, where it is figured as a cow recumbent in a boat with head surmounted by a star; and again, immediately following, as the goddess **Sothis,** accompanied by the goddess Anget, with two urns from which water is flowing, emblematic of the inundation at the rising of the star. But in the earlier temple service of Denderah it was **Isis Sothis,** at Philae **Isis Sati,** or **Satit,** and, for a long time in Egypt's mythology, the resting-place of the soul of that goddess, and thus a favorable star. Plutarch made distinct reference to this; although it should be noted that the word Isis at times also indicated anything luminous to the eastward heralding sunrise. Later it was **Osiris,** brother and husband of Isis, but this word also was applied to any celestial body becoming invisible by its setting. Thus its titles noticeably changed in the long period of Egypt's history.

As **Thoth,** and the most prominent stellar object in the worship of that country,—its ḥeliacal rising was in the month of Thoth,—it was in some way associated with the similarly prominent sacred ibis, also a symbol of Isis and Thoth, for, in various forms, the bird and star appear together on Nile monuments, temple walls, and zodiacs.

Sirius was worshiped, too, as **Sihor,** the Nile Star, and, even more commonly, as **Sothi** and **Sothis,** its popular Graeco-Egyptian name, the **Brightly Radiating One,** the **Fair Star of the Waters;** but in the vernacular was **Sept, Sepet, Sopet,** and **Sopdit; Sed,**[1] and **Sot,**—the Σήθ of Vettius Valens.

Upon this star was laid the foundation of the Canicular, Sothic, or Sothiac Period named after it, which has excited the attention and puzzled the minds of historians, antiquarians, and chronologists. Lockyer has an admirable discussion of this in his *Dawn of Astronomy.*

Sir Edwin Arnold writes of it in his *Egyptian Princess:*

> And even when the Star of Kneph has brought the summer round,
> And the Nile rises fast and full along the thirsty ground;

for the Egyptians always attributed to the Dog-star the beneficial influence of the inundation that began at the summer solstice; indeed, some have said that the Aethiopian Nile took from Sirius its name Siris, although others consider the reverse to be the case. Minsheu, who dwells much on this, ends thus: "*Some thinke that the* Dog-starre *is called* Sirius, *because at the time the* Dogge-starre *reigneth,* Nilus *also overfloweth* as though the water *were led by that Starre.*" Indeed, it has been fancifully asserted that its canine title originated in Egypt, "because of its supposed watchful care over the interests of the husbandman; its rising giving him notice of the approaching overflow of the Nile."

Caesius cited for it **Solechin** as from that country, signifying the Starry Dog, and derived from the Egypto-Greek word Σολεκήν.

[1] According to Mueller, this Sed, or Shed, of the hieroglyphic inscriptions appeared in Hebrew as El Shaddar.

Perhaps it is the ancient importance of this Dog on the Nile that has given the popular name, the **Egyptian X,** to the figure formed by the stars Procyon and Betelgeuze, Naos and Phaet, with Sirius at the vertices of the two triangles and the centre of the letter. On our maps Sirius marks the nose of the Dog.

The Phoenicians are said to have known it as **Hannabeah,** the Barker.

The astronomers of China do not seem to have made as much of Sirius as did those of other countries, but it is occasionally mentioned, with other stars in Canis Major, as **Lang Hoo**; and Reeves quoted for it **Tseen Lang,** the Heavenly Wolf. Their astrologers said that when unusually bright it portended attacks from thieves.

Some have called it the **Mazzārōth** of the *Book of Job;* others the **H·aṣīl** of the Hebrews; but this people also knew it as **Sihor,** its Egyptian name, and Ideler thinks that the adoration of the Sᵉērīm, or "Devils" of the Authorized Version of our *Bible*, the "He Goats" of the Revision, which, as we see in *Leviticus* xvii, 7, was specially prohibited to the Jews, may have had reference to Sirius and Procyon, the **Two Sirii** or **Shi'rayān,** that must have been well known to them in the land of their long bondage as worshiped by their taskmasters.

The culmination of this star at midnight was celebrated in the great temple of Ceres at Eleusis, probably at the initiation of the Eleusinian mysteries; and the Ceans of the Cyclades predicted from its appearance at its heliacal rising whether the ensuing year would be healthy or the reverse. In Arabia, too, it was an object of veneration, especially by the tribe of Ḳais, and probably by that of Kodhā'a, although Muḥammād expressly forbade this star-worship on the part of his followers. Yet he himself gave much honor to some "star" in the heavens that may have been this.

In early astrology and poetry there is no end to the evil influences that were attributed to Sirius.

Homer wrote, in Lord Derby's translation,

> The brightest he, but sign to mortal man
> Of evil augury.

Pope's very liberal version of the same lines,—

> Terrific glory! for his burning breath
> Taints the red air with fevers, plagues and death,—

seems to have been taken from the *Shepheard's Kalendar* for July:

> The rampant Lyon hunts he fast with dogge of noysome breath
> Whose balefull barking brings in hast pyne, plagues and dreerye death.

Spenser, however, was equally a borrower, for we find in the *Aeneid:*

The dogstar, that burning constellation, when he brings drought and diseases on sickly mortals, rises and saddens the sky with inauspicious light;

and in the 4th *Georgic:*

Jam rapidus torrens sitientes Sirius Indos
Ardebat coelo,

rendered by Owen Meredith in his *Paraphrase* on Vergil's Bees of Aristaeus:

Swift Sirius, scorching thirsty Ind,
Was hot in heaven.

Hesiod advised his country neighbors, "When Sirius parches head and knees, and the body is dried up by reason of heat, then sit in the shade and drink,"— advice universally followed, even till now, although with but little thought of Sirius. Hippocrates made much, in his *Epidemics* and *Aphorisms*, of this star's power over the weather, and the consequent physical effect upon mankind, some of his theories being current in Italy even during the last century; while the result of all physic depended upon the sign of the zodiac in which the sun chanced to be. Manilius wrote of Sirius:

from his nature flow
The most afflicting powers that rule below.

But these expressions as to the hateful character of the Dog-star may have been induced in part from the evil reputation of the dog in the East.

Its heliacal rising, 400 years before our era, corresponded with the sun's entrance into the constellation Leo, that marked the hottest time of the year, and this observation, originally from Egypt, taken on trust by the Romans, who were not proficient observers, and without consideration as to its correctness for their age and country, gave rise to their *dies canicularíae*, the dog days, and the association of the celestial Dog and Lion with the heat of midsummer. The time and duration of these days, although not generally agreed upon in ancient times, any more than in modern, were commonly considered as beginning on the 3d of July and ending on the 11th of August, for such were the time and period of the unhealthy season of Italy, and all attributed to Sirius. The Greeks, however, generally assigned fifty days to the influence of the Dog-star. Yet even then some took a more correct view of the matter, for Geminos wrote:

It is generally believed that Sirius produces the heat of the dog days; but this is an error, for the star merely marks a season of the year when the sun's heat is the greatest.

But he was an astronomer.

The idea prevailed, however, even with the sensible Dante in his "great scourge of days canicular"; while Milton, in *Lycidas*, designated it as "the swart star." And the notion holds good with many even to the present time. This character doubtless is indicated on the Farnese globe, where the Dog's head is surrounded with sun-rays.

But Pliny took a kinder view of this star, as in the "xii. chapyture of the xi. booke of his naturall hystorie," on the origin of honey:

> This coometh from the ayer at the rysynge of certeyne starres, and especially at the rysynge of *Sirius*, and not before the rysynge of *Vergiliae* (which are the seven starres cauled *Pleiades*) in the sprynge of the day;

although he seems to be in doubt whether "this bee the swette of heaven, or as it were a certeyne spettyl of the starres." This idea is first seen in Aristotle's *History of Animals*. So, too, in late astrology wealth and renown were the happy lot of all born under this and its companion Dog. Our modern Willis wrote in his *Scholar of Thebet ben Khorat:*

> Mild Sirius tinct with dewy violet,
> Set like a flower upon the breast of Eve.

When in opposition Sirius was supposed to produce the cold of winter.

It has been in all history the brightest star in the heavens, thought worthy by Pliny of a place by itself among the constellations, and even seen in broad sunshine with the naked eye by Bond at Cambridge, Massachusetts, and by others at midday with very slight optical aid; but its color is believed by many to have changed from red to its present white. This question recently has been discussed, by See in the affirmative and Schiaparelli in the negative, at a length not allowing repetition here, the weight of argument, however, seeming to be against the admission of any change of color in historic times.

Aratos' term ποικίλος, applied to the Dog, is equally appropriate to Sirius now in the sense of many-colored or changeful, and is an admirable characterization, as one realizes when watching this magnificent object coming up from the horizon on a winter evening. Tennyson, who is always correct as well as poetical in his astronomical allusions, says in *The Princess:*

> the fiery Sirius alters hue
> And bickers into red and emerald;

this, of course, being largely due to its marked scintillation; and Arago gave **Barāḳish** as an Arabic designation for Sirius, meaning Of a Thousand

Colors; and said that as many as thirty changes of hue in a second had been observed in it.[1]

Sirius, notwithstanding its brilliancy, is by no means the nearest star to our system, although it is among the nearest; only two or three others having, so far as is yet known, a smaller distance. Investigations up to the present time show a parallax of 0″.39, indicating a distance of 8.3 light years, nearly twice that of *α* Centauri.

Some are of the opinion that the apparent magnitude of Sirius is partly due to the whiteness of its tint and its greater intrinsic brilliancy; and that the red stars, Aldebaran, Betelgeuze, and others, would appear much brighter than now if of the same color as Sirius; rays of red light affecting the retina of the eye more slowly than those of other colors. The modern scale of magnitudes that makes this star —1.43,— about 9½ times as bright as the standard 1st-magnitude star Altair (*α* Aquilae),— would make the sun —25.4, or 7000 million times as bright as Sirius; but, taking distance into account, we find that Sirius is really forty times brighter than the sun.

Its spectrum, as type of the Sirian in distinction from the Solar, gives name to one of the four general divisions of stellar spectra instituted by Secchi from his observations in 1863–67; these two divisions including nearly $\frac{11}{12}$ of the observed stars. Of these about one half are Sirian of a

> brilliantly white colour, sometimes inclining towards a steely blue. The sign manual of hydrogen is stamped upon them with extraordinary intensity

by broad, dark shaded lines which form a regular series.

It is found by Vogel to be approaching our system at the rate of nearly ten miles a second, and, since Rome was built, has changed its position by somewhat more than the angular diameter of the moon.

It culminates on the 11th of February.

The celebrated Kant thought that Sirius was the central sun of the Milky Way; and, eighteen centuries before him, the poet Manilius said that it was "a distant sun to illuminate remote bodies," showing that even at that early day some had knowledge of the true character and office of the stars.

Certain peculiarities in the motion of Sirius led Bessel in 1844, after ten years of observation, to the belief that it had an obscure companion with which it was in revolution; and computations by Peters and Auwers led Safford to locating the position of the satellite, where it was found as pre-

[1] Montigny's scintillometer has marked as many as seventy-eight changes in a second in various white stars standing 30° above the horizon, though a somewhat less number in those of other colors.

dicted on the 31st of January, 1862, by the late Alvan G. Clark,[1] at Cambridgeport, Mass., while testing the 18½-inch glass now at the Dearborn Observatory. It proved to be a yellowish star, estimated as of the 8½ magnitude, but difficult to be seen because of the brilliancy of Sirius, and then 10″ away; this diminishing to 5″ in 1889; and last seen and measured by Burnham at the Lick Observatory before its final disappearance in April, 1890. Its reappearance was observed from the same place in the autumn of 1896 at a distance of 3″.7, with a position angle of 195°. It has a period of 51½ years, and an orbit whose diameter is between those of Uranus and Neptune; its mass being ⅓ that of Sirius and equal to that of our sun, although its light is but $\frac{1}{10000}$ of that of its principal. So that it may be supposed to be approaching non-luminous solidity,— one of Bessel's "dark stars."

It is remarkable that Voltaire in his *Micromegas* of 1752, an imitation of *Gulliver's Travels*, followed Dean Swift's so-called prophetic discovery of the two moons of Mars by a similar discovery of an immense satellite of Sirius, the home of his hero. Swift, however, owed his inspiration to Kepler, who more than a century previously wrote to Galileo:

> I am so far from disbelieving in the existence of the four circumjovial planets, that I long for a telescope to anticipate you, if possible, in discovering two round Mars (as the proportion seems to me to require), six or eight round Saturn, and perhaps one each round Mercury and Venus.

Other stars are shown by the largest glasses in the immediate vicinity of Sirius, two additional having very recently been discovered by Barnard at the Yerkes Observatory.

β, 2.3, white.

Murzim, generally but less correctly **Mirzam,** and occasionally **Mirza,** is from **Al Murzim,**[2] the Announcer, often combined by the Arabs with β Canis Minoris in the plural **Al Mirzamāni,** or as **Al Mirzamā al Shi'rayain,** the two Sirian Announcers; Ideler's idea of the applicability of this title being that this star announced the immediate rising of the still brighter Sirius.

Buttmann asserted that it also was **Al Kalb,** the Dog, running in front

[1] His death occurred on the 9th of June, 1897, in the sixty-fifth year of his age, just after the completion and successful installation of the 40-inch glass in the Yerkes Observatory, the greatest of his many great lenses, and the last, excepting the 24-inch for Mr. Percival Lowell.

[2] Literally the Roarer, and so another of the many words in the Arabic tongue for the lion, of which that people boasted of having four hundred.

of Sirius, but this must have been from early times in the Desert. In our maps it marks the right fore foot of the Dog.

The Chinese called it **Kuen She,** the Soldiers' Market.

γ, 4.5, is Burritt's **Muliphen** that properly belongs to δ and to stars in Columba; but the *Century Atlas* has it **Mirza.**

It is **Isis** with Bayer, which Ideler confirms, but Grotius applied the title to the adjacent μ, adding, however, *nisi potius quarta sit*, thus referring to γ.

Montanari said that it entirely disappeared in 1670, and was not again observed for twenty-three years, when it reappeared to Miraldi, and since has maintained a steady lustre, although faint for its lettering.

It marks the top of the Dog's head.

δ, 2.2, light yellow,

is the modern **Wezen,** from **Al Wazn,** Weight, "as the star seems to rise with difficulty from the horizon"; but Ideler justly calls this an astonishing star-name.

It also was one of the **Muḥlifaïn** particularly described under Columba.

The Chinese knew η and κ of Canis Major, with stars in Argo, as **Hoo She,** the Bow and Arrow.

Gould thought δ variable. It lies near the Dog's hind quarter, and has a 7.5-magnitude companion 2′ 45″ away, readily seen with an opera-glass.

ϵ, Double, 2 and 9, pale orange and violet.

Adara, Adhara, Adard, Udara, and **Udra** are from **Al ʽAdhārā,** the Virgins, applied to this star in connection with δ, η, and o; perhaps from the Arabic story of Suhail. It has also been designated **Al Zara,** with probably the same signification, although this form is erroneous.

The component stars are 7″.5 apart, at a position angle of 160°.6.

ζ, 3, light orange.

Furud is either from **Al Furud,** the Bright Single Ones, or, perhaps by a transcriber's error, from **Al Ḳurūd,** the Apes, referring to the surrounding small stars with some of those of Columba; Ideler thought the latter derivation more probable. Al Sufi mentioned these as **Al Agribah,** the Ravens.

ζ marks the toe of the right hind foot.

η, 2.4, pale red.

Aludra is from **Al ʽAdhrā,** the singular of **Al ʽAdhārā,** and one of that group. This title has been universal from the days of Arabian catalogues and globes to our modern lists.

Smyth wrote in his notes on η, "Well may Hipparchus be dubbed the Praeses of ancient astronomers!" for that great man used this star, then at 90° of right ascension, as convenient in astronomical reckoning.

μ, a double, of 4.7 and 8th magnitudes, 2″.9 apart, yellow and blue, was known as **Isis** by Grotius, although he admitted that γ might have been the one referred to by this title.

o^1, a red star of the 4th magnitude, and π, a double, of 5th and 10th magnitudes, with other small stars in the body of the Dog, were the Chinese **Ya Ke,** the Wild Cock.

Bayer's star-lettering for this constellation ended with o, but Bode added others down to ω.

★

The Dog's-precursor, too, shines bright beneath the Twins.

Brown's *Aratos.*

Canis Minor, the Lesser Dog,

is **der Kleine Hund** of the Germans; **le Petit Chien** of the French; and **il Cane Minore** of the Italians; Proctor, ignoring La Lande, strangely altered it to **Felis.**

It was not known to the Greeks by any comparative title, but was always προκύων, as rising before his companion Dog, which Latin classic writers transliterated **Procyon,** and those of late Middle Ages as **Prochion** and **Procion.** Cicero and others translated this into **Antecanis,**— sometimes **Anticanis,— Antecedens Canis, Antecursor, Praecanis, Procanis,** and **Procynis**; or changed to plain **Canis.** To this last from the time of Vitruvius, perhaps before him, the Romans added various adjectives; *septentrionalis*, from its more northerly position than that of Canis Major; *minor*, *minusculus*, and *parvus*, in reference to its inferior brightness; *primus*, as rising first; and *sinister*, as on the left hand, in distinction from the Canis *dexter* on the right. Lucan described both of the Dogs as *semi deosque Canes.*

It was also **Catellus** and **Catulus,** the Puppy.

Horace wrote of it,

> Jam Procyon furit,

which Mr. Gladstone rendered,

> The heavens are hot with Procyon's ray,

as though it were the **Canicula,** and he was followed by others in this; indeed, Pliny began the dog days with its heliacal rising on the 19th of July, and strangely said that the Romans had no other name for it.

With mythologists it was Actaeon's dog, or one of Diana's, or the Egyptian **Anubis;** but popularly **Orion's 2d Hound,** often called **Canis Orionis,** and thus confounded as in other ways with the Sirian asterism. Hyginus had **Icarium Astrum,** referring to the dog Maera; Caesius, **Erigonius** and **Canis virgineus** of the same story, but identified by Ovid with Canis Major; and Firmicus, **Argion,** that perhaps was for Ulixes' dog Ἄργος. It also was considered as representing Helen's favorite, lost in the Euripus, that she prayed Jove might live again in the sky.

It shared its companion's much mixed, degenerate nomenclature, as in the 1515 *Almagest's* "Antecedens Canis *et est* **Alsehere Ascemie Algameisa**"; while the industrious Bayer as usual had some strange names for it. Among these are **Fovea,** a Pit, that Caesius commented much upon, but little to our enlightenment; and Συκάμινος, or **Morus,** the Sycamine tree, the equivalent of one of its Arabic titles. His **Aschemie** and **Aschere,** as well as Chilmead's **Alsahare alsemalija,** and mongrel words from the foregoing *Almagest*, etc., can all be detected in their original **Al Shiʽrā al Shāmiyyah,** the Bright Star of Syria, thus named because it disappeared from the Arabs' view at its setting beyond that country.

We also find **Al Jummaizā,** their Sycamine, although some say that this should be **Al Ghumaiṣāʽ,** the Dim, Watery-eyed, or Weeping One; either from the fact that her light was dimmer than that of her sister Al Shiʽrā, or from the fable connected with Suhail and his marriage to Al Jauzah and subsequent flight, followed by Al Shiʽrā below the Milky Way, where she remained, the other sister, Al Ghumaiṣāʽ, being left in tears in her accustomed place, or it may be from a recollection of the Euphratean title for Procyon,—the **Water-dog.** Bayer wrote the word **Algomeiza;** Riccioli, **Algomisa** and **Algomiza;** and others, **Algomeysa, Algomyso, Alchamizo,** etc. Thus the Two Dog-stars were the Arabs' **Al Ukhtā Suhail,** the Sisters of Canopus. Still another derivation of the name is from **Al Ghamūs,** the Puppy; but this probably was a later idea from the Romans.

Also borrowing from them, the Arabians called it **Al Kalb al Asghar,** the Lesser Dog,—Chilmead's **Alcheleb Alasgar,** Riccioli's **Kelbelazguar,**—and **Al Kalb al Mutaḳaddim,** the Preceding Dog.

In Canis Minor lay a part of **Al Dhirā' al Asad al Maḳbuḍah,** the Contracted Fore Arm, or Paw, of the early Lion; the other, the Extended Paw, running up into the heads of Gemini.

Like its greater neighbor, Procyon foretold wealth and renown, and in all astrology has been much regarded. Leonard Digges[1] wrote in his *Prognostication Everlasting of Right Good Effect,* an almanac for 1553,—

Who learned in matters astronomical, noteth not the great effects at the rising of the starre called the Litel Dogge.

Caesius made it the **Dog of Tobias,** in the *Apocrypha,* that Novidius had claimed for Canis Major; but Julius Schiller imagined it the **Paschal Lamb.**

Who traced out the original outlines of Canis Minor, and what these outlines were, is uncertain, for the constellation with Ptolemy contained but two recorded stars, and no 'αμόρφωτοι; and even now Argelander's map shows only 15, although Heis has 37, and Gould 51.

Canis Minor lies to the southeast from the feet of Gemini, its western border over the edge of the Milky Way, and is separated by Monoceros from Canis Major and Argo.

α, Binary, 0.4, and 13, yellowish white and yellow.

Procyon, varied by **Procion** and **Prochion,**—Προκύων in the original,—has been the name for this from the earliest Greek records, distinctly mentioned by Aratos and Ptolemy, and so known by all the Latins, with the equivalent **Antecanis.**

Ulug Beg designated it as **Al Shi'rā al Shāmiyyah,** shortened to **Al Shāmiyyah;** Chrysococca transcribing this into his Low Greek Σιαὴρ Σιαμὴ, and Riccioli into **Siair Siami;** all of these agreeing with its occasional English title the **Northern Sirius.** The *Alfonsine Tables* of 1521 quote it as **Aschere, Aschemie** *et* **Algomeysa;** those of 1545, as **prochion** & **Algomeyla.**

It thus has many of its constellation's names; in fact, being the *magna pars* of it, probably itself bore them before the constellation was formed.

[1] It was this Digges who, nearly fifty years before Galileo, wrote of the telescope as though it were an instrument with which he was familiar,—perhaps from Roger Bacon's writings of 350 years before him.

Jacob Bryant insisted that its title came to Greece from the Egyptian **Pur Cahen.**

Euphratean scholars identify it with the **Kakkab Paldara, Pallika,** or **Palura** of the cylinders, the Star of the Crossing of the Water-dog, a title evidently given with some reference to the River of Heaven, the adjacent Milky Way; and Hommel says that it was the **Kak-shisha** which the majority of scholars apply to Sirius.

Dupuis said that in Hindu fables it was **Singe Hanuant**; and Edkins that it, or Sirius, was the Persian **Vanand.**

Reeves' Chinese list gives it as **Nan Ho,** the Southern River, in which β and η were included.

With the natives of the Hervey Islands it was their goddess **Vena.**

In astrology, like its constellation, it portended wealth, fame, and good fortune. Procyon culminates on the 24th of February.

Elkin determined its parallax as $0''.341$, making its distance from our system about 9½ light years; and, according to Vogel, it is approaching us at a speed of nearly six miles a second. Gould thinks it slightly variable.

Its spectrum is on the border between Solar and Sirian.

It is attended by several minute companions that have long been known; but in November, 1896, Schaeberle of the Lick Observatory discovered a 13th-magnitude yellowish companion, about $4''.6$ away, at a position angle of $318^{\circ}.8$, that may be the one predicted by Bessel in 1844 as explaining its peculiar motion,— a motion resembling that of Sirius, which astronomers had found to be moving in an oval orbit entirely unexplained until the discovery of its companion by Alvan G. Clark in 1862. Barnard, at the Yerkes Observatory in 1898, makes the close companion of Procyon $4''.83$ away, at a position angle of 326°.

The period of revolution of this most magnificent system is about forty years, in an orbit slightly greater than that of Uranus, the combined mass being about six times that of our sun and earth, and the mass of the companion equaling that of our sun. Its light is three times greater.

β, 3.5, white.

Gomeisa is from the Ghumaiṣā' of the constellation, changed in the *Alfonsine Tables* to **Algomeyla,** and by Burritt to **Gomelza.**

Occasionally it has been **Al Gamus,** from another of the Arabians' titles for the whole; and **Al Murzim,** identical with the name of β Canis Majoris, and for a similar reason,— as if announcing the rising of the brightest star of the figure. The Arabs utilized this, with Procyon, to mark the terminal points of their short Cubit, or Ell, **Al Dhirā'**, their long Cubit being the line between Castor and Pollux of Gemini. This same word appears in the title of one of the moon stations in that constellation.

β has some close companions of the 10th and 12th magnitudes.

ζ, θ, o, and π were the Chinese **Shwuy Wei**, a Place of Water, a designation that may have been given them from their nearness to the River of Heaven, the Galaxy.

> Thy Cold (for thou o'er Winter Signs dost reign,
> Pull'st back the *Sun*, and send'st *us* Day again)
> Makes Brokers rich.
>
> Thomas Creech's translation of Manilius' *Poeticon Astronomicon.*

Capricornus,

next to the eastward from Sagittarius, is our **Capricorn,** the French **Capricorne,** the Italian **Capricorno,** and the German **Steinbock,**— Stone-buck, or Ibex,— the Anglo-Saxon **Bucca** and **Buccan Horn.**

The common Latin name was varied by the **Caper** of Ausonius, **flexus Caper** of Manilius, **Hircus corniger** of Vergil, **hircinus Sidus** of Prudens, **Capra** and **aequoris Hircus,** the Sea Goat; while Minsheu's "**Capra** *illa* **Amalthea**" indicates that it was identified by some with the goat usually assigned to Auriga. All this, doubtless, was from oriental legends, perhaps very ancient, which made Capricorn the nurse of the youthful sun-god that long anticipated the story of the infant Jupiter and Amalthea. The Latin poets also designated it as **Neptuni proles,** Neptune's offspring; **Pelagi Procella,** the Ocean Storm; **Imbrifer,** the Rain-bringing One; **Signum hiemale,** and **Gelidus,** because then at the winter solstice, the equivalent Ἀθαλπής appearing with the Greeks, which Riccioli repeated as **Athalpis.**

Aratos called it Αἰγοκέρως, the Horned Goat, to distinguish it from the Αἴξ of Auriga, as did Ptolemy, but Ionic writers had Αἰγοκέρευς; and this word, Latinized as **Aegoceros,** was in frequent use with all classical authors who wrote on astronomy. The *Arabo-Latin Almagest* of 1515 turned this into **Alcaucurus,** explained by *habens cornua hirci;* and Bayer mentioned

Alcantarus. Eratosthenes knew it as Πάν and 'Αιγι-Πᾶν, the Goat-Footed Pan, half fishified, Smyth said, by his plunge into the Nile in a panic at the approach of the monster Typhon; the same story being told of Bacchus, so that he, too, always was associated with its stars.

In Persia it was **Bushgali, Bahi** or **Vahik,** and **Goi;** in the Pahlavi tongue, **Nahi;** in Turkey, **Ughlak;** in Syria, **Gadjo;** and in Arabia, **Al Jady,** usually written by us **Giedi;** all meaning the Goat, or, in the latter country, the Bādan, or Ibex, known to zoölogists as *Capra beden.* Burritt's **Tower of Gad,** at first sight presumably Hebrew, would seem rather to be a bungled translation [1] from the Arabic, and in no way connected with the Jewish tribe. Riccioli had **Elgedi, Elgeudi,** and **Gadio.**

Very frequent mention was made of this constellation in early days, for the Platonists held that the souls of men, when released from corporeity, ascended to heaven through its stars, whence it was called the **Gate of the Gods;** their road of descent having been through Cancer. But some of the Orientals knew it as the **Southern Gate of the Sun,** as did the Latins in their **altera Solis Porta.** Berōssōs is reported by Seneca to have learned from the old books of Sargon [2] that the world would be destroyed by a great conflagration when all the planets met in this sign.

Numa Pompilius, the second mythical king of Rome, whose date has been asserted as from 715 to 673 B. C., began the year when the sun was in the middle of Capricorn, and when the day had lengthened by half an hour after the winter solstice.

In astrology, with Taurus and Virgo, it was the **Earthly Trigon,** and black, russet, or a swarthy brown, was the color assigned to it; while, with Aquarius, it was the **House of Saturn,** as that planet was created in this constellation, and whenever here had great influence over human affairs; as Alchabitus asserted, in the *Ysagogicus* of 1485, *caput et pedes habet;* and it always governed the thighs and knees. It also was regarded as under the care of the goddess Vesta, and hence **Vestae Sidus.** Ampelius singularly associated it with the burning south wind Auster, and Manilius said that it reigned over France, Germany, and Spain; in later times it ruled Greece, India, Macedonia, and Thrace, Brandenburg and Mecklen-

[1] The Arabic word *Burj* signifies both Constellation and Tower, or Fortress.

[2] This Sargon has been considered the almost mythical founder of the first Semitic empire, 3850 B. C., but inscriptions recently unearthed at Nuffar, and only deciphered in 1896 at Constantinople by Professor Herman V. Hilprecht of the University of Pennsylvania, make it evident that Babylonia was an important kingdom at least three or four millenniums before him. Sargon's astronomical work, the *Illumination of Bel,* in 72 books, was compiled by the priests of that god, and translated into Greek by Berōssōs about 260 B. C. Fragments of this last work still remain to us.

burg, Saxony and Wilna, Mexico and Oxford. Manilius also wrote of it as in our motto, and

> at Caesar's Birth Serene he shone.

The almanac of 1386 has: "Whoso is borne in Capcorn schal be ryche and wel lufyd"; in 1542 the Doctor, as Arcandum was called, showed that a man born under it would be a great gallant, would have eight special illnesses, and would die at sixty; and according to Smyth it was "the very pet of all constellations with astrologers, having been the fortunate sign under which Augustus and Vespasian were born," although elsewhere, in somewhat uncourtly style, he quotes: "prosperous in dull and heavy beasts." It also appears to have been much and favorably regarded by the Arabians, as may be seen in their names for its chief stars, and in the character assigned by them to its lunar mansions. But these benign qualities were only occasional, caused probably by some lucky combination with a fortunate sign, as is known only to the initiated, for its general reputation was the reverse; and, in classical days, when coincident with the sun, it was thought a harbinger of storms and so ruler of the waters,—Horace's

> tyrannus Hesperiae Capricornus undae.

Aratos had clearly showed this long before:

> Then grievous blasts
> Break southward on the sea, when coincide
> The *Goat* and sun; and then a heaven-sent cold.

Ovid expressed much the same opinion in connection with the story of Acaetes; but ages before them this seems to have been said of it on Euphratean tablets.

Caesius and Postellus are authority for its being **Azazel**, the Scapegoat of *Leviticus;* although Caesius also mentioned it as **Simon Zelotes,** the Apostle. Suetonius in his *Life of Augustus*, and Spanheim in his *De Nummis*, said that Capricorn was shown on silver coins of that emperor, commemorating the fact that it was his natal sign; and it always has been regarded in astrology as the **Mansion of Kings.** It is seen, too, on a coin found in Kent, struck by the British prince Amminius, and was the most frequent of the zodiacal figures on uranographic amulets of the 14th and 15th centuries, "worn as a kind of astral defensive armor."

Its figuring generally has been consistent, and as we now see it, with the head and body of a goat, or ibex, ending in a fish's tail. Manuscripts from the 2d to the 15th century show it thus; a Syrian seal of 187 B. C. has it in the same way; as also an early Babylonian gem, surmounted, not inappropriately, by the crescent moon, for Capricorn was a nocturnal sign; and the same figure is on a fragment of a Babylonian planisphere, now in the British Museum, supposed to be of the 12th century B. C. So that this may be considered its original form, in full agreement with its amphibious character, and with some resemblance, in the grouping of the chief stars, to a goat's horns and a fish's tail. From this figuring Camões, in *Os Lusiadas* of 1572, called it the **Semi-Capran Fish,** as it now is with us the **Goat-Fish** and the **Sea Goat.** Still at times it has been a complete goat-like animal, and was so considered by Aratos, Eratosthenes, and Ptolemy, as by the more modern Albumasar, Kazwini, Ulug Beg, and in occasional mediaeval manuscripts. It was thus shown on some Egyptian zodiacs; although on that of Denderah it appears in its double form, where "an ibis-headed man rides on Capricornus, under which sign Sirius rose anti-heliacally"; the ibis being sacred to Isis, with which Sirius was identified. Still differently, a silver bowl from Burmah engraved with the Brahmin zodiac, probably copied from original sources, makes the Fish entire in Capricorn, and omits the Goat; while Jensen says that in Babylonia the Goat and Fish, both complete, were occasionally used together for the constellation.

Jewish Rabbis asserted that the tribe of Naphtali adopted this sign as their banner emblem,—"Naphtali is a hind let loose,"—as if Capricorn were a deer, or antelope; others ascribed it to Benjamin, or to Reuben; but Aquarius more fitly represented the latter.

Some connect the sign in Egyptian astronomy with **Chnum, Chnemu, Gnoum,** or **Knum,** the God of the Waters, associated with the rising of the Nile and worshiped in Elephantine at the Cataracts, this divinity bearing goat's, not ram's, horns. Others have said that it was the goat-god **Mendes**; and La Lande cited the strange title **Oxirinque** from the Greek adjective descriptive of a **Swordfish,** our constellation sometimes being thus shown, when it was considered the cause of the inundation. In Coptic Egypt it was Ὀπέυτυς, *Brachium Sacrificii*; and Miss Clerke says that it was figured in that country as a **Mirror,** emblematic of life.

Earlier Hindu names were **Mriga** and **Makara,**—the Cingalese **Makra** and the Tamil **Makaram,** an Antelope; but occasionally it was shown with a goat's head upon the body of a hippopotamus, signifying some amphibious creature, and a later term was **Shī-shu-mara** or **Sim-shu-mara,** the Crocodile, although this originally was marked by stars of Draco. Varāha Mihira took his title for it, **Akokera,** from the Greeks; and it was the last in order of the zodiacal signs of India, as on the Euphrates. In the Aztec calendar it appeared as **Cipactli,** with a figure like that of the narwhal.

It was the zodiacal **Bull,** or **Ox,** of Chinese astronomy, that later became **Mo Ki,** the Goat-Fish. Williams says that, with stars of Sagittarius, it was **Sing Ki,** the Starry Record, and with a part of Aquarius **Hiuen hiau**; while in very early days, with Aquarius and Sagittarius, it was the **Dark Warrior,** etc., the so-called Northern one of the four large divisions of the zodiac. Flammarion asserts that Chinese astronomers located among its stars a conjunction of the five planets 2449 B. C.

Sayce, Bosanquet, and others think that they have without doubt identified it with the Assyrian **Munaχa,** the Goat-Fish; and we see other probable names in **Shah** or **Shahu,** the Ibex, and in **Niru,** the Yoke, this last perhaps a popular one. Brown gives for it the Akkadian **Su-tul** of the same meaning; and another possible title, resembling the early Hindu, was **Makhar,** claimed also for Delphinus. It seems likewise to have been known as the **Double Ship.** Jensen says that "the amphibious Ia Oannes of the Persian Gulf was connected with the constellation Capricornus"; Sayce, that a cuneiform inscription designates it as the **Father of Light,**—a title which, astronomically considered, could not have been correct except about 15000 years ago, when the sun was here at the summer solstice; that "the goat was sacred and exalted into this sign"; and that a robe of goatskins was the sacred dress of the Babylonian priests. So that, although we do not know when Capricornus came into the zodiac, we may be confident that it was millenniums ago, perhaps in prehistoric days. It was identified with the 10th Assyrian month Dhabitu, corresponding to December–January.

Its symbol, ♑, usually is thought to be τρ, the initial letters of τράγος, Goat, but La Lande said that it represents the twisted tail of the creature; and Brown similarly calls it "a conventional representation of a fish-tailed goat." Indeed it is not unlike the outline of these stars on a celestial globe.

The sun is in the constellation from the 18th of January to the 14th of February, when, as Dante wrote in the *Paradiso*,

> The horn of the celestial goat doth touch the sun;

and Milton mentions the latter's low elevation during this time,

> Thence down amain
> . . .
> As deep as Capricorn.

The title Tropic of Capricorn, originating from the fact that when first observed the point of the winter solstice was located here, now refers to the sign and not to the constellation, this solstice at present being 33° to the westward, in the figure of Sagittarius, near its star μ.

Capricorn is, after Cancer, the most inconspicuous in the zodiac, and chiefly noticeable for the duplicity of its *lucida*.

Argelander charted 45 naked-eye stars within its borders; and Heis 63.

α^1, Double, 3.2 and 4.2, yellow.
α^2, Triple, 3, 11.5, and 11.5, pale yellow, ash, and lilac.

These are the **Prima** and **Secunda Giedi,** or plain **Algedi,** from the Arabian constellation title Al Jady.

Other titles, **Dabih** and the degenerated **Dschäbbe** and **Dshabeh,** applied to them, but more commonly to β, have been traced by some to Al Jabbah, the Forehead, although the stars are nearer the tip of the horn; but the names undoubtedly come from **Al Saʽd al Dhābiḥ,** the Lucky One of the Slaughterers, the title of the 20th *manzil* (of which these *alphas* and β were the determinant point), manifestly referring to the sacrifice celebrated by the heathen Arabs at the heliacal rising of Capricorn. And of similar signification was the Euphratean **Shak-shadi** and the Coptic **Eupeutōs**, or **Opeutus,** for the same lunar asterism of those peoples.

Brown thinks that α, then seen only as a single star, with β and ν was known by the Akkadians as **Uz,** the Goat; and as **Enzu** in the astronomy of their descendants; while Epping is authority for the statement that this, or perhaps β, marked the 26th ecliptic asterism of the Babylonians, **Qarnu Shahū,** the Horn of the Goat. Brown also says that α represented the 8th antediluvian king Amar Sin,— 'Αμέμψινος.

In Hipparchos' time the two *alphas* were but 4′ apart, and it was not till towards Bayer's day that they had drifted sufficiently away from each other to be readily separated by the naked eye. Their distance in 1880 was 6¼′, and this is increasing by 7″ in every hundred years.

They culminate on the 9th of September.

Smyth described a minute blue companion of α^2 which he caught "in little evanescent flashes, so transient as again to recall Burns's snow-flakes on a stream"; and mentioned Sir John Herschel's suggestion that this might shine by reflected light. Alvan G. Clark doubled this in 1862, the distance being 1″.2, and the position angle 239°.

β^1, and β^2, 2.5 and 6, each double, orange yellow and sky blue.

Dabih Major and **Dabih Minor** are the names of this so-called double, but telescopically multiple, star, taken from the title of the *manzil* of which, with α, it formed part.

These *betas*, with α, ν, o, π, and ρ farther to the south, were the 20th *sieu* of China, **Nieu,** or **Keen Nieu,** the Ox, anciently **Ngu,** or **Gu,** themselves being the determinants. The lunar asterism was in some way intimately connected in religious worship with the rearing of the silkworm in that country.

The two stars mark the head of the Goat, the components 205″ distant from each other, and each very closely double. The duplicity of β^1 was first recognized in 1883 by Barnard from its behavior at an occultation by the moon, this discovery being soon verified and measured by Professor Young, Hough, and other observers.

γ, 3.8.

Nashira is from **Al Sa'd al Nashirah,** the Fortunate One, or the Bringer of Good Tidings, which the early Arabs applied to this when taken with δ. Smyth gave it as **Sa'dubnáshirah;** and the *Standard Dictionary* repeats this as **Saib' Nasch-rú-ah!**

Bayer had the later **Deneb Algedi,** the Tail of the Goat, that is more proper for δ; the *Alfonsine Tables* of 1521, **Denebalchedi,** which has degenerated to **Scheddi;** and the fine wall star-map of Doctor Ferdn. Reuter, **Deneb Algethi;** but this is erroneous, and a confusion with the Arabian title for the constellation Hercules.

γ marked the 27th Babylonian ecliptic asterism, **Mahar sha hi-na Shahū,** the Western One in the Tail of the Goat.

With δ, ϵ, κ and stars in Aquarius and Pisces it was the Chinese **Luy Pei Chen,** the Intrenched Camp.

δ, 3.1.

Deneb Algedi is the transcription by Ulug Beg's translator of **Al Dhanab al Jady,** the Tail of the Goat; changed to **Scheddi** in some lists,— a name also found for γ.

Ideler said that these stars were **Al Muḥibbain,** the Two Friends, an Arabic allegorical title for any two closely associated objects; but Beigel differed with him as to this, and wrote it **Al Muhanaim,** the Two Bending Stars,— in the flexure of the tail,— for "moral beings are foreign to the nomad sky."

It marked the 28th ecliptic constellation of Babylonia, **Arkat sha hi-na Shahū,** the Eastern One in the Tail of the Goat.

5° to the eastward is the point announced by Le Verrier[1] as the position

[1] Flammarion, who was intimate with Le Verrier, thinks that the latter never had the curiosity to observe his planet through the telescope, strangely content with his mathematical achievement! And it is interesting to know that Doctor Galle, in his 85th year, in 1896 received the congratulations of the astronomical world upon the 50th anniversary of the finding of Neptune.

of his predicted new planet,— Neptune,— where Galle, first assistant of the celebrated Encke at the Berlin Observatory, under Le Verrier's direction, visually discovered it on the 23d of September, 1846. It had been suspected by Bouvard in 1821, and seen six times from France and England just previous to its discovery, but without knowledge of its character.

ζ, η, θ, and ι, 4th- and 5th-magnitude stars on the body, were respectively **Yen, Chow, Tsin,** and **Tae,** names of old feudal states in China.

λ, 5.4, with ξ Aquarii and others near by, was **Tien Luy Ching,** the Heavenly Walled Castle; and μ, 5.24, was **Kuh,** Weeping. λ and μ mark the extreme end of the tail.

ν, 4.7, was Kazwini's **Al Shat,** the Sheep that was to be slaughtered by the adjacent Dhābiḥ, the stars β.

The following also seem to be named only in China: v, 5.3, marked **Loo Sieu,** the Lace-like asterism; ϕ, 5.3, and χ, 5.3, taken together were **Wei,** the name of one of the old feudal states; ψ, 4.3, was **Yue,** a Battle-ax; while the 5th-magnitudes A, *b*, and *m* also bore titles from feudal times of the states **Tsoo, Tsin,** and **Chaou.**

Bayer gave A, *b*, and *c* as *Tres ultimae* Deneb Algedi; but Heis puts A in the right fore arm, *b* in the belly, and *c* — Flamsteed's 46 — outside of and beyond the tail, in the ribs of Aquarius, thus showing a change of figuring in the past three centuries.

★

A place where Cassiopea sits within
Inferior light, for all her daughter's sake.

Mrs. Browning's *Paraphrases on Nonnus.*

Cassiopeia, or Cassiope,

more correctly **Cassiepeia,** although variously written, is one of the oldest and popularly best known of our constellations, and her throne, "the shinie Casseiopeia's chair" of Spenser's *Faerie Queen*, is a familiar object to the most youthful observer. It also is known as the **Celestial W** when below the pole, and the **Celestial M** when above it.

Hyginus, writing the word **Cassiepia,** described the figure as bound to her seat, and thus secured from falling out of it in going around the pole head downward,— this particular spot in the sky having been selected by the queen's enemies, the sea-nymphs, to give her an effectual lesson in humility, for a location nearer the equator would have kept her nearly upright. Aratos said of this:

> She head foremost like a tumbler sits.

Her outstretched legs also, for a woman accustomed to the fashions of the East, must have added to her discomfort.

Euripides and Sophocles, of the fifth century before our era, wrote of her, while all the Greeks made much of the constellation, knowing it as *Κασσιέπεια* and Ἡ *τον θρόνου*, She of the Throne. But at one time in Greece it was the **Laconian Key,** from its resemblance to that instrument, the invention of which was attributed in classical times to that people;[1] although Pliny claimed this for Theodorus of Samos in Caria, 730 B. C., whence came another title for our stars, **Carion.** The learned Huetius (Huet, bishop of Avranches and tutor of the dauphin Louis XV) more definitely said that this stellar key represented that described by Homer as sickle-shaped in the wardrobe door of Penelope:

> A brazen key she held, the handle turn'd,
> With steel and polish'd elephant adorned;

and Aratos wrote of the constellation:

> E'en as a folding door, fitted within
> With key, is thrown back when the bolts are drawn.

But even Ideler did not understand this simile, although the outline of the chief stars well shows the form of this early key.

The Romans transliterated the Greek proper name as we still have it, but also knew Cassiopeia as **Mulier Sedis,** the Woman of the Chair; or simply as **Sedes,** qualified by *regalis* or *regia;* and as **Sella** and **Solium.** Bayer's statement that Juvenal called it **Cathedra mollis** was an error from a misreading of the original text. Hyde's title **Inthronata** has been repeated by subsequent authors; and **Cassiopeia's Chair** is the children's name for it now.

The Arabians called it **Al Dhāt al Kursiyy,** the Lady in the Chair,—Chilmead's **Dhath Alcursi,**—the Greek proper name having no signification to them; but the early Arabs had a very different figure here, in no way connected with the Lady as generally is supposed,—their **Kaff al Ḥadib,**

[1] Locks and keys, however, were used at the siege of Troy; have been found in Egyptian catacombs and sculptured on the walls of the Great Temple of Karnak; disinterred from the palaces of Khorsabad near Nineveh; and twice mentioned in our *Old Testament,* as early as Ehud's time in the *Book of Judges,* iii, 24 and 25.

the large Hand Stained with Henna, the bright stars marking the finger-tips; although in this they included the nebulous group in the left hand of Perseus. Chrysococca gave it thus in the Low Greek Χεὶρ βεβαμένη; and it sometimes was the **Hand of,** *i. e.* next to, **the Pleiades,** while Smyth said that in Arabia it even bore the title of that group, **Al Thurayya,** from its comparatively condensed figure.

The early Arabs additionally made **Two Dogs** out of Cassiopeia and Cepheus, from which may have come Bayer's **Canis**; but his **Cerva,** a Roe, is not explained, although La Lande asserted that the Egyptian sphere of Petosiris had shown a **Deer** to the north of the Fishes. Al Tizini imagined a **Kneeling Camel** from some of its larger stars, whence the constellation's name **Shuter** found with Al Naṣr al Dīn, and common for that animal in Persia.

The *Alfonsine Tables* and *Arabo-Latin Almagest* described the figure as *habens palmam delibutam*, Holding the Consecrated Palm, from some early drawing that is still continued; but how the palm, the classic symbol of victory and Christian sign of martyrdom, became associated with this heathen queen does not appear. Similarly La Lande cited **Siliquastrum,** the name for a tree of Judaea, referring to the branch in the queen's hand.

Bayer's Hebrew title for it, **Aben Ezra,** was by a misreading of Scaliger's notes.

La Lande quoted **Harnacaff** from the *Metamorphoses of Vishnu*, but the later Hindus said **Casyapi,** evidently from the classical word.

Grimm gives the Lithuanian **Jostandis,** from Josta, a Girdle, although without explanation.

As the figure almost wholly lies in the Milky Way, the Celts fixed upon it as their Llys Don, the Home of Don, their king of the fairies and father of the mythical character Gwydyon,[1] who gave name to that great circle.

Schiller's *Wallenstein*, as versified by Coleridge, has

> That one
> White stain of light, that single glimmering yonder,
> Is from Cassiopeia, and therein
> Is Jupiter—

a blunder on the part of the translator that has puzzled many, as "therein" should be "beyond" or "in that direction," but even then what did the poet have in mind?

In early Chinese astronomy our constellation was **Ko Taou** according to Williams, although Reeves limited that title to the smaller ν, ξ, o, and π, with

[1] Gwydyon has been identified with the classical Hermes-Mercury, the reputed inventor of writing, a practitioner in magic and builder of the rainbow.

the definition of a Porch-way; but later on its prominent stars were **Wang Liang,** a celebrated charioteer of the Tsin Kingdom about 470 B. C.

As a stellar figure in Egypt Renouf identified it with the **Leg,** thus mentioned in the *Book of the Dead*, the Bible of Egypt, that most ancient ritual, 4000 years old or more:

> Hail, leg of the northern sky in the large visible basin.

And in some constellated form its stars unquestionably were well known on the Euphrates with the rest of the Royal Family, and shown there on seals.

The earthly Cassiopeia ought to have been black, and is so described by Milton in his verses of *Il Penseroso* on

> That starr'd Ethiop Queen that strove
> To set her beauty's praise above
> The Sea-nymphs;

while Landseer with the same idea called her **Cushiopeia,** the Queen of Cush, or Kush, but the *Leyden Manuscript* makes her of fair complexion, lightly clad, upright and unbound in a very uncomfortable chair; and such is the general representation. But in the 17th-century reconstruction of sky figures in the interests of religion, our Cassiopeia became **Mary Magdalene**; or **Deborah** sitting in judgment under her palm tree in Mount Ephraim; or **Bathsheba,** the mother of Solomon, worthy to sit on the royal throne.

The astrologers said that it partook of the nature of Saturn and Venus.

Professor Young gives the word *Bagdei* as a help to memorizing the order of the chief components from their letters β, α, γ, δ, ϵ, ι; the last being the uppermost when the figure is on the horizon, hanging head downwards.

Cassiopeia lies between Cepheus, Andromeda, and Perseus, Argelander cataloguing 68 stars here, but Heis, 126; and the constellation is rich in clusters.

α, Multiple and slightly variable, 2.2 to 2.8, pale rose.

Schedar is first found in the *Alfonsine Tables*, and was **Schedir** with Hevelius; **Shadar, Schedar, Shedar, Sheder, Seder, Shedis, Zedaron,** etc., elsewhere; and all supposed to be from **Al Sadr,** the Breast, which the star marks in the figure. Some, however, have asserted that they are from the Persian Shuter for the constellation.

Ulug Beg called it **Al Dhāt al Kursiyy** from the whole, which Riccioli changed to **Dath Elkarti.**

Smyth said that it was known as **Lucida Cassiopea,**— a matter-of-fact statement, as the brightest star in any sky figure is the *lucida*.

Birt noticed its variability in 1831, which is now determined as in a period of about 79 days, although irregular.

It culminates on the 18th of November.

Burnham has discovered two additional faint companions, the nearest 17''.5 away: the companion first known, a smalt blue star, having been found by Sir William Herschel, in 1781, 63'' away.

α, β, η, and κ were the Chinese **Yūh Lang,** or **Wang Leang.**

β, 2.4, white.

Caph, Chaph, or **Kaff,** on the upper right-hand corner of the chair, are from the Arabic title of the constellation; but Al Tizini designated the star as **Al Sanām al Nākah,** the Camel's Hump, referring to the contemporaneous Persian figure.

With α Andromedae and γ Pegasi, as the **Three Guides,** it marks the equinoctial colure, itself exceedingly close to that great circle; and, being located on the same side of the pole as is Polaris, it always affords an approximate indication of the latter's position with respect to that point. This same location, 32° from the pole, and very near to the prime meridian, has rendered it useful for marking sidereal time. When above Polaris and nearest the zenith the astronomical day begins at 0 hours, 0 minutes, and 0 seconds; when due west the sidereal time is 6 hours; when south and nearest the horizon, 12 hours, and when east, 18 hours; this celestial clock-hand thus moving on the heavenly dial contrary to the motion of the hands of our terrestrial clocks, and at but one half the speed.

Beta's parallax, 0''.16, indicates a distance of 20 light years.

Just north of it is an especially bright patch in the Milky Way.

> When first **Al Aaraf** knew her course to be
> Headlong thitherward o'er the starry sea.
>
> Edgar Allan Poe's *Al Aaraf.*

About 5° to the west-northwest of Caph, 1½° distant from κ, and forming a parallelogram with Caph, γ and α, appeared, in 1572, a famous *nova* visible in full daylight and brighter than Venus at perigee.

Poe's name for it is from the Arabians' Al Orf,— in the plural Al Arāf,— their temporary abode of spirits midway between H ven and Hell, and so applicable to this temporary star. This object was known for two centuries after its appearance as the **Stranger,** or the **Pilgrim, Star,** and the **Star in the Chayre,** but by us as **Tycho's Star,** although it was first noticed by Schuler at Wittenberg in Prussia, on the 6th of August; again at Augsburg by Hainzel, and at Winterthür, Switzerland, by Lindauer, on the 7th of November; and on the 9th by Cornelius Gemma, who called it the **New Venus.** Maurolycus began its systematic study at Messina on the 8th, while Tycho did not see it till the 11th, at the time of its greatest brilliancy; but his published account of it in 1602, in his *Astronomiae Instauratae Proegymnasmata*, has caused his name to be identified with it. Its lustre began to wane in the following December, and it was inserted in the *Rudolphine Tables* as "Nova anni 1572" of the 6th magnitude, to which it had at that time decreased. It disappeared entirely in March, 1574, so far as could then be known.

This *nova* is said to have incited Tycho to the compilation of his star-catalogue, as that of seventeen centuries earlier may have been the occasion of the catalogue of Hipparchos. At all events, it created a great commotion in its time, and induced Beza's celebrated prediction of the second coming of Christ,[1] as it was considered a reappearance of the Star of Bethlehem. The statement that this star appeared in 945 and 1264 rests upon the very doubtful authority of the Bohemian astrologer Cyprian Leowitz, and is not credited by our modern astronomers; although Williams asserts that a large comet was seen in the latter year near Cassiopeia. The reddish 10½-magnitude, known as B Cassiopeiae, singularly variable in its light, is now to be seen 0′.8 from the spot assigned by Argelander to the star of 1572, and is thought possibly to be identical with it.

The Chinese recorded Tycho's *nova* as **Ko Sing,** the Guest Star.

γ, Binary, 2 and 11, brilliant white,

in Cassiopeia's girdle, was the Chinese **Tsih,** a Whip.

This was the first star discovered to contain bright lines in its spectrum, —by Secchi in 1886,—and so is of much interest to astronomers. The spectrum is peculiarly variable, as also is its light.

The components are 2″.1 apart, at a position angle of 255°.2, and there has been no change in angle or distance since measured by Burnham in 1888. A telescope of high power shows several minute companions.

[1] In the same way the comet of 1843 confirmed the Millerites in their belief in the immediate destruction of the world.

δ, 3,

is **Ruchbah,** sometimes **Rucba** and **Rucbar,** from **Al Rukbah,** the Knee.

It was utilized by Picard in France, in 1669, in determining latitudes during his measure of an arc of the meridian,— the first use of the telescope for geodetic purposes.

ε, of 3.6 magnitude, nearer the foot, also has borne the title **Ruchbah.**

ζ, of the 4th, and λ, of the 5th magnitude, marking the face, were the Chinese **Foo Loo,** a By-path.

η, Binary, 4 and 7.5, orange and violet,

very near *a*, is one of the finest objects in the sky for a moderate-sized telescope; and, although unnamed, it is worth noting that the components were 5″ apart in 1892, at a position angle of 193°, their period being about 200 years. The parallax is 0″.15 according to Struve; or 0″.45 according to Davis' measures of Rutherfurd's photographs. It is certainly a neighbor, and probably the nearest to us of all the stars in this constellation.

θ, 4.4, and μ, Triple, 5.1, 10.5, and 11, deep yellow, blue, and ruddy.

The Arabians knew these as **Al Marfiḳ,** the Elbow, where they lie; and the *Century Cyclopedia* gives **Marfak** as a present title for either star.

μ has the great proper motion of 3″.8 annually, a rate that will carry it around the heavens in 300,000 years.

★

The ramping Centaur!

. . . .

The Centaur's arrow ready seems to pierce
Some enemy; far forth his bow is bent
Into the blue of heaven.

John Keats' *Endymion.*

Centaurus, the Centaur,

is from the Κένταυρος that Aratos used, probably from earlier times, for it was a universal title with the Greeks; but he also called it Ἱππότā Φήρ, the Horseman Beast, the customary term for a centaur in the Epic and Aeolic dialects. This, too, was the special designation of the classical Pholos, son of Silenus and Melia, and the hospitable one of the family, who died in consequence of exercising this virtue toward Hercules. Apollodorus tells us that the latter's gratitude caused this centaur's transformation to the sky as our constellation, with the fitting designation Ἐυμενής, Well-disposed.

Eratosthenes asserted that the stellar figure represented *Χείρων*, a title that, in its transcribed forms **Chiron** and **Chyron**, was in frequent poetical use in classical times, and is seen in astronomical works even to Ideler's day. This has appropriately been translated the Handy One, a rendering that well agrees with this Centaur's reputation. He was the son of Chronos and the ocean nymph Philyra, who was changed after his birth into a Linden tree, whence **Philyrides** occasionally was applied to the constellation; although a variant story made him **Phililyrides**, the son of Phililyra, the Lyre-loving, from whom he inherited his skill in music. He was imagined as of mild and noble look, very different from the threatening aspect of the centaur Sagittarius; and Saint Clement of Alexandria wrote of him that he first led mortals to righteousness. His story has been thought in some degree historic, even by Sir Isaac Newton. As the wisest and most just of his generally lawless race he was beloved by Apollo and Diana, and from their teaching became proficient in botany and music, astronomy, divination, and medicine, and instructor of the most noted heroes in Grecian legend. Matthew Arnold wrote of him in *Empedocles on Etna:*

> On Pelion, on the grassy ground,
> Chiron, the aged Centaur lay,
> The young Achilles standing by.
> The Centaur taught him to explore
> The mountains where the glens are dry
> And the tired Centaurs come to rest,
> And where the soaking springs abound.
> . . . . .
> He told him of the Gods, the stars,
> The tides.

Indeed, he was the legendary inventor of the constellations, as we see in Dyer's poem *The Fleece:*

> Led by the golden stars as Chiron's art
> Had marked the sphere celestial;

and the father of Hippo, mentioned by Euripides as foretelling events from the stars.

The story of Pholos is repeated for Chiron: that, being accidentally wounded by one of the poisoned arrows of his pupil Hercules, the Centaur renounced his immortality on earth in favor of the Titan Prometheus, and was raised to the sky by Jove. His name and profession are yet seen in the mediaeval medicinal plants *Centaurea*, the Centaury, and the still earlier *Chironeion*.

Prometheus evidently inherited Chiron's astronomical attainments, as well as his immortality, for Aeschylus, who thought him the founder of civilization and "full of the most devoted love for the human race," made him say in *Prometheus Bound:*

> I instructed them to mark the stars,
> Their rising, and, a harder science yet,
> Their setting.

The conception of a centaur's figure with Homer, Hesiod, and even with Berōssōs, probably was of a perfect human form, Pindar being the first to describe it as semi-ferine, and since his day the human portion of the Centaur has been terminated at the waist and the hind quarters of a horse added. William Morris thus pictured him in his *Life and Death of Jason:*

> at last in sight the Centaur drew,
> A mighty grey horse trotting down the glade,
> Over whose back the long grey locks were laid,
> That from his reverend head abroad did flow;
> For to the waist was man, but all below
> A mighty horse, once roan, but now well-nigh white
> With lapse of years; with oak-wreaths was he dight
> Where man joined unto horse, and on his head
> He wore a gold crown, set with rubies red,
> And in his hand he bare a mighty bow,
> No man could bend of those that battle now.

Some ancient artists and mythologists changed these hind quarters to those of a bull, thus showing the **Minotaur,** and on the Euphrates it was considered a complete **Bull.** The Arabians drew the stellar figure with the hind parts of a Bear, but adopted the Greek title in their **Al Kentaurus,** that has been considered as the original of the otherwise inexplicable **Taraapoz,** used in Reduan's *Commentary* for our constellation.

Some of the Centaur's stars, with those of Lupus, were known to the early Arabs as **Al Ḳaḍb al Karm,** the Vine Branch; and again as **Al Shamārīlī,** the broken-off Palm Branches loaded with dates which Kazwini described as held out in the Centaur's hands. This degenerated into **Asemarik,** and perhaps was the origin of Bayer's word **Asmeat.** He also had **Albeze;** and Riccioli, **Albezze** and **Albizze,**— unintelligible unless from the Arabic **Al Wazn,** Weight, that was sometimes applied to α and β.

Hyde is our authority for another title (from Albumasar), **Birdun,** the Pack-horse.

Ptolemy described the figure with Lupus in one hand, and the Thyrsus in the other, marked by four 4th-magnitude stars, of which only two can now be found; this **Thyrsus** being formed, Geminos said, into a separate constellation by Hipparchos as *θυρσόλογκος*,— in the Manitius text as *θύρσος*,— and Pliny wrote of it in the same way, but their selection of such small stars seems remarkable.

The Centaur faces the east, and the Farnese globe shows him pointing with left hand to the Beast and the adjacent circular Altar; but in the *Hyginus* of 1488 the Beast is in his outstretched hands, the Hare on the spear, and a canteen at his waist; the *Alfonsine Tables* have the Thyrsus in his right hand and Lupus held by the fore foot in his left, which was the Arabian idea. The *Leyden Manuscript* gives a striking delineation of him with shaven face, but with heavy mustache (!), bearing the spear with the Hare dangling from the head, and a Kid, instead of the Beast, held out in his hands towards the Altar, the usual libation carried in the canteen. Bayer shows the Centaur with Lupus; Burritt has him in a position of attack, with the spear in his right hand and the shield on his left arm, the Thyrsus and vase of libation depicted on it; Grotius calling this portion of the constellation **Arma.** The *Century Dictionary* illustrates a Bacchic wand with the spear.

In Rome the constellation was **Centaurus,** the *duplici Centaurus imagine* of Manilius, and the **Geminus biformis** of Germanicus; **Minotaurus; Semi Vir,** the Half Man, and **Semi Fer,** the Half Beast; **Pelenor** and **Pelethronius** from the mountain home of the centaurs in Thessaly; **Acris Venator,** the Fierce Hunter; and Vergil had **Sonipes,** the Noisy-footed. The *Alfonsine Tables* designated it as **Sagittarius** *tenens pateram seu crateram* to distinguish it from the other Sagittarius with the more appropriate bow.

Robert Recorde, in 1551, had the **Centaure Chiron,** but Milton, in 1667, wrote Centaur for the zodiac figure, as so many others have done before and since his day; in fact, Sagittarius undoubtedly was the original Centaur and from the Euphrates, the Centaur of the South probably being of Greek conception. But in the classical age confusion had arisen among the unscientific in the nomenclature of the two figures, this continuing till now; much that we find said by one author for the one appearing with another author for the other. During the 17th century, however, distinction was made by English authors in calling this the **Great Centaure.**

In some mediaeval Christian astronomy it typified **Noah,** but Julius Schiller changed the figure to **Abraham with Isaac;** and Caesius likened it to **Nebuchadrezzar** when "he did eat grass as oxen."

This is one of the largest constellations, more than 60° in length, its centre about 50° south of the star Spica below Hydra's tail; but Aratos located it entirely under the Scorpion and the Claws, an error that Hipparchos criticized. It shows in the latitude of New York City only a few of its components in the bust, of which θ, a variable 2d-magnitude on the right shoulder, is visible in June about 12° above the horizon when on the meridian, and 27° southeast from Spica, with no other star of similar brightness in its vicinity. It was this that Professor Klinkerfues of Göttingen mentioned in his telegram to the Madras Observatory, on the 30th of November, 1872, in reference to the lost Biela comet which he thought had touched the earth three days previously and might be found in the direction of this star.

ι on the left shoulder, a 2½-magnitude, is about 11° west of θ.

Gould's list contains 389 naked-eye stars in this constellation.

One of the remarkable nebulae of the heavens, N. G. C., 3918, was discovered here by Sir John Herschel, who called it the **Blue Planetary,** "very like Uranus, only half as large again."

A 7th-magnitude *nova* that appeared in Centaurus between the 14th of June and the 8th of July, 1895, has changed since its discovery to a gaseous nebula, as has been the case with recent *novae* in Auriga, Cygnus, and Norma.

α, Binary, 0.2 and 1.5, white and yellowish.

Baily's edition of Ulug Beg's catalogue gives this as **Rigil Kentaurus,** from **Al Rijl al Kentaurus,** the Centaur's Foot; describing it as on the toe of the right front hoof, and Bayer so illustrated it. Chrysococca had the synonymous πους κονrουρος; and our *Century Dictionary* retains **Rigel,** although this is better known for the bright star in Orion. Burritt located on the left fore hoof a 4th-magnitude star that he wrongly lettered a; and above the pastern our 1st-magnitude, also lettered a, with the title **Bungula,** which I find only with him and the *Standard Dictionary*. He gives no explanation of this, nor can I trace it further; it may be a word specially coined by Burritt from β and *ungula,* the hoof, although even in this the letter is wrong.

Ideler said that a and β also have been the Arabic **Ḥaḍar,** Ground, and **Wazn,** Weight, as is explained at the star β; but he seemed at a loss as to the proper assignment of these words, although inclining to **Ḥaḍar** for β.

These two stars were among the much discussed **Al Muhlifaïn** described at γ Argūs and δ Canis Majoris.

Alpha's splendor naturally made it an object of worship on the Nile, and its first visible emergence from the sun's rays, in the morning at the autumnal equinox, has been connected by Lockyer with the orientation of at least nine temples in northern Egypt dating from 3800 to 2575 B. C., and of several in southern Egypt from 3700 B. C. onward. As such object of worship it seems to have been known as **Serk-t.**

It bore an important part, too, in southern China as the determinant of the stellar division **Nan Mun,** the South Gate.

α lies in the Milky Way, 60° south of the celestial equator, culminating with Arcturus, but is invisible from north of the 29th parallel. It is of the greatest interest to astronomers, being, so far as is now known, the nearest to our system of all the stars, although more than 275,000 times the distance of the earth from the sun,— 92,892,000 miles,— and 100 millions of times the distance from the earth to the moon,— 238,840 miles. Its parallax, first taken at the Cape of Good Hope by Henderson in 1839, and later by Gill and Elkin, and now fixed at 0''.75, shows a distance equal to that traveled by light in $4\frac{1}{3}$ years.

We can better realize the immensity of this distance from Professor Young's statement that if the line from the earth to the sun's centre be represented as 215 feet long, one to this star would be 8000 miles; and from Sir John Herschel's illustration :

> to drop a pea at the end of every mile of a voyage on a limitless ocean to the nearest fixed star, would require a fleet of 10,000 ships of 600 tons burthen, each starting with a full cargo of peas.

The nicety of parallactic observation, too, is shown by the fact that "an angle of 2'' is that in which a circle of $\frac{6}{10}$ of an inch in diameter would be seen at the distance of a mile."

Were our sun removed to the distance of α Centauri, its diameter of 866,400 miles would subtend an angle of only $\frac{1}{143}$ of a second of arc, of course utterly inappreciable with the largest telescope; and if seen from that star, would appear as a 2d-magnitude near the chair of Cassiopeia.

α was first discovered to be double by Richaud at Pondicherry, India, in 1689; but there seems discrepancy in the magnitudes respectively attributed to the components. Early astronomers thought the lesser star, α^1, a 4th-magnitude; even recently Gould has estimated it as $3\frac{1}{2}$; yet Miss Clerke writes, "the lesser, though emitting only $\frac{1}{3}$ as much light as its neighbour, is still fully entitled to rank as of the 1st magnitude"; all of which may indicate an increase of brilliancy since its observation began. Together they give nearly four times as much light as the sun, while their mass is double that of the latter.

The period of orbital revolution is about eighty-one years; the position angle in 1897, 208°; and they now are 21''.5 apart,—about 2700 millions of miles,—and yet connected! This distance is increasing.

Their proper motion, 3''.7 annually, or about 446 millions of miles across the line of vision, will carry them to the Southern Cross in 12,000 years.

The spectrum of α^2, the larger star, is midway between the Sirian and Solar.

β, 1.2.

Burritt located this near the right fore leg, calling it **Agena,** but gave no meaning or derivation of the word, and I have not found it elsewhere; Bayer placed it on the left hind quarter.*

Ḥaḍar and **Wazn,** Ground and Weight, seem to have been applied without much definiteness to α and β of this constellation, and to stars in Argo, Columba, and Canis Major, probably on account of their proximity to the horizon; the meridian altitude of β, 1000 years ago at Cairo, in 30° of north latitude, being only 4°. Hyde, however, said that α and γ were the stars referred to by these Arabic titles.

The Chinese call β **Mah Fuh,** the Horse's Belly.

This and α are the **Southern Pointers,** *i. e.* towards the Southern Cross, often regarded as the Cynosure of the southern hemisphere.

The Bushmen of South Africa knew them as **Two Men that once were Lions**; and the Australian natives as **Two Brothers** who speared Tchingal to death, the eastern stars of the Cross being the spear points that pierced his body.

γ, 2.4, that Bayer placed on the right fore foot, with τ, 4.4, were the early Chinese **Koo Low,** an Arsenal Tower; and δ, 2.8, was the later **Ma Wei,** the Horse's Tail.

The early ϵ, ζ, ν, and ξ^2, the four *Dictis a nautis Croziers* of Halley's catalogue, are the Southern Cross; ζ probably being Al Tizini's **Al Nā'ir al Baṭn al Kentaurus,** the Bright One in the Centaur's Belly.

θ, Double and variable, 2.2 to 2.7 and 14.3, red and bluish,

appears in the *Century Cyclopedia* as **Chort,** an error from the editor's writing *Centauri* for *Leonis*, this letter and title really belonging to θ Leonis, on the hind quarter of the Lion near the Ribs, that the Arabic **H·ārātan** signifies. θ in this constellation marks the left shoulder of the figure.

Harvard observers at Arequipa have reported an 8th-magnitude com-

*Allen, who correctly deciphered the meaning of Bungula (p. 152), failed to recognize Agena as a similar compound, *a* and *gena*, the knee.—Donald H. Menzel, 1963.

panion 3″ away, at a position angle of 180°. See does not find this at the Lowell observatories; but in 1897 discovered the companion noted in the heading, about 70″ away, at a position angle of 128°.6.

In China κ was **Ke Kwan,** a Cavalry Officer; μ, ν, and φ were **Wei,** the Balance; *i, g, k,* ψ, and A, with another adjacent, were **Choo,** a Pillar; and some small stars near the foot of the Cross were **Hae Shan,** the Sea and the Mountain.

The letter ω was applied by Bayer to a hazy 4th-magnitude star in *imo dorso* of the human part of the figure, which Halley, in 1677, inserted in his catalogue as a nebula; but at Feldhausen, on the Cape of Good Hope, the better telescope of Sir John Herschel showed it as "a noble globular cluster, beyond all comparison the richest and largest in the heavens." This appears absolutely round, 20′ in diameter, and contains many thousands of 13th- to 15th-magnitude stars; while its uniform structure indicates that it may be among the youngest of its class. It is the N. G. C. 5139, and has been splendidly photographed by Bailey at Arequipa, showing 6336 stars, among which he finds 122 variables.

It comes to the meridian on the 1st of June, about 36° south of Spica, but is invisible from north of the 34th parallel.

★

Kepheus is like one who stretches forth both hands.

Brown's *Aratos.*

Cepheus,

the French **Céphée** and the Italian **Cefeo,** is shown in royal robes, with one foot on the pole, the other on the solstitial colure, his head marked by a triangle, the 4th-magnitudes δ, ε, and ζ; γ and κ, near the knees, forming an equilateral triangle with Polaris; and almost universally has been drawn as Aratos described in the motto. Some see in his stars a large **K** open towards Cassiopeia,—ε, ζ, ξ, β, and κ, with ν and γ. Achilles Tatios, probably of our 5th century, claimed that the constellation was known in Chaldaea twenty-three centuries before our era, when the earthly King was recognized in that country's myths as the son of Belos, of whom Pliny wrote, *Inventor hic fuit sideralis scientiae.*

In Greek story, like so many other stellar personages, Cepheus was connected with the Argonautic expedition.

The figure bore our title among all early astronomers and classic authors, but Germanicus added **Iasides** from the 'Ιασίδαο of Aratos; Nonnus had 'Ανὴρ βασιλήιος from his royal station, which became **Vir regius** and even **Regulus.** Others said that he was the aged **Nereus** and thus also **Senex aequoreus,** and others strangely called it **Juvenis aequoreus.**

Cantans, Sonans, and **Vociferans** show early confusion with the not far distant Boötes; while **Dominus solis, Flammiger, Inflammatus,** and **Incensus** are fiery epithets that do not seem appropriate for so faint a figure, unless originating from the fable that the tables of the Sun were spread in Aethiopia, the land where Cepheus reigned when on earth. Some one, however, has suggested that they are from the fact that his head is surrounded and illuminated by the Milky Way, although itself in an entirely bare spot in that great circle of light. This appeared in Horace's lines:

> Clarus occultum Andromedae pater
> Ostendit ignem.

Cepheus is an inconspicuous constellation, but evidently was highly regarded in early times as the father of the Royal Family, and his story well known in Greek literature of the 5th century before Christ. The name Κηφεύς, compared by Brown to Khufu of Great Pyramid fame, was the source of many queer titles from errors in Arabic transcription—first into **Ḳifaūs, Ḳikaūs, Kankaus**; later into **Fikaus, Fifaus,** and **Ficares,** or **Phicares,** its usual designation in Persia, and **Phicarus.** Chilmead suggested that Phicares was a Phoenician title equivalent to Flammiger, and identical with Πυρκᾶεύς, the Fire-kindler, which, transliterated as **Pirchaeus,** has been used for these stars. Later on in astronomical literature we find **Caicans, Ceginus, Ceichius, Chegnius, Chegninus, Cheguinus,** and **Chiphus,** some of which also are seen for Boötes.

The later Hindus knew Cepheus as **Capuja,** adopted from Greece; but Hewitt claims that with their prehistoric ancestors it represented **Kapi,** the Ape-God, when its stars α and γ were the respective pole-stars of 21000 and 19000 B. C.

Dunkin derives our title from the Aethiopic Hyk, a King, but the connection with Aethiopia probably can only be allowed by considering that country the Asian Aethiopia, for our Cepheus is unquestionably of Euphratean origin. Still Bayer's illustration of it is that of a typical African.

In China, somewhere within this constellation's boundaries, was the **Inner Throne of the Five Emperors.**

Arabian astronomers translated Inflammatus into **Al Multahab**; but the nomads knew Cepheus, or at least some of its stars, as **Al Aghnām,** the Sheep, and thus associated with the supposed **Fold,** a large figure around the pole very visible traces of which appear in the nomenclature of components of this and other circumpolar constellations. Bayer specified certain of these,—η, θ, γ, κ, π, and ρ,—as the **Shepherd,** his **Dog,** and the **Sheep**; but Smyth alluded to the whole of Cepheus as the **Dog,** Cassiopeia being his mate. Riccioli quoted from Kircher, as to these, the Arabic "**Raar, Kelds** & **San**: *nempe Pastorem*, *Canem*, *Oves*," more correctly transcribed **Rāiʻ, Kalb,** and **Shām.**

A translator of Al Ferghani's[1] *Elements of Astronomy* called the constellation **Al Radif,** the Follower, which may have come by some misunderstanding from the near-by Al Ridf in the tail of the Swan, for Cepheus does not seem ever to have been known by any such title. The early Arabs' **Ḳidr,** the Pot, was formed by the circle of small stars from ζ and η on the hand of our figure extending to the wing of the Swan.

In the place of Cepheus, Caesius wished to substitute **King Solomon,** or **Zerah,** the Aethiopian, whom King Asa overthrew, as told in the *2d Book of the Chronicles*, xiv, 9–12; but Julius Schiller said that it should be **Saint Stephen.**

Argelander gives 88 naked-eye components; Heis, 159.

α, 2.5, white.

Alderamin, from **Al Deraimin** of the *Alfonsine Tables* of 1521, originally was **Al Dhirāʻ al Yamīn,** the Right Arm, but it now marks that shoulder. Bayer wrote it "**Aderaimin** *corrupte* Alderamin"; Schickard, **Adderoiaminon;** Assemani, **Alderal jemīn;** while elsewhere we find **Al Derab, Al Deraf, Alredaf,** and **Alredat.** Kazwini mentioned it as **Al Firḳ,** but, although thus found on the Borgian globe, Ideler thinks it a mistake of that author, as a single star cannot represent a Flock, which Al Firḳ signifies. Ulug Beg more appropriately called α, β, and η **Al Kawākib al Firḳ,** the Stars of the Flock, although by this last word a Herd of Antelopes may be intended.

α culminates on the 27th of September.

It will be the Polaris of the year 7500; while midway between it and α Cygni lies the north polar point of the planet Mars.

[1] This author was Aben al Khethīr of Fergana in Sogdiana, prominent in 9th-century astronomy and much quoted from the 16th to the 18th centuries as Alfergan, Alferganus, Alfragani, and Alfraganus. His work, a valuable one for its day, was translated with notes by Golius (the Dutch Jakob Gohl), and published after the latter's death in 1669.

β, Double, 3.3 and 8, white and blue.

Alfirk is now current for this star, although originally given to *α*; and **Ficares** is occasionally seen, from one of the degenerated names for the whole constellation that also may have been applied by the Arabs to others of its brighter stars.

The components are about 14″ apart, and the position angle is 251°.

γ, 3.5, yellow.

Errai of the Palermo and **Er Rai** of other catalogues, but sometimes **Arrai**, is from **Al Rāi'**, the Shepherd, a title indigenous to Arabia.

In China it was **Shaou Wei,** a Minor Guard.

γ now marks the left knee of the King, and will be the pole-star of 2600 years hence.

δ, ε, ν, and ζ, of about the 4th magnitude, were the Chinese **Tsaou Foo,** a charioteer of Mu Wang, the 5th emperor of the Chow dynasty, 536 B. C.

δ is a noted double, the yellow and blue components 41″ apart, at a position angle of 192°. The smaller is of the 7th magnitude, but the larger varies from 3.7 to 4.9 in a period of $10\frac{2}{3}$ days. This was discovered by Goodricke [1] in 1784; and Belopolsky thinks it a spectroscopic binary, the period of revolution equaling the period of variation.

From its neighborhood radiate the **Cepheid** meteors, visible from the 10th to the 28th of June.

Surrounding δ, ε, ζ, and λ, which mark the King's head, is a vacant space within the southern edge of the Milky Way similar to the Coal-sacks of the Northern and Southern Cross.

η and θ, 4th-magnitude stars on and near the right wrist, mark **Al Ḳidr.**

κ, a double star, 4.4 and 8.5, is the Chinese **Shang Wei,** the Higher Guard. The components are yellow and blue, 7″.5 apart, at a position angle of 124°.

μ, Irregularly variable, 4 (?) to 5 (?), garnet,

about 5° east of the head of Cepheus, is Sir William Herschel's celebrated **Garnet Star,** and so entered by Piazzi in the *Palermo Catalogue*, yet strangely omitted from Flamsteed's list, perhaps owing to its variability. This, suspected by Hind in 1848, was confirmed by Argelander.

[1] John Goodricke of York, England, is still remembered in the astronomy of the last century as a diligent and successful observer of variable stars, although he was a deaf-mute and died at the early age of 22 years.

It is one of the deepest-colored stars visible to the naked eye, and comparison with the near-by α will show its peculiar tint, which, however, sometimes changes to orange.

ξ, Binary, 4.5 and 7, blue.

Kazwini called this **Al Ḳurḥaḥ,** an Arabic word that Ideler translated as a white spot, or blaze, in the face of a horse; but thinking this not a proper stellar name, suggested **Al Ḳirdah,** the Ape. He seems here, however, to have forgotten Al Hiḳ'ah of Orion, of the same meaning as that to which he objected.

The components are 7″ apart, and their position angle is 285°.

ρ, a 5th-magnitude, was **Al Kalb al Rā'i,** the Shepherd's Dog, guarding the Flock shown by α, β, and η; k, h, and v, with others between the feet and Polaris, were **Al Aghnām,** the Sheep, apparently separated from the Flock.

υ^1 and υ^2, 5th-magnitude stars, are given by Bayer, under the title **Castula,** as from Nonius, equivalent to Ταινία, the Front of the Garment, which they mark.

Sundry small members of this constellation and Camelopardalis were the Chinese **Hwa Kae,** the State Umbrella.

★

To Cerberus too a place is given —
His home of old was far from heaven.

Quoted in Smyth's *Bedford Catalogue.*

Cerberus

is the Italian **Cerbero,** Secchi associating it with **Ramo,** the Branch, and the French combining both in the title **Rameau et Cerbere.**

This sub-constellation, a former adjunct of Hercules, but now entirely disregarded by astronomers, is supposed to have originated with Hevelius in his *Firmamentum Sobiescianum*, although Flammarion asserts that it was on the sphere of Eudoxos with the Branch. The 4th- to 5th-magnitude stars that Hevelius assigned to it are Flamsteed's 93, 95, 96, and 109, lying half-way between the head of Hercules and the head of the Swan.

The royal poet James I designated the infernal Cerberus as "the thrie headed porter of hell," and the heavenly one has been so figured, although with serpents' darting tongues; but the abode and task of the creature would seem to render very inappropriate his transfer to the sky, so that it probably was only made for the purpose of mythological completeness, as the death of this watch-dog of Hades fitly rounded out the circle of Hercules' twelve labors.

Others have said that the figure typified the serpent destroyed by the Hero while it was infesting the country around Taenarum, the *Μέτωπον* of Greece, the modern Cape Matapan.

Some of the stars of Cerberus were known in China as **Too Sze,** the Butcher's Shop; and others as **Meen Too,** a Cloth Measure.

★

The south wind brings her foe
The Ocean beast.

Brown's *Aratos.*

Cetus, the Whale, or Sea Monster,

is the French **Baleine,** the Italian **Balaena,** and the German **Wallfisch.**

This constellation has been identified, at least since Aratos' day, with the fabled creature sent to devour Andromeda, but turned to stone at the sight of the Medusa's head in the hand of Perseus. Equally veracious additions to the story, from Pliny and Solinus, are that the monster's bones were brought to Rome by Scaurus, the skeleton measuring forty feet in length and the vertebrae six feet in circumference; from Saint Jerome, who wrote that he had seen them at Tyre; and from Pausanias, who described a near-by spring that was red with the monster's blood. But the legend in which Cetus figured seems to have been current on the Euphrates long before our era; and, descending to Euripides and Sophocles, appeared in their dramas, as also in much subsequent literature.

For its stellar title the Greeks usually followed Aratos and Eratosthenes in Κῆτος, but they also had 'Ορφίς, 'Ορφός, and 'Ορφώς, some species of cetacean; and the equivalent Πρῆστις and Πρίστις,[1] from πρῆθειν, to blow or spout, the common habit of the animal. The last word, variously transliterated, was common for the constellation with Roman authors, appearing as **Pristis, Pristix,** and **Pistrix,** qualified by the adjectives *auster*, *Nereia*, *fera*, *Neptunia*, *aequorea*, and *squammigera*. Cetus, however, has been the usual title from the days of Vitruvius, varied by **Cete** with the 17th-century astronomical writers, although the stellar figure is unlike any whale known to zoölogy.

The *Harleian*[2] and *Leyden Manuscripts* show it with greyhound head, ears, and fore legs, but with a long, trident tail; the whole, perhaps, modeled after the ancient bas-relief of Perseus and Andromeda in the Naples Museum. It is found thus on the Farnese globe, and this figuring may have given rise to, or originated from, the early title that La Lande cited, **Canis Tritonis,** his own **Chien de Mer.** But the *Hyginus* of 1488 has a dolphin-like creature with proboscis and tusks, all imitated in the edition of 1535 by Micyllus; and Dürer still further varied the shape of the head and front parts.

Thus in these, as, in fact, in all delineations, it has been a strange and ferocious marine creature, in later times associated with the story of Andromeda, and at first, perhaps, was the Euphratean Tiāmat, of which other forms were Draco, Hydra, and Serpens; indeed, some have thought that our Draco was Andromeda's foe because of its proximity to the other characters of the legend. But as an alternative signification of the word Κῆτος is Tunny,[3] also a signification of Χελιδόνιας, applied to the Northern Fish of the zodiac, it is not unlikely that the latter figure should be substituted in the story for the time-honored Whale.

Cetus is sometimes represented swimming in the River Eridanus, although usually as resting on the bank with fore paws in the water; its head, directly under Aries, marked by an irregular pentagon of stars, and its body stretching from the bend in Eridanus to that in the Stream from the Urn. It occupies a space of 50° in length by 20° in breadth, and so is one of the most extended of the sky figures; yet it shows no star larger than of the 2d magnitude, and only one of that lustre.

[1] This word is seen in more modern days in the Physetere that Rabelais used.

[2] This is the famous No. 647 of the Harleian Collection of manuscripts in the British Museum, from Robert Harley, the first earl of Oxford. It is an illuminated copy of Cicero's translation of the *Phainomena*, and has been reproduced and annotated by Ottley in the 26th volume of *Archaeologia* for 1834, its editor supposing it to be from the 2d or 3d century. Verses from Manilius are inscribed within the figure outlines.

[3] This tunny, the horse-mackerel of our American coast and the *Albacora thynnus* of ichthyology, is found in the Mediterranean up to 1000 pounds' weight.

Argelander enumerates 98 stars in the constellation, and Heis 162.

The 1515 *Almagest* and the *Alfonsine Tables* called it **Balaena,** but Firmicus said **Belua,** the Beast or Monster, a more appropriate name than ours. Bayer mentioned it as **Draco,** and drew it so, but without wings; he also cited for it **Leo, Monstrum marinum, Ursus marinus, Orphas,** and **Orphus**; and Grotius quoted **Gibbus,** Humped, from anonymous writers.

The Arabian astronomers of course knew the Greek constellation and called it **Al Ḳeṭus,** from which have come **Elketos, Elkaitos,** and **Elkaitus**; but their predecessors, who had not heard of the Royal Family and its foe, separated these stars into three very different asterisms. Those in the head, α, γ, δ, λ, μ, ξ^1, and ξ^2, were **Al Kaff al Jidhmah,** the Part of a Hand, from a fancied resemblance to their Stained Hand, our Cassiopeia; η, θ, τ, ζ, and υ, in the body of our Cetus, were **Al Na'āmāt,** the Hen Ostriches; and the four in a straight line of 3° length across the tail, all lettered ϕ, were **Al Niṭhām,** the Necklace.

The biblical school of the 17th century of course saw here the **Whale that swallowed Jonah**; and commentators on that great astronomical poem, the *Book of Job*, have said that it typified the **Leviathan** of which the Lord spoke to the patriarch. Julius Schiller thought it "**SS. Joachim and Anna.**"

The **Easy Chair** has popularly been applied to it from the arrangement of its chief stars, the back of the chair leaning towards Orion.

Although an old constellation, Cetus is by no means of special interest, except as possessing the south pole of the Milky Way and the Wonderful Star, the variable Mira; and from the fact that it is a condensation point of nebulae directly across the sphere from Virgo, also noted in this respect.

α, 2.9, bright orange.

Menkar of the *Alfonsine Tables* of 1521, Scaliger's **Monkar,** and now sometimes **Menkab,** from **Al Minḥar,** the Nose, still is the popular, but inappropriate name, for it marks the Monster's open jaws. It is the prominent star in the northeastern part of the constellation, and culminates on the 21st of December.

Al Kaff al Jidhmah, found on the Borgian globe, is Ulug Beg's and Al Tizini's designation for it, taken from that for all the stars in the head; but modern lists apply this solely to γ.

In astrological days it portended danger from great beasts, disgrace, ill fortune, and illness to those born under its influence.

In China α, γ, δ, λ, μ, ν, o, ξ^1, and ξ^2, were **Tseen Kwan,** Heaven's Round Granary.

The other 'neath the dusky Monster's tail.
Brown's *Aratos.*

β, 2.4, yellow.

Deneb Kaitos is from the Arabian **Al Dhanab al Ḳaiṭos al Janūbīyy,** the Tail of the Whale towards the South, *i. e.* the Southern Branch of the Tail. Chrysococca synonymously had 'Οὖρα τοῦ Καίτου, arbitrarily formed from the Arabic; and the *Alfonsine Tables* of 1521 called it **Denebcaiton.**

Very differently it was the Arabs' **Al Ḍifdiʽ al Thānī,** the Second Frog, that we see in the present **Difda,** Latinized as **Rana Secunda;** the star Fomalhaut being Al Ḍifdi' al Awwal, the First Frog.

In China it was **Too Sze Kung,** Superintendent of Earthworks.

Although below it in lettering, this star is now brighter than α, yet both were registered γ—*i. e.* of the 3d magnitude—by Ptolemy; and Miss Clerke asserts that this inversion of brilliancy took place during the last century. It is nearly 40° southwest from α, culminating on the 21st of November.

One third of the way towards β Andromedae is a group of unnamed stars from which Smyth said that a new asterism, **Testudo,** was proposed.

γ, Double, 3.5 and 7, pale yellow and blue.

Al Kaff al Jidhmah is the Arabs' name for the whole group marking the Whale's head, but in modern lists is exclusively applied to this star.

The components are 2″.5 apart, at a position angle of 290°.

ε, of the 5th magnitude, with π, was a part of the **Ostrich's Nest** that mainly lay in Eridanus; and, with π, ρ, and σ, also was Al Sufi's **Al Sadr al Ḳaiṭos,** the Whale's Breast.

Notwithstanding its lettering, it is the faintest of these four stars.

ε, ρ, and σ were the Chinese **Tsow Kaou,** Hay and Straw.

ζ, 3.9, topaz yellow,

is **Baten Kaitos,** the Arabian **Al Baṭn al Ḳaiṭos,** the Whale's Belly, although the star is higher up in the body. The *Alfonsine Tables* had **Batenkaiton** and **Batenel Kaitos;** and Chilmead, **Boten.**

In astrology it portended falls and blows.

It forms, with the 5th-magnitude χ, a very coarse naked-eye double; and itself has a 7½-magnitude companion 3′ 6″ distant.

η, 3.6, yellow.

Deneb and **Dheneb** are names for this star, especially in English lists, maps, and globes; but incorrectly, as η, on the Heis *Atlas*, lies at the base of the tail, and in Bayer's and Argelander's on the Monster's flank, while there are two others, β and ι, so named in the proper location. Still, although a misnomer, the title seems to be generally recognized. The *Century Cyclopedia* extends it as **Deneb Algenubi.** This error in name has led to another, for the star has been mistaken for the Rana Secunda of the Arabs, the Second Frog, the Arabs' Al Ḍifdiʿ al Thānī,— β Ceti.

ι, 3.6, bright yellow,

is another **Deneb Kaitos** to which the Arabians added **Al Shamāliyy** as being in the Northern branch of the tail, although Heis places it in the Southern. From this Arabic adjective the *Standard Dictionary* very unsatisfactorily gives **Schemali** simply as the star's title. With η, θ, ν, τ, and stars in the modern Fornax, it made up the Chinese asterism **Tien Yuen,** Heaven's Temporary Granary.

λ, of about 4½ magnitude, is occasionally called **Menkar,** and, as it exactly marks the Nose of Cetus, the title would seem more appropriate than it is to α; but it was applied by the Arabs to both.

o, Variable, 1.7 to 9.5, flushed yellow.

Mira, Stella Mira, and **Collum Ceti** are all titles for this **Wonderful Star** in the Whale's neck, the show object in the heavens as a variable of long period and typical of its class.

It was first noticed as a 3d-magnitude on the 13th of August, 1596, and again on the 15th of February, 1609, by David Fabricius, an amateur astronomer and disciple of Tycho Brahē; but its true character was not ascertained till 1638 by Phocylides Holwarda of Holland,— the first established record of a variable star.

Bayer lettered it in 1603 as of the 4th magnitude, evidently at a time of its diminished brilliancy and without knowledge of its variability; Hevelius, having observed it from 1659 to 1682, inserted it in his *Prodromus* as the *Nova in Collo Ceti;* and Flamsteed, numbering it 68, described it as *in pectore nova* and of the 6th magnitude on the 18th of October, 1691, and again on the 28th of September, 1692.

"This was singular in its kind till that in Collo Cygni was discovered; and the attention it excited among astronomers is detailed in the *Historiola Mirae Stellae*" of Hevelius in 1662; thus virtually naming it and "commemorating the amazement excited by the detection of stellar periodicity."

Its period, fixed by Bouillaud in 1667 as 333 days, is now given as 331, but this is subject to extreme irregularities,—at various times it has not been seen at all with the naked eye for several years consecutively,—and its maxima and minima are even more irregular. While it has been known almost to equal Aldebaran in its light, as it did under Herschel's observations on the 6th of November, 1779, Chandler gives its maximum as from 1.7 to 5, and its minimum from 8 to 9.5. It thus sometimes sends out at its maximum fifteen hundredfold more light than at its minimum, and "after three centuries of notified activity gives no sign of relaxation." It is generally at its brightest for about a fortnight; the increase occupying about seven weeks and the decrease about three months. The maximum of 1897 occurred about the 1st of December, when it was a little below the 3d magnitude.

Sir William Herschel wrote of it in 1783 as being of a deep garnet color like μ Cephei.

The spectrum is of Secchi's 3d type, with extremely brilliant hydrogen lines at the time of maximum.

Mira lies almost exactly on the line joining γ and ζ, a little nearer the former star.

ϕ^1, ϕ^2, ϕ^3, and ϕ^4, 5th- to 6th-magnitude stars, were the Arabs' **Al Nithām.** In China they were **Tien Hwan,** Heaven's Sewer. It was near these that Harding of Lilienthal discovered the minor planet Juno, on the 2d of September, 1804, the 3d of these objects found.

c and *y*, small stars near τ, were the Chinese **Foo Chih,** the Ax and Skewer.

★

Chamaeleon,

the German **Chamäleon,** the French **Caméléon,** and the Italian **Camaleonte,** is a small and unimportant constellation below Carina, Octans separating it from the south pole. It was first published and figured by Bayer among his new constellations from observations by navigators of the preceding century. Pontanus, in Chilmead's *Treatise*, included it with Musca as "the **Chamaeleon with the flie**"; but Julius Schiller entirely changed its character by combining it with Apus and Musca in his biblical **Eve.**

None of its stars seem to be named except in China, where some of the larger were **Seaou Tow,** a small Measure or Dipper, that our α, θ, η, ι, ε, μ^2, and μ^1 well show.

Gould gives 50 naked-eye components from 4.2 to the 7th magnitude. The constellation culminates about the 1st of May.

Circinus, the Pair of Compasses,

formed by La Caille, lies close to the front feet of the Centaur, south from Lupus and Norma, its inventor appropriately associating it with the latter.

It is the German **Zirkel,** the French **Compas,** and the Italian **Compasso.**

Gould catalogues in it 48 stars down to the 7th magnitude; *a*, its *lucida*, being of only 3.5.

The constellation culminates about the middle of June.

★

> Others underneath the hunted *Hare*,
> All very dim and nameless roll along.
>
> Brown's *Aratos*.

Columba Noae, Noah's Dove,

now known simply as **Columba,** is the **Colombe de Noé** of the French, **Colomba** of the Italians, and **Taube** of the Germans, lying south of the Hare, and on the meridian with Orion's Belt.

Although first formally published by Royer in 1679, and so generally considered one of his constellations, it had appeared seventy-six years before correctly located on Bayer's plate of Canis Major, and in his text as *recentioribus Columba;* one of these "more recent" being Petrus Plancius, the Dutch cosmographer and map-maker of the 16th century, and instructor of Pieter Theodor. While these are the first allusions to Columba in modern times, yet the following from Caesius may indicate knowledge of its stars,[1] and certainly of the present title, seventeen centuries ago. Translating from the *Paedagogus* of Saint Clement of Alexandria, he wrote:

> Signa sive insignia vestra sint Columba, sive Navis coelestis cursu in coelum tendens sive Lyra Musica, in recordationem Apostoli Piscatoris.

Still it was not recognized by Bartschius twenty-one years after Bayer, nor by Tycho, Hevelius, or Flamsteed; but Halley gave it, in the same year as Royer, with ten stars; and our Gould, two centuries later in Argentina, increased the number to seventeen. It was made up from the southwestern

[1] But the faintness of this constellation is against the probability of such use, and would imply that some other, and more noticeable, sky-group was known as a Dove, possibly Coma Berenices.

outliers of Canis Major, near to the Ship,—Noah's Ark,—and so was regarded as the attendant Dove.

Smyth wrote of its modern formation, and of its nomenclature in Arab astronomy:

Royer cut away a portion of Canis Major, and constructed Columba Noachi therewith in 1679. The part thus usurped was called Muliphein, from *al-muhlifeïn*, the two stars sworn by, because they were often mistaken for Soheïl, or Canopus, before which they rise: these two stars are now α and β Columbae. Muliphein is recognized as comprehending the two stars called *Ḥaḍ'ár*, ground, and *al-wezn*, weight.

Reference already has been made to Al Muḥlifaïn at the stars γ, ζ, and λ Argūs, δ Canis Majoris, and α Centauri.

α, 2.5.

Phaet, Phact, and **Phad** are all modern names for this, perhaps of uncertain derivation, but said to be from the Ḥaḍar already noted under the constellation.

The Chinese call it **Chang Jin,** the Old Folks.

Although inconspicuous, Lockyer thinks that it was of importance in Egyptian temple worship, and observed from Edfū and Philae as far back as 6400 B. C.; but that it was succeeded by Sirius about 3000 B. C., as α Ursae Majoris was by γ Draconis in the north. And he has found three temples at Medinet Habu, adjacent to each other, yet differently oriented, apparently toward α, 2525, 1250, and 900 years before our era: all these to the god Amen. He thinks that as many as twelve different temples were oriented to this star; but the selection of so faint an object for so important a purpose would seem doubtful.

Phaet is 33° south of ε Orionis, the central star in the Belt, and culminates on the 26th of January.

β, 2.9.

Wezn, or **Wazn,** is from **Al Wazn,** Weight.

With α it was among the disputed **Al Muḥlifaïn**; and Al Tizini additionally called both stars **Al Aghribah,** the Ravens, a title that Hyde assigned to a group in Canis Major.

Chilmead's *Treatise* has this brief description of Columba:

11 *Starres: of which there are two in the backe of it of the second magnitude, which they call the* **Good messengers,** *or bringers of good newes: and*

those in the right wing are consecrated to the Appeased Deity, *and those in the left, to the Retiring of the waters in the time of the Deluge.*

Heis locates α and β in the back; υ^2 in the right wing, and ϵ in the left. θ and κ were included by Kazwini in the Arabic figure **Al Kurud,** the Apes.

In China they were **Sun,** the Child; λ being **Tsze,** a Son; and the near-by small stars, **She,** the Secretions.

★

The streaming tresses of the Egyptian queen.
William Cullen Bryant's *The Constellations.*

Not Berenice's locks first rose so bright,
The heavens bespangling with dishevell'd light.
Pope's *Rape of the Lock.*

Coma Berenices, Berenice's Hair,

the **Chevelure** of the French, **Chioma** of the Italians, and the **Haupthaar** of the Germans, lies southwest from Cor Caroli.

It seems to have been first alluded to by Eratosthenes as **Ariadne's Hair** in his description of Ariadne's Crown; although subsequently, in his account of Leo, he mentioned the group as Πλόκαμος Βερενίκης 'Ευεργέτιδος. But for nearly 2000 years its right to a place among the constellations was unsettled, for it has been the ἀμόρφωτοι behind the Lion's tail, or connected with Virgo, or partly recognized as an asterism by itself. Tycho, however, set the question at rest in 1602 by cataloguing it separately, adopting the early title as we have it now.

Aratos, perhaps, alluded to it, although indefinitely, in the 146th line of the *Phainomena:*

Each after each, ungrouped, unnamed, revolve;

but, of course, did not give its name, for he wrote under the 2d Ptolemy (Philadelphus), whereas it was not known till about 243 B. C., in the reign of the 3d (Euergetes), the brother and husband of Berenice, whose amber hair we now see in the sky figure. It was the happy invention of this constellation by Conon that consoled the royal pair after the theft of the tresses from the temple of Arsinoë Aphrodite at Zephyrium. Some versions of the story turned the lady's hair into a hair-star, or comet.

The scholiast on Aratos, however, referred to it, as did Callimachus, the latter calling it Βόστρυκος Βερενίκης;[1] and his poem on it, now lost, was imitated 200 years later by Catullus, in one of his most beautiful odes, describing it as

> the consecrated spoils of Berenice's yellow head, which the divine Venus placed, a new constellation, among the ancient ones, preceding the slow Boötes, who sinks late and reluctantly into the deep ocean.

The beautiful and touching legend of the Sudarium of Veronica, with its *vera icon*, has been associated with our constellation from the similarity in words, some supposing the saint to have been the Herodian Bernice,—in Latin Beronica,—converted to Christianity through her sympathy for the Saviour's sufferings. Lady Eastlake has fully told this story in her continuation of Mrs. Jameson's *History of our Lord.*

Hyginus had Βερενίκης πλόκαμος; and Ptolemy, simple πλόκαμος for three of its stars among the ἀμόρφωτοι of Leo, calling it νεφελοειδής συστροφή, a cloudy condensation. This was rendered **Al Atha** by Reduan, or, as Golius printed it, **Al Ultha,** literally a Mixture.

Manilius did not mention Coma, although he wrote 250 years after Conon; nor of course did the versifiers of Aratos, at least by name, as the figure is not distinctly specified in the *Phainomena.*

Crines and **Crines Berenices** are found in classical times; Flamsteed has the plural **Comae Berenices,** and La Lande **Capilli. Cincinnus** appears on Mercator's globe of 1551, but there consists of only one star and two nebulae; and the *Latin Almagest* of the same year wrote *Convolutio nubilosa quae cincinnus vocatur*, with this marginal note, all for Coma's stars as *informes* of Leo: *Plocamos graecē, latinē vero cincinnus, hoc est, caesaries & coma virginis, Berenices fortasse crinis qui ā Poeta Calimacho in astra relatus est. Sed cincinnum barbari tricam vocant.* The *Almagest* of 1515 already had **Trica,** describing it as *nubilosa* and *luminosa;* but Bayer

[1] The word Berenice, sometimes Beronice, is from Βερενίκη, the Macedonian form of the purer Greek Φερενίκη, Victory-bearing; and is the Βερνίκη, or Bernice, of the *New Testament*, the name of the notorious daughter and wife of the Agrippas. From it some philologists derive the Italian *Vernice*, the French *Vernis*, the Spanish *Barniz*, and our *Varnish*, all from the similar amber color of the lady's hair; Βερενίκη having later become the Low Greek word for amber.

changed this to **Tricas, Tericas,** and **Triquetras,** taking these probably from the Low Greek τριχες, which doubtless is the origin of our word "tresses."

Pliny wrote in the *Historia Naturalis: nec* [*cernit*] *Canopum Italia et quem vocant Berenices crinem*, which Bostock and Riley correctly translated, in 1855, "nor can we, in Italy, see the star Canopus, or Berenice's Hair"; but Holland had rendered this, in 1601, "neither hath Italy a sight of Canopus, named also Berenices Hair," from which mistranslation it was long inferred that the southern heavens contained another sky group bearing this same title. And this blunder has been perpetuated, even in Doctor Murray's *New English Dictionary*, which defines the word as the name "formerly of the southern star Canopus," citing as authority the foregoing passage from Holland. Pliny's statement as to the invisibility of Coma from Italy of course was incorrect then as now.

Julius Schiller asserted that the constellation represented the *Flagellum Christi.*

Thompson writes in his *Glossary*, p. 134, that

It has been suggested by Landseer, *Sabaean Researches*, p. 186, from the study of an Assyrian symbolic monument, that the stars which Conon converted into the Coma Berenices (Hygin. *P. A.* ii, 24, cf. Ideler, *Sternnamen*, p. 295) and which lie in Leo opposite to the Pleiades in Taurus, were originally constellated as a Dove; and that this constellation, whose first stars rise with the latest of those of Argo, and whose last rise simultaneously with the hand of the Husbandman, links better than the Pleiad into the astronomical Deluge-myth. The case rests on very little evidence, and indeed is an illustration of the conflicting difficulties of such hypotheses: but it is deserving of investigation, were it only for the reason that the Coma Berenices contains seven visible stars (Hygin.), and the Pleiad six, a faint hint at a possible explanation of the lost Pleiad.

Serviss, who has some beautiful stellar similes, says that it is a

curious twinkling, as if gossamers spangled with dewdrops were entangled there. One might think the old woman of the nursery rhyme who went to sweep the cobwebs out of the sky had skipped this corner, or else that its delicate beauty had preserved it even from her housewifely instinct.

In *Hudibras* the constellation was **Berenice's periwig**; while another old-fashioned name has been **Berenice's Bush,** found in Thomas Hill's *Schoole of Skil* of 1599, but even then rendered classic in its use by Chaucer and Spenser; and Smyth says that there has been a name still homelier.

Bayer also mentioned **Rosa,** a Rose, or a Rose Wreath; but he figured it on his plate of Boötes as a **Sheaf of Wheat,** in reference to the Virgo Ceres close by; indeed, Karsten Niebuhr, at Cairo in 1762, heard it called **Al Huzmat,** the Arabic term for that object, or for a Pile of Fruit, Grain, or Wood. The Dresden globe has it as an **Ivy Wreath,** or, just as probably, a **Distaff** held in the Virgin's hand, which has been designated *Fusus vel Colus*, *Fila et Stamina*, the Distaff, Thread, and Woof; or perhaps the **Caduceus** of Mercury, placed here when Coma was a part of Virgo and this latter constellation the astrological house of that planet.

But very differently in early Arabia it was **Al Ḥauḍ,** the Pond, into which the Gazelle, our Leo Minor, sprang when frightened at the lashing of the Lion's tail; although some of the Desert observers claimed that this Pond lay among the stars of the neck, breast, and knees of the Greater Bear; and Lach substituted it for the Gazelle in our location of Leo Minor. The Arabian astronomers knew Coma as **Al Halbah,** or **Al Ḍafīrah,** the Coarse Hair, or Tuft, in the tail of the Lion of the zodiac, thus extending that figure beyond its present termination at the star Denebola.

Coma probably was known in early Egypt as the **Many Stars.**

The Chinese had several names here; the *lucida* being **Hing Chin;** *u* and *w* in the Reeves list, **Chow Ting,** the Imperial Caldron of the Chow dynasty; a small group toward Virgo, **Woo Choo How;** *a*, *b*, *c*, *d*, *e*, and *f*, **Lang Wei,** Official Rank; *p*, **Lang Tseang,** a General, and *v*, **Shang Tseang,** a Higher General;[1] while **Tsae Ching,** the Favorite Vassal, was the title for Bode's 2629. This abundant nomenclature, in so faint a figure, shows great interest on the part of the Chinese in this beautiful little group.

Argelander numbers thirty-six stars here, Heis extending this to seventy; and Chase, of the Yale Observatory, has taken measures of thirty-two of these. The constellation culminates about the middle of May.

Although it is not easy for the casual observer to locate any of the individual stars except the *lucida*, three have been lettered — α, β, γ — that Baily claimed for Flamsteed's 7, 15, and 23. Of these Fl. 15, an orange star, is generally supposed to be the Arabian **Al Ḍafīrah,** from Ulug Beg's name for the whole that he located among the *informes* of Leo. Hyde cited some ancient codices as applying to Fl. 21, toward the south, the title **Kissīn,** a species of Ivy, Convolvulus, or perhaps the climbing Dog-rose. This appeared with Ulug Beg, evidently from Ptolemy's κίσσινος, but Ideler said that it was intended to mark *c*, *g*, and *h*, and Baily, that it was for Fl. 21 or 23.

There evidently is much uncertainty as to the lettering and numbering of Coma's stars; and it seems remarkable that such minute objects should bear individual names.

[1] Some of these letters may be from Flamsteed, as he applied *a*, *b*, *c*, *d*, *e*, *f*, *g*, and *h* to a small portion — the centre — of the constellation; but Baily, his editor, has rejected them as being only a temporary arrangement.

Near Fl. 6 is the **Pin-wheel Nebula**, N. G. C. 4254, 99 M., one of the pyrotechnics of the sky; while Fl. 31 closely marks the pole of the Milky Way, more exactly in right ascension 12° 40′ and north polar distance 28°; the southern pole lying in Cetus.

★

> . . . other few,
> Below the Archer under his forefeet,
> Led round in circle roll without a name.
>
> Brown's *Aratos*.

Corona Australis, the Southern Crown,

often qualified by other synonymous adjectives, *austrina*, *meridiana*, *meridionalis*, or *notia*, is an inconspicuous constellation, although accepted in Ptolemy's time as one of the ancient forty-eight. On modern maps its location is close to the waist of Sagittarius, on the edge of the Milky Way.

The Germans know it as the **Südliche Krone**; the French, as the **Couronne Australe**; and the Italians, as the **Corona Australe.**

Aratos did not mention it by name unless in his use of the plural Στεφάνοι for both of the Crowns; yet doubtless had it in mind when he wrote of the Δινωτοί Κύκλω in our motto. His scholiast and Geminos had 'Ουρανίσκος, the Canopy; Δευτέρος, the Second; and Δευτέρα Κύκλα, the Second Circle. Hipparchos is said to have known it as Κηρυκεῖον, the Caduceus, or Herald's Wand of Peace, but this is not found in his *Commentary*. Ptolemy called it Στεφάνος νοτίος, the Southern Wreath.

Germanicus rendered the supposed reference in the *Phainomena* as *Corona sine honore—i. e.* without any such noteworthy tradition as is connected with the Northern Crown; commenting upon which, Grotius said that this author, as well as Cicero and Avienus, understood Aratos to refer to the southern figure; and added that this was the **Centaur's Crown**, those personages frequently being represented as wearing such. This idea doubtless originated from the outspreading sun-rays, in crown-like form, around the heads of the Gandharvas, the Aryan celestial horses that probably were the forerunners of the Centaurs. It was thus appropriately associated with the centaur Sagittarius and took the title **Corona Sagittarii**.

Manilius did not allude to it; but others of the classical poets thought it the Crown that Bacchus placed in the sky in honor of his mother Semele; or one in commemoration of the fivefold victory of Corinna over Pindar in their poetical contest; and some considered it the early **Bunch of Arrows** radiating from the hand of the Archer, often imagined as a wheel. This idea was expressed in its titles Τροχός Ἰξίονος and **Rota Ixionis,** the Wheel of Ixion, perhaps from the latter's relationship to the centaur Pholos.

Albumasar called it **Coelum,** while **Coelulum** and **parvum Coelum,** the Little Sky, *i.e.* Canopy, are from the *Satyricon*,[1] the encyclopaedic writings of the Carthaginian Martianus Mineus Felix Capella of the 5th century, in the 8th book of which he treats of astronomy.

La Lande cited **Sertum australe,** the Southern Garland, and **Orbiculus Capitis**; Proctor, Brown, and Gore of the present day have **Corolla,** the Little Crown, but this was used 250 years ago by Caesius, who also gave **Spira australis**, the Southern Coil, and said that its stars represented the **Crown of Eternal Life** promised in the *New Testament.* Julius Schiller, however, went back a millennium before our era to the **Diadem of Solomon.**

Al Sufi is our authority for the Arabs' **Al Ḳubbah,** literally the Tortoise, but secondarily the Woman's Tent, or traveling apartment, from its form; and it was **Al Ḣibā'**, the Tent, and Kazwini's **Al Udḥā al Na'ām,** the Ostrich's Nest, for the same reason; the birds themselves being close by in what now are the Archer and the Eagle. **Al Fakkah,** the Dish, was borrowed from the Northern Crown, but among the later Arabians it was **Al Iklīl al Janūbiyyah,** their equivalent for our title; Chilmead giving this as **Alachil Algenubi**; Riccioli, **Elkleil Elgenubi**; and Caesius, **Aladil Algenubi.**

The Chinese knew it by the figure current in early Arabia—**Peē,** the Tortoise.

Bayer illustrated Corona as a typical wreath, but without the streaming ribbons of its northern namesake, and the original *Alfonsine Tables* show a plain heart-shaped object with no semblance to the name. Gould assigns to it forty-nine stars, many more than even Heis does to its much more celebrated and noticeable counterpart in the north. Its *lucida*, the 4th-magnitude α, at the eastern edge of the constellation, is **Alfecca meridiana** in the Latin translation of Reduan's *Commentary.* It culminates on the 13th of August.

[1] This was a popular text-book centuries ago, and noticeable even by us, as it contains a very clear statement of the heliocentric system, probably from Hicetas of Syracuse, 344 B.C.; and may have led Copernicus, who quoted him in 1543, to his own conclusions on the subject.

> Looke! how the crowne which Ariadne wore
> Upon her yvory forehead, . . .
> Being now placed in the firmament,
> Through the bright heavens doth her beams display,
> And is unto the starres an ornament,
> Which round about her move in order excellent.
>
> Spenser's *Faerie Queen.*

Corona Borealis, the Northern Crown,

is the French **Couronne Boréale,** the German **Nördliche Krone,** and the Italian ancestral **Corona.**

It was the only stellar crown known to Eratosthenes and the early Greeks, but they called it Στέφανος, a Wreath; and their successors, who had begun to locate the Southern Crown, added to this title of the original the distinguishing πρῶτος and βόρειος to show its priority and its northern position. The Latins adopted the Greek name and adjectives in **Corona borea, borealis,** and **septentrionalis**; and further knew it as the **Crown of Vulcan** fashioned *ex auro et indicis gemmis;* or **of Amphitrite,** probably from its proximity in the sky to the Dolphin associated with that goddess. But generally it was **Ariadnaea Corona, Corona Ariadnae, Corona Ariadnes, Cressa Corona, Corona Gnosida, Corona Cretica** and **Gnossis,** varied by **Minoia Corona** and **Minoia Virgo** found with Valerius Flaccus and Germanicus, and **Ariadnaea Sidus** with Ovid; these classical designations referring to Ariadne, or to her father Minos, king of Crete, and to her birthplace in that island, at Gnosos, where Theseus married her. When deserted by him she became the wife of Liber Bacchus, and so took his name Libera; while the crown that Theseus — or, as some said, the goddess Venus — had given her was transferred to the sky, where it became our Corona; and, as early as the 3d century B. C., Apollonius Rhodius wrote in his *Argonauticae:*

> Still her sign is seen in heaven,
> And midst the glittering symbols of the sky
> The starry crown of Ariadne glides.

Keats changed this in his *Lamia* to **Ariadne's tiar;** and others made it the **Coiled Hair of Ariadne** as companion to the Streaming Tresses of Berenice. Some authors, however,— Ovid among them in his *Fasti,*— said that Ariadne herself became the constellation; and Mrs. Browning, in her Paraphrases from Nonnus of *How Bacchus comforts Ariadne:*

> Or wilt thou choose
> A still surpassing glory? — take it all —
> A heavenly house, Kronion's self for kin.

This legend of Ariadne and her Crown seems to have been first recorded by Pherecydes early in the 5th century before Christ.

Dante, referring to Ariadne's descent, called these stars **la Figliuola di Minoi,** the poet giving much prominence to her father,[1] who "was so renowned for justice as to be called the Favorite of the Gods, and after death made Supreme Judge in the Infernal Regions."

In all ages Corona has been a favorite, popularly as well as in literature, and few of our stellar groups have had as many titles, although the English of the Middle Ages usually wrote its wearer's name "Adrian" and "Adriane."

Chaucer had this strange passage on the constellation:

> And in the sygne of Taurus men may se
> The stonys of hire coroune shyne clere;

but this seems unintelligible, unless from some confusion in the poet's mind with the location of Koronis of the Hyades. These, however, lie in the heavens just opposite the Crown, and Skeat ingeniously suggests that Chaucer may have meant that when the Sun was in Taurus the Crown was specially noticeable in the midnight sky, as is exactly the case.

"England's Arch Poet," Edmund Spenser, wrote in the *Shepheard's Kalendar*[2] of 1579:

> And now the Sunne hath reared up his fierie footed teme,
> Making his way between the Cuppe and golden Diademe;

one of the early titles of Corona being **Diadema Coeli.**

The **Wreath of Flowers,** occasionally seen for it, is merely the early signification of the words *Στέφανος* and Corona.

Oculus was another name of the constellation — a term common in poetry and post-Augustan prose for any celestial luminary; and Prudens[3] called it **Maera,** the Shining One.

As the *ardens corona* of the *Georgics*, Vergil included it with the Pleiades as a calendar sign, May translating the passage:

[1] Dante furnished him "with a tail (*colla coda*), thus converting him, after the mediaeval fashion, into a Christian demon." It was a long tail, too, for we read:

> Who bore me unto Minos, who entwined eight times his tail about his stubborn back.

[2] It may not be generally known that this was first published as the *Twelve Aeglogues, Proportionable to the Twelve monethes*.

[3] Aurelius Clemens Prudentius, the Latin Christian poet of our 4th century.

> But if thou plow to sowe more solid graine,
> A wheat or barley harvest to obtaine:
> First let the morning Pleiades be set,
> And Ariadne's shining Coronet,
> Ere thou commit thy seed to ground, and there
> Dare trust the hope of all the following yeare.

Columella, in a similar connection, called it **Gnosia Ardor Bacchi,** and **Naxius Ardor,** from Naxos, where Ariadne had been deserted by Theseus; and specially designated its *lucida* as *clara stella.*

Its stars were favored also by the astrologers, Manilius expressing this in:

> Births influenc'd then shall raise fine Beds of Flowers,
> And twine their creeping Jasmine round their Bowers;
> The Lillies, Violets in Banks dispose,
> The Purple Poppy, and the blushing Rose:
> For Pleasure shades their rising Mounts shall yield,
> And real Figures paint the gawdy Field:
> Or they shall wreath their Flowers, their Sweets entwine,
> To grace their Mistress, or to Crown their Wine.

Bayer said of it *Azophi* **Parma,** by which he meant that Al Sufi called it a Shield; but the majority of Arabian astronomers rendered the classical title by **Al Iklīl al Shamāliyyah,** which degenerated into **Acliluschemali** and **Aclushemali,** and appeared with Ulug Beg as plain **Iklīl.**

But in early Arabia there was a different figure here, **Al Fakkah,** the Dish, which Ulug Beg's translator gave as **Phecca,** and others as **Alphaca, Alfecca, Alfacca, Foca, Alfeta,** and **Alfelta**; while Riccioli said **Alphena** *Syrochaldaeis;* and Schickard, **Alphakhaco.**

Hyde quoted **Ḳaṣʽat al Sālik,** and **Ḳaṣʽat al Masākīn,** the Pauper's Bowl; and the Persians had the same in their **Kāsah Darwīshān,** the Dervish's Platter, or **Kāsah Shekesteh,** the Broken Platter, because the circle is incomplete. Bullialdus Latinized some of these titles in his **Discus parvus confractus,** evidently taken from Chrysococca's Πινάκιν κεκλασμένον, a Small Broken Dish, which, however, should read Πινάκιον.

The *Alfonsine Tables* have **Malfelcarre,** "of the Chaldaeans," Riccioli's **Malphelcane,** considered by Ideler a degenerate form of the Arabic **Al Munīr al Fakkah,** the Bright One of the Dish; though Buttmann derived it from **Al Malf al Khatar,** the Loop of the Wreath, or the Junction of the Crown; and Scaliger suggested **Al Malif al Kurra,** of somewhat similar meaning, more correctly written **Al Milaff al Kurrah.** Bayer said **Malphelcarre** *quod est sertum pupillae,* the Circle of the Pupil of the Eye; and, although he did not explain this, may have written better than he knew, for Pupilla is the Latin equivalent of Κόρη, which, as a proper name, was a title for Persephone. In La Lande's *Astronomie* Dupuis devoted much space to his identification of this goddess, the Latin Proserpina, with the Chaldaean Phersephon, taking the title from Phe'er, Crown, and Serphon, Northern. Thus, if Dupuis be correct, the origin of the figure, as well as of the name, may lie far back of Cretan days.

The Hebrews are said to have called it **'Aṭārōth,** the Crown,—perhaps of the Semitic queen Cushiopeia; and the Syrians, **Ashtaroth,** their Astarte, the 'Αφροδίτη of the Greeks and the Venus of the Latins; but all this seems doubtful, as also is Ewald's conjecture that it was the biblical Mazzārōth.

Blake quotes from Flammarion, **Vichaca,** but without explanation.

Reeves catalogued it as the Chinese **Kwan Soo,** a Cord.

In Celtic story Corona was **Caer Arianrod,** the House of Arianrod or Ethlenn, the sister of Gwydyon and daughter of Don, the Fairy King, this name bearing a singular resemblance to that of the classical owner of the Crown.

The Shawnee Indians knew it as the **Celestial Sisters,** the fairest of them being the wife of the hunter White Hawk, our Arcturus.

Caesius said that it represented the **Crown that Ahasuerus placed upon Esther's head,** or the **golden one of the Ammonite King** of a talent's weight, or the **Crown of Thorns** worn by the Christ.

The *Leyden Manuscript* shows it as a laurel wreath, and thus, or as a typical crown, it appears on the maps. In the *Firmamentum Firmianum,* a work of 1731, in honor of the persecuting bishop of Salzburg, of the Firmian family, the figuring is that of the **Corona Firmiana,** with a stag's antlers from the coat of arms of that family. But an exception to the rule may be noted in an illustration, in the original *Alfonsine Tables,* of a plain three-quarter circle, entirely unlike either crown or wreath. Proctor suggested that in the earliest astronomy it may have formed the right arm of Boötes.

It is interesting to the astronomer from its many close binaries, and is a favorite object with youthful observers, who generally know it as **Ariadne's Crown.** It certainly is much more like that for which it is named than usually is the case with our sky figures; and it is equally suggestive to the Australian native of the **Woomera,** our Boomerang, his idea of Corona's stars.

Here appeared very suddenly, 58′ south of ε, on the 12th of May, 1866, the celebrated **Blaze Star** as a 2d-magnitude visible to the naked eye for only eight days, declining, with some fluctuations, to the 10th magnitude at the rate of half a magnitude a day, but rising again to the 8th, where it still remains as T Coronae, a pale yellow, slightly variable star. Although called a *nova,* Argelander had already mapped it on the 18th of May, 1855, and again noted it on the 31st of March, 1856, probably at its normal magnitude. It was the first temporary star to be "studied by the universal chemical method"— the spectroscope.

Near its place the **Variabilis Coronae,** now lettered R, was discovered by Pigott in 1795, still varying from 5.8 to 13, but with much irregularity.

Professor Young repeats the *βαγδει* of Cassiopeia as a help to the memory in locating the stars of this constellation. The extreme northern one is *θ*, but then follow in order *β*, *α*, *γ*, *δ*, *ε*, *ι*. They form an almost perfect semicircle 20° northeast of Arcturus.

Argelander gives a total of 27 stars visible to the naked eye; and Heis, 31.

> One plac'd i' th' front above the rest displays
> A vigorous light, and darts surprizing rays —
> The Monument of the forsaken Maid.
>
> Creech's *Manilius.*

α, 2.4, brilliant white.

Alphecca, the **Alphaca** of Burritt's *Atlas* of 1835, was Ulug Beg's **Al Nā'ir al Fakkah,** the Bright One of the Dish, this Nā'ir being equivalent to the Latin word *lucida.*

Bayer asserted that the Arabs knew this star as **Pupilla,** which also appears in the nomenclature of the constellation, with a possible clue to its derivation; but as the word belongs to Lyra, and is certainly not Arabic, we may have to recur to first principles for its origin in the classical *Papilla.*

Munir, found with Bayer as of the "Babylonians,"— by whom he probably intended those gifted in astrology,— is from the Arabs, and synonymous with their Nā'ir. Chilmead gave this as **Munic.**

In Vergil's *Georgics* it was **Gnosia Stella Coronae.**

Gemma and **Gemma Coronae** were not used in classical times, but are later titles, perhaps from Ovid's *gemmasque novem* that Vulcan combined with his *auro* to make Ariadne's Crown; but Spence said, in his *Polymetis,* that the word should be taken in its original meaning of a Bud, referring to the unopened blossoms and leaves of the floral crown, thus agreeing with the early idea of the figure. The **Gema** occasionally seen unquestionably is from an early type omission.

Alphecca is the central one of the seven brightest members of the group, and in modern times has been **Margarita Coronae,** the Pearl of the Crown, occasionally transformed into **Saint Marguerite.** It marks the loop, or knot, of the ribbon along which are fastened the buds, flowers, or leaves of the wreath shown in early drawings with two long out-streaming ends.

The spectrum is of Secchi's Solar type; and the star is receding from our system at the rate of about twenty miles a second. It has a distant 8th-magnitude companion, and culminates on the 28th of June.

It marks the radiant point of the **Coronids,** the meteor shower visible from the 12th of April to the 30th of June.

β, a 4th-magnitude northwest from Alphecca, is **Nusakan** in the 2d edition of the *Palermo Catalogue*, derived from the Masākīn of the constellation.

γ, η, and σ, although unnamed, are all interesting binary stars.

★

> Till, rising on my wings, I was preferr'd
> To be the chaste Minerva's virgin bird.
>
> Joseph Addison's translation of Ovid's *Metamorphoses*.

Corvus

was the **Raven** in Chaucer's time, and the Germans still have **Rabe**; but the French follow the Latins in **Corbeau,** as the Italians do in **Corvo,** and we in the **Crow.**

Although now traversed by the 20th degree of south declination, 2000 years ago it lay equally on each side of the celestial equator. It contains only 15 naked-eye stars according to Argelander,—26 according to Heis,—yet was a noted constellation with the Greeks and Romans, and always more or less associated with the Cup and with the Hydra, on whose body it rests. Ovid said of this combination in his *Fasti:*

> Continuata loco tria sidera, Corvus et Anguis,
> Et medius Crater inter utrumque jacet;

but while always so drawn, the three constellations for a long time have been catalogued separately.

The Greeks called it Κόραξ, Raven; and the Romans, Corvus. Manilius designating it as **Phoebo Sacer Ales,** and Ovid as **Phoebeïus Ales,** mythology having made the bird sacred to Phoebus Apollo in connection with his prophetic functions, and because he assumed its shape during the conflict of the gods with the giants.

Ovid, narrating in the *Metamorphoses* the story of Coronis, and of her unfaithfulness to Apollo,[1] said that when the bird reported to his master this unwelcome news he was changed from his former silver hue to the present black, as Saxe concludes the story:

> Then he turned upon the Raven,
> "Wanton babbler! see thy fate!
> Messenger of mine no longer,
> Go to Hades with thy prate!
>
> "Weary Pluto with thy tattle!
> Hither, monster, come not back;
> And — to match thy disposition —
> Henceforth be thy plumage black!"

This story gave rise to the stellar title **Garrulus Proditor.**

Another version of the legend appears in the *Fasti* — viz., that the bird, being sent with a cup for water, loitered at a fig-tree till the fruit became ripe, and then returned to the god with a water-snake in his claws and a lie in his mouth, alleging the snake to have been the cause of his delay. In punishment he was forever fixed in the sky with the Cup and the Snake; and, we may infer, doomed to everlasting thirst by the guardianship of the Hydra over the Cup and its contents. From all this came other poetical names for our Corvus — **Avis Ficarius,** the Fig Bird; and **Emansor,** one who stays beyond his time; and a belief, in early folk-lore, that this alone among birds did not carry water to its young.

Florus called it **Avis Satyra,** the Bird of the Satyrs, and **Pomptina,** from the victory of Valerius when aided by a raven on the Pontine Marsh.

This bird and an ass appear together on a coin of Mindaon, which is interpreted as a reference to the almost simultaneous setting of the constellations Corvus and Cancer, for the ass always has been associated with the latter in the Ὄνοι, or Asini, of its stars.

The Raven of Rome and Greece became **Al Ghurāb** in Arabia; but in earlier days four of its stars were **Al ʽArsh al Simāk al ʽAzal,** the Throne of the Unarmed One, referring to the star Spica. These naturally have been considered β, γ, δ, and η; but Firuzabadi, as interpreted by Lach, said that they were θ, κ, ψ, and g; and the same stars were **Al ʽAjz al Asad,** the Rump of the ancient Lion. Other early titles for the whole were **Al Ajmāl,**

[1] It may be noted here that Apollo and Coronis were even still more closely connected with astronomy in being the parents of Aesculapius, who afterwards became the Serpent-holder Ophiuchus.

the Camel, and **Al Ḣibā'**, the Tent; this last generally qualified by **Yamaniyyah,** the Southern, to distinguish it from that in Auriga. Instead of Ajmāl, Hyde quoted, from the *Mudjïzat*, **Ahmal,** or **Ḥamal,** the Ram, but this does not seem probable here.

As these stars were utilized by the Arabs in forming their exaggerated Asad, so also were they by the Hindus in the immense **Praja-pāti,** of which they marked the hand,—this title being duplicated for Orion, and much better known for that constellation. The head of the figure was marked by Citrā, our Spica, and the thighs by the two Viçākhas, α and β Librae; while the Anuradhas, β, δ, and π Scorpii, formed Praja-pāti's standing-place. Incongruously enough, they considered Nishtya, or Svati,—our star Arcturus,—as the heart; but as this was far out of the proper place for that organ, Professor Whitney substituted ι, κ, and λ Virginis of the *manzil* and *sieu*.

The *Avesta* mentions a stellar Raven, **Eorosch**; but how, if at all, this coincided with ours is unknown; although Hewitt thinks that our Corvus, under the title **Vanant,** marked the western quarter of the earliest Persian heavens.

Nor is the reason for the association of Corvus with Hydra evident, although there is a Euphratean myth, from far back of classical days, making it one of the monster ravens of the brood of Tiāmat that Hydra represented; and upon a tablet appears a title that may be for Corvus as the **Great Storm Bird,** or **Bird of the Desert,** to which Tiāmat gave sustenance, just as Aratos described Κόραξ pecking the folds of the Hydra. The prominent stars of Corvus have otherwise been identified with the Akkadian **Kurra,** the Horse.

The Hebrews knew it as **'Ōrebh,** or **Ōrev,** the Raven; and the Chinese, as a portion of their great stellar division the **Red Bird,** while its individual stars were an **Imperial Chariot** ruling, or riding upon, the wind.

In later days it has been likened to **Noah's Raven** flying over the Deluge, or alighting on Hydra, as there was no dry land for a resting-place; or one of those that fed the prophet Elijah; but Julius Schiller combined its stars with those of Crater in his **Ark of the Covenant.**

α, 4.3, orange.

Al Chiba is from the Desert title for the whole Arabic figure; but Ulug Beg and the Arabian astronomers designated it as **Al Minḣar al Ghurāb,** the Raven's Beak.

Reeves said that it was the Chinese **Yew Hea,** the Right-hand Linch-pin.

Although lettered first, it now is so much less brilliant than the four following stars that some consider it as having decreased since Bayer's day, and perhaps changed in color, for Al Sufi called it red.

β, a ruddy yellow 3d-magnitude star, seems unnamed except in China, where it is **Tso Hea,** the Left-hand Linch-pin; but under this title were included γ, δ, and η.

γ, 2.3.

Gienah is from Ulug Beg's **Al Janāḥ al Ghurāb al Aiman,** the Right Wing of the Raven, although on modern charts it marks the left. **Algorab,** given in the *Alfonsine Tables* to this star, is now usually applied to δ.

γ is the brightest member of the constellation, and some Chinese authorities said that it alone marked their 11th *sieu.* It culminates on the 10th of May.

δ, Double, 3.1 and 8.5, pale yellow and purple.

Algorab, the generally received modern title, is from the *Palermo Catalogue;* Proctor has **Algores.** It is on the right wing, and at the upper left corner of the square. The components are 24″ apart; but, owing to its color, the smaller is not readily distinguishable. The position angle is 210°.

All the foregoing stars, ε being added, constituted the 11th *nakshatra,* **Hasta,** the Hand, with Savitar, the Sun, as its presiding divinity; δ marking the junction with **Citrā,** the next lunar station.

The 11th *sieu,* **Tchin,** the Cross-piece of a chariot, anciently **Kusam,** contained β, γ, δ, and ε; but, according to some authorities, only γ. This, however, always was the determining star.

ζ, a 6th-magnitude double, almost on the limit of invisibility, strangely seems to have borne a name in China,— **Chang Sha,** a Long Sand-bank.

Al Bīrūnī said that with β, γ, and δ it marked the hind quarters of the monstrous early Lion.

★

. . . the generous Bowl
Of Bacchus flows, and chears the thirsty Pole.
Creech's *Manilius.*

Crater, the Cup,

is the French **Coupe,** the German **Becher,** and the Italian **Tazza,** formed by several 4th- and 5th-magnitude stars above the Hydra's back, just westward from Corvus, and 30° south of Denebola, in a partly annular form opening to the northwest. This was long considered a part of the threefold constellation **Hydra et Corvus et Crater**; but modern astronomers catalogue it separately, Argelander assigning to it 14 stars, and Heis extending the number to 35.

In early Greek days it represented the Κάνθαρος, or Goblet, of Apollo, but universally was called Κρατήρ, which in our transliterated title obtained with all Latins, Cicero writing it **Cratera**; while Manilius described it as **gratus Iaccho Crater,** so using the mystic, poetical name often applied to Bacchus. In ancient manuscripts it appears as **Creter.** The Greeks also called it Κάλπη, a Cinerary Urn; 'Αργεῖον, 'Υδρεῖον, and 'Υδρία, a Water-bucket.

The Romans additionally knew it as **Urna, Calix,** or **Scyphus,** and, poetically, as **Poculum,** the Cup, variously, of Apollo, Bacchus, Hercules, Achilles, Dido, Demophoön, and Medea; its association with this last bringing it into the long list of Argonautic constellations.

Hewitt connected it with the **Soma-cup** of prehistoric India; and Brown with the **Mixing-bowl** in the Euphratean myth of Istar-Kirke, referring to the words of the prophet Jeremiah:

> Babylon hath been a golden cup in the Lord's hand.

But any connection here would seem doubtful, although the Jews knew it as **Cōs,** a Cup. Hewitt also identifies it with "the Akkadians' **Mummu Tiāmut,** the chaos of the sea, the mother of heaven and earth, and the child of Tiāmut, the mother (*mut*) of living things (*tia*)"; but all this better suits Corvus.

It was known in England two or three centuries ago as the **Two-handed Pot**; and Smyth tells us of a small ancient vase in the Warwick collection bearing an inscription thus translated:

> Wise ancients knew when Crater rose to sight,
> Nile's fertile deluge had attained its height;

although Egyptian remains thus far show no allusion to the constellation.

In early Arabia it was **Al Ma'laf,** the Stall,—a later title there for the Praesaepe of Cancer; but when the astronomy of the Desert came under Greek influence it was **Al Bāṭiyah,** the Persian **Badiye,** and the **Al Batinah** of Al Achsasi, all signifying an earthen vessel for storing wine. Another title, **Al Kās,** a Shallow Basin,—**Alhas** in the Alfonsine lists,—has since been turned into **Alker** and **Elkis**; but Scaliger's suggestion of **Alkes** generally has been adopted, although now applied to the star α. These same *Tables* Latinized it as **Patera,** and as **Vas,** or **Vas aquarium.**

Riccioli's strange **Elvarad** and **Pharmaz** I cannot trace to their origin.

Its more conspicuous stars, with χ and others in Hydra, twenty-two in all, formed the 10th *sieu*, **Yh, Yih,** or **Yen,** Wings or Flanks; and the whole constellation may have been the Chinese **Heavenly Dog** shot at by **Chang,** the divinity of the 9th *sieu* in Leo, which also bore that god's name.

Caesius said that Crater represented the **Cup of Joseph** found in Benjamin's sack, or **one of the stone Water-pots of Cana,** or the **Cup of Christ's Passion**; others called it the **Wine-cup of Noah,** but Julius Schiller combined some of its stars with a part of Corvus as the **Ark of the Covenant.**

Astrologically it portended eminence to those born under its influence.

α, 4.1, orange.

Alkes is our title from Scaliger, but it also has been **Alker,** and in the *Alfonsine Tables* **Alhes**: all from Al Kās of the constellation.

The Latin designation for it — **Fundus vasis** — well describes its position at the base of the Cup.

Since it is the only named star in the figure, and the first lettered, it may have been brighter 300 years ago; but δ, a 3.9-magnitude, is now the *lucida*.

α has several optical companions, and culminates on the 20th of April, about 32° nearly due south from β Leonis.

β, of 4.4 magnitude, at the southern edge of the base, was one of Al Tizini's **Al Sharāsīf,** the Ribs,— *i. e.* of the Hydra,— and the first of the set.

★

The four that glorify the night!
Ah! how forget when to my ravish'd sight
The Cross shone forth in everlasting light!

Samuel Rogers' *The Voyage of Columbus.*[1]

Crux, the Cross,

is the German **Kreuz,** the Italian **Croce,** the French **Croix** and, in the 1776 edition of Flamsteed's *Atlas*, **Croisade.** With us it is the **Southern Cross.**

It was unknown to the ancients by its present title, its four chief stars being noted by Ptolemy as a part of the Centaur, which now surrounds it on three sides. As such Bayer outlined it over the hind feet, lettering it ϵ, ζ, ν,

[1] In this poem Rogers makes the great discoverer bring the telescope into use a century before its invention!

and ξ Centauri; but these now are α, β, γ, and δ Crucis,—the 1.3-magnitude *lucida* at the foot, the 2d-magnitude γ at the top, with β and δ, the early ξ and ν, as the transverse: these last, respectively, of 1.7 and 3.4 magnitudes. A fifth star, ϵ, of the 4th magnitude, between α and δ, somewhat interferes with the regularity of the figure; and there are forty-nine others visible to the naked eye within the constellation boundaries.

The statement that it was mentioned by Hipparchos probably is erroneous, although he distinctly alluded to its β as of the Centaur; but Pliny may have known it as **Thronos Caesaris** in honor of the emperor Augustus; yet it was then invisible from Italy, though plainly visible from Alexandria, where it may have been thus named by some courtly astronomer. And Al Bīrūnī wrote that a star could be seen from Multan in India, in 30° 12′ of north latitude, "which they call **Sūla**," the Beam of Crucifixion. This, if a reference to the Cross, is a striking anticipation of the modern figure. Hewitt, repeating this title as **Ṣhūla,** claimed it for the south pole of Hindu astronomers.

Whittier said, in his *Cry of a Lost Soul:*

> The Cross of pardon lights the tropic skies;

which is correct for our day, as it is not now entirely visible above 27° 30′ of north latitude. It was last seen on the horizon of Jerusalem —31° 46′ 45″ —about the time that Christ was crucified. But 3000 years previously all its stars were 7° above the horizon of the savages along the shores of the Baltic Sea, in latitude 52° 30′.

Its invention as a constellation is often attributed to Royer as of 1679, but it had been the theme of much description for nearly two centuries before him, and we know that it was illustrated by Mollineux of England, in 1592, on his celestial globe, with others of the new southern figures; and Bayer drew it over the hind legs of the Centaur, giving it in his text as *modernis crux, Ptolemaeo pedes Centauri.* Bartschius had it separately in 1624, and Caesius catalogued it in 1662 as though well known; hence it seems remarkable that it was only outlined over the Centaur in the Flamsteed *Atlas.*

Crux lies in the Milky Way,—here a brilliant but narrow stream three or four degrees wide,—and is noticeable from its compression as well as its form, being only 6° in extent from north to south, and less in width, the upper star a clear orange in color, and the rest white; the general effect being that of a badly made kite rather than of a cross. So that, notwithstanding all the poetry and romance associated with it,—perhaps owing to these,—it usually disappoints those from northern latitudes who see it for the first time.

For twelve centuries, from Pliny to Dante, we find no allusion to its stars till that great poet, turning from his contemplation, in the *Purgatorio*, of Venus "veiling the Fishes,"

> posi mente
> Al altro polo e vidi quatro stelle
> Non viste mai fuor che alla prima gente,

in which Baron Alexander von Humboldt, in his *Examen Criticum*, insists that he refers to the Cross; while Longfellow, translating the passage

> and fixed my mind
> Upon the other pole and saw four stars
> Ne'er seen before save by the primal people,

calls it an acknowledged reference to the same, figuring, as it were, the cardinal virtues, Justice, Prudence, Fortitude, and Temperance, attributes of Cato as the Guardian of Purgatory, claiming that

> We here are Nymphs and in the Heaven are Stars.

Later on in the same canto we read again of Cato:

> The rays of the four consecrated stars
> Did so adorn his countenance with light.

But this reference to the "primal people" is not, Barlow says in his *Study of Dante*, to our first parents, as Cary's translation has it, but to the early races of mankind, who 5000 years ago could see the Cross from latitudes very much higher even than that of Italy. In the same passage Dante alludes to its local invisibility in his apostrophe to the northern heavens:

> O! thou septentrional and widowed site
> Because thou art deprived of seeing these!

and in the 8th canto calls them *Le quatro chiare stelle.*

Whence Dante learned all this we do not know, for it was not till 200 years later that we have any published account of the constellation; but that he paid great attention to the heavens is evident from his frequent and intelligent allusions to them throughout the *Divine Comedy*. He was, too, a man of erudition as well as of imagination and poetical genius,— Carlyle called him the spokesman of ten silent centuries,— and may have seen some of the Arabic celestial globes, on at least one of which — probably the Borgian of 1225 — we know that the stars of the Centaur were represented; and he doubtless had frequent opportunities of intercourse with learned travelers,[1] or some of the many returned voyagers among his own adventurous countrymen, worthy successors to their ancient neighbors the Phoenicians. This should be sufficient to account for these allusions without attributing them to prophetic inspiration. And here, although in no way connected with the Cross, I would call attention to a fact pleasing to star-lovers — viz., "the beautiful and endless aspiration, so artistically and silently suggested by Dante, in closing each part of his poem with the word *stelle*."

The *Inferno* ends with:

> Thence we came forth to rebehold the stars;

the *Purgatorio*:

> Pure and disposed to mount unto the stars;

and the *Paradiso*:

> The love which moves the sun and the other stars.

Note, too, the poet's perhaps unconscious advance in astronomical knowledge beyond his contemporaries in associating the sun with the stars.

Vespucci, on his third voyage in 1501, called to mind the passages from Dante, insisting that he himself was the first of Europeans to see the Four Stars, but did not use the title of the Cross, and called them **Mandorla.**[2] Vasco da Gama said of it in the *Lusiadas*:

> A group quite new in the new hemisphere,
> Not seen by others yet;

while nearly four centuries after him, in our day, Lord Lytton (Owen Meredith) has something similar in his *Queen Guenevere*:

> Then did I feel as one who, much perplext,
> Led by strange legends and the light of stars
> Over long regions of the midnight sand
> Beyond the red tract of the Pyramids,
> Is suddenly drawn to look upon the sky,
> From sense of unfamiliar light, and sees,
> Reveal'd against the constellated cope,
> The great cross of the South.

Writers of the 16th century made frequent mention of it in their accounts of southern navigation; Corsali saying in 1517, as translated by Eden:

[1] Marco Polo was his contemporary.

[2] This, literally "an Almond," is the word used in Italian art for the *vescica piscis*, the oblong glory, surrounding the bodies of saints ascending to heaven.

Above these [the Magellanic Clouds] appeareth a marveylous crosse in the myddest of fyve notable starres which compasse it abowt (as doth Charles Wayne the northe pole) with other starres whiche move with them abowt .xxx. degrees distant from the pole, and make their course in .xxiiii. houres. This crosse is so fayre and beutiful, that none other hevenly sygne may be compared to it as may appear by this fygure.[1]

A. The pole Antartike. B. The Crosse.

Subsequently, in 1520, Pigafetta, the companion of Magellan, mentioned it as **El Crucero**, and *una croce maravigliosa* used for the determination of altitudes, saying that Dante first described it; Pedro Sarmiento de Gamboa called it **the Star Crucero** and **the Stars of Crucero**; Blundevill, in 1574,

[1] I use this "fygure" not for its artistic excellence, but as illustrating the early ignorance of locations and magnitudes of southern stars. The Clouds here especially are misplaced with respect to the pole.

Crosier and, very differently, the **South Triangle**, but this was twenty-nine years before Bayer gave this title to other stars. Eden also cited the **Crossiers** and **Crosse Stars**; Chilmead, **Crusero** and **Crusiers**; Sir John Narborough, **Crosers**; and Halley, in 1679, **Crosiers**.

A century before Halley, the Portuguese naturalist Cristoval d'Acosta, writing the title **Cruzero**,— the old Spanish **Cruciero**,— termed the Cross the **Southern Celestial Clock**; and as such it has served a useful purpose for nearly 400 years. Von Humboldt, in his *Voyage to the Equinoctial Regions of the New Continent*, alluding to the Portuguese and Spaniards, wrote:

> A religious sentiment attaches them to a constellation the form of which recalls the sign of the faith planted by their ancestors in the deserts of the New World; —

a thought which Mrs. Hemans beautifully expressed in her *Cross of the South* where the Spanish traveler says:

> But to thee, as thy lode-stars resplendently burn
> In their clear depths of blue, with devotion I turn,
> Bright Cross of the South! and beholding thee shine,
> Scarce regret the loved land of the olive and vine.
> Thou recallest the ages when first o'er the main
> My fathers unfolded the ensign of Spain,
> And planted their faith in the regions that see
> Its imperishing symbol ever blazoned in thee.

Von Humboldt adds:

> The two great stars, which mark the summit and the foot of the Cross, having nearly the same right ascension, it follows that the constellation is almost perpendicular at the moment when it passes the meridian. This circumstance is known to the people of every nation situated beyond the Tropics or in the southern hemisphere.
>
> It has been observed at what hour of the night, in different seasons, the Cross is erect or inclined.
>
> It is a time piece, which advances very regularly nearly four minutes a day, and no other group of stars affords to the naked eye an observation of time so easily made.
>
> How often have we heard our guides exclaim in the savannahs of Venezuela and in the desert extending from Lima to Truxillo, "Midnight is past, the Cross begins to bend." How often these words reminded us of that affecting scene when Paul and Virginia, seated near the source of the river of Lataniers, conversed together for the last time, and when the old man, at the sight of the Cross, warns them that it is time to separate, saying, "*la Croix du Sud est droite sur l'horizon.*"

Von Humboldt thought it remarkable that these so striking and well-defined stars should not have been earlier separated from the large ancient constellation of the Centaur, especially since Kazwini and other Muhammadan astronomers took pains to discover crosses elsewhere in the sky; and he said that the ancient Persians, who knew the Cross well, celebrated a feast by its name, their descendants, to whom it was lost by precession, finding its successor in the Dolphin.

The Pareni Indians of his day made much of the stars of the Cross, calling them **Bahumehi**, after one of their principal fishes.

Lockyer alludes to it as the **Pole-star of the South**, which it may be when on the meridian, as the most prominent constellation in the vicinity of the pole, although its base star is nearly 28° from that point, about four and one half times the length of the Cross. But this idea is an old one; Minsheu's *Guide* having, at the word "Cruzero," *Quatuor stella poli*, *Foure starres crossing;* and Sarmiento, even earlier, had much the same, but asserted that, "with God's help," he was enabled to select another pole-star nearer the true point.

In modern China it has been **Shih Tsze Kea**, the equivalent of our word.

The five stars are shown on postage stamps of Brazil,—Camões' Realms of the Holy Cross,—surrounded by twenty-one stars symbolizing the twenty-one states, and some of the coins bear the same. But this name for that country was not new with the poet, for it was given by the discoverer Cabral, on the 1st of May, 1500; and the fine *Ptolemaeus* printed at Rome in 1508, with the first engraved map of the new continent, carries as its title for South America, *Terra sancte crucis*.

Partly within the constellation's boundaries, and at the point of the nearest approach of the Milky Way to the south pole, is the pear-shaped **Coal-sack, or Soot-bag,** 8° in length by 5° in breadth, containing only one star visible to the naked eye, and that very small, although it has many that are telescopic, and a photograph taken at Sydney in 1890 shows about as many in proportion as in the surrounding region. This singular vacancy was first formally described by Peter Martyr, although observed in 1499 by Vicente Yañez Pinzon, and designated by Vespucci as **il Canopo fosco,** and perhaps alluded to by Camões. Narborough wrote of it in 1671 as "a small black cloud which the foot of the Cross is in"; but before him it was **Macula Magellani,** Magellan's Spot, and fifty years ago Smyth mentioned it as the **Black Magellanic Cloud.** Froude described it in his *Oceana* as "the inky spot—an opening into the awful solitude of unoccupied space." A native Australian legend, which "reads almost like a Christian parable," says that it was "the embodiment of evil in the shape of an Emu, who lies in wait at the foot of a tree, represented by the stars of the Cross, for an opossum driven by his persecutions to take refuge among its branches."

The Peruvians imagined it a heavenly **Doe** suckling its fawn.

Although this is the most remarkable of those "curious vacancies through which we seem to gaze out into an uninterrupted infinity," there are many other such in the heavens; an extended list of forty-nine being given by Sir John Herschel in his *Observations at the Cape of Good Hope*, and an abbreviated one by Espin in *Webb's Celestial Objects*.

α, Triple, 1, 2, and 6.

Acrux, in Burritt's *Atlas*, probably is a word of his own coining from α Crucis. Al Tizini defined its position as near the ankle-bone of the right hind foot of the Centaur, in which Bayer's plate agrees, lettering it ζ.

It was discovered to be double by some Jesuit missionaries sent by King Louis XIV to Siam in 1685; and another companion, of the 6th magnitude, is 60″ away. The two larger stars are 5″ apart, with a position angle of 120°.

α lies 2° east of the equinoctial colure, and, at its culmination, touches the horizon in latitude 27° 30′ on the 13th of May, due south from Corvus.

γ, the uppermost star, is on the horizon of the Lowe Observatory, at an elevation of 3700 feet, in latitude 34° 20′. Gould thinks it variable, for it has been variously estimated, even by the same observer, as from 1.8 to 2.4.

Around the 6½-magnitude κ is the celebrated cluster of colored stars, N. G. C. 4755, occupying one forty-eighth of a square degree of space; the central and principal one being of a deep red, surrounded by about 130 others, green, blue, and of various shades; but Miss Clerke writes:

It must be confessed that, with moderate telescopic apertures, it fails to realize the effect of colour implied by Sir John Herschel's [its discoverer] comparison to "a gorgeous piece of fancy jewellery." A few reddish stars catch the eye at once; but the blues, greens and yellows belonging to their companions are pale tints, more than half drowned in white light.

Gould, however, called it exquisitely beautiful.

★

Custos Messium, the Harvest-keeper,

is the German **Erndtehüter**, and the Italian **Mietitore.** La Lande published this on his globe of 1775, forming it from some inconspicuous stars not far from the pole, between the Camelopard, Cassiopeia, and Cepheus.

His alternative title, **Le Messier,** Smyth said was "in poorish punning compliment to his friend, the 'Comet ferret,'" as King Louis XV had called him, who for thirty years had been the gatherer and keeper of the harvest of comets, and the discoverer of twelve between the years of 1794 and 1798. This title also may have been induced by the fact that the two neighboring royal personages were rulers of an agricultural people, and the Giraffe an animal destructive to the grain-fields; all perhaps selected because the Phoenicians are said to have imagined a large **Wheat Field** in this part of the sky.

Its inventor was the enthusiastic astronomer who would spend nights on the Pont Neuf over the Seine, explaining the wonders of the variable Algol to all whom he could interest in the subject, and whose seclusion in his observatory, amid the turmoil of the French Revolution, enabled him to "thank his stars" that he had escaped the fate of so many of his friends.

Custos has now passed out of the recognition of astronomers.

★

Those deathless odalisques of heaven's hareem,
The Stars, unveil; a lonely cloud is roll'd
Past by the wind, as bears an azure stream
A sleeping swan's white plumage fringed with gold.

Adam Mickiewicz' *Polish Evening Hymn.*

Cygnus, the Swan,

that modern criticism says should be **Cycnus**, lies between Draco and Pegasus. The French know it as **Cygne**; the Italians as **Cigno**; the Spaniards as **Cisne**; and the Germans as **Schwan.**

It was Κύκνος with Eratosthenes, but usually Ὄρνις with other Greeks, by which was simply intended a **Bird** of some kind, more particularly a **Hen**; although the αιόλος of Aratos may indicate that he had in view the "quickly flying swan"; but, as this Greek adjective also signifies "varied," it is possible that reference was here made to the Bird's position in the Milky Way, in the light and shade of that great circle. With this idea, Brown renders it "spangled." Aratos also described it as ἠρόεις, "dark," especially as to its wings, an error which Hipparchos corrected.

When the Romans adopted the title that we now have, our constellation became the mythical swan identified with Cycnus, the son of Mars, or of the Ligurian Sthenelus; or the brother of Phaëthon, transformed at the river

Padus and transported to the sky.[1] Associated, too, with Leda, the friend of Jupiter and mother of Castor, Pollux, and Helena, it was classed among the Argonautic constellations, and **Helenae Genitor,** with other names derived from the well-known legend, was applied to it.

Popularly the constellation was **Ales, Avis,** and **Volucris,** a Bird,— **Ales Jovis, Ales Ledaeus,** and **Avis Veneris,**— while **Olor,** another word for the Swan, both ornithological and stellar, has been current even to modern times. **Phoebi Assessor** is cited by La Lande, the bird being sacred to that deity; and **Vultur cadens** is found for it, but this was properly Lyra's title. As the bird of Venus it also has been known as **Myrtilus,** from the myrtle sacred to that goddess; and it was considered to be **Orpheus,** placed after death in the heavens, near to his favorite Lyre.

Our Cygnus may have originated on the Euphrates, for the tablets show a stellar bird of some kind, perhaps **Urakhga,** the original of the Arabs' Rukh, the Roc, that Sindbad the Sailor knew. At all events, its present figuring did not originate with the Greeks, for the history of the constellation had been entirely lost to them, as had that of the mysterious Engonasin, — an evident proof that they were not the inventors of at least some of the star-groups attributed to them.

In Arabia, although occasionally known as **Al Ṭā'ir al Ardūf,** the Flying Eagle, Chilmead's **Altayr,** or as **Al Radīf,** it usually was **Al Dajājah,** the Hen, and appears as such even with the Egyptian priest Manetho, about 300 B. C., this degenerating into the **Adige, Adigege, Aldigaga, Addigagato, Degige, Edegiagith, Eldigiagich,** etc., of early lists, some of these even now applied to its brightest star.

Scaliger's **Al Ridhadh,** for the constellation, which degenerated to **El Rided,** perhaps is the origin of our Arided for the *lucida*, but its signification is uncertain, although the word is said to have been found in an old Latin-Spanish-Arabic dictionary for some sweet-scented flower.

Hyde gives **Kathā** for it, the Arabic **Al Kaṭāt,** a bird in form and size like a pigeon; indeed, Al Ṣufi's translator, Schjellerup, defined the latter's title for it, **Al Ṭā'ir,** as *le pigeon de poste;* but Al Kaṭāt is now the Arabs' word for a common gallinaceous game-bird of the desert, perhaps the mottled partridge.

The *Alfonsine Tables*, in the recent Madrid edition, supposed to be a reproduction of the original, illustrate their **Galina** by a forlorn Hen instead

[1] While Cygnus was thus prominent in myth and the sky, the swan was especially so in ancient ornithology, and the subject of many fables, where its "hostility" to other birds and to beasts was made much of; but in these Thompson sees astronomical symbolism, as already has been alluded to under Aquila.

of a Swan, with the bungled Arabic title **altayr aldigeya,** although elsewhere they say *Olor: Hyparcus Cygnum vocat;* the *Arabo-Latin Almagest* of 1515 had **Eurisim:** *et est volans: et jam vocatur gallina. et dicitur eurisim quasi redolens ut lilium ab ireo;* the *Alfonsine Tables* of 1521 have **Hyresym;** *et dicitur quasi redolens ut lilium: et est volans: et jam vocatur gallina;* Bayer wrote of it, *quasi Rosa redolens Lilium;* Riccioli, *quasi Galli rosa;* and contemporaries of this last author wrote **Hirezym** and **Hierizim.** Ideler's comments on all this well show the roundabout process by which some of our star-names have originated, and are worthy quotation entire:

They have, moreover, made use of the translated Greek Ὄρνις, as is shown by the Borgian Globe, on which is written **Lūrnis,** or **Urnis** (for the first letter is not connected with the second, so that we have both readings). It is most probable that from this *Urnis* originated the *Eurisim* in the foregoing rare title. Probably the translator found in the Arabic original the, to him, foreign word *Urnis.* He naturally surmised that it was Greek, only he did not know its proper signification. On the other hand, the plant Ἐρύσιμον (*Erysimum officinale,* Linn.) occurred to him, which the Romans called Ireo (see Pliny, *Hist. Nat.* xviii, 10, xxii, 25), and this recalled the richly scented Iris or Sword Lily (*Iris florentina,* Linn.), and so, as it seems to me, he traced the thought through a perfectly natural association of ideas to his beautiful *Eurisim, quasi redolens, ut lilium ab ireo.* At the same time I believe I have here struck the trail of the title Albireo, which has never yet been satisfactorily explained. This is given to the star on the beak,—β,— by Bayer and in our charts. It seems to me to be nothing more than the above *ab ireo,* which came to be turned into an Arabic star-name by means of an interpolated *l.*

The early **Gallina** continued in use by astronomers even to the last century.

Cygnus usually is shown in full flight down the Milky Way, the Stream of Heaven, "uppoised on gleaming wings"; but old drawings have it apparently just springing from the ground.

Caesius thought that the constellation represented the **Swan** in the Authorized Version of *Leviticus* xi, 18, the **Timshēmath** of the Hebrews; but this is a Horned Owl in the Revision, or may have been an Ibis. Other Christians of his time saw here the **Cross of Calvary, Christi Crux,** as Schickard had it, Schiller's **Crux cum S. Helena;** these descending to our day as the **Northern Cross,** well known to all, and to beginners in stellar observations probably better than by the stars' true title. Lowell was familiar with it, and thus brings it into his *New Year's Eve, 1844:*

Orion kneeling in his starry niche,
The Lyre whose strings give music audible
To holy ears, and countless splendors more,
Crowned by the blazing Cross high-hung o'er all;

and Smith, in *Come Learn of the Stars:*

> Yonder goes Cygnus, the Swan, flying southward,—
> Sign of the Cross and of Christ unto me.

This Cross is formed by α, γ, η, and β, marking the upright along the Galaxy, more than 20° in length, ζ, ϵ, γ, and δ being the transverse.

These last also were an Arab asterism, **Al Fawāris,** the Riders; α and κ sometimes being added to the group.

The Chinese story of the Herdsman, or Shepherd, generally told for our Aquila, and of his love for the skilful Spinster, our Lyra, occasionally includes stars in Cygnus.

While interesting in many respects, it is especially so in possessing an unusual number of deeply colored stars, Birmingham writing of this:

> A space of the heavens including the Milky Way, between Aquila, Lyra, and Cygnus, seems so peculiarly favored by red and orange stars that it might not inaptly be called the Red Region, or the Red Region of Cygnus.

Argelander locates 146 naked-eye members of the constellation, and Heis 197, its situation in the Galaxy accounting for this density. Of these stars Espin gives a list of one hundred that are double, triple, or multiple. The **Lace-work Nebula,** N. G. C. 6960, also lies within its borders.

We find among classical authors Ἰκτῖνος, Miluus, Milvus, and Mylvius, taken from the *Parapegmata*, and, even to modern days, supposed to be titles for our Cygnus, Aquila, or some unidentified sky figure; but Ideler showed that by these words reference probably was made to the Kite, the predaceous bird of passage annually appearing in spring, and not to any stellar object.

α, 1.4, brilliant white.

Deneb is from **Al Dhanab al Dajājah,** the Hen's Tail, which has become **Denebadigege, Denebedigege, Deneb Adige,** etc.

In the *Alfonsine Tables* **Arided** appears, and is still frequently seen for this star, as **Al Ridhādh** and **El Rided** formerly were for the constellation. Referring to this last title, Caesius termed α **Os rosae,** the German **Rosemund,** although he also designated it as **Uropygium,** the Pope's Nose of our Thanksgiving dinner-tables.

α also, and correctly enough, is **Aridif,** from **Al Ridf,** the Hindmost; but Bayer changed it to **Arrioph,** and Cary to **Arion.**

Bayer gave **Gallina** as an individual title.

Mr. Royal Hill says that this and the three adjacent bright stars in the figure are known as the **Triangles.**

Deneb has no sensible proper motion, and hence has been considered as deserving the term, generally inappropriate, of a "fixed star"; but spectroscopic investigations made at Greenwich seemed to show motion at the rate of thirty-six miles a second toward the earth, and so only apparently stationary. Such motion, Newcomb says, would eventually carry it at some time,—probably between 100,000 and 300,000 years hence,—past our system at about $\frac{1}{100}$ part of its present distance, making it the nearest and the brightest of the earth's neighbors. But Vogel's recent and more trustworthy measures at Potsdam give its rate as about five miles a second.

Elkin estimated its parallax in 1892 as $0''.047$,—practically insensible. Its spectrum is Sirian.

Photographs by Doctor Max Wolf, of Heidelberg, in June, 1891, show that it and γ are involved in one vastly extended nebula.

It rises in the latitude of New York City at sunset on the 12th of May, culminating on the 16th of September, and lies so far to the north that it is visible at some hour of every clear night throughout the year.

β, Double,—perhaps binary, 3.5 and 7, topaz yellow and sapphire blue.

Albireo, the now universal title, is in no way associated with Arabia, but apparently was first applied to the star from a misunderstanding as to the words *ab ireo* in the description of the constellation in the 1515 *Almagest.* **Albirco** in the *Standard Dictionary* undoubtedly is from a type error, as also may be **Abbireo, Alberio,** and **Albeiro,** which occasionally are used.

The Arabians designated β as **Al Minḣar al Dajājah,** the Hen's Beak, where it is still located on our maps. Riccioli wrote this **Menkar Eldigiagich**; and also had **Hierizim.**

β is one of the show objects of the sky, and Miss Clerke, calling its colors golden and azure, says that it presents "perhaps the most lovely effect of colour in the heavens." Being $35''$ apart, the components can readily be resolved by a field-glass. The system, if binary, has a very long period of revolution, as yet undetermined, the present position angle being $56°$.

Close to β appeared a *nova* on the 20th of June, 1670, described by the Carthusian monk Anthelmus of Dijon. This disappeared after two years of varying brilliancy, but may still exist as a 10th- to 11th-magnitude variable, discovered, in the supposed location, by Hind in 1852.

In the neck of the Swan, not far from β, is the variable χ^2, ranging from 4.5 to 13.5 in 406 days. Sometimes, at its maximum, it is of only the 6th magnitude.

γ, 2.7, is **Sadr,**—incorrectly **Sudr,**—from **Al Sadr al Dajājah,** the Hen's Breast, and one of the **Fawāris** of the Arabs.

Reeves said that in China it was **Tien Tsin,** the name of a city; but this generally was given to the group of four stars, α, β, γ, and δ.

γ is in the midst of beautiful streams of small stars, itself being involved in a diffused nebulosity extending to α; while the space from it to β perhaps is richer than any of similar extent in the heavens. Espin asserts that around γ and the horns of Taurus seem to centre the stars showing spectra of the fourth type. Its own spectrum is Solar. According to observations at Potsdam, it is in motion toward us at the rate of about four miles a second.

ϵ, 2.6, yellow,

on the right wing, is **Gienah,** from the Arabic **Al Janāḥ,** the Wing.

Between α, γ, and this star is the **Northern Coal-sack,** an almost vacant space in the Milky Way; another, still more noticeable and celebrated, coincidently being located in the Southern Cross.

6° to the northeast from ϵ is 61 Cygni, with a parallax of 0″.5, and thus, so far as we now know, the nearest star to us in the northern heavens, with the exception of La Lande 21185 Ursae Majoris. If the distance from the earth to the sun be considered as one inch, that to this star would be about seven and one half miles. It also is remarkable for its great proper motion toward the star σ,—5″.16 annually,—near to which it probably will be in 15,000 years. 4000 years ago it was near ϵ.

It is a double 6th-magnitude, and may be binary, the components 20″ apart, with a position angle of 121° in 1890. It was the first star successfully observed for parallax,—by Bessel between the years 1837 and 1840.

ζ and ρ, with two other adjacent small stars, were the Chinese **Chay Foo,** a Storehouse for Carts.

π^1, 4.8,

is **Azelfafage,** possibly a corrupted form of **Adelfalferes,** from **Al Ṭhīlf al Faras,** the Horse's Foot or Track; and, to quote Ideler,

> It follows either that the foot of Pegasus [now marked by π Pegasi] extended to this star, or that in this region was supposed to be located the feet of the Stallion which, as we shall see farther on, some Arab astronomer introduced between Pegasus and the Swan.

Or the title may be, as seems more probable, from **Al 'Azal al Dajājah,** the Tail of the Hen, which it exactly marks. It is sometimes **Azelfafge;** but

Bayer, with whom the word apparently first occurs, had "**Azelfage** *id est* **Tarcuta.**" [1]

π^1, with about twenty other stars in Cygnus, Andromeda, and Lacerta, was comprised in the early Chinese **Tang Shay,** the Dragon.

P, or Fl. 34, a 5th-magnitude, located at the base of the Swan's neck, is one of the few so-called gaseous stars having bright lines in their spectra. It was discovered by Janson, as a *nova* of the 2d magnitude, on the 18th of August, 1600; was numbered 27 in Tycho's catalogue, with the designation of *nova anni* 1600 *in pectore Cygni;* and Kepler thought it worthy of a monograph in 1606. Christian Huygens, the Dutch astronomer of the 17th century, called it the **Revenante of the Swan,** from its extraordinary light changes; but these now seem to have ceased.

ω^3, Double, 5½ and 10, pale red,

is **Ruchba** from **Al Rukbah al Dajājah,** the Hen's Knee; but the three stars ω now mark the tertiaries of the left wing.

The components of ω^3 are 56″.3 apart, at a position angle of 86°.3; and other minute stars are in the same field.

*

. . . the Delphienus heit
Up in the aire.

King James I, in *Ane schort Poeme of Tyme.*

Delphinus, the Dolphin,

is **Dauphin** in France, **Delfino** in Italy, and **Delphin** in Germany: all from the Greek Δελφίς and Δελφίν, transcribed by the Latins as **Delphis** and **Delphin.** This last continued current through the 17th century, and in our day was resumed by Proctor for his reformed list. Chaucer, in the *Hous of Fame*, had **Delphyn,** and later than he it was **Dolphyne.**

It now is one of the smallest constellations, but originally may have included the stars that Hipparchos set off to form the new Equuleus; and in all astronomical literature has borne its present title and shape, with many and varied stories attached, for its namesake was always regarded as the most remarkable of marine creatures.

[1] What is this last? It seems to have escaped comment by all of the authorities.

In Greece it also was Ἱερὸς Ἰχθύς, the Sacred Fish, the creature being of as much religious significance there as a fish afterwards became among the early Christians; and it was the sky emblem of philanthropy, not only from the classical stories connected with its prototype, but also from the latter's devotion to its young. It should be remembered that our stellar Dolphin is figured as the common cetacean, *Delphinus delphis*, of Atlantic and Mediterranean waters, not the tropical *Coryphaena* that Dorado represents.

Ovid, designating it as *clarum sidus*, personified it as **Amphitrite,** the goddess of the sea, because the dolphin induced her to become the wife of Neptune, and for this service, Manilius said, was "rais'd from Seas" to be

> The Glory of the Floud and of the Stars.

From this story the constellation was known as **Persuasor Amphitrites,** as well as **Neptunus** and **Triton.**

With Cicero it appeared as **Curvus,** an adjective that appropriately has been applied to the creature's apparent form in all ages[1] down to the "bended dolphins" in Milton's picture of the Creation. Bayer's **Currus** merely is Cicero's word with a typographical error, for he explained it, *Ciceroni ob gibbum in dorso;* but he also had **Smon** *nautis*, and Riccioli **Smon** *barbaris*, which seems to be the *Simon*, Flat-nosed, of old-time mariners, quoted by Pliny for the animal.

Another favorite title was **Vector Arionis,** from the Greek fable that attributed to the dolphin the rescue of Arion on his voyage from Tarentum to Corinth — a variation of the very much earlier myth of the sun-god Baal Hamon. Hence comes Henry Kirke White's

> lock'd in silence o'er Arion's star,
> The slumbering night rolls on her velvet car.

In continuation of the Greek story of Arion and his Lyre appears Μουσικὸν ζώδιον, the **Musicum signum** of the Latins; or this may come from the fact mentioned in Ovid's *Fasti* that the constellation was supposed to contain nine stars, the number of the Muses, although Ptolemy prosaically catalogued 10; Argelander, 20; and Heis, 31.

Riccioli and La Lande cited **Hermippus** for Delphinus, and **Acetes** after the pirate-pilot who protected Bacchus on his voyage to Naxos and Ariadne; while to others it represented **Apollo** returning to Crissa or piloting Castalius from Crete.

[1] Huet, in his notes on Manilius, quoted many examples of the use of this term by the Latins, and said *Perpetuam hoc Delphinum Epitheton.*

The Hindus, from whom the Greeks are said to have borrowed it,—although the reverse of this may have been the case,—knew it as **Shī-shu-māra,** or **Sim-shu-māra,** changed in later days to **Zizumara,** a Porpoise, also ascribed to Draco. And they located here the 22d *nakshatra*, **Çravishthā,** Most Favorable, also called **Dhanishthā,** Richest; the Vasus, Bright or Good Ones, being the regents of this asterism, which was figured as a Drum or Tabor; β marking the junction with Catabishaj.

Brown thinks that it may have been the Euphratean **Makhar,** although Capricorn also claimed this.

Al Bīrūnī, giving the Arabic title **Al Ḳa'ūd,** the Riding Camel, said that the early Christians—the Melkite[1] and Nestorian sects—considered it the **Cross of Jesus** transferred to the skies after his crucifixion; but in Kazwini's day the learned of Arabia called α, β, γ, and δ **Al 'Uḳūd,** the Pearls or Precious Stones adorning **Al Ṣalīb,** by which title the common people knew this Cross; the star ε, towards the tail, being **Al 'Amūd al Ṣalīb,** the Pillar of the Cross. But the Arabian astronomers adopted the Greek figure as their **Dulfīm,** which one of their chroniclers described as "a marine animal friendly to man, attendant upon ships to save the drowning sailors."

The *Alfonsine Tables* of 1545 said of Delphinus, *Quae habet stellas quae sapiunt naturam*, a generally puzzling expression, but common in the 1551 translation of the *Tetrabiblos*, where it signifies stars supposed to be cognizant of human births and influential over human character,—*naturam*. Ptolemy, as is shown in these *Four Books*, was a believer in the genethliacal influence of certain stars and constellations, of which this seems to have been one specially noted in that respect.

Delphinus lies east of Aquila, on the edge of the Milky Way, occupying, with the adjoining aqueous figures, the portion of the sky that Aratos called the **Water.** It culminates about the 15th of September.

Caesius placed here the **Leviathan** of the 104th *Psalm;* Novidius, the **Great Fish** that swallowed Jonah; but Julius Schiller knew some of its stars as the **Water-pots of Cana.** Popularly it now is **Job's Coffin,** although the date and name of the inventor of this title I have not been able to learn.

The Chinese called the four chief stars and ζ **Kwa Chaou,** a Gourd.

α, 4, pale yellow; β, Binary, 4 and 6, greenish and dusky.

The strange names **Sualocin** and **Rotanev** first appeared for these stars in the *Palermo Catalogue* of 1814, and long were a mystery to all, and

[1] These Melkites, or Royalists as the name indicates, were of the Greek Church, whose spiritual head now is the Czar, the royal head of Russia, and successor of the Byzantine Patriarch.

seemingly a great puzzle to Smyth, which he perhaps never solved, although he was very intimate with the staff of the Palermo Observatory. Webb, however, discovered their origin by reversing the component letters, and so reading *Nicolaus Venator*, the Latinized form of Niccolo Cacciatore, the name of the assistant and successor of Piazzi. But Miss Rolleston, in her singular book *Mazzaroth*, considered in some quarters as of authority, wrote that they are derived, α from the

> Arabic Scalooin, swift (*as the flow of water*);

and β from the

> Syriac and Chaldee Rotaneb, or Rotaneu, *swiftly running* (*as water in the trough*).

For no part of this scholarly (!) statement does there seem to be the least foundation. Burritt gave these titles as **Scalovin** and **Rotanen.**

α may be variable to the extent of half a magnitude in fourteen days.

β is a very close pair, 0″.68 apart in 1897, at a position angle of 357°, with the rapid orbital period of about twenty-six years. Another companion, purple in color and of the 11th magnitude, 6″ away, has lately been discovered by See, and so β may be ternary; while two other stars of the 10th and 13th magnitudes are about 30″ away.

γ is a beautiful double of 4th and 5th magnitudes, 11″ apart, with a position angle of 270°; but, if binary, their motion is extremely slow. The components are golden and bluish green, and a fine object for small glasses.

ε, a 4th-magnitude, although lying near the dorsal fin of our present figure, bears the very common name **Deneb,** from **Al Dhanab al Dulfīm,** the Dolphin's Tail. But in Arabia it also was **Al ʽAmūd al Ṣalīb,** as marking the Pillar of the Cross. In China it was **Pae Chaou,** the Rotten Melon.

The comparative brilliancy of β, γ, δ, and ε has been variously estimated — a fact which the observations of Gould at Albany in 1858, and at Cordoba in 1871–74, prove to be occasioned by variability, within moderate limits, of all four.

★

Dorado, the Goldfish,

first published by Bayer among his new southern figures, is still thus known in Germany and Italy, but the French say **Dorade**; and Flammarion has **Doradus,** perhaps from confusion with its supposed genitive case. The word is from the Spanish, and refers not to our little exotic cyprinoid, but to the large *coryphaena* of the tropical seas, of changing colors at death. On the planisphere in Gore's translation of *l'Astronomie Populaire* it is strangely rendered **Gold Field**; and **Craver,** in the Colas' list of the *Celestial Handbook* of 1892, is equally erroneous. Chilmead mentions it as the **Gilthead fish,** but this, in ichthyology, was a very different fish, the *Crenilabrus melops* of British coasts.

Caesius combined its stars with the Greater Cloud and the Flying Fish to form his Old Testament figure of **Abel the Just.**

The alternative title **Xiphias,** the Swordfish, I first find in the *Rudolphine Tables* of 1627; Halley used it, in addition to Dorado, in his catalogue of 1679; Flamsteed gave both names in his edition of Sharp's catalogue; and the modern Stieler's planisphere still has **Schwerdtfisch.** *Xiphias*, however, had appeared in astronomy in the first century of our era, for Pliny applied it to sword-shaped comets, as Josephus did to that "which for a year (!) had hung over Jerusalem in the form of a sword,"—possibly Halley's comet of A. D. 66.

The *Rudolphine Tables* and Riccioli catalogued here 6 stars of 4th and 5th magnitudes, but Gould 42 from 3.1 to 7.

The head of Dorado marks the south pole of the ecliptic, so that, according to Caesius, the constellation gave its name to that point as the Polus Doradinalis. Within 3° of this pole is the very remarkable nebula 30 Doradūs, that Smyth called the **True Lover's Knot,** although now known as the **Great Looped Nebula,** N. G. C. 2070, described by Sir John Herschel as an assemblage of loops and one of the most extraordinary objects in the heavens,—"the centre of a great spiral."

ε appears in Reeves' list as **Kin Yu,** but this star being only a 5th-magnitude, and these words signifying a Goldfish, they doubtless were designed for the whole figure introduced into China by the Jesuits.

ζ, a 5th-magnitude, bears the Chinese title **Kaou Pih.**

★

With vast convolutions Draco holds
Th' ecliptic axis in his scaly folds.
O'er half the skies his neck enormous rears,
And with immense meanders parts the Bears.

Erasmus Darwin's *Economy of Vegetation.*

Draco, the Dragon,

the German **Drache,** the Italian **Dragone,** and the French **Dragon,** was Δράκων with the Greeks—indeed this has been the universal title in the transcribed forms of the word. Classic writers, astronomers, and the people have known it thus, although Eratosthenes and Hipparchos called it Ὄφις, and in the Latin *Tables*, as with some of the poets, it occasionally appeared, with the other starry snakes, as **Anguis, Coluber, Python,** and **Serpens.** From the latter came **Aesculapius,** and perhaps **Audax.**

It was described in the *Shield of Hercules*, with the two Dogs, the Hare, Orion, and Perseus, as

> The scaly horror of a dragon, coiled
> Full in the central field;

and mythologists said that it was the Snake snatched by Minerva from the giants and whirled to the sky, where it became **Sidus Minervae et Bacchi** or the monster killed by Cadmus at the fount of Mars, whose teeth he sowed for a crop of armed men.

Julius Schiller, without thought of its previous character, said that its stars represented the **Holy Innocents of Bethlehem**; others, more consistently, that it was the **Old Serpent,** the tempter of Eve in the Garden; Caesius likened it to the **Great Dragon** that the Babylonians worshiped with Bel; and Olaus Rudbeck,[1] the Swedish naturalist of about 1700, said that his countrymen considered it the ancient symbol of the **Baltic Sea**; but he also sought to show that Paradise was located in Sweden!

Delitzsch asserted that a Hebrew conception for its stars was a **Quiver**; but this must have been exceptional, for the normal figure with that people was the familiar Dragon, or a sea monster of some kind. Renan thought that the allusion of Job to "the crooked serpent" in our *Authorized Version* is to this, or possibly to that of Ophiuchus; but the Dragon would seem to be the most probable as the ancient possessor of the pole-star, then, as ours now is, the most important in the heavens; while this translation of the original is specially appropriate for such a winding figure. The Reverend Doctor Albert Barnes renders it "fleeing," and Delitzsch, "fugitive"; but the *Revised Version* has "swift," a very unsuitable epithet for Draco's slow motion, yet applicable enough to the more southern Hydra.

Referring to Draco's change of position in respect to the pole from the effect of precession, Proctor wrote in his *Myths and Marvels of Astronomy*:

> One might almost, if fancifully disposed, recognize the gradual displacement of the Dragon from his old place of honour, in certain traditions of the downfall of the great Dragon whose "tail drew the third part of the stars of heaven," alluded to in *The Revelation* xii, 4;

and the conclusion of that verse, "did cast them to the earth," would show a possible reference to meteors.

[1] Rudbeck perhaps was "the sagacious Swede" of whom the Pope speaks in Browning's *The Ring and the Book*.

In Persia Draco was **Azhdehā,** the Man-eating Serpent, occasionally transcribed **Hashteher;** and, in very early Hindu worship, **Shī-shu-māra,** the Alligator, or Porpoise, which also has been identified with our Delphinus.

Babylonian records allude to some constellation near the pole as a **Snail** drawn along on the tail of a Dragon that may have been our constellation; while among the inscriptions we find **Sīr,** a Snake, but to which of the sky serpents this applied is uncertain. And some see here the dragon **Tiāmat,**[1] overcome by the kneeling sun-god Izhdubar or Gizdhubar, our Hercules, whose foot is upon it. Rawlinson, however, said that Draco represented **Hea** or **Hoa,** the third god in the Assyrian triad, also known as **Kim-mut.**

As a Chaldaean figure it probably bore the horns and claws of the early typical dragon, and the wings that Thales utilized to form the Lesser Bear; hence these are never shown on our maps. But with that people it was a much longer constellation than with us, winding downwards and in front of Ursa Major, and, even into later times, clasped both of the Bears in its folds; this is shown in manuscripts and books as late as the 17th century, with the combined title **Arctoe et Draco.** It still almost incloses Ursa Minor. The usual figuring is a combination of bird and reptile, *magnus et tortus*, a **Monstrum mirabile** and **Monstrum audax,** or plain **Monstrum** with Germanicus. Vergil had **Maximus Anguis,** which,

> after the manner of a river, glides away with tortuous windings, around and through between the Bears; —

a simile that may have given rise to another figure and title, found in the *Argonauticae*,—**Ladon,** from the prominent river of Arcadia, or, more probably, the estuary bounding the Garden of the Hesperides, which, in the ordinary version of the story, Draco guarded, "the emblem of eternal vigilance in that it never set." Here he was **Coluber** *arborem conscendens*, and **Custos Hesperidum,** the Watcher over the golden fruit; this fruit and the tree bearing it being themselves stellar emblems, for Sir William Drummond wrote:

> a fruit tree was certainly a symbol of the starry heavens, and the fruit typified the constellations;

and George Eliot, in her *Spanish Gypsy:*

[1] This notable creation of Euphratean mythology was the personification of primeval chaos, hostile to the gods and opposed to law and order; but Izhdubar conquered the monster in a struggle by driving a wind into its opened jaws and so splitting it in twain. Cetus, Hydra, and the Serpent of Ophiuchus also have been thought its symbols. Its representation is found on cylinder seals recently unearthed.

The stars are golden fruit upon a tree
A'll out of reach.

Draco's stars were circumpolar about 5000 B. C., and, like all those similarly situated,— of course few in number owing to the low latitude of the Nile country,— were much observed in early Egypt, although differently figured than as with us. Some of them were a part of the **Hippopotamus,** or of its variant the **Crocodile,** and thus shown on the planisphere of Denderah and the walls of the Ramesseum at Thebes. As such Delitzsch says that it was **Hes-mut,** perhaps meaning the Raging Mother. An object resembling a ploughshare held in the creature's paws has fancifully been said to have given name to the adjacent Plough.

The hieroglyph for this Hippopotamus was used for the heavens in general; while the constellation is supposed to have been a symbol of **Isis Hathor, Athor,** or **Athyr,** the Egyptian Venus; and Lockyer asserts that the myth of Horus which deals with the Hor-she-shu, an almost prehistoric people even in Egyptian records, makes undoubted reference to stars here; although subsequently this myth was transferred to the Thigh, our Ursa Major. It is said that at one time the Egyptians called Draco **Tanem,** not unlike the Hebrew **Tannīm,** or Aramaic **Tannīn,** and perhaps of the same signification and derived from them.

The Egyptian **Necht** was close to, or among, the stars of Draco; but its exact location and boundaries, how it was figured, and what it represented, are not known.

Among Arabian astronomers **Al Tinnīn** and **Al Thuʽbān** were translations of Ptolemy's *Δράκων*; and on the Borgian globe, inscribed over β and γ, are the words **Alghavil Altannin** in Assemani's transcription, the Poisonous Dragon in his translation, assumed by him as referring to the whole constellation. That there was some foundation for this may be inferred from the traditionary belief of early astrologers that when a comet was here poison was scattered over the world. Bayer cited from Turkish maps **Etanin,** and from others **Aben, Taben,** and **Etabin**; Riccioli, **Abeen** *vel* **Taeben**; Postellus, **Daban**; Chilmead, **Alanin**; and Schickard, **Attanino. Al Shujāʽ,** the Snake, also was applied to Draco by the Arabians, as it was to Hydra; and **Al Ḥayyah,** the Snake, appeared for it, though more common for our Serpens, with which word it was synonymous.

Bayer had **Palmes emeritus,** the Exhausted Vine Branch, that I do not find elsewhere; but the original is probably from the Arabs for some minor group of the constellation.

Williams mentions a great comet, seen from China in 1337, which passed through **Yuen Wei,** apparently some unidentified stars in Draco. The creature itself was the national emblem of that country, but the Dragon of the Chinese zodiac was among the stars now our Libra: Edkins writes that Draco was **Tsï Kung**, the Palace of the Heavenly Emperor, adding, although not very clearly, that this palace

is bounded by the stars of Draco, fifteen in number, which stretch themselves in an oval shape round the pole-star. They include the star **Tai yi**, ξ, o, σ, s, of Draco, which is distant about ten degrees from the tail of the Bear and twenty-two from the present pole. It was itself the pole in the Epoch of the commencement of Chinese astronomy.

Draco extends over twelve hours of right ascension, and contains 130 naked-eye components according to Argelander; 220, according to Heis: but both of these authorities extend the tail of the figure, far beyond its star λ, to a 4th-magnitude under the jaws of Camelopardalis,— much farther than is frequently seen on the maps.

α, 3.6, pale yellow.

Thuban and **Al Tinnin** are from the Arabic title for the whole of Draco, and **Azhdeha** from the Persian.

It is also **Adib, Addib, Eddib, Adid, Adive,** and **El Dsib**, all from **Al Dhi'bah,** the Hyaenas, that also appears for the stars ζ, η, and ι, as well as for others in Boötes and Ursa Major. Al Tizini called it **Al Dhīlī**, the Male Hyaena.

Among seamen it has been the **Dragon's Tail,** a title explained under γ.

In China it was **Yu Choo,** the Right-hand Pivot; the space towards ι being **Chung Ho Mun.**

Sayce says that the great astrological and astronomical work compiled for the first Sargon, king of Agade, or Akkad, devoted much attention to this star, then marking the pole, as **Tir-An-na,** the Life of Heaven; **Dayan Same,** the Judge of Heaven; and **Dayan Sidi,** the Favorable Judge,— all representing the god **Caga Gilgati,** whose name it also bore. Brown applies these titles to Wega of the Lyre, the far more ancient pole-star,— but this was 14,000 years ago!— and cited for α Draconis **Dayan Esiru,** the Prospering Judge, or the Crown of Heaven, and **Dayan Shisha,** the Judge Directing, as having the highest seat amongst the heavenly host. About 2750 B. C. it was less than 10′ from the exact pole, although now more than 26°; and as it lies nearly at the centre of the figure, the whole constellation then visibly swung around it, as on a pivot, like the hands of a clock, but in the reverse direction.

The star could be seen, both by day and night, from the bottom of the central passage[1] of the Great Pyramid of Cheops (Knum Khufu) at Ghizeh, in 30° of north latitude, as also from the similar points in five other like structures; and the same fact is asserted by Sir John Herschel as to the two pyramids at Aboussseir.

Herschel considered that there is distinct evidence of Thuban formerly being brighter than now, as its title from its constellation, and its lettering, would indicate; for with Bayer it was a 2d-magnitude,— in fact the only one of that brilliancy in his list of Draco,— and generally so in star-catalogues previous to two centuries ago. It culminates on the 7th of June.

β, probably Binary, 3 and 14, yellow.

Rastaban and **Rastaben** are from **Al Rās al Thuʽbān,** the Dragon's Head,— Schickard's **Raso tabbani.**

In early Arab astronomy it was one of **Al ʽAwāïd,** the Mother Camels, γ, μ, ν, and ξ completing the figure, which was later known as the **Quinque Dromedarii.** From the Arabic word comes another modern name, **Alwaid,** unless it may be from a different conception of the group as **Al ʽAwwād,** the Lute-player. Still other Desert titles were **Al Rāḳis,** the Dancer, or Trotting Camel, now given to μ; and it formed part of **Al Ṣalīb al Wāḳiʽ,** the Falling Cross, β and ξ forming the perpendicular, γ, μ, and ν the transverse; and thus designated as if slanting away from the observer to account for the paucity of stars in the upright.

Asuia, current in the Middle Ages and since, was from Al Shujāʽ, and often has been written **Asvia,** the letter *u* being mistakenly considered the early *v*. The companion, 4″ away, at a position angle of 13°.4, was discovered by Burnham.

β and γ, 4° apart, near the solstitial colure, have been known as the **Dragon's Eyes,** incorrect now, although Proctor thought them so located in the original figuring of a front view of Draco. Modern drawings place them on the top of the head.

In China they were **Tien Kae.**

γ, Double, 2.4 and 13.2, orange.

Eltanin, also written **Ettanin, Etannin, Etanim, Etamin,** etc., is from Ulug Beg's **Al Rās al Tinnīn,** the Dragon's Head, applied to this, as it also

[1] This passage, 4 feet by 3½ feet in diameter and 380 feet long, was directed northward to this star, doubtless by design of the builder, from a point deep below the present base, at an inclination of 26° 17′ to the horizon. At the time of its building, perhaps four millenniums before our era, the Southern Cross was entirely visible to the savage Britons.

is to α; Riccioli wrote it **Ras Eltanim.** The word Tinnīn is nearly synonymous with Thuʻbān, and Bayer mentioned **Rastaben** as one of its titles, the Alfonsine **Rasaben,** and now **Rastaban** in the *Century Cyclopedia;* but in early Arabic astronomy it was one of the Herd of Camels alluded to at β.

Firuzabadi referred to a Rās al Tinnīn and Dhanab al Tinnīn in the heavens, the Dragon's Head, and Tail; but these have no connection with our Draco, reference being there made solely to the ascending and descending nodes in the orbits of the moon and planets known to Arabian astronomers under these titles. Primarily, however, these were from India, and known as Rahu and Kitu. This idea seems to have originated from the fact that the moon's undulating course was symbolized by that of the stellar Hydra; and had the latter word been used instead of "Dragon," the expression would now be better understood. But it was familiar to seamen as late as the 16th century, for "the head and tayle of the Dragon"[1] appears in Eden's *Dedication*, of 1574, to Sir Wyllyam Wynter; and even now the symbols, ☊ for the ascending node and ☋ for the descending, are used in text-books and almanacs.

γ has been a notable object in all ages. It was observed with a telescope by Doctor Robert Hooke in the daytime in 1669 while endeavoring to determine its parallax, but his result afterwards was found to be due to the effect of aberration. Subsequently this star was used by Bradley for the same purpose, although unsuccessfully; but, on the other hand, it gave him his great discovery of the aberration of light,[2] of which Hooke of course was ignorant.

Millenniums before this, however, it was of importance on the Nile, as it ceased to be circumpolar about 5000 B. C., and a few centuries thereafter became the natural successor of Dubhe (α Ursae Majoris), which up to that date had been the prominent object of Egyptian temple worship in the north. γ was known there as **Isis,** or **Taurt Isis,**— the former name applied at one time to Sirius,—and it marked the head of the Hippopotamus that was part of our Draco. Its rising was visible about 3500 B. C. through the central passages of the temples of Hathor at Denderah and of Mut at Thebes; Canopus being seen through other openings toward the south at the same date. And Lockyer says that thirteen centuries later it became the orientation point of the great Karnak temples of Rameses and Khons at Thebes, the passage in the former, through which the star was

[1] The nodical month also is called the Dracontic, or Draconitic.

[2] The date of this discovery has been variously given as from 1726 to 1729, although it was first called to Bradley's attention on the 21st of December, 1725, by an unexplained discordance in his observations; but it took some time for him to complete this explanation.

observed, being 1500 feet in length; and that at least seven different temples were oriented toward it. When precession had put an end to this use of these temples, others are thought to have been built with the same purpose in view; so that there are now found three different sets of structures close together, and so oriented that the dates of all, hitherto not certainly known, may be determinable by this knowledge of the purpose for which they were designed. Such being the case, Lockyer concludes that Hipparchos was not the discoverer of the precession of the equinoxes, as is generally supposed, but merely the publisher of that discovery made by the Egyptians, or perhaps adopted by them from Chaldaea.

He also states that **Apet, Bast, Mut, Sekhet,** and **Taurt** were all titles of one goddess in the Nile worship, symbolized by γ Draconis.

It is interesting to know that the Boeotian Thebes, the City of the Dragon, from the story of its founder, Cadmus, shared with its Egyptian namesake the worship of this star in a temple dedicated, so far as its orientation shows, about 1130 B. C.: a cult doubtless drawn from the parent city in Egypt, and adopted elsewhere in Greece, as also in Italy in the little temple to Isis in Pompeii. Here, however, the city authorities interfered with this star-worship in one of their numerous raids on the astrologers, and bricked up the opening whence the star was observed.

γ lies almost exactly in the zenith of Greenwich, in fact, has there been called the **Zenith-star;** and, being circumpolar, descends toward the horizon, but, without disappearing, rises easterly, and thus explains the poet's line:

the East and the West meet together.

It was nearer the pole than any other bright star about 4000 years ago.

Its minute companion, 21″ distant, at a position angle of 152°, was discovered by Burnham.

δ, 3.1, deep yellow,

is the **Nodus secundus** of several catalogues, as marking the 2d of the four Knots, or convolutions, in the figure of the Dragon.

Al Tizini called it **Al Tāis,** the Goat, as the prominent one of the quadrangle, δ, π, ρ, and ε, which bore this title at a late period in Arabic indigenous astronomy; although that people generally gave animal names only to single stars. The **Jais,** which is found in various lists, maps, and globes, would seem to be a typographical error, or an erroneous transliteration of the original Arabic. δ also may have been one of Firuzabadi's two undetermined stars **Al Tayyasān,** the Two Goatherds.

δ, ε, π, ρ, and σ were the Chinese **Tien Choo**, Heaven's Kitchen.

ζ, a 3d-magnitude, was **Al Dhi'bah,** that we have also seen for *a*.

The Chinese knew it as **Shang Pih,** the Higher Minister.

Half-way between it and δ, within 7′ of the planetary nebula N. G. C. 6543, is the north pole of the ecliptic; the south pole being in the head of Dorado. Denning considers ζ the radiant point of the meteor streams of the 19th of January and of the 28th of March.

η, a double 2d- and 8th-magnitude, deep yellow and bluish star, was known in China as **Shang Tsae,** the Minor Steward.

The components are about 5″ apart, and the position angle is 143°.1.

ζ and η together were **Al Dhī'bain,** the **Duo Lupi** of early works, the Two Hyaenas or Wolves, lying in wait for the Camel's Foal, the little star **Al Ruba',** protected by the Mother Camels, the larger stars in our Draco's head. They also were **Al 'Auhaḳān,** the Two Black Bulls, or Ravens, the Arabic signifying either of these creatures; but this last word likewise appears for ω and *f*, and for χ and ψ; all of these titles being from Arabia's earliest days.

θ, a 4.3-magnitude, is **Hea Tsae,** the Lowest Steward; while the smaller stars near it were **Tien Chwang.**

ι, 3.6, orange.

Smyth mentioned this as **Al Ḍhiba'** of the Dresden globe and of Ulug Beg, but Kazwini had called it **Al Dhīlī,** the Male Hyaena, from which comes **Ed Asich,** its usual title now, the **Eldsich** of the *Century Cyclopedia.*

In China it was **Tso Choo,** the Left Pivot.

It marks the radiant point of the **Quadrantid** meteors of the 2d and 3d of January, so called from the adjacent Mural Quadrant.

A 9th-magnitude pale yellow companion is 2′ distant.

λ, 4.1, orange.

Giansar and **Giauzar** are variously derived: either from Al Jauzā', the Twins,— a little star is in close proximity,— or from Al Jauzah, the Central One, as it is nearly midway between the Pointers and Polaris; or, and still better, from the Persian Ghāuzar,— Al Bīrūnī's Jauzahar of Sāsānian origin,— the Poison Place, referring to the notion that the nodes, or points where the moon crosses the ecliptic, were poisonous because they "happened to be called the Head and Tail of the Dragon." This [illegible]ngular idea descended into comparatively modern times, and, although these points are far removed from Draco, still obtains in the name for λ. **Juza** is another popular title.

It also has been known as **Nodus secundus,** the Second Knot, possibly because thus located on some drawings; yet it is far removed from δ, which usually bears that name.

In China it was **Shang Poo,** or **Shaou Poo.**

Although the last lettered star in the figure, it lies at a considerable distance from the end, as figured on the atlases of Heis and Argelander.

μ, Binary, 5 and 5.1, brilliant white and pale white.

Al Rāḳis, from Ulug Beg's catalogue, turned into **Arrakis** and **Errakis,** generally has been thought to signify the Dancer, perhaps to the neighboring Lute-player, the star β; but here probably the Trotting Camel, one of the group of those animals located in this spot. Ideler added for it **Al Rāfad,** the Camel Pasturing Freely, that the original, differently pointed, may mean. The little star in the centre of the group of Camels, β, γ, μ, ν, and ξ, is named **Al Ruba'** on the Borgian globe, although almost invisible; but did not appear in the catalogues till Piazzi's time, except with Julius Schiller in his *Coelum Stellatum Christianum* of 1627, where it is the 37th star in his constellation of the Holy Innocents.

Assemani mentioned μ as **Al Ca'ab,** the Little Shield or Salver, but gave no reason for this, and its inappropriateness renders the claim very doubtful.

In modern drawings it marks the nose or tongue of Draco.

The components are 2″.5 apart, with a position angle of 165°; and their period is long, although not yet accurately determined.

ν, on the Dragon's head, already mentioned in connection with β, γ, μ, and ξ, is an interesting double for a small telescope. The components are each of 4.6 magnitude, about 62″ apart, with a position angle of 313°.

According to Wagner's determination of the parallax,— not yet, however, confirmed,— they are near neighbors to us, at a distance of about eleven light years.

ξ, 3.8, yellow,

was one of the Herd of Camels; but its modern individual name, **Grumium,** is the barbarism found for it in the *Almagest* of 1515, an equivalent of *γένυς* used by Ptolemy for the Dragon's under jaw. The word is now seen in the Italian *grugno* and the French *groin.*

Bayer followed Ptolemy in calling the star **Genam.**

Proctor thought that it marked Draco's darted tongue in the earliest representations of the figure,— unless ι Herculis were such star; while Denning considers it the radiant point of the meteor stream seen about the 29th of May,— the **Draconids.**

σ, 6.5, in the second coil northeast from δ, is **Alsafi,** corrupted from **Athāfi,** erroneously transcribed from the Arabic plural **Athāfiyy,** by which the nomads designated the tripods of their open-air kitchens; one of these being imagined in σ, τ, and υ. Uthfiyyah is the singular form. It probably is one of the nearest stars to our system,— about thirteen light years away according to Brunowski's unconfirmed determination.

ϕ, a 4th-magnitude double, was the Chinese **Shaou Pih,** the Minor Minister; and χ, of slightly greater brilliancy, was **Kwei She.**

ψ^1 and ψ^2, 4.3 and 5.2, pearly white and yellow.

Dsiban, from **Al Dhībain** (the Arabs' title for ζ and η), has been given by some to this pair, and Lach thought that with χ it also was **Al ʽAuhaḳān,** which we similarly find for ζ and η.

In China it was **Niu She,** the Palace Governess, or a Literary Woman.

The components of ψ^1 are about 30″ apart, with a position angle of 15°.

ω, 4.9, and f, 5.1.

These dim stars, between ζ and the group ϕ, χ, and ψ, were **Al Aṭhfār al Dhīb,** the Hyaena's claws, stretched out to clutch the Camel's Foal. They thus appear with Ulug Beg and on the Dresden globe; but elsewhere occasionally were known as **Al ʽAuhaḳān,** a designation shared with ζ and η, and with ϕ and χ. They also sometimes were **Al Dhīli,** the Wolf.

There seems to be confusion, and some duplication, in the nomenclature of Draco's stars, but their many titles show the great attention paid to the constellation in early days.

★

. . . the flaming shoulders of the Foal of Heav'n.

Omar Khayyám's *Rubáiyát.*

Equuleus, the Foal,

that modern Latin critics would turn into **Eculeus,** lies half-way between the head of Pegasus and the Dolphin, marked by the trapezium of 4th- to 5th-magnitude stars,— α, β, γ, and δ,— although Argelander catalogues nine others, and Heis twelve down to 6.7 magnitude. Thus "the flaming shoulders" of our motto are lacking here, and the reference may be to Pegasus, to which the characterization certainly is more appropriate.

The Germans call it **Füllen,** the Filly, and **Kleine Pferd,** which with us is the **Little Horse,** the French **Petit Cheval,** and the Italian **Cavallino.**

Hood wrote of it about 1590:

> This constellation was named of almost no writer, saving *Ptolomee* and *Alfonsus* who followith Ptolomee, and therefore no certain tail or historie is delivered thereof, by what means it came into heaven;

but we know that Geminos mentioned it as having been formed by Hipparchos, its stars till then lying in the early Dolphin. Still Hipparchos did not allude to it in his *Commentary*, nor did Hyginus, Manilius, or Vitruvius, a century after him.

Ptolemy catalogued it as Ἵππου Προτομή, this last word equivalent to our Bust for the upper part of an animal figure; but with later astronomers it was **Equus primus** and **prior,** as preceding Pegasus in rising; while from its inferior size come our own title and **Equulus, Equiculus,** and **Equus Minor.** Gore's translation of *l'Astronomie Populaire*, following Proctor, has **Equus,** the larger Horse being **Pegasus.**

Ptolemy's idea of the incompleteness of the figure was repeated in the **Equi Sectio, Equi Praesectio, Sectio equina, Sectio Equi minoris, Semi-perfectus,** and **Praesegmen** of various authors and Latin versions of the *Syntaxis* and of the *Alfonsine Tables;* the *Almagest* of 1551 gave **Praecisio Equi.**

Chrysococca's *Tables* had Κεφαλή Ἵππου, the **Equi Caput** of some Latin writers, and the **Horse's Head** of our day.

The Arabians followed Ptolemy in calling it **Al Kiṭ'ah al Faras,** Part of a Horse, Chilmead's **Kataat Alfaras; Al Faras al Thānī,** the Second Horse, alluding either to its inferior size, or to the time of its adoption as a constellation; and **Al Faras al Awwal,** the First Horse, in reference to its rising before Pegasus. From the first of these comes the modern **Kitalpha,** sometimes applied to the constellation, and generally to the brightest star. Riccioli's **Elmac Alcheras** certainly is a barbarism,— not unusual, however, with him; but La Lande's rarely used **Hinnulus,** a Young Mule, has more to commend it.

With the Hindus it was another of their **Açvini,** the Horsemen, although their figuring resembled ours.

Some of the mythologists said that the constellation represented **Celeris,** the brother of Pegasus, given by Mercury to Castor; or **Cyllarus,** given to Pollux by Juno; or the creature struck by Neptune's trident from the earth when contesting with Minerva for superiority; but it also was connected with the story of Philyra and Saturn. Caesius, in modern times, associated it with the **King's Horse** that Haman hoped for, as is told in the *Book of Esther;* and Julius Schiller, with the **Rosa mystica.**

The constellation comes to the meridian on the 24th of September.

α, 3.8,

is **Kitalpha,** from the Arabian name for the whole figure, strangely turned by Burritt into **Kitel Phard.** Stieler has **Kitalphar.**

With β it was the Chinese **Sze Wei.**

δ, Triple and binary, 5, 5, and 10, topaz yellow and pale sapphire.

The two largest stars form a system noted as the quickest in orbital revolution of all known binaries except κ Pegasi, and perhaps the 7th-magnitude Ll. 9091 in Orion, on the border of Taurus. Its period is about 11½ years, and the components are so close that they can be separated only by the largest telescopes; their maximum distance apart every seven years is but 0″.44, this occurring in 1897, their position angle being 208°.

ε is another triple, much resembling δ in character; the component stars, 5.7, 6.2, and 7.1 in magnitude, are 1″.3, and 10″.4 apart, the colors of the first two yellowish, the last ashy white.

*

Equuleus Pictoris, the Painter's Easel,

was formed, and thus named, by La Caille, but also has been called **Pluteum Pictoris;** astronomers know it as **Pictor.** It is the **Chevalet du Peintre,** or the **Palette,** of the French; the **Pittore** of the Italians; and the **Malerstaffelei** of the Germans.

The constellation lies just south of Columba, between Canopus and the south pole of the ecliptic in Dorado, La Caille assigning to it 14 stars, of from 3½ to 5½ magnitudes; but Gould catalogued 67 down to the 7th.

Near its ε, and close to Columba, Kapteyn recently has discovered an 8.2-magnitude orange-yellow star having a proper motion of 8″.7 annually, thus much exceeding that of Goombridge's 1830 Ursae Majoris, hitherto the **Flying Star.**

. . . amnis, quod de coelo exoritur sub solio Jovis.

Plautus' *Trinummus.*

. . . the starry Stream.
For this a remnant of Eridanos,
That stream of tears, 'neath the gods' feet is borne.

Brown's *Aratos.*

The River Eridanus,

the French **Eridan,** the Italian **Eridano,** and the German **Fluss Eridanus,** is divided into the Northern and the Southern Stream; the former winding from the star Rigel of Orion to the paws of Cetus; the latter extending thence southwards, southeast, and finally southwest below the horizon of New York City, 2° beyond the *lucida* Achernar, near the junction of Phoenix, Tucana, Hydrus, and Horologium. Excepting Achernar, however, it has no star larger than a 3d-magnitude, although it is the longest constellation in the sky, and Gould catalogues in it 293 naked-eye components.

Although the ancients popularly regarded it as of indefinite extent, in classical astronomy the further termination was at the star θ in 40° 47′ of south declination; but modern astronomers have carried it to about 60°.

With the Greeks it usually was ὁ Ποταμός, the River, adopted by the Latins as **Amnis, Flumen, Fluvius,** and specially as **Padus** and **Eridanus;** this last, as 'Ερıδανός, having appeared for it with Aratos and Eratosthenes. Geographically the word is first found in Hesiod's Θεογονία for the Phasis[1] in Asia, celebrated in classic history and mythology,

That rises deep and stately rowls along

into the Euxine Sea near the spot where the Argonauts secured the golden fleece.

Other authors identified our Eridanus with the fabled stream flowing into the ocean from northwestern Europe,—a stream that always has been a matter of discussion and speculation (indeed, Strabo called it "the nowhere existing"),—or with Homer's Ocean Stream flowing around the earth, whence the early titles for these stars, **Oceanus** and the **River of Ocean.** They also have been associated with the famous little brook under the Acropolis; with the Ligurian Bodencus—the Padus of ancient, and the Po

[1] This is the modern Rion, or Rioni, the Fasch of the Turks; this last title being a general appellation in early Oriental geography for all rivers, perhaps from the Sanskrit Phas, Water, or Was, still seen in the German Wasser.

of modern, Italy,—famous in all classical times as the largest of that country's rivers, Vergil's *Rex fluviorum;* with the Ebro of Spain; with the Granicus of Alexander the Great; with the Rhenus and the Rhodanus,—our Rhine and Rhone; and with the modern Radaune, flowing into the Vistula at Danzig.

Some of these originals of our River, especially the Padus, were seats of the early amber trade, thus recalling the story of the Heliades, whose tears, shed at the death of their brother Phaëthon, turned into amber as they fell into "that stream of tears" on which that unfortunate was hurled by Jove after his disastrous attempt to drive the chariot of the sun. This was a favorite theme with poets, from Ovid, in the *Metamorphoses*, to Dean Milman, in *Samor*, and the foundation of the story that the river was transferred to the sky to console Apollo for the loss of his son.

But none of these comparatively northern streams suit the stellar position of our Eridanus, for it is a southern constellation, and it would seem that its earthly counterpart ought to be found in a corresponding quarter. In harmony with this, we know that Eratosthenes and the scholiasts on Germanicus and Hyginus said that it represented the Nile, the only noteworthy river that flows from the south to the north, as this is said to do when rising above the horizon. Thus it was **Nilus** in the *Alfonsine Tables*, the edition of 1521 saying, *Stellatio fluvii id est Eridanus sive Gyon sive Nilus;* **Gyon**[1] coming from the statement in *Genesis* ii, 13:

> the name of the second river is Gihon: the same is it that compasseth the whole land of Cush;

this latter being misunderstood for the Nile country instead of the Asiatic Kush that was unquestionably intended by the sacred writer. La Lande cited **Mulda**, equivalent to another title for the stellar Eridanus,—Μέλας, Black,—and so again connected with Egypt, whose native name, Khem, has this same meaning, well describing the color of the fertile deposit that the Nile waters leave on the land. This became the Latin **Melo**, an early name for the Nile, as it also was for the constellation.

This allusion to the Nile recalls the ancient wide-spread belief that it and the Euphrates were but different portions of the same stream; and Brown, in his monograph *The Eridanus*, argues that we should identify the Euphrates with the sky figure. He finds his reasons in the fact that both are frequently alluded to, from very early days to the classical age, as The River,

[1] The word Sihor for the Nile, in our Authorized Version of *Jeremiah* ii, 18, is Γηων in the *Septuagint*, Josephus also using it in his Ἰουδαϊκή Ἀρχαιολογία, or *Jewish Antiquities*, in referring to the Nile as one of the four great branches of the River of Paradise.

the Euphrates originally being Pura or Purat, the Water, as the Nile was, and even now is, Ioma or Iauma, the Sea; that they resemble each other as long and winding streams with two great branches; that each is connected with a Paradise — Eden and Heaven; that the adjoining constellations seem to be Euphratean in origin; and that each is in some way associated with the Nile, and each with the overthrow of the sun-god.

There is much in the Euphratean records alluding to a stellar stream that may be our Eridanus,— possibly the Milky Way, another sky river; yet it is to the former that the passage translated by Fox Talbot possibly refers:

> Like the stars of heaven he shall shine; like the River of Night he shall flow;

and its title has been derived from the Akkadian Aria-dan, the Strong River. George Smith thinks that the heavenly Eridanus may have been the Euphratean **Erib-me-gali.**

Its hither termination at the star Rigel gave it the title **River of Orion,** used by Hipparchos, Proclus, and others; and Landseer wrote:

> the stars now constellated as Erydanus were originally known in different countries by the names of **Nile, Nereus,** and **Ocean,** or **Neptune.**

Riccioli cited for it **Vardi,** and a Moorish title, according to Bayer, was **Guad,**— the 1720 edition of the *Uranometria* has **Guagi,**— all these from the Arabic *wādī*, and reminding us of the Wādī al Kabīr, the Great River, the Spaniards' Guadalquivir; but the common designation among the Arabians was **Al Nahr,** the River, transcribed **Nar** and **Nahar,**— Chilmead's **Alvahar**; this Semitic word, occasionally written Nahal, also having been adduced as a derivation of the word Nile.

Assemani quoted **Al Kaff Algeria** from the Borgian globe for stars in the bend of the stream; but Ideler claimed these for Al Kaff al Jidhmah of Cetus.

Caesius thought our Eridanus the sky representative of the **Jordan,** or of the **Red Sea,** which the Israelites passed over as on dry land.

Old illuminated manuscripts added a venerable river-god lying on the surface of the stream, with urn, aquatic plants, and rows of stars; for all of which the *Hyginus* of 1488 substitutes the figure of a nude woman, with stars lining the lower bank. Bayer's illustration is quite artistic, with reeds and sedge on the margins. The monster Cetus often is depicted with his fore paws, or flippers, in the River.

α, 0.4, white.

Achernar is from **Al Ākhir al Nahr,** the End of the River, nearly its present position in the constellation, about 32° from the south pole; but the title was first given to the star now lettered θ, the farthest in the Stream known by Arabian astronomers. For α Bayer had **Acharnar** *pro* **Acharnahar** *vel* **Acharnarim,** and **Enar**; Caesius, **Acarnar**; Riccioli, **Acarnaharim** and **Acharnaar**; Scaliger, **Acharnarin**; Schickard, **Achironnahri**; while **Achenar** and **Archarnar** are still occasionally used.

This star is supposed to be one of Dante's **Tre Facelle,** notwithstanding its invisibility from Italy.

Chinese astronomers knew it as **Shwuy Wei.**

Ptolemy did not mention it, although he could have seen it from the latitude of Alexandria, 31° 11′,—a fact, among others, which argues that his catalogue was not based upon original observations, but drawn from the now lost catalogue of Hipparchos, compiled at Rhodes, more than 5° further north, from which place Achernar was not visible.

It culminates on the 4th of December, due south of Baten Kaitos.

β, 2.9, topaz yellow.

Cursa, 3° to the northwest of Rigel in Orion, is the principal star in this constellation, seen from the latitude of New York City.

The word is from **Al Kursiyy al Jauzah,** the Chair, or Footstool, of the Central One, *i. e.* Orion, formed by β, λ, and ψ Eridani with Orionis, and regarded as the support of his left foot; but in the earlier astronomy of the nomads it was one of **Al Udḥā al Na'ām,** the Ostrich's Nest, that some extended to o^1 and o^2.

The *Century Cyclopedia* gives **Dhalim** as an alternative title, undoubtedly from **Al Ṭhalīm,** the Ostrich; but, although used for β by several writers, this better belongs to θ.

The Chinese called β **Yuh Tsing,** the Golden Well.

γ^1, 3, yellow.

Zaurac and **Zaurak** are from the Arabic **Al Nā'ir al Zauraḳ,** the Bright Star of the Boat; but Ideler applied this early designation to the star that now is α of our Phoenix.

With δ, ϵ, η, and others near, it made up the Chinese **Tien Yuen,** the Heavenly Park.

η, 3.7, pale yellow.

Azha is supposed to have been the **Azḥā** of Al Sufi, and the equivalent **Ashiyane** of the Persians, and was known by Kazwini as **Al Udḥiyy,** being chief among the stars of the Ostrich's Nest, which the word signifies. The other components were ζ, ρ, and σ; but this last, the 17th of Ptolemy, is not now to be identified in the sky, although it may be one of the three stars ρ displaced by proper motion since Ptolemy's time.

Near η, towards τ, are some other stars — ε and π Ceti among them — which in early days were included in the Nest, but later were set apart by Al Sufi as **Al Sadr al Ḳetus,** the Breast of the Whale.

θ, Double, 3 and 5.25.

Achernar was the early name for this at the then recognized end of the stream, Halley saying of it, *ultima fluminis in veteri catalogo,* referring to Tycho's work, of which his own was a supplement. Various forms of its title are given under α, but **Acamar,** from the *Alfonsine Tables,* is peculiar to θ.

Ulug Beg called it **Al Ṭhalīm,** the Ostrich, but Hyde rendered this the Dam, as if blocking the flow of the stream to the south.

Bullialdus, in his edition of Chrysococca's work, had it 'Αὖλαξ, the Furrow, equivalent to the *sulcus* used by Vergil to denote the track of a vessel, appropriate enough to a star situated in the Stream of Ocean; and Riccioli distinctly gave **Sulcus** for it in his *Astronomia Reformata.*

It is the solitary star visible from the latitude of New York City in early winter evenings, low down in the south, on the meridian with Menkar of the Whale; but Baily said that its brilliancy has probably lessened since Ptolemy's time, for the latter designated it by α — *i. e.* of the 1st magnitude.

Between it and Fomalhaut lie many small stars, not mentioned by Ptolemy, that Hyde said were **Al Zibāl**; but Al Sufi had already called them **Al Ri'āl,** the Little Ostriches.

ι, κ, ϕ, and χ, of about the 4th magnitude, were another **Tien Yuen** of the Chinese, different from that marked by γ; ι and κ are the lowest in the constellation visible from the latitude of New York.

μ and ω, 4th-magnitude stars lying westward of β, were **Kew Yew** in China; Reeves including under this title b and the stars of the Sceptre.

o^1, 4.1, clear white.

In early Arabia this was **Al Baiḍ,** the Egg, from its peculiarly white color, as well as from its position near the Ostrich's Nest. Modern lists generally write it **Beid.**

Situla, the Urn, also has been used for it, although there is no apparent applicability here, and the title is universally recognized for κ Aquarii.

o^2, Triple, 4, 9.1, and 10.8, orange and sky blue,

is the **Keid** of modern lists, Burritt's **Kied,** from **Al Ḳaid,** the Egg-shells, thrown out from the nest close by.

The Abbé Hell used it in the construction of his constellation Psalterium.

Its duplicity was discovered by Sir William Herschel in 1783, and in 1851 Otto Struve found the smaller star itself double and a binary of short period. The system is remarkable from its great proper motion of 4″.1 annually. The two larger stars are 83″ apart, at a position angle of 108°, and the smaller 4″ apart, at an angle of 111°. The parallax by Elkin indicates a distance of twenty light years.

τ^2, 4, yellow.

Angetenar of the *Alfonsine Tables,* now the common title, the **Argentenar** of Riccioli and **Anchenetenar** of Scaliger, is from **Al Ḣināyat al Nahr,** the Bend in the River, near which it lies; Ideler transcribing this as **Al Anchat al Nahr.** This is one of Bayer's nine stars of the same letter lying just above Fornax; he said of them, *sibi mutuo succedentes novem.*

See found, in 1897, a 14.9-magnitude bluish star, about 52″ away, at a position angle of 128°.3.

υ^1–υ^7

mark another series of seven stars called in Bayer's text **Beemim** and **Theemim.** This last, used by Bode and now in current use, is perhaps the Arabic Al Tau'amān and the Jews' Tĕōmīm, the Twins, from the pairs υ^1, υ^2, and υ^3, υ^4. Grotius thought it derived either from the foregoing or from an Arabic term for two medicinal roots; but Ideler's suggestion that it is from the Hebrew Bamma'yīm, In the Water, would seem more reasonable, although we have but few star-names from Judaea, and he intimated that it might be a distorted form of Al Ṭhalīm, the Ostrich. The *Almagest* of 1515 has **Beemun;** and the *Standard Dictionary,* **The.eʿ.nim.**

★

Felis, the Cat,

a word which Latin lexicographers now write **Faelis,** was formed by La Lande from stars between Antlia and Hydra, and first published in his *Bibliographie Astronomique* of 1805. Its inventor said of it:

I am very fond of cats. I will let this figure scratch on the chart. The starry sky has worried me quite enough in my life, so that now I can have my joke with it.

In *Die Gestirne*, thé 2d edition of Bode's maps, it appears as **Katze,** with twenty stars; but, except with Secchi, who included it as **Gatto** in his planisphere of 1878, it has long been discontinued in the catalogues and charts.

Proctor assigned this title to Canis Minor, but no one has followed him in this change.

★

Fornax Chemica, or Fornax Chymiae, the Chemical Furnace,

was formed by La Caille from stars within the southern bend of the River; but modern astronomers, by whom it is still recognized, have abbreviated the title to **Fornax.**

The Chinese know it as **Tien Yu,** Heaven's Temporary Granary.

Bode changed the early name in 1782 to **Apparatus chemicus,** and translated it as the **Chemische Apparat, Chymische Ofen,** and **l'Apparat Chimique,** an alteration in honor of the celebrated chemist Antoine Laurent Lavoisier. These titles, however, have fallen into disuse.

Gould assigns to it 110 stars, from 3.6 to 7th magnitudes.

a, the *lucida*, is a double of 4th and 7th magnitudes, 3″ apart, with a position angle of 320°, and may be binary. It comes to the meridian on the 19th of December.

★

Frederici Honores.

In 1787 Bode formed, and in 1790 published in the *Jahrbuch*, this minor constellation as **Friedrich's Ehre,**—Frederick's Glory, Burritt's **Gloria Frederica,** and Miss Clerke's **Gloria Frederici,**—in honor of the great Frederick II of Prussia, who had died in 1786.

It was made up from thirty-four stars in the space between Cepheus, Andromeda, Cassiopeia, and the Swan, where Royer, in 1679, had attempted to replace the earlier Lacerta of Hevelius by his Sceptre and Hand of Justice. But he borrowed for his new creation from the northern hand of Andromeda, which he moved to a more easterly position, entirely indifferent to the fact that it had been "stretched out there for 3000 years." Bode's figure was thus described:

Below a Nimbus, the sign of royal dignity, hang, wreathed with the imperishable Laurel of fame, a Sword, Pen and an Olive Branch, to distinguish this ever to be remembered monarch, as hero, sage and peacemaker.

It is now seldom mentioned, and has been discarded from the charts, while Lacerta maintains its position in this much occupied spot.

*

Then both were cleans'd from blood and dust
To make a heavenly sign;
The lads were, like their armour, scour'd,
And then hung up to shine;
Such were the heavenly double-Dicks,
The sons of Jove and Tyndar.

John Grubb, in Percy's *Reliques of Ancient English Poetry.*

Gemini, the Twins.

The conception of a sky couple for these stars has been universal from remote antiquity, but our Latin title dates only from classical times, varied by **Gemelli,** which still is the Italian name. The Anglo-Saxons knew them as **ge Twisan,** and the Anglo-Normans as **Frère;** the modern French as **Gémeaux,** and the Germans as **Zwillinge,** Bayer's **Zwilling.**

While on earth these Twins were sons of Leda, becoming, after their transfer to the sky, **Geminum Astrum, Ledaei Fratres, Ledaei Juvenes,** and **Ledaeum Sidus;** Dante calling their location **Nido di Leda,** the Nest of Leda. Cowley, the contemporary of Milton, wrote of them as the **Ledaean Stars,** and Owen Meredith of our day as

The lone **Ledaean lights** from yon enchanted air.

They also were **Gemini Lacones,**— Milton's **Spartan Twins** and William Morris' **Twin Laconian Stars; Spartana Suboles** from their mother's home, and **Cycno generati** from her story; **Pueri Tyndarii, Tyndarides, Tyndaridae,** and Horace's **clarum Tyndaridae Sidus,** from Tyndarus, their supposed father; while the **Oebalii** and **Oebalidae** of Ovid, Statius, and Valerius Flaccus are from their grandfather, Oebalus, king of Sparta. Manilius called them **Phoebi Sidus** as being under Apollo's protection.

Individually they were **Castor and Pollux,**— Dante's and the Italians' **Castore e Polluce; Apollo and Hercules, Triptolemus and Iasion, Theseus and Pirithoüs.** Horace wrote **Castor fraterque magni Castoris;** Pliny, **Castores;** and Statius had **alter Castor** from their alternate life and death that the modern James Thomson repeated in the Summer of his *Seasons:*

Th' alternate Twins are fix'd.

But Welcke gave an astronomical turn to these titles by seeing in the first **Astor,** the Starry One, and in Pollux **Polyleukes,** the Lightful.

With the Greeks they were Δίδυμοι, the Twins,— Riccioli's **Didymi,**— originally representing two of the Pelasgian Κάβειροι, but subsequently the Boeotian Διόσκυροι,—**Dioscuri** in Rome,—the Sons of Zeus; as also **Amphion and Zethus,** Antiope's sons, who, as Homer wrote, were

Founders of Thebes, and men of mighty name,

strikingly shown on the walls of the Spada Palace in Rome, and with the Farnese Bull now in the Naples Museum. Plutarch called them Ἄνακες, Lords,— Cicero's **Anaces,**— and Σιώ, the Two Gods of Sparta; Theodoretus, Ἐφέστιοι, the Familiar Gods; others, **Dii Samothraces,** from the ancient seat of worship of the Cabeiri; and **Dii Germani,** the Brother Gods.

In India they always were prominent as **Açvini,** the **Ashwins,** or Horsemen, a name also found in other parts of the sky for other Hindu twin deities; but, popularly, they were **Mithuna,** the Boy and Girl, the Tamil **Midhunam,** afterwards changed to **Jituma,** or **Tituma,** from the Greek title.

A Buddhist zodiac had in their place a **Woman** holding a golden cord.

Some of the Jews ascribed them to the tribe of Benjamin, although others more fitly claimed them for Simeon and Levi jointly, the Brethren. They called them **Teōmīm;** the Tyrians, **Tome;** and the Arabian astronomers, **Al Tau'amān,** the Twins; but in early Desert astronomy their two bright stars formed one of the fore paws of the great ancient Lion; although they also were **Al Burj al Jauzā',** the Constellation of the Twins. From this came Bayer's **Algeuze,** which, however, he said was *unrecht,* thus making Riccioli's **Elgeuzi** and **Gieuz** equally wrong. Hyde adopted another form of the word,— Jauzah, the Centre,— as designating these stars' position *in medio coeli,* or in a region long viewed as the centre of the heavens; either because they were a zenith constellation, or from the brilliancy of this portion of the sky. Julius Pollux, the Egypto-Greek writer of our second century, derived the title from Jauz, a Walnut, as mentioned in his *Onomasticon.* But there is much uncertainty as to the stellar signification and history of this name, as will be further noticed under Orion.

The 1515 *Almagest* has the inexplicable **Alioure,** said to be from some early edition of the *Alfonsine Tables.*

The Persians called the Twins **Du Paikar,** or **Do Patkar,** the Two Figures; the Khorasmians, **Adhupakarik,** of similar meaning; and Riccioli wrote that they were the "Chaldaean" **Tammech.**

Kircher said that they were the Κλύσος, or *Claustrum Hori,* of the Egyptians; and others, that they represented the two intimately associated gods, **Horus the Elder,** and **Horus the Younger,** or **Harpechruti,**—the Harpocrates of Greece.

The Twins were placed in the sky by Jove, in reward for their brotherly love so strongly manifested while on earth, as in the verses of Manilius:

> Tender Gemini in strict embrace
> Stand clos'd and smiling in each other's Face;

and were figured as Two Boys, or Young Men, drawn exactly alike:

> So like they were, no mortal
> Might one from other know;

or as Two Infants, **Duo Corpuscula.** But Paulus Venetus and other illustrators of Hyginus showed **Two Angels,** and the Venetian edition of Albumasar of 1489 has two nude seated figures, a Boy and a Girl, with arms outstretched upon each other's shoulders.

The *Leyden Manuscript* shows two unclad boys with Phrygian caps, each surmounted by a star and Maltese cross; one with club and spear, the other with a stringed instrument. Bayer had something similar, Pollux, however, bearing a peaceful sickle.

Caesius saw here the **Twin Sons of Rebecca,** or **David and Jonathan;** while other Christians said that the stars together represented **Saint James the Greater;** or, to go back to the beginning of things, **Adam and Eve,** who probably were intended by the nude male and female figures walking hand in hand in the original illustration in the *Alfonsine Tables.* A similar showing appears, however, on the Denderah planisphere of 1300 years previous.

The Arabians drew them as **Peacocks,** from which came a mediaeval title, **Duo Pavones;** some of the Chaldaeans and Phoenicians, as a **Pair of Kids** following Auriga and the Goat, or as **Two Gazelles;** the Egyptians, as **Two Sprouting Plants;** and Brown reproduces a Euphratean representation of a couple of small, naked, male child-figures, one standing upon its head and the other standing upon the former, feet to feet; the original Twins being the sun and moon, when the one is up the other is generally down;

a variant representation showing the positions reversed and the figures clothed.

Another symbol was a **Pile of Bricks,** referring to the building of the first city and the fratricidal brothers — the Romulus and Remus of Roman legend; although thus with a very different character from that generally assigned to our Heavenly Twins. Similarly Sayce says that the Sumerian name for the month May–June, when the sun was in Gemini, signified "Bricks" (?).

In classical days the constellation was often symbolized by two stars over a ship; and having been appointed by Jove as guardians of Rome, they naturally appeared on all the early silver coinage of the republic from about 269 B. C., generally figured as two young men on horseback, with oval caps, surmounted by stars, showing the halves of the egg-shell from which they issued at birth. On the *denarii*, the "pence" of the good Samaritan, they are in full speed as if charging in the battle of Lake Regillus, and the *sestertii* and *quinarii* have the same; but even before this, about 300 B. C., coins were struck by the Bruttii of Magna Graecia, in Lower Italy, that bore the heads of the Twins on one side with their mounted figures on the other. The coins of Rhegium had similar designs, as had those of Bactria.

For their efficient aid in protecting their fellow Argonauts in the storm that had nearly overwhelmed the Argo, the Gemini were considered by the Greeks, and even more by the Romans, as propitious to mariners, Ovid writing in the *Fasti:*

Utile sollicitare sidus utrumque rati,

which moral John Gower, the friend of Chaucer, rendered:

A welcome couple to a vexed barge;

and Horace, in his *Odes*, as translated by Mr. Gladstone:

So Leda's twins, bright-shining, at their beck
Oft have delivered stricken barks from wreck.

In *The Acts of the Apostles*, xxviii, 11, we read that the Twin Brothers were the "sign," or figurehead, of the ship in which Saint Paul and his companions embarked after the eventful voyage that had ended in shipwreck on Malta; or, as Tindale rendered it in 1526:

a ship of Alexandry, which had wyntred in the Yle, whose badge was Castor and Pollux,—

the Greek Alexandria, and Ostia, the harbor of Rome, specially being under the tutelage of the Twins, who were often represented on either side of the bows of vessels owned in those ports.

The incident of the storm in the history of the Twins seems to have associated them with the electrical phenomenon common in heavy weather at sea, and well known in ancient times, as it is now. Pliny described it at length in the *Historia Naturalis*, and allusions to it are frequent in all literature; the idea being that a double light, called Castor and Pollux, was favorable to the mariner. Horace designated this as *Fratres Helenae, lucida sidera*, rendered by Mr. Gladstone "Helen's Brethren, Starry Lights"; Rabelais wrote:

> He had seen Castor at the main yard arm;

and our Bryant:

> resplendent cressets which the Twins
> Uplifted in their ever-youthful hands.

A single light was "that dreadfull, cursed, and threatening meteor called Helena,"—the sister of the Twins that brought such ill luck to Troy.

In modern times these lights are known as Composant, Corposant, and Corpusant, from the Italian Corpo Santo; Pigafetta ending one of his descriptions of a dangerous storm at sea with "God and the Corpi Santi came to our aid"; and as the Fire of Saint Helen, Saint Helmes, or Telmes —San Telmo of Spain; or of San Anselmo, Ermo, Hermo, and Eremo, from Anselmus, or Erasmus, bishop of Naples, martyred in Diocletian's reign. Ariosto wrote of it, *la disiata luce di Santo Ermo;* and in Longfellow's *Golden Legend* the Padrone exclaims:

> Last night I saw Saint Elmo's stars,
> With their glittering lanterns all at play
> On the tops of the masts and the tips of the spars,
> And I knew we should have foul weather to-day.

The phenomenon also has been called Saint Anne's Light; and some one has dubbed it Saint Electricity. In recent centuries, with seamen of the Latin races, it has been Saint Peter and Saint Nicholas; the former from his walking on the water, and the latter from the miracles attributed to him of stilling the storm on his voyage to the Holy Land when he restored to life the drowned sailor, and again on the Aegean Sea. These miracles have made Nicholas the patron saint of all Christian maritime nations of the south of Europe, and famous everywhere. In England alone 376 churches are dedicated to him,—more than to that country's Saint George.

In Eden's translation from Pigafetta's account of his voyage with Magellan, 1519–1522, we read that when off the coast of Patagonia the navigators

were in great daungiour by tempest. But as soon as the three fyers cauled saynte Helen, saynte Nycolas, and saynt Clare, appered uppon the cabels of the shyppes, suddeynely the tempest and furye of the wyndes ceased . . . the which was of such comfort to us that we wept for joy.

This Saint Clare is from Clara d'Assisi, the foundress of the order of Poor Clares in the 13th century, by whose rebuke the infidel Saracens were put to flight when ravaging the shores of the Adriatic. Von Humboldt mentioned in *Cosmos* another title, San Pedro Gonzalez, probably Saint Peter of Alcantara, another patron saint of sailors, "walking on the water through trust in God."

A few words as to Pigafetta may be not uninteresting. His work is described in Eden's *Decades* as

A briefe declaration of the vyage or navigation made abowte the worlde. Gathered owt of a large booke wrytten hereof by Master Antonie Pygafetta Vincentine [*i. e.* from Vincenza], Knyght of the Rhodes and one of the coompanye of that vyage in the which, Ferdinando Magalianes a Portugale (whom sum caule Magellanus) was generall Capitayne of the navie.

Pigafetta was knighted after his return to Seville in the ship *Victoria* that Transilvanus wrote was "more woorthye to bee placed amonge the starres then that owlde Argo." And it was from Eden's translation of this "large booke" that Shakespeare is supposed to have taken his Caliban of the *Tempest*, whose "dam's god, Setebos," was worshiped by the Patagonians. Indeed Caliban himself seems to have been somewhat of an astronomer, for he alludes to Prospero as having taught him how

To name the bigger light, and how the less,
That burn by day and night.

The Gemini were invoked by the Greeks and Romans in war as well as in storm. Lord Macaulay's well-known lines on the battle of Lake Regillus, 498 B. C., one of his *Lays of Ancient Rome*, have stirred many a schoolboy's heart, as Homer's *Hymn to Castor and Pollux* did those of the seamen of earliest classical days. Shelley has translated this last:

Ye wild-eyed muses! sing the Twins of Jove,
. . . .
. . . mild Pollux, void of blame,
And steed-subduing Castor, heirs of fame.
These are the Powers who earth-born mortals save

> And ships, whose flight is swift along the wave.
> When wintry tempests o'er the savage sea
> Are raging, and the sailors tremblingly
> Call on the Twins of Jove with prayer and vow,
> Gathered in fear upon the lofty prow,
> And sacrifice with snow-white lambs, the wind
> And the huge billow bursting close behind,
> Even then beneath the weltering waters bear
> The staggering ship—they suddenly appear,
> On yellow wings rushing athwart the sky,
> And lull the blasts in mute tranquillity,
> And strew the waves on the white ocean's bed,
> Fair omen of the voyage; from toil and dread,
> The sailors rest rejoicing in the sight,
> And plough the quiet sea in safe delight.

They seem to have been a common object of adjuration among the Romans, and, indeed, as such have descended to the present time in the boys' "By Jiminy!" while the caricature of 1665, *Homer A la Mode*, had, as a common expression of that day, "O Gemony!" And theatre-goers will recall the "O Gemini!" of Lucy in Sheridan's *Rivals*.

Astrologers assigned to this constellation guardianship over human hands, arms, and shoulders; while Albumasar held that it portended intense devotion, genius, largeness of mind, goodness, and liberality. With Virgo it was considered the **House of Mercury,** and thus the **Cylenius tour** of Chaucer; and a fortunate sign, ruling over America, Flanders, Lombardy, Sardinia, Armenia, Lower Egypt, Brabant, and Marseilles; and, in ancient days, over the Euxine Sea and the river Ganges. High regard, too, was paid to it in the 17th century as being peculiarly connected with the fortunes of the south of England and the city of London; for the Great Plague and Fire of 1665 and 1666 occurred when this sign was in the ascendant, while the building of London Bridge and other events of importance to the city were begun when special planets were here. But two centuries previously it was thought that whoever happened to be born under the Twins would be "ryght pore and wayke and lyf in mykul tribulacion." Chinese astrologers asserted that if this constellation were invaded by Mars, war and a poor harvest would ensue.

Ampelius assigned to it the care of Aquilo, the North Wind, the Greek Boreas that came from the north one third east.

Its colors were white and red like those of Aries, and it was the natal sign of Dante, who was born on the 14th of May, 1265, when the sun entered it for the first time in that year. He made grateful acknowledgment of this in the *Paradiso:*

> O glorious stars, O light impregnated
> With mighty virtue, from which I acknowledge
> All of my genius, whatsoe'er it be;

and called them *gli Eterni Gemelli.* How like this is to Hesiod's reference to the Muses!

> To them I owe, to them alone I owe,
> What of the seas, or of the stars, I know.

The sign's symbol, ♊, has generally been considered the Etrusco-Roman numeral, but Seyffert thinks it a copy of the Spartans' emblem of their Twin Gods carried with them into battle. Brown derives it from the cuneiform 𒈦, the ideograph of the Akkad month Kas, the Twins, the Assyrian Simānu, corresponding to parts of our May and June when the sun passed through it. The constellation was certainly prominent on the Euphrates, for five of its stars marked as many of the ecliptic divisions of that astronomy.

The Gemini were the **Ape** of the early Chinese solar zodiac, and were known as **Shih Chin;** Edkins, calling it **Shi Ch'en,** says that this title was transferred to it from Orion. Later on the constellation was known as **Yin Yang,** the Two Principles; and as **Jidim,** an important object of worship.

The Reverend Mr. William Ellis wrote, in his *Polynesian Researches*, that the natives of those islands knew the two stars as Twins, Castor being **Pipiri** and Pollux **Rehua;** and the whole figure **Na Ainanu,** the Two Ainanus, one Above, the other Below, with a lengthy legend attached; but the Reverend Mr. W. W. Gill tells the same story, in his *Myths and Songs of the South Pacific*, as belonging to stars in Scorpio. The Australian aborigines gave them a name signifying Young Men, while the Pleiades were Young Girls; the former also being **Turree** and **Wanjil,** pursuing Purra, whom they annually kill at the beginning of the intense heat, roasting him by the fire the smoke of which is marked by Coonar Turung, the Great Mirage. The Bushmen of South Africa know them as **Young Women,** the wives of the eland, their great antelope.

Aristotle has left an interesting record of the occultation, at two different times, of some one of the stars of Gemini by the planet Jupiter, the earliest observation of this nature of which we have knowledge, and made probably about the middle of the 4th century B. C.

The southern half of the constellation lies within the Milky Way, α and β, on the north, marking the heads of the Twins between Cancer and Auriga, and noticeably conspicuous over setting Orion in the April sky.

Argelander enumerates 53 naked-eye stars, and Heis 106.

Starry Gemini hang like glorious crowns
Over Orion's grave low down in the west.
Tennyson's *Maud.*

α, Binary, 2.7 and 3.7, bright white and pale white.

Castor, Ovid's **Eques,** the Horseman of the Twins, and the mortal one as being the son of Tyndarus, is the well-known name for this star, current for centuries; but in later Greek days it was Ἀπόλλων, and **Apollo** with the astronomers even through Flamsteed's time.

It will be remembered that till toward the Christian era this name for the god of day was the title of the planet Mercury when morning star,[1] its rapid orbital movement and nearness to the sun preventing its earlier identification with the evening star,[1] which was designated, as now, after the god of thieves and darkness. In Percy's *Reliques* Mercury is described as "the nimble post of heaven"; Goad, in 1686, called it

a squirting lacquey of the sun, who seldom shows his face in these parts, as if he were in debt;

while this same quick motion induced the alternative word of the chemists for quicksilver, as well as for the very uncomfortable human temperament that Byron described:

a mercurial man
Who fluttered over all things like a fan.

Notwithstanding, however, the supposed difficulty of seeing Mercury,— Copernicus died regretting that he had never observed it, although this was doubtless partly due to his high latitude and the mists arising from the Vistula at Thorn,—the canon Gallet, whom La Lande styled Hermophile, saw it 100 times, and Baily said that Hevelius observed it 1100 times! Indeed, it is easily visible in the latitude of New York City for several days, at its elongation, if one knows where to look for it.

But to return to our star Castor.

It was Ἀπέλλων in the Doric dialect, which degenerated into **Afelar, Aphellon, Aphellan, Apullum, Aphellar,** and **Avellar**; the **Avelar** of Apian[2] of the 16th century subsequently appearing as **Anelar,** the Alfonsine **Anhelar.**

[1] As morning and evening star in Egypt it was Set and Horus; in India, Buddha and Rauhinya; and in Greece Ἐρόεις, the Lovely One, and Στίλβων, the Sparkling One. Its earliest observation, reported by Ptolemy as from Chaldaea, was on the 15th of November, 265 B. C., the planet then being between β and δ Scorpii.

[2] This Apian was Pieter Bienewitz, whose surname was Latinized, after the fashion of his day, into Apianus; *apis*, our word *bee*, taking the place of the German *biene.*

Caesius had the synonymous **Phoebus,** and also cited **Theseus,** but this should rather be applied to β as another title of the original Hercules. Bayer gave **Rasalgeuze**; and Riccioli, **Algueze** *vel potius* **Elgiautzi,** but these also better belong to β.

The Babylonians used Castor to mark their 11th ecliptic constellation, **Mash-mashu-Mahrū,** the Western One of the Twins; while with Pollux the two constituted **Mas-tab-ba-gal-gal,** the Great Twins. In Assyria they were **Mas-mas** and **Tuāmu,** the Twins, although that country knew other twin stars here as well as elsewhere in the sky. As an object of veneration Castor was **Tur-us-mal-maχ,** the Son of the Supreme Temple; but in astrology, everywhere, it has been a portent of mischief and violence.

When the Arabians adopted the Greek figures they designated this star as **Al Rās al Taum al Muḳaddim,** the Head of the Foremost Twin; but, according to Al Tizini, the early and indigenous term was **Al Awwal al Dhirāʽ,** the First in the Paw or Forearm. Reference was made by this to the supposed figure of the enormous early Lion, the nomads' **Asad,** the Outstretched Forearm of which α and β marked as **Al Dhirāʽ al Mabsuṭāt.** This extended still further over Gemini, the other, the Contracted one, **Al Maḳbūḍah,** running into Canis Minor. The rest of this monstrosity included Cancer, part of our Leo, Boötes, Virgo, and Corvus, as was mentioned by Kazwini, and commented on by Ideler, who sharply criticized mistakes in its construction. Al Bīrūnī also described this ancient figure, especially complaining of the many errors and much confusion in the Arab mind as to the nomenclature of the two stars, although he himself used titles for them generally applied only to Sirius and Procyon. Ideler and Beigel attributed this exaggerated and incongruous formation to blunders of misunderstanding and transcription by early writers and copyists. Indeed, the former asserted that the whole was the creation of grammarians who knew nothing of the heavens, and arbitrarily misrepresented older star-names.

The two bright stars were the 5th *manzil*, **Al Dhirāʽ,** and the 5th *nakshatra*, **Punarvarsū,** the Two Good Again; Aditi, the sky goddess, mother of the Adityas, being the presiding divinity, and β marking the junction with Pushya, the next *nakshatra*. They also constituted the 5th *sieu*, **Tsing,** a Well, or Pit, anciently **Tiam,** although this was extended to include ϵ, d, ζ, λ, ξ, γ, ν, and μ, Biot making the last the determinant star.

α and β also were a distinct Chinese asterism, **Ho Choo,** and with γ and δ were **Pih Ho.**

As marking lunar stations, Brown thinks them the Akkadian **Supa,** Lustrous; the Coptic **Pimafi,** the Forearm; the Persian **Taraha,** the Sogdian **Ghamb,** and the Khorasmian **Jiray,**— these last three titles signifying the

Two Stars. Hyde wrote that the Copts knew it as Πιμάι, or Πιμάιντεκεων, the Forearm of the Nile; κεων being for Gihon, a name for that river.

Castor is 7° north of the ecliptic, but, although literally heading the constellation, is now fainter than its companion, and astronomers generally are agreed that there has been inversion of their brilliancy during the last three centuries. It culminates on the 23d of February.

It is among

> those double stars
> Whereof the one more bright
> Is circled by the other,

viewed by the Self-indulgent Soul of Tennyson's *Palace of Art;* and Sir John Herschel called it the largest and finest of all the double stars in our hemisphere; while the rapid revolution of its two components first convinced his father of the existence of binary systems. But Bradley had already noticed a change of about 30° in their angle of position between 1718 and 1759, and "was thus within a hair's breadth of the discovery of their physical connection," afterwards predicted, in 1767, by the Reverend John Michell, and positively made in 1802 by Sir William Herschel, who coined the word "binary" now applied to this class of stars. Burnham wrote in 1896 that we have only 36 pairs whose orbits can be said to be well determined, and about 230 other pairs probably binary systems; and there are 1501 other pairs, within 2″ of space between the components, from which the foregoing number may be increased; as well as other pairs now known only as having a common proper motion.[1] Of course the stars observed till now have been almost entirely in the northern heavens,— within 120° of the pole,—so that these numbers may be largely added to as astronomers turn their attention to the southern skies with this object in view.

The orbit of Castor is such, however, that the observations of even a century do not enable us to calculate its size or period with any certainty; but the period certainly is long,— probably between 250 and 1000 years. The components at present are about 5″.7 apart, equal to the angle subtended by a line an inch long at the distance of half a mile. Their position angle is about 227°.

The spectrum is of the Sirian type, and, according to the Potsdam observers, the star is approaching us at the rate of 18.5 miles a second. In 1895 Belopolsky announced that the larger star, like Spica, is a spectro-

[1] In a note from Professor Burnham, of the 19th of July, 1898, in regard to these figures, he says: "The statements I made a couple of years ago about binary systems will hold good generally at this time. . . . So far as well-determined orbits are concerned, I do not think anything could be added to the estimate I made."

scopic binary, completing its revolution in less than three days around the centre of gravity between it and an invisible companion, with a velocity of about 15½ miles a second.

Burnham thinks that the 9.5-magnitude star, 73″ distant, forms, with the two larger, one vast physical system.

In 1888 Barnard found five new nebulae within 1° of Castor.

β, 1.1, orange,

is **Pollux,** formerly **Polluces,** the Greek Πολυδευκής; Ovid's **Pugil,** the Pugilist of the Two Brothers, and the immortal one as being son of Zeus.

As companion of 'Απόλλων, this was Ἡρακλῆς and Ἡρακλέης, descending to Flamsteed's day as **Hercules,** and degenerating, in early catalogues, into **Abrachaleus,** that Caesius derived from the Arabic Ab, Father, and the Greek word; this being contracted by some to **Aracaleus,** by Grotius to **Iracleus,** by Hyde to **Heraclus,** and by Riccioli to **Garacles.** All these are queer enough, as are some of Castor's titles; but what shall be said of Riccioli's **Elhakaac,** that he attributes to the Arabs for α and β jointly, and **Ketpholtsuman** for β alone, and with no clue to their origin!

It was the early Arabs' **Al Thānī al Dhirā'**, the Second in the Forearm; but the later termed it **Al Rās al Taum al Mu'ah·h·ār,** the Head of the Hindmost Twin, and **Al Rās al Jauzā'**, the Head of the Twin,—the Alfonsine **Rasalgense** and **Rasalgeuze,** that elsewhere is **Rasalgauze.** Riccioli cited **Elhenaat,** but this he also more properly gave to γ.

β was the determinant of the 12th Babylonian ecliptic asterism **Mashmashu-arkū,** the Eastern One of the Twins; and individually **Mu-sir-kes-da,** the Yoke of the Inclosure.

It lies 12° north of the ecliptic, the zodiac's boundary line running between it and Castor; and Burnham has found five faint companions down to 13.5 magnitude.

Elkin gives its parallax as 0″.057; and Scheiner, its spectrum as Solar; its rate of recession from us being about one mile a second.

It is one of the lunar stars made use of in navigation; and, in astrology, differed from its companion in portending eminence and renown.

Ptolemy characterized β as ὑπόκιρρος, a favorite word with him for this star-tint, and generally supposed to signify "yellowish" or "reddish," Bayer correctly following the former in his *subflava;* but the *Alfonsine Tables* of 1521 translated it *quae trahit ad aerem, et est cerea.* Miss Clerke, somewhat strongly, says "fiery red."

The two *lucidae* probably bore the present title of the constellation long antecedent to the latter's formation; they certainly were the **Mas-mas,** or Twins, of the Assyrians, independent of the rest of the figure.

As a convenient measuring-rod it may be noted that α and β stand 4½° apart; and this recalls an early signification of their *manzil* title, Al Dhirā', the Arabs' Ell measure of length that the stars were said to indicate. This naturally became the dual **Al Dhirā'ān** that also was used on the Desert for other similar pairs of stars.

γ, 2.2, brilliant white.

Almeisan, Almisan, Almeisam, and **Almisam** are from **Al Maisan,** the Proudly Marching One, its early Arabic name, which Al Firuzabadi, however, said was equally applicable to any bright star.

Riccioli called it **Elhenaat,** but **Alhena** is now generally given to it, from **Al Han'ah,** the 4th *manzil*, γ, μ, ν, η, and ξ, in the feet of the Twins. This word, usually translated a Brand, or Mark, on the right side of a camel's, or horse's, neck, was defined by Al Bīrūnī as Winding, as though the stars of this station were winding around each other, or curving from the central star; and they were **Al Nuḥātai,** the dual form of Al Nuḥāt, a Camel's Hump, itself a curved line. Some Arabic authority found in them, with χ^1 and χ^2 of Orion, the **Bow** with which the Hunter is shooting at the Lion.

In Babylonia γ marked the 10th ecliptic constellation, **Mash-mashu-sha-Risū,** the Twins of the Shepherd (?), and, with η, probably was **Mas-tab-ba-tur-tur,** the Little Twins; and, with η, μ, ν, and ξ, all in the Milky Way, may have been the Babylonian lunar mansion **Khigalla,** the Canal, and the equivalent Persian **Rakhvad,** the Sogdian **Ghathaf,** and the Khorasmian **Gawthaf.**

δ, Double, 3.8 and 8, pale white and purple.

Wasat and **Wesat** are from **Al Wasat,** the Middle, *i. e.* of the constellation; but some have referred this to the position of the star very near to the ecliptic, the central circle.

In China it was **Ta Tsun,** the Great Wine-jar.

The components are 7″ apart, with a position angle of 203°, and may form a binary system.

Just north of δ lies the radiant point of the **Geminids,** visible early in October; another stream of meteors bearing the same title appearing from the northeastern border of the constellation and at its maximum on the 7th of December.

ε, Double, 3.4 and 9.5, brilliant white and cerulean blue.

Mebsuta is from **Al Mabsuṭāt,** the Outstretched, from its marking the extended paw of the early Arabic Lion, but now it is on the hem of Castor's tunic. Burritt had it **Melucta** in his *Geography*, and **Mebusta** in his *Atlas;* Professor Young, following English globes, has **Meboula;** and elsewhere we find **Menita, Mesoula,** and **Mibwala.**

ε, δ, λ, and others near by, were the Chinese **Tung Tsing.**

ζ, Variable, 3.7 to 4.5, pale topaz.

Mekbuda is from **Al Maḳbūḍah,** Contracted, the Arabic designation for the drawn-in paw of the ancient Asad; but some, with less probability, derive it from **Al Mutakabbidah,** a Culminating Star.

Its variations, discovered by J. F. Julius Schmidt at Athens in 1847, have a period of about ten days, but Chandler says that definitive investigations are not completed. Lockyer thinks it also a spectroscopic binary.

η, Binary and variable, 3.2 to 3.7, and 9.

Propus is from the Πρόπους of Hipparchos and Ptolemy, indicating its position in front of Castor's left foot, and is its universal title, with the equivalent **Praepes.** Riccioli wrote it Πρόπος, and Flamsteed gave both Πρόπους and **Propus;** but Tycho had applied this last to the star Fl. 1 among the *extras* of Gemini. This position of η similarly made it the **Pish Pai** of the Persians.

Bassus and Hyginus said **Tropus,** Turn, referring to the apparent turning-point of the sun's course at the summer solstice, which now is more precisely marked by the star *y* just eastward from η; and Flamsteed also had Τρόπος.

Flammarion's assertion that Hipparchos knew η as a distinct constellation, **Propus,** does not seem well founded.

Tejat prior is from **Al Tahāyī,** an anatomical term of Arabia by which it was known in early days; a name also applied to stars in the head of Orion. The Arabs included it with γ and μ in their **Nuḥātai;** the Chinese knew it as **Yuë,** a Battle-ax; and in Babylonia it marked the 8th ecliptic constellation, **Maru-sha-pu-u-mash-mashu,** the Front of the Mouth of the Twins.

It portended lives of eminence to all born under its influence.

The variability of η was discovered by Schmidt in 1865, and its period is now considered as 229–231 days; in 1881 Burnham found it double, the components 1″.08 apart, and likely to prove an interesting binary system.

Near this star Sir William Herschel discovered the planet Uranus on the 13th of March, 1781. He thought it a comet, and its discovery as such was communicated to the Royal Astronomical Society on the 26th of April. Its true nature, however, first suspected by Maskelyne, was announced in the succeeding year by Lexell of Saint Petersburg and by La Place; and Herschel then published it on the 7th of November, 1782, as the Georgium Sidus, thus following Galileo, who, till he knew their true nature, had named Jupiter's satellites Sidera Cosmiana and Sidera Medicea, after his patron the 2d Cosmo di Medici, and Tardé, who had called the sun-spots Borbonica Sidera. Continental astronomers designated the planet as Herschel, and this in a much varied orthography, strangely erroneous considering the fame of its discoverer. We find it thus with La Lande in 1792; indeed, Herschel appeared as an alternative title in our text-books as late as fifty years ago; but Bode suggested the present Uranus to conform to the mythological nomenclature of the other planets, and because the name of the oldest god was specially applicable to the oldest — as the most distant — body then known in our system.

Uranus, however, had been observed and noted as a star twenty-two times previously by various observers; these are called "the ancient observations"; and Miss Clerke writes: "There is, indeed, some reason to suppose that he had been detected as a wandering orb by savage 'watchers of the skies' on the Pacific long before he swam into Herschel's ken."[1]

The 4th-magnitude θ, and ι, ν, τ, and ϕ, collectively were **Woo Chow Shih,** or **Woo Choo How,** the Seven Feudal Princes of China.

ι is **Propus** in the *Standard Dictionary*, although it lies between the shoulders of the Twins.

μ, Double, 3.2 and 11, crocus yellow and blue,

occasionally has been known as **Tejat posterior,** and sometimes as **Nuḥātai,** from the *manzil* of that title of which it formed a part.

The *Century Dictionary* and *Cyclopedia* apply to it the **Pish Pai** seen for η, yet appropriate enough for this similarly situated star; but in Flamsteed's edition of Tycho's catalogue we distinctly read of it, *dicta* **Calx,** the Heel.

It marked the 9th ecliptic constellation of Babylonia as **Arkū-sha-pu-u-mash-mashu,** the Back of the Mouth of the Twins.

[1] The Burmans, too, thought that there was an 8th planet, Rahū, but invisible; and the Hindus named other imaginary planets Kethu, Rethu, and Kulican; and figured Sani, their god Saturn, with a circle around him of intertwined serpents ages before Galileo's day; although this has had a very different explanation.

In China it was included with Castor and others in the *sieu* **Tsing.**

The components are 80″ apart, at a position angle of 79°.

ξ, a 4th-magnitude, was Al Bīrūnī's **Al Zirr,** the Button.

χ, a 5th-magnitude, with μ Cancri, was the Chinese **Tseih Tsing,** Piled-up Fuel.

★

Globus Aerostaticus, vel Aetherius, the Balloon,

was formed by La Lande in 1798, but, like most of his stellar creations, seems to have passed out of the recognition of science.

It lay east of the Microscope, between the tail of the Southern Fish and the body of Capricorn.

Bode published it in his *Die Gestirne* as the **Luft Ballon,** Ideler's **Luft Ball,** with twenty-two stars; and Father Secchi still had it in his maps as the Italian **Aerostáto.** With the French it was the **Ballon Aérostatique.**

★

Proxima sideribus numinibusque feror.

Flavius Avianus' *15th Fable.*

Grus, the Crane,

is one of the so-called Bayer groups, **la Grue** of the French and Italians, **der Kranich** of the Germans; and the title is appropriate, for Horapollo, the grammarian of Alexandria, about A. D. 400, tells us that the crane was the symbol of a star-observer in Egypt, presumably from its high flight as described in our motto.

Caesius, who carried his biblical symbols even to the new constellations, imagined this to be the **Stork in the Heaven** of *Jeremiah* viii, 7, although the Crane occurs in the same verse; but Julius Schiller combined it with Phoenix in a representation of **Aaron the High Priest.**

The Arabians included its stars in the Southern Fish, Al Sufi giving its α, β, δ, θ, ι, and λ as unformed members of that constellation.

The components, with the exception of the *lucida*, form a gentle curve southwest from this Fish, and among them are stars noted in astronomy.

One hundred and seven are catalogued by Gould as being visible to the naked eye.

α, marking the body of the bird, is the conspicuous 2d-magnitude southwest from Fomalhaut when the latter culminates in autumn evenings, itself coming to the meridian on the 11th of October. It was Al Tizini's **Al Nā'ir,** the Bright One, *i. e.* of the Fish's tail, when that constellation extended over the stars of our Grus. The Chinese knew it as **Ke.**

β, a 2.2-magnitude red star, was Al Tizini's Rear One at the end of the tail of his Fish, thirty-five minutes of arc to the eastward from α. It is in the left wing of the Crane.

γ, a 3d-magnitude, was the same author's **Al Dhanab,** the Tail itself, but now marks the eye in the bird's figure.

π^1, a 6.7-magnitude deep crimson star, and its somewhat brighter white companion, π^2, are like "little burnished discs of copper and silver, seen under strong illumination."

The alternative title for the stars of Grus,

Phoenicopterus, the Flamingo,

is now seldom, if ever, used, nor can I find any record of its inventor, or date of its adoption as a constellation name. Chilmead's *Treatise* contains this reference to it:

> *The* Phoenicopter *we may call the Bittour* [the old English word for Bittern].
>
> . . . .
>
> *The* Spaniards *call it* Flamengo: *and it is described with the wings spread abroad, and as it were striking with his bill at the South Fish, in that part where he boweth himselfe. This Asterisme consistith of 13 Starres: of which, that of the second magnitude in his head is called, the* Phoenicopters Eye: *and it hath two other Stars also of the same magnitude, one in his backe, and the other in his left wing. And those two which are in the middle of his necke,* Paulus Merula in his first booke of his Cosmography, *calleth his Collar or Chaine.*

The absence of our titles in the foregoing description would show that the **Bittern,** or **Flamingo,** was the popular English figuring and title in the early part of the 17th century.

★

> Hercules with flashing mace.
>
> Bryant's *The Constellations.*

Hercules,

stretching from just west of the head of Ophiuchus to Draco, its eastern border on the Milky Way, is one of the oldest sky figures, although not known to the first Greek astronomers under that name,—for Eudoxos had Ἐνγούνασι; Hipparchos, Ἐνγόνασι, *i. e.* ὁ ἐν γόνασι καθήμενος, Bending on his Knees; and Ptolemy, ἐν γόνασιν. Aratos added to these designations Ὀκλάζων, the Kneeling One, and Εἴδωλον, the Phantom, while his description in the *Phainomena* well showed the ideas of that early time as to its character:

> . . . like a toiling man, revolves
> A form. Of it can no one clearly speak,
> Nor to what toil he is attached; but, simply,
> *Kneeler* they call him. Labouring on his knees,
> Like one who sinks he seems; . . .
> . . . And his right foot
> Is planted on the twisting *Serpent's* head.

But all tradition even as to

> Whoe'er this stranger of the heavenly forms may be,

seems to have been lost to the Greeks, for none of them, save Eratosthenes, attempted to explain its origin, which in early classical days remained involved in mystery. He wrote of it, ὀυτός, φασὶν, Ἡρακλῆς ἐστίν, standing upon the Ὄφις, our Draco; and some modern students of Euphratean mythology, associating the stars of Hercules and Draco with the sun-god Izhdubar[1] and the dragon Tiāmat, slain by him, think this Chaldaean myth the foundation of that of the classical Hercules and the Lernaean Hydra. Izhdubar is shown on a cylinder seal of 3000 to 3500 B. C., and described in that country's records as resting upon one knee, with his foot upon the Dragon's head, just as Aratos says of his Ἐνγόνασι, and as we have it now. His well-known adventures are supposed to refer to the sun's passage through the twelve zodiacal signs, appearing thus on tablets of the 7th century before Christ. This myth of several thousand years' antiquity may have been adopted by Greece, and the solar hero changed into Hercules with his twelve familiar labors.

This constellation is said to have been an object of worship in Phoenicia's most ancient days as the sky representative of the great sea-god **Melkarth.** Indeed, it has everywhere been considered of importance, judging from its abundant nomenclature and illustration, for no other sky group seems to have borne so many titles.

The usual Greek name was transliterated **Engonasi, Engonasis,** and **Engonasin** down to the days of Bullialdus, with whom it appeared in the queer

[1] Izhdubar was identified with Nimrod, and known, too, as Gizdhubar, Gilgamesh, or Gi-il-ga-mes, the Γίλγαμος of Aelian. He was aided in his exploits by his servant-companion, the first Centaur, Ea-bani, or Hea-bani, the Creation of Ea.

combination of Greek and Roman letters Ό εn Ίonacín; but the poets translated it as **Genuflexus, Genunixus,** and **Geniculatus; Ingeniculatus** with Vitruvius; **Ingeniclus** and **Ingeniculus** with Firmicus; while **Ingenicla Imago** and **Ignota Facies** appear in Manilius,— his familiar line,

Nixa venit species genibus, sibi conscia causae,

being liberally translated by Creech,

Conscious of his shame
A constellation kneels without a name.

We see with other authors the synonymous **Incurvatus in genu, Procidens, Prociduus, Procumbens in genua,** and **Incumbens in genibus; Defectum Sidus** and **Effigies defecta labore;** and the *Tetrabiblos* of 1551 had **Qui in genibus est.**

It also was **Saltator,** the Leaper; Χάρωψ, the Keen-eyed One; Κορυνήτης and Κορυνηφόρος, the equivalents of **Clavator** and **Claviger,** the Club-bearer of the Latins: all applied to the constellation in early days, from classical designations of the hero **Hercules,** whose own name has now become universal for it. Although we first find this in the *Catasterisms,* Avienus asserted that it was used by Panyasis, the epic poet of 500 B. C., and uncle of Herodotus, perhaps to introduce into the heavens another Argonaut. The **Nessus** of Vitruvius came from the story of Deïanira, the innocent cause of Hercules' death, when, as in the *Death of Wallenstein,*

Soared he upward to celestial brightness;

Nisus, from the city of Nisa; **Malica, Melica, Melicartus,** and **Melicerta,** from the name of its king, known later as **Palaemon,**— although some refer these to the title of the great god of Phoenicia, Melkarth, the King of the City; and **Aper,** from the Wild Boar slain at Elis. It was **Cernuator,** the Wrestler, from the hero's skill; **Caeteus, Ceteus,** and **Cetheus,** as son of Lycaon, and so uncle or brother of Kallisto, who, as Ursa Major, adjoined this constellation; indeed, it was even known as **Lycaon** himself, weeping over Kallisto's transformation. Ovid's **Alcides** was a common poetical title, either from 'Αλκή, Strength, or from Alcaeus, Hercules' grandfather; while **Almannus** and **Celticus** came from the fact that a similar hero was worshiped by the Germans and Celts, themselves noted for strength and daring deeds, and said to have been descended from Hercules. The unexplained **Pataecus** and **Epipataecus** are from Egypt; **Maceris,** from Libya; while **Desanaus, Desanes,** and **Dosanes,** or **Dorsanes,** are said to be of Hindu origin.

Other titles are **Ixion,** laboring at his wheel, perhaps because Hercules also labored; or from the radiated object shown on Euphratean gems, a supposed representation of the solar prototype of Hercules, which in later times may easily have been regarded as a wheel; **Prometheus,** bending in chains on Caucasus; **Thamyris,** sad at the loss of his lyre; **Amphitryoniades,** from the supposed sire of Hercules; **Heros Tirynthius,** from the place where he was reared; and **Oetaeus,** from the mountain range of Thessaly whence he ascended the funeral pyre. The **Sanctus** that has appeared as a title is properly **Sancus,** the Semo Sancus, of Sabine-Umbrian-Roman mythology, identified with Hercules. **Theseus** was a name for this constellation, from the similar adventures of the originals; **Mellus** and **Ovillus** trace back to the Malum and Ovis in the myth of the Apples, or Sheep, of the Hesperides, with which the story of Hercules is connected,— different ideas, but both from *μῆλον* with this double signification; although La Lande thought that reference was made to the skin of the lion thrown over the hero's shoulder. We also occasionally see **Diodas, Manilius, Orpheus,** and **Trapezius,** the exact connection of which with our sky figure is not certain.

The 4th edition of the *Alfonsine Tables* singularly adds **Rasaben,** from the neighboring Draco's Al Rās al Thu'bān.

Bayer erroneously quoted Γνύξ ἐριπών, on Bended Knee, as if from Homer; and gave Ἔιδωλον ἄπευθος, the Unknown Image, and **Imago laboranti similis.** He also cited the Persians' **Ternuelles,** which Beigel suggested might be from their mistaken orthography of the word Hercules; and Hyde added another term, from that people, in **Ber zanū nisheste,** Resting on his Knees, a repetition of the earliest idea as to the figure.

Flammarion states that he found our modern title first mentioned in an edition of Hyginus of 1485,— but he had not read Eratosthenes; and some say that even this Hercules of Hyginus was really designed for the adjacent Ophiuchus.

The modern Italians' **Ercole** is like their Roman predecessors' abbreviated name for the deity, who was one of their most frequent objects of adjuration.

Our stellar figure generally has been drawn with club and lion-skin, the left foot on Draco and the right near Boötes, the reversal of these by Aratos being criticized by Hipparchos; but the Farnese globe shows a young man, nude and kneeling; while the *Leyden Manuscript* very inappropriately drew it as a young boy, erect, with a short star-tipped shepherd's crook, bearing a lion's skin and head. Bayer shows the strong man kneeling, clothed in the lion's skin, with his "all brazen" club and the Apple Branch.

This last he called **Ramus pomifer,** the German **Zweig,** placing it in the right hand of Hercules, on the edge of the Milky Way; but this even then was an old idea, for the Venetian illustrator of Hyginus in 1488 showed, in the constellation figure, an Apple Tree with a serpent twisted around its trunk. Argelander followed Bayer's drawing, but Heis transfers the Branch to the left hand, with two vipers as a reminder of the now almost forgotten stellar Cerberus with serpents' tongues, which Bayer did not know. The French and Italians, who give more prominence to these adjuncts of Hercules than do we, have combined them in a sub-constellation **Rameau et Cerbere** and **Ramo e Cerbero.** In all this, as well as in some of the titles of the Hercules constellation and of Draco, reappears the story of the Golden Fruits of the Hesperides with their guardian dragon.

It may have been the serpent and apples in our picturing of the constellation that aided Miss Rolleston to her substitution of the biblical **Adam** for the mythological Hercules. Others, however, changed the latter to **Samson** with the jawbone of an ass; and Julius Schiller multiplied him into the **Three Magi.**

The Arabians turned the classical Saltator, or Leaper, into **Al Raḳis,** the Dancer;[1] as also Ἐνγόνασι into **Al Jāthiyy a'la Rukbataihi,** the One who Kneels on both Knees; this subsequently degenerating into **Elgeziale rulxbachei, Alcheti hale rechabatih, Elzegeziale,** and **Elhathi.** It also has often appeared as **Alchete** and **Alcheti**; as **Algethi,** and, in the 1515 *Almagest* and *Alfonsine Tables* of 1521, as **Algiethi** *incurvati super genu ipsius.*

Argelander catalogues 155 naked-eye stars in Hercules, and Heis 227.

Between ζ and η, two thirds of the way from ζ, is N. G. C. 6205, 13 M., the finest cluster in the northern heavens. Halley discovered this in 1714 and thought it a nebula, whence its early title, the **Halley Nebula**; but it is remarkable that it was not sooner seen, for it is visible by the unaided eye, although only 8′ in diameter. Herschel's estimate that it contains 14,000 stars is so high that some regard it as a typographical error for 4000; the number counted by Harvard observers is 724, outside of the nucleus. Miss Clerke records an opinion that it may be 558,000 millions of miles in diameter, and distant from us sixty-five light years; but we have as yet no certain determination of either size or distance. Burnham notes one of its central stars as double, an infrequent occurrence in compressed clusters; and Campbell of the Lick Observatory writes:

[1] The foregoing Dancer, Beigel said, was in the East merely a posture-maker, which the configuration of these stars plainly shows, and hence this title is appropriate. It seems to have wandered to the near-by Draco for the faint μ, although with a different signification,— the Trotting Camel.

In the Hercules cluster the stars are perhaps very little denser than the streams of nebulous matter in which they are situated, and hence their density is [*i. e.* may be] only something a thousand millionth part of that of the sun.

Bailey finds no variables in it.

In the early days of Arab astronomy a space in the heavens, coinciding with parts of Hercules, Ophiuchus, and Serpens, was the **Rauḍah,** or Pasture, the Northern Boundary of which, the **Nasaḳ Shāmiyy,** was marked by the stars β and γ Herculis, the Syrians' **Row of Pearls,** with β and γ Serpentis in continuation of the Pasture line; while δ, α, and ε Serpentis, with δ, ε, ζ, and η Ophiuchi, formed the Southern Boundary, the **Nasaḳ Yamaniyyah.** The group of stars now known as the **Club of Hercules** was the **Sheep within the Pasture.**

α, Double and both irregularly variable, 3.1 to 3.9 and 5 to 7, orange red and bluish green.

Ras Algethi, also **Ras Algathi,** on Malby's globe **Ras Algothi,** is from **Al Rās al Jāthīyy,** the Kneeler's Head; but it often is **Ras Algeti,** sometimes **Ras Algiatha,** and the *Standard Dictionary* has **Ras Algetta.** It was **Rasacheti** with Chilmead. Riccioli's **Ras Elhhathi** and **Ras Alhathi** probably came from **Ras Alheti** of the first three editions of the *Alfonsine Tables;* but in the 4th edition very incorrectly appeared **Rasaben** for both the star and the constellation, probably taken from the neighboring Al Rās al Thuʿbān of Draco;—all Arabian translations of the Greek names.

The nomads' title for it was **Al Kalb al Rāʿi,** the Shepherd's Dog, that our α shared with the adjoining *lucida* of Ophiuchus, 5° distant.

The Chinese called it **Ti Tso,** the Emperor's Seat; and **Tsin.**

Some small stars in Hercules, near α, were included with ι and κ Ophiuchi in the asterism **Ho,** one of the measures of China.

This is a beautiful pair, but apparently not binary, for there has been no certain change in the last century. The components are 4″.8 apart, at a position angle of 119°. Its variability, discovered by Sir William Herschel in 1795, is now described by Chandler as shown by "very irregular oscillations in periods of two to four months." It is one of the most noted of Secchi's 3d type with banded spectra.

α culminates on the 23d of July.

β, 2.8, pale yellow.

Korneforos and **Kornephoros** are from the Κορυνηφόρος which we have seen applied to the whole figure. Burritt has **Kornephorus** *vel* **Rutilicus,**

perhaps the diminutive of *rutilus*, "golden red," or "glittering," an adjective applied to Arcturus; but this term is by no means appropriate for β. The *Arabo-Latin Almagest* of 1515 reads *rutillico*, adding *propinque cillitico*, this last unintelligible unless explained by the Basel edition of 1551 as *penes axillam seu scapulam;* so that we may perhaps consider the alternative title to be from the barbarism used to show the star's position on the shoulder of the figure. Indeed, Bayer said of it, *Rutilicum barbari dicunt.* Ideler, however, asserted his belief that it was from *rutellum*, the diminutive of *rutrum*, a sharp instrument of husbandry or war, in Roman times, that Hercules in some early representations, especially on the Arabic globes, is carrying. The *Century Cyclopedia* gives **Rutilico** as a rarely used name.

β was the Chinese **Ho Chung**, In the River, while the 4th-magnitude γ was **Ho Keen**, Between the River.

Its spectrum is like that of the sun, and the star is approaching our system at the rate of about 22 miles a second.

ζ, 3.1 and 6.5, is a remarkable binary with a period of only 34½ years, the distance between the stars ranging from 0″.6 to 1″.7. According to Belopolsky, it is approaching us at the rate of nearly forty-four miles a second,— the greatest velocity of approach or recession so far ascertained.

θ, 4.1, with adjacent small stars, was **Tien Ke**, Heaven's Record.

κ, Double, 4.8 and 7, light yellow and pale garnet.

Marfak, Mirfak, Marsia, Marfic, and **Marsic** are all found for this star,— as for λ Ophiuchi; but it properly is **Marfik,** from **Al Marfiḳ,** the Elbow; the titles written with the letter *s* probably coming from early confusion with the letter *f*. The Dorians similarly called it Κύβιτον, the Elbow.

In China, with two other stars near by, it was **Tsung Tsing,** an Ancestral Star.

Ptolemy and the Arabian astronomers located it on the right elbow, but Smyth on the left; Heis places it in the right hand, as did Bayer; while Burritt has Marsic in the proper place, but letters it χ.

λ, 4.8, deep yellow.

Masym, Maasym, Maasim, Mazym, Mazim, and **Masini** are from the Arabic **Miʿṣam,** the Wrist, although Ptolemy as well as most of the stellar map-makers located ο on that part of the figure; but Bayer, probably by an oversight, gave the title to λ, not far from the left shoulder, and hence the mistake which still survives. Burritt applied Masym to this lettered star at the elbow, and duplicated it at the one on the hand, omitting the letter; but this title had appeared in the *Latin Almagest* of 1515 and the *Alfonsine Tables* of 1521, not as a proper name, but simply indicative of the position of the star o, which, though now unnamed, should bear that title instead of λ. The same word is used in those works to describe the positions of θ and η Aurigae in the similar location, but is there written **Mahasim.** The *Century Cyclopedia*, by a misprint for λ, uses Masym for χ Herculis in the left hand of the giant.

λ also was **Chaou,** one of the early feudal states of China.

> The Sun flies forward to his brother Sun;
> The dark Earth follows wheel'd in her ellipse.
>
> Tennyson's *The Golden Year.*

Although Johann Tobias Mayer of Göttingen seems to have been the pioneer, in 1760, in the efforts to ascertain the direction of the sun's motion among the stars, yet Sir William Herschel was the first successful investigator as to this, about 1806, and he settled upon the vicinity of λ as the objective point of our solar system, the **Apex of the Sun's Way;** and his determination was, in a great measure, confirmed by later astronomers.

Some recent observations, however, change this: either to ν of this constellation, to the group of small stars four or five degrees north of west from ν, to the immediate vicinity of Wega in the Lyre, or to the neighborhood of Arided, near the tail of the Swan,— yet all in the same general quarter of the heavens. Thirty-five separate determinations of this Apex, made from 1783 to 1892, locate it variously between 227° 18′ and 289° of right ascension, and between 14° 26′ and 53° 42′ in north declination; the weight of authority being in favor of some point[1] in Hercules near the boundary between it and Lyra. The *velocity* of the sun's motion is found by Potsdam computers of spectroscopic observations to be from 7½ to 11¼ miles a second; this is more reliable than the value deduced by other methods.

The **Sun's Quit,** the point in the heavens opposite to the Apex, according to Todd, lies about midway between the stars Sirius and Canopus.

μ^1, a 4th-magnitude triple, half-way between Wega of the Lyre and α Herculis, was the Chinese **Kew Ho,** the Nine Rivers.

The distance between the large star and its 9th-magnitude companion is

[1] Professor Young thinks the Apex in about 267° of right ascension and 31° of declination, but that the data are not yet sufficient to give a very close determination of either the sun's speed or direction, since the problem is embarrassed by the probability of systematic motions among the stars themselves. Results so far obtained are to be regarded only as rather rough approximations.

31″; while the companion itself is a close binary with a period of about 45 years, the distance seldom exceeding 1″.

ν and ξ, of the 4th magnitude, with the small *b*, were the Chinese **Chung Shan,** the Middle Mountain. Some recent investigations place here the Apex of the Sun's Way.

ω, a 4th-magnitude double, by some early transcriber's error, is now given as **Cujam,** from Caiam, the accusative of Caia, the word used by Horace for the Club of Hercules, which is marked by this star. **Gaiam, Guiam,** and **Guyam,** frequently seen, are erroneous. In Burritt's *Atlas* the star is wrongly placed within the uplifted right arm.

The **Club of Hercules** is supposed to have been a separate constellation with Pliny.

★

Horologium Oscillatorium, the Pendulum Clock,

lies to the eastward of Achernar,—*a* of Eridanus,—and north of Hydrus.

In France it is **Orloge**; in Italy, **Orologio**; and in Germany, **Pendeluhr.**

Although shown on the maps, it is rarely mentioned; and the only object in it known to be of special interest is a variable star, detected by Harvard observers in Peru, changing in light from 9.7 to 12.7 in a period of about three hundred days. Gould catalogues 68 stars down to the 7th magnitude; *a*, the *lucida*, being 3.8.

Whitall had on his planisphere a figure, which he entitled **Horoscope,** between "Chemica Fornar" and "Caela Sculptoris," but no Horologium. His title is undoubtedly for our constellation, as it occupies Horologium's place.

★

> Close by the *Serpent* spreads; whose winding Spires
> With order'd stars resemble scaly Fires.
>
> Creech's *Manilius.*

Hydra, the Water-snake,

is the French **Hydre,** the German **Grosse Wasserschlange,** and the Italian **Idra,** and may be classed among the Argonautic constellations, as it was said to represent the **Dragon of Aetes.**

Its stars are now well defined under this single title, but anciently were described, with their riders Corvus and Crater, as Ovid wrote:

Anguis, Avis, Crater, sidera juncta micant.

This continued to the 18th century, Flamsteed and other early astronomers making of them even four divisions, **Hydra, Hydra et Crater, Hydra et Corvus,** and **Continuatio Hydrae. Nepa** and **Nepas,** originally African words for the terrestrial crab and scorpion, seem also to have been used for this constellation in classic times.

Aratos called it Ὕδρη; Eratosthenes, Hipparchos, and Geminos, Ὕδρος, the **Hydros** of Germanicus, while others wrote it Ὕδρα; but Eratosthenes again had it all under Κόραξ, and Hipparchos also used Δράκων.

In Low Latin it has been **Hidra, Idra,** and **Ydra**; and, in the *Almagest* of 1551, **Hydrus** in the masculine, which, correct enough before Bayer's day, would now confound it with the new southern figure. Riccioli, and Hyde in his translation of Ulug Beg's catalogue, had it thus, showing its continuance till then as a common title, although often written **Idrus** and **Idrus aquaticus,** as well as changed to **Serpens aquaticus.**

Other names, also used for the northern Dragon, have been **Draco, Asiua,** and **Asuia,** or **Asvia,** which Bayer referred to as *ἀσούγια non ἀσβία*; but these are not Greek words, and doubtless are from **Al Shujā',** the Snake, transformed, as only the late mediaeval astronomical writers and their immediate successors could transform classical and Arabic terms into their Low Latin and Greek; Chilmead wrote it **Alsugahh.** Still another conception and title may be seen in the *Arabo-Latin Almagest's Stellatio Ydre: et est species serpentium: et jam nominatur Asiua. secur';* where the last word, if an abbreviation for *securis*, "ax," seems not inappropriate when taking the western half of Hydra for a somewhat crooked handle, and Corvus for the ax-head. The **Asina,** or She Ass, which La Lande mentioned, is probably a continuation of some early type error in the barbarous Asiua.

Coluber, the Snake, and **Echidna,** the Viper, also obtain for Hydra, with the adjectives **Furiosus, Magnanimus,** and **Sublimatus,** here used as proper nouns, as they were for Orion. The Arabians similarly called it **Al Ḥayyah,** another of their words for a snake,—**El Havic** in Riccioli's *New Almagest.*

Its representation has generally been as we have it, but the *Hyginus* of 1488 added a tree in whose branches the Hydra's head is resting; probably a recollection of the dragon that guarded the apple-trees of the Hesperides, although this duty really belonged to our Draco; and at times it has been shown as three-headed. Map-makers have always figured it in its present form, the Cup resting midway on its back, with the Raven pecking at one of its folds; Hydra preventing the latter's access to the Cup in punishment for its tattling about Coronis, or for its delay in Apollo's service. The minor constellation **Turdus,** or **Noctua,** only recently has been added to it.

Those who saw biblical symbols among the stars called Hydra the **Flood**; Corvus, **Noah's Raven**; and Crater, the **Cup** "out of which the patriarch sinned"; but Julius Schiller said that the whole represented the **River Jordan.**

The 7th *sieu*, **Lieu,** a Willow Branch, or **Liu,** a Circular Garland,— was the creature's head, 15° south of Praesaepe, δ being the determinant, and formed the beak of the Red Bird; it governed the planets and was worshiped at festivals of the summer solstice as an emblem of immortality.

Here, too, was the 7th *nakshatra*, **Āçleshā,** or **Āçreshā,** the Embracer, figured as a **Wheel,** with Sarpas, the Serpents, as presiding divinities; ϵ marking the junction with the *nakshatra* Maghā.

The 8th *sieu*, **Sing,** a Star, anciently **Tah,** was formed by α, σ, and τ, with others smaller lying near them, α being the determinant. This asterism constituted the neck of the Red Bird, and, Edkins asserts, was also known as the **Seven Stars.**

The 9th *sieu* consisted of κ, υ^1, υ^2, λ, μ, ϕ, and another unascertained, and was called **Chang,** or **Tchang,** a Drawn Bow,— Brown says "anciently **Tjung,** the Archer,"— υ^1 being the determinant; the god Chang using this bow to slay the Sky Dog, our Crater. The stars between Corvus and Crater were **Kien Mun,** and those between γ Hydrae and Spica of the Virgin were **Tien Mun,** Heaven's Gate. These lie beyond the outlines of the Virgin's robe on the Heis map, but on Burritt's are included in the tip of her left wing.

Hydra is supposed to be the snake shown on a uranographic stone from the Euphrates, of 1200 B. C., "identified with the source of the fountains of the great deep," and one of the several sky symbols of the great dragon Tiāmat. Certain stars near, or perhaps in the tip of Hydra's tail and in Libra, seem to have been the Akkadian **En-te-na-mas-luv,** or **En-te-na-mas-mur,** the Assyrian **Etsen-tsiri,** the Tail-tip.

Theon said that the Egyptians considered it the sky representative of the **Nile,** and gave it their name for that river.

After Al Sufi's day, in our 10th century, the figure was much lengthened, and now stretches for nearly 95° in a winding course from Cancer to Scorpio; this well agreeing with the fable of its immense marine prototype, the Scandinavian Kraken. Conrad Gesner, the 16th-century naturalist, gave an illustration of this in its apparently successful attack upon the ship Argo.

The constellation cannot be seen in its entirety till Crater is on the meridian. Argelander enumerates in it 75 stars; Heis, 153.

For an unknown period its winding course symbolized that of the moon; hence the latter's nodes are called the Dragon's Head and Tail. When a comet was in them poison was thought to be scattered by it over the world; but these fanciful ideas are now associated with Draco.

Al Sufi mentioned an early Arab figure, **Al H'ail,** the Horse, formed from stars some of which now belong to our Hydra, but more to Leo and Sextans.

> The Water-serpent's gleaming bend.
>
> Brown's *Aratos.*

α, 2, orange.

Alphard, Alfard, and **Alpherd,**—**Alphart** in the *Alfonsine Tables* and **Pherd** with Hyde,—are from **Al Fard al Shujā'**, the Solitary One in the Serpent, well describing its position in the sky. Caesius gave **Alpharad,** which on the Reuter wall-map was **Alphrad**; and a still more changed title is **Alphora.** The Arabs also knew *a* as **Al Faḳār al Shujā'**, the Backbone of the Serpent; but Ulug Beg changed this to **Al 'Unk al Shujā'**, the Serpent's Neck; and it shared the Suhel of other bright stars as **Suhel al Fard,** and **Suhel al Shām,** the Solitary, and the Northern, Suhail.

Tycho first called it **Cor Hydrae,** the Hydra's Heart,—Riccioli's **Kalb Elhavich** and **Kalbelaphard,**—which, with the alternative **Collum Hydrae,** the Hydra's Neck, is current even now.

In China it determined the 8th *sieu*, and was the prominent star of the Red Bird that combined the seven lunar divisions of the southern quarter of the heavens. Its longitude is said to have been ascertained there in the 19th century before our era, but the statement may be questionable; as also that it was observed passing the meridian at sunset on the day of the vernal equinox during the time of the emperor Yao, about 2350 B. C. It culminates on the 26th of March.

β and ξ were the Chinese **Tsing Kew,** the Green Hill.

δ, ε, ζ, η, ρ, and σ, 3d to 5th magnitudes, on the head, were Ulug Beg's **Min al Az'al,** Belonging to the Uninhabited Spot.

ε is a remarkable triple,—an 8th-magnitude 3½″ from a 3.8-magnitude, the latter divided by Schiaparelli, in 1892, into two of nearly equal brightness 0″.2 apart,—which probably form a rapid ternary system.

ι, a 4th-magnitude, was the Chinese **Ping Sing,** a Tranquil Star.

κ, a 5th-magnitude, and the stars of about the same brilliancy extending from it to β, with β Crateris, were Al Sufi's **Al Sharāsīf,** the Ribs.

σ, 4.6, was Ulug Beg's **Al Minliar al Shujā'**, the Snake's Nose.

τ^1, 4.9, flushed white, and τ^2, 4.6, lilac, with ι and the 5th-magnitude A, form the curve in the neck, Ptolemy's Καμπή; but Kazwini knew them as **'Ukdah,** the Knot.

★

Hydrus,

first published by Bayer, must not be confounded with the ancient Hydra. It lies between Horologium and Tucana; the head adjoining the polar Octans, the tail almost reaching the magnificent star Achernar of Eridanus.

The French know the figure as **l'Hydre Mâle;** and the Germans as **der Kleine Wasserschlange.**

Out of this, with Tucana and the Lesser Cloud, Julius Schiller made his biblical constellation **Raphael.**

The Chinese formed from the stars of Hydrus, with others surrounding it, four of their later asterisms: **Shay Show,** the Serpent's Head, marked by ε and ζ; **Shay Fuh,** the Serpent's Belly, towards Tucana; **Shay We,** the Serpent's Tail, entirely within the boundaries of Hydrus; and **Foo Pih,** of unknown signification, marked by γ, a red 3.2-magnitude, specially mentioned by Corsali in his account of the Magellanic Clouds.

In it Gould catalogues 64 stars from 2.7 to 7th magnitudes.

The 2.7-magnitude *lucida* β, in the tail, is of a remarkably clear yellow hue, and the nearest conspicuous star to the south pole, although 12° distant.

★

Indus, the Indian,

is the German **Indianer,** the Italian **Indiano,** and the French **Indien;** La Lande giving the alternative **Triangle Indien,** probably from the general outline of its chief stars.

It is one of Bayer's new constellations, south of the Microscope, between Grus and Pavo, and, although generally supposed to represent a typical American Indian, its publisher drew it as a far more civilized character, yet nude, with arrows in both hands, but no bow. Flamsteed's *Atlas* has a similar figuring. Julius Schiller, however, went much further back in point of time and joined it with Pavo as the patriarch **Job.**

Indus, or its *lucida* a, was **Pe Sze** in China, where it also was known as the **Persian,** a title from the Jesuit missionaries.

Gould assigned to it 84 naked-eye stars, from 3.1 to 7th magnitudes; but none of these are specially noticeable except the 6.3 γ, which may be a variable, and ϵ, with the unusually large proper motion of $4''.6$ annually, a rate of speed that will carry it to the south pole in 50,000 years.

★

Lacerta, the Lizard,

is the French **Lézard,** the Italian **Lucertola,** and the German **Eidechse,**— Bode's **Eidexe,**— extending from the head of Cepheus to the star π at the left foot of Pegasus, its northern half lying in the Milky Way.

This inconspicuous constellation was formed by Hevelius from outlying stars between Cygnus and Andromeda, this special figure having been selected because there was not space for any of a different shape. But he drew "a strange weasel-built creature with a curly tail," heading the procession of his offerings to Urania illustrated in his *Firmamentum Sobiescianum* of 1687. Flamsteed's picture is more like a greyhound, but equally uncouth; that by Heis is typically correct.

Its inventor gave it the alternative title of **Stellio,** the Stellion, a newt with star-like dorsal spots found along the Mediterranean coast. Somewhat coincidently its stars, with those in the eastern portion of Cygnus, were combined by the early Chinese in their **Flying Serpent.**

Hevelius catalogued 10 components; Argelander, 31; and Heis, 48. They come to the meridian about the middle of April. It has no named star, and its *lucida*, a, is only of 3.9 magnitude.

β, 4.5, marks the radiant point of the **Lacertids,** a minor meteor stream visible through August and September.

Before the Lizard was formed, Royer introduced here, in 1679, the

Sceptre and Hand of Justice,

commemorating his king, Louis XIV; and a century later Bode substituted the Frederici Honores, in honor of his sovereign Frederick the Great; but Lacerta has held its place, while Royer's figure has been entirely forgotten, and Bode's nearly so.

> In pride the Lion lifts his mane
> To see his British brothers reign
> As stars below.
>
> Edward Young's *Imperium Pelagi.*

Leo, the Lion,

is **Lion** in France, **Löwe** in Germany, and **Leone** in Italy. In Anglo-Norman times it was **Leun.** It lies between Cancer and Virgo, the bright Denebola 5° north of the faint stars that mark the head of the latter constellation; but Ptolemy extended it to include among its ἀμόρφωτοι the group now Coma Berenices.

In Greek and Roman myth this was respectively Λέων and Leo, representing the **Nemean Lion,** originally from the moon, and, after his earthly stay, carried back to the heavens with his slayer Hercules, where he became the poet's **Nemeaeus; Nemeas Alumnus; Nemees Terror; Nemeaeum Monstrum;** and, in later times, **No Animal Nemaeo truculento** of Camões. It also was **Cleonaeum Sidus,** from Cleonae, the Argolic town near the Nemean forest where Hercules slew the creature; **Herculeus;** and **Herculeum Astrum.** But the Romans commonly knew it as Leo, Ovid writing **Herculeus Leo** and **Violentus Leo.**

Bacchi Sidus was another of its titles, that god always being identified with this animal, and its shape the one usually adopted by him in his numerous transformations; while a lion's skin was his frequent dress. But Manilius had it **Jovis et Junonis Sidus,** as being under the guardianship of these deities; and appropriately so, considering its regal character, and especially that of its *lucida.*

The Egyptian king Necepsos, and his philosopher Petosiris, taught that at the Creation the sun rose here near Denebola; and hence Leo was **Domicilium Solis,** the emblem of fire and heat, and, in astrology, the **House of the Sun,** governing the human heart, and reigning in modern days over Bohemia, France, Italy, and the cities of Bath, Bristol, and Taunton in England, and our Philadelphia. In ancient times Manilius wrote of it as ruling over Armenia, Bithynia, Cappadocia, Macedon, and Phrygia. It was a fortunate sign, with red and green as its colors; and, according to Ampelius, was in charge of the wind Thrascias mentioned by Pliny, Seneca, and Vitruvius as coming from the north by a third northwest. Ancient physicians thought that when the sun was in this sign medicine was a poison, and even a bath equally harmful (!); while the weather-wise said that thunder foretold sedition and deaths of great men. The adoption of this animal's form for a zodiac sign has fancifully been attributed to the fact that when the sun was among its stars in midsummer the lions of the desert left their accustomed haunts for the banks of the Nile, where they could find relief from the heat in the waters of the inundation; and Pliny is authority for the statement that the Egyptians worshiped the stars of Leo because the rise of their great river was coincident with the sun's entrance among them. For the same reason the great Androsphinx is said to have been sculptured with Leo's body and the head of the adjacent Virgo; although Egyptologists maintain that this head represented one of the early kings, or the god Harmachis. Distinct reference is made to Leo in an inscription on the walls of the Ramesseum at Thebes, which, like the Nile temples generally, was adorned with the animal's bristles; while on the planisphere of Denderah its figure is shown standing on an outstretched serpent. The Egyptian stellar Lion, however, comprised only a part of ours, and in the earliest records some of its stars were shown as a **Knife,** as they now are as a **Sickle.** Kircher gave its title there as *Πιμεντεκέων*, *Cubitus Nili.*

The Persians called it **Ser** or **Shīr**; the Turks, **Artān**; the Syrians, **Aryō**; the Jews, **Aryē**; and the Babylonians, **Arū,**—all meaning a Lion; the last title frequently being contracted to their letter equivalent to our **A.**

It was the tribal sign of Judah, allotted to him by his father Jacob as recorded in *Genesis* xlix, 9, and confirmed by Saint John in *The Revelation* v, 5; Landseer suggesting that this association was from the fact that Leo was the natal sign of Judah and so borne on his signet-ring given to Tamar.

Christians of the Middle Ages and subsequently, who figured biblical characters throughout the heavens in place of the old mythology, called it **one of Daniel's lions**; and the apostolic school, **doubting Thomas.**

On Ninevite cylinders Leo is depicted as in fatal conflict with a bull, typifying the victory of light over darkness; and in Euphratean astronomy it was additionally known as **Gisbar-namru-sa-pan,** variously translated, but by Bertin as the Shining Disc which precedes Bel; the latter being our Ursa Major, or in some way intimately connected therewith. Hewitt says that it was the Akkadian **Pa-pil-sak,** the Sceptre, or the Great Fire; and Sayce identifies it with the Assyrian month Abu, our July–August, the Fiery Hot; Minsheu assigning as the reason for this universal fiery character of the constellation, "because the sunne being in that signe is most raging and hot like a lion."

Thus throughout antiquity the animal and the constellation always have been identified with the sun,—indeed in all historic ages till it finally appears on the royal arms of England, as well as on those of many of the early noble families of that country. During the 12th century it was the only animal shown on Anglo-Norman shields.

As a zodiacal figure it was of course entirely different from the ancient Asad of Arabia, that somewhat mythical Lion extending from Gemini over our Cancer, Leo, Virgo, Libra, and parts of other constellations, both north and south of the zodiac; but the later Arabians also adopted Ptolemy's Leo and transferred to it the **Asad** of the early constellation. This appeared in the various corrupted forms cited by Bayer,— **Alasid, Aleser, Asis, Assid,** and others similar, of which Assemani gives a long list; Schickard adding **Alasado** and **Asedaton;** and Riccioli, specially mentioning **Asid** and **Ellesed,** cautioned his readers against the erroneous **Alatid** and **Alezet.**

Early Hindu astronomers knew it as **Asleha,** and as **Sinha,** the Tamil **Simham;** but the later, influenced by Greece and Rome, as **Leya,** or **Leyaya,** from the word Leo. It contained the 8th *nakshatra*, **Maghā,** Mighty, or Generous; as also the 9th and 10th, **Pūrva,** and **Uttara, Phalgunī,** the Former, and the Latter, Phalgunī, a word of uncertain meaning,— perhaps the Bad One,— the single station being represented by a **Fig-tree,** and the combined by a **Bed** or **Couch.**

Nearly the same stars were included in the 8th, 9th, and 10th *manazil* of Arabia as **Al Jabhah,** the Forehead; **Al Zubrah,** the Mane; and **Al Ṣarfah,** the Turn.

Of the *sieu*, however, none appear in Leo, the Chinese having adopted, instead, stations among the stars of Hydra and Crater, so that many infer that their lunar asterisms were original with themselves. In the later native solar zodiac of China the Lion's stars were the **Horse,** and in the earlier a part of the Red Bird; while Williams says that they also were **Shun Ho,** the Quail's Fire; but in the 16th century the Chinese formally adopted our Leo, translating it as **Sze Tsze.** The space between it and Virgo was **Tae Wei,** or **Shaou Wei,** and the western half of Leo, with Leo Minor, was regarded as a **Yellow Dragon** mounting upwards, marked by the line of ten stars from Regulus through the Sickle. It also was another of the **Heavenly Chariots** of imperial China.

Its symbol, ♌, has been supposed to portray the animal's mane, but seems more appropriate to the other extremity; the *Hyginus* of 1488 and the *Albumasar* of 1489 showing this latter member of extraordinary length, twisting between the hind legs and over the back, the *Hyginus* properly locating the star Denebola in the end; but the *International Dictionary*, in a more scholarly way, says that this symbol is a corruption of the initial letter of Λέων. Lajard's *Culte de Mithra* mentions the hieroglyph of Leo as among the symbols of Mithraic worship, but how their Lion agreed, if at all, with ours is not known.

One of the sultans of Koniyeh, ancient Iconium, put the stellar figure on his coins.

Its drawing has generally been in a standing position, but, in the *Leyden Manuscript*, in a springing attitude, with the characteristic **Sickle** fairly represented. Young astronomers know the constellation by this last feature in the fore parts of the figure, the bright Regulus marking the handle; its other stars successively being η, γ, ζ, μ, and ϵ. Nor is this a recent idea, for Pliny is thought to have given it separately from Leo in his list of the constellations; but not much could have been left of the Lion after this subtraction except his tail.

These same Sickle stars were a lunar asterism with the Akkadians as **Gis-mes,** the Curved Weapon; with the Khorasmians and Sogdians as **Khamshish,** the Scimetar; but with the Copts as **Titefui,** the Forehead.

The sun passes through Leo from the 7th of August to the 14th of September. Argelander catalogues in it 76 stars, and Heis 161.

In Leo and Virgo lay the now long forgotten asterism **Fahne,** of which Ideler wrote:

> The Flag is a constellation of the heavens, one part in Leo and one part in Virgo. Has many stars. On the iron [the arrowhead of the staff] in front one, on the flag two, on every fold of the flag one.

This is illustrated in the 47th volume of *Archaeologia*, and it appeared as a distinct constellation in a 15th-century German manuscript, perhaps the original of the work of 1564 from which Ideler quoted. Brown repeats a Euphratean inscription, "The constellation of the *Yoke* like a flag floated," although he claims no connection here, and associates the Yoke with Capricorn.

> Il Petto del lione ardente.
>
> Dante's *Paradiso.*

α, Triple, 1.7, 8.5, and 13, flushed white and ultramarine.

Regulus was so called by Copernicus, not after the celebrated consul of the 1st Punic war, as Burritt and others have asserted, but as a diminutive of the earlier **Rex,** equivalent to the βασιλίσκος of Ptolemy. This was from the belief that it ruled the affairs of the heavens,—a belief current, till three centuries ago, from at least 3000 years before our era. Thus, as **Sharru,** the King, it marked the 15th ecliptic constellation of Babylonia; in India it was **Maghā,** the Mighty; in Sogdiana, **Magh,** the Great; in Persia,

Miyan, the Centre; among the Turanian races, **Masu,** the Hero; and in Akkadia it was associated with the 5th antediluvian King-of-the-celestial-sphere, Amil-gal-ur, 'Αμεγάλαρος. A Ninevite tablet has:

If the star of the great lion is gloomy the heart of the people will not rejoice.

In Arabia it was **Malikiyy,** Kingly; in Greece, *βασιλισκός ἀστήρ*; in Rome, **Basilica Stella;** with Pliny, **Regia;** in the revival of European astronomy, **Rex;** and with Tycho, **Basiliscus.**

So, too, it was the leader of the **Four Royal Stars** of the ancient Persian monarchy, the **Four Guardians of Heaven.** Dupuis, referring to this Persian character, said that the four stars marked the cardinal points, assigning **Hastorang,** as he termed it, to the North; **Venant** to the South; **Tascheter** to the East; and **Satevis** to the West: but did not identify these titles with the individual stars. Flammarion does so, however, with Fomalhaut, Regulus, and Aldebaran for the first three respectively, so that we may consider Satevis as Antares. This same scheme appeared in India, although the authorities are not agreed as to these assignments and identifications; but, as the right ascensions are about six hours apart, they everywhere probably were used to mark the early equinoctial and solstitial colures, four great circles in the sky, or generally the four quarters of the heavens. At the time that these probably were first thought of, Regulus lay very near to the summer solstice, and so indicated the solstitial colure.

Early English astrologers made it a portent of glory, riches, and power to all born under its influence; Wyllyam Salysbury, of 1552, writing, but perhaps from Proclus:

The Lyon's herte is called of some men, the Royall Starre, for they that are borne under it, are thought to have a royall nativitie.

And this title, the **Lion's Heart,** has been a popular one from early classical times, seen in the *Καρδία λεόντος* of Greece and the **Cor Leonis** of Rome, and adopted by the Arabians as **Al Kalb al Asad,** this degenerating into **Kalbelasit, Kalbeleced, Kalbeleceid, Kalbol asadi, Calb-elez-id, Calb-elesit, Calb-alezet,** and **Kale Alased** of various bygone lists. Al Bīrūnī called it the **Heart of the Royal Lion,** which "rises when Ṡuhail rises in Al Ḥijāz."[1]

Bayer and others have quoted, as titles for Regulus, the strange **Tyberone** and **Tuberoni Regia;** but these are entirely wrong, and arose from a misconception of Pliny's *Stella Regia appellata Tuberoni in pectore Leonis*,

[1] The province containing Mecca, Medina, and Jiddah, and reaching to Tehama, the low land bordering on the Red Sea.

rendered "the star called by Tubero the Royal One in the Lion's breast"; Holland's translation reading:

> The cleare and bright star, called the **Star Royal**, appearing in the breast of the signe Leo, *Tubero*[1] mine author saith.

Naturally sharing the character of its constellation as the **Domicilium Solis,** in Euphratean astronomy it was **Gus-ba-ra,** the Flame, or the Red Fire, of the House of the East; in Khorasmia, **Achir,** Possessing Luminous Rays; and throughout classical days the supposed cause of the summer's heat, a reputation that it shared with the Dog-star. Horace expressed this in his *Stella vesani Leonis.*

It was of course prominent among the lunar-mansion stars, and chief in the 8th *nakshatra* that bore its name, **Maghā,** made up by all the components of the Sickle; and it marked the junction with the adjoining station Pūrva Phalgunī; the Pitares, Fathers, being the regents of the asterism, which was figured as a House. In Arabia, with γ, ζ, and η of the Sickle, it was the 8th *manzil,* **Al Jabhah,** the Forehead. In China, however, the 8th *sieu* lay in Hydra; but the astronomers of that country referred to Regulus as the **Great Star in Heen Yuen,** a constellation called after the imperial family, comprising α, γ, ε, η, λ, ζ, χ, ν, o, ρ, and others adjacent and smaller reaching into Leo Minor. Individually it was **Niau,** the Bird, and so representative of the whole quadripartite zodiacal group.

In addition to the evidence, from its nomenclature, of the ancient importance of this star is the record, although perhaps questionable, of an observation of its longitude 1985 years before the time of Ptolemy; and of a still earlier one in Babylonia, 2120 B. C., Regulus then being in longitude 92° 30′, but now over 148°. Its position, and that of Spica, observed by Hipparchos, when compared with the earlier records are said to have revealed to him the phenomenon of the precession of the equinoxes. It was then in longitude 119° 50′. Smyth wrote of it:

> The longitude of Regulus has, through successive ages, been made a datum-step by the best astronomers of all nations.

This is the faintest of the so-called 1st-magnitude stars, with but $\frac{1}{13}$ of the brightness of Sirius. It has a spectrum of the Sirian type, and is approaching the earth at the rate of 5½ miles a second. Elkin has determined its parallax as 0″.089. It lies very close to the ecliptic, almost covered by the sun on the 20th of August; and, as one of the lunar stars, is much observed in navigation. It culminates on the 6th of April.

[1] This was Lucius Tubero, the intimate literary friend of Cicero.

The companion, about 3′ away, described "as if steeped in indigo," was discovered by Winlock to be itself closely double, 3″.3 apart, at a position angle of 88°.5.

β, 2.3, blue.

Denebola — sometimes **Deneb** — is the modern name for this star, abbreviated from **Al Dhanab al Asad,** the Lion's Tail, the Greek 'Αλκαία; Bayer gave it as **Denebalecid** and **Denebaleced**; Chilmead, as **Deneb Alased**; and Schickard, as **Dhanbol-asadi.** Riccioli omitted the first syllable of the original, and called the star **Nebolellesed, Nebollassid** "of the Nubian astrologers," and **Alazet** *apud Azophi,* his title for Al Sufi. Elsewhere it is **Nebulasit** and **Alesit**; the *Alfonsine Tables* have **Denebalezeth** and the very appropriate **Dafira,** from the similar Arabic term for the tuft of coarse hair at the end of the tail in which the star lies. Proctor called it **Deneb Aleet,** and there may be other degenerated forms of the original. Kazwini cited **Al Aḳtāb al Asad,** the Viscera of the Lion, or **Al Ḳatab,** a Small Saddle: inappropriate names, Ideler said, and inferred that they should be **Al Ḳalb,** which in the course of time might have wandered here from Regulus, the genuine Ḳalb, or Heart, of the Lion.

It marked the 10th *manzil,* **Al Ṣarfah,** the Changer, *i. e.* of the weather, given by Ulug Beg as the star's individual title; and Al Bīrūnī wrote of it: "The heat turns away when it rises, and the cold turns away when it disappears." Chilmead cited **Asumpha,** which he attributed to Alfraganus; Baily called this **Serpha**; and Hyde changed it to **Mutatrix.**

With the 4th-magnitude Fl. 93, it constituted the 10th *nakshatra,* **Uttara Phalgunī,** and was the junction star with the adjacent Hasta; the regents of this and the next asterism, the Pūrva Phalgunī, being the Adityas, Āryaman and Bagha. Al Bīrūnī, however, said that Hindu astronomers pointed out to him a star in Coma Berenices as forming the lunar station with Denebola; and they claimed that the great scientific attainments of Varāha Mihira were due to his birthday having coincided with the entrance of the moon into Uttara Phalgunī.

The Chinese knew it, with four small neighboring stars, as **Woo Ti Tso,** the Seat of the Five Emperors, surrounded by twelve other groups, variously named after officers and nobles of the empire.

In Babylonian astronomy it marked the 17th ecliptic constellation, **Zibbat A.,** the Tail of the Lion, although Epping gives this with considerable doubt as to its correctness. Other Euphratean titles are said to have been **Lamash,** the Colossus; **Sa.** Blue, the Assyrian **Samu**; and **Mikid-isati,** the

Burning of Fire, which may be a reference to the hot season of the year when the sun is near it.

The Sogdians and Khorasmians had a similar conception of it, as shown in their titles **Widhu** and **Widhayu,** the Burning One; but the Persians called it **Avdem,** the One in the Tail. Hewitt writes of it as, in India, the **Star of the Goddess Bahu,** the Creating Mother.

With θ, it was the Coptic **Asphulia,** perhaps the Tail; but Kircher had a similar Ἄσπολια, in Virgo, as from Coptic Egypt.

Denebola was of unlucky influence in astrology, portending misfortune and disgrace, and thus opposed to Regulus in character as in position in the figure.

Its spectrum is Sirian, and it is approaching our system at the rate of about twelve miles a second. It comes to the meridian on the 3d of May, and, with Arcturus and Spica, forms a large equilateral triangle, as also another similar with Arcturus and Cor Caroli, these, united at their bases, constituting the celebrated **Diamond of Virgo.**

Several small stars, some telescopic, in its immediate vicinity, are the **Companions of Denebola.**

γ, Double and perhaps binary, 2.2 and 3.5, bright orange and greenish yellow.

Smyth wrote of this that it

> has been improperly called Algieba, from *Al jeb-bah,* the forehead; for no representation of the Lion, which I have examined, will justify that position,—

a well-founded criticism, although as, after Regulus, it is the brightest member of the *manzil* Al Jabbah, it may have taken the latter's title. The star, however, is on the Lion's mane, the Latin word for which, **Juba,** distinctly appeared for γ with Bayer, Riccioli, and Flamsteed. Hence it is not at all unlikely that **Algieba,**— also written **Algeiba,**— is from the Latin, Arabicized either by error in transcription or by design.

Sir William Herschel discovered its duplicity in 1782, and Kitchiner asserted that this and α Lyrae are the only stars upon which he ventured to use his high telescopic power of 6450. In 1784 he saw both components of γ white, and in 1803 he announced their binary (?) character. They now are 3″.7 apart, at a position angle of 114°; and according to Doberck have a period of revolution of about 402.62 years, although this is very uncertain, for " since the first reliable measures of distance the change to this time is only 12°."

γ is in approach toward us at the rate of about twenty-four miles a second, the greatest velocity toward our system of any star noted by the Potsdam observers, yet only half that of ζ Herculis as determined at Poulkowa. Its spectrum is Solar.

δ, Coarsely triple, 2.7, 13, and 9, pale yellow, blue, and violet.

Zosma and **Zozma** are from ζῶσμα, an occasional form of ζῶμα, the Girdle, found in the *Persian Tables;* but its propriety as a stellar title is doubtful, for the star is on the Lion's rump, near the tail.

Ulug Beg very correctly termed it **Al Ṭhahr al Asad,** the Lion's Back, which has become **Duhr** and **Dhur** of modern catalogues.

With θ, on the hind quarter, it constituted the 9th *manzil*, **Al Zubrah,** the Mane, and itself bears this name as **Zubra,**— strange titles for star and station so far away from that feature of the animal. δ and θ also were **Al Kāhil al Asad,** the Space between the Shoulders of the Lion; and **Al H·arātān,** sometimes transcribed **Chortan,** and translated the Two Little Ribs, or the two Khurt, or Holes, penetrating into the interior of the Lion; but all these seem as inapplicable as are the other titles.

In India they marked the corresponding *nakshatra*, **Pūrva Phalgunī,** δ being the junction star between the two Phalgunī asterisms.

On the Euphrates they were **Kakkab Kua,** the constellation of the god Kua, the Oracle; and in Egypt, according to Hewitt, **Mes-su,** the Heart of Su. In Sogdiana they were **Wadha,** the Wise; in Khorasmia, **Armagh,** the Great; and with the Copts **Pikhōrion,** the Shoulder.

In China δ was **Shang Seang,** the Higher Minister of State.

Its spectrum is Sirian, and the star is approaching our system at the rate of about nine miles a second.

Flamsteed observed it and 6 Virginis on the 13th of December, 1690, with the object which nearly a century later proved to be the planet Uranus. He made record of the observation, but without any thought of having seen a hitherto unknown member of our system.

ϵ, 3.3, yellow.

The Arabians designated this as **Al Rās al Asad al Janūbiyyah,** the Southern Star in the Lion's Head; but by us it is practically unnamed, although the *Century Cyclopedia* says "rather rarely **Algenubi.**" With μ, it was **Al Ashfār,** the Eyebrows, near to which they lie.

It marked the 14th ecliptic constellation of Babylonia, **Rishu A.,** the Head of the Lion.

The Chinese knew these two stars as **Tsze Fe**; while ε, individually, was **Ta Tsze,** the Crown Prince.

ζ, Double, 3.7 and 6,

is Burritt's **Adhafera, Aldhafara,** and **Aldhafera,** by some confusion perhaps with Al Ashfār of the near-by ε and μ. It is on the crest of the mane, and was one of the *manzil* Al Jabhah; sometimes taking the latter's name, as in Baily's edition of Ulug Beg.

From a point a little to the west of ζ, and not much farther from γ,[1] issue the **Leonids,** the meteor stream of November 9th to 17th, its maximum now occurring on the 13th–14th, which about every thirty-three years has furnished such wonderful displays, the last in 1866 and the next due in 1899.

Their first noticed appearance may have been in the year 137, since which date the stream has completed fifty-two revolutions. According to Theophanes of Byzantium, the shower was seen from there in November, 472; but the late Professor Newton, our deservedly great authority on the whole subject of meteors, commenced his list of the Leonids with their appearance on the 13th of October, 902, the Arabian Year of the Stars, during the night of the death of King Ibrahim ben Ahmad, and added:

It will be seen that all these showers are at intervals of a third of a century, that they are at a fixed day of the year, and that the day has moved steadily and uniformly along the calendar at the rate of about a month in a thousand years.

Oppolzer's and Leverrier's observations showed the identity of their orbit with that of Tempel's comet, I of 1866; and they are supposed to have entered our system by some comparatively recent action, as they still come in shoals and are not lengthened out in a continuous line. It was suggested by Leverrier, and confirmed by Adams, that Uranus may have produced this effect early in the year 126 of our era.

Apparently the most remarkable showers in the long Leonid history were the one observed by Von Humboldt and his companion Bonpland on the 12th of November, 1799, from Venezuela, and by various other observers throughout the western hemisphere; and that of November 13, 1833, splendidly seen from this country. The lesser one of the 13th–14th of November, 1866, was more especially noticeable from the Old World, and others, remarkable yet gradually declining, were annually seen from 1867 to 1869.

These meteors appear at an elevation of from sixty-one to ninety-six miles, during the latter part of the night, at a speed of forty-four miles a

[1] When first observed the radiant point was in Cancer.

second,[1] and generally are characterized by a greenish, or bluish, tint, with vivid and persistent trains. It probably was to them that Milton alluded in his

> Swift as a shooting star
> In Autumn thwarts the night.

The stream seems to be lengthening, and consequently thinning out, so that the great displays of long period may eventually cease, while the annual may become more brilliant than now.

Many other meteor streams are visible about the same time as the Leonids, Mr. W. F. Denning having given a list of sixty-eight; the brightest of these, the **Ursids,** being often mistaken by the casual observer for the Leonids, as their radiant, near μ Ursae Majoris, is less than 20° distant from the radiant in Leo.

θ, 3.5,

in the *manzil* Al Zubrah, shares with δ the title **Al Ḥ·arātān,** Al Bīrūnī saying that "when they rise Suhail is seen in Al Izak,"— wherever this may be. The *Century Cyclopedia* gives **Chort** as the individual name, from the combined title. Ulug Beg substituted the 5th-magnitude Fl. 72 for δ as the second member of the *manzil*, his translator placing them *in coxis*, "in the hips," as does the Heis *Atlas*.

In China it was **Tsze Seang,** the Second Minister of State.

ι, Binary and perhaps variable, 4.6 and 7.4, yellowish — possibly varying.

Reeves mentioned this as **Tsze Tseang,** the Second General.

The lesser star is suspected of change in color and in brilliancy down to the 9th magnitude. The components now are about 2″.6 apart, at a position angle of 57°.

κ, Double, 4.8 and 10.5, yellow and blue.

This was designated by Ulug Beg as **Al Minḥar al Asad,** the Lion's Nose, still correct for it as laid down on the Heis *Atlas*, although now never used as a star-title.

The components are 3″ apart, at a position angle of 203°.8.

[1] It is owing to this great velocity that no Leonid has ever been known to reach the earth's surface, its substance being dissipated by the intense heat occasioned by the resistance of the atmosphere.

λ, 4.8, red.

Alterf is from **Al Tarf,** the name for the 7th *manzil,* which it formed with ξ Cancri. The word has generally been rendered the Glance, *i. e.* of the Lion's eye, although on modern maps the star lies in the open mouth, where Ptolemy located it. But it also had the secondary meaning of the Extremity, still more appropriate here, and so understood by Ideler.

μ, 4.3, orange,

and ε were **Al Ashfār,** the Eyebrows; but, singly, the Arabians designated μ as **Al Rās al Asad al Shamāliyy,** the Lion's Head towards the South, which, by abbreviation, has become **Rasalas** in modern lists; and sometimes, but very insufficiently, plain **Alshemali.** Al Nasr al Dīn mentioned ε and μ as "a whip's length apart," a common expression for measurement among the Arabs, here indicating a little more than 2°.

π, a 5th-magnitude red star, was the Chinese **Yu Neu,** the Honorable Lady.

ρ, a 4th-magnitude, marked the 16th ecliptic constellation of Babylonia, **Maru-sha-arkat-Sharru,** that Epping translated the Fourth Son (or the Four-Year-Old Son) behind the King.

σ, 4.1, is the Chinese **Shang Tseang,** the Higher General.

χ, a 5th-magnitude, with *c* and *d,* was **Ling Tae,** a Wonderful Tower, and ψ, a double of the 6th and 10th magnitudes, bright orange and bluish white in color, was **Tsew Ke,** a Wine-flagon, but this included ξ and ω Leonis with κ and ξ Cancri.

★

Each after each, ungrouped, unnamed, revolve.

Brown's *Aratos.*

Leo Minor, the Lesser Lion,

is the French **Petit Lion,** the German **Kleine Löwe,** and the Italian **Leoncino.** Proctor arbitrarily changed the title to **Leaena,** the Lioness.

It was formed by Hevelius from eighteen stars between the greater Lion and Bear, in a long triangle with a fainter line to the south, and thus named because he said it was "of the same nature" as these adjoining constellations. Argelander assigned to it 21 components, and Heis 40.

Aratos is supposed to have alluded to these "ungrouped, unnamed" stars under the hind paws of Ursa Major; and Ptolemy had some of them among the ἀμόρφωτοι of his Λέων. Ideler surmised that they were the Arabs' **Al Ṭhibā' wa-Aulāduhā,** the Gazelle with her Young, shown in this location on the Borgian globe; but Lach, that they were **Al Haud,** the Pond, into which the Gazelle sprang, as noted under Coma Berenices.

The Chinese made two asterisms of it,—**Nuy Ping,** an Inner Screen, and **Seaou Wei**; but also included our Lesser Lion with the Greater in their still greater **Dragon** mounting to the highest heavens, and in yet another figure, the **State Chariot.**

The Denderah planisphere located here the zodiacal Crab, but whether by design, or in error, is unknown; although some see in the Lesser Lion's stars, with others from the Bear's feet, a well-marked Scarab that was Egypt's idea of Cancer. This was in a part of the sky thought to have been sacred to the great god Ptah.

Fl. 46, 4.

To the *lucida* Hevelius applied the adjective **Praecipua,** Chief, which Piazzi inserted as a proper name in the *Palermo Catalogue.* Burritt mentioned it, under the letter *l*, as the **Little Lion,** from its being the principal star in the figure.

It culminates on the 14th of April.

In Smyth's *Bedford Catalogue* we read that Praecipua has three distant companions,—7½, pale gray; 13, reddish; and 12, of violet tint.

★

Behind him Sirius ever speeds as in pursuit, and rises after,
And eyes him as he sets.

Poste's *Aratos.*

Lepus, the Hare,

the German **Hase,** the Portuguese **Lebre,** the Italian **Lepre,** and the French **Lièvre,** is located just below Orion and westward from his Hound.

It was Λαγῶς among the Greeks — Λαγωός in the Epic dialect,— Aratos characterizing its few and faint stars by the adjective γλαυκός. With the Greeks of Sicily, the country noted in early days for the great devastations by hares, the constellation was Λέπορις, whence came the fanciful story that our Hare was placed in the heavens to be close to its hunter, Orion. Riccioli enlarged upon this in his *Almagestum Novum*:

Quia Orion in gratiam Dianae, quae leporino sanguine gaudebat, plurimum venatu leporis gauderet.

Among the Romans it was simply **Lepus**, often qualified by the descriptive *auritus*, "eared"; *dăsȳpus*, "rough-footed"; *levipes*, "light-footed"; and *velox*, "swift."

The Arabians adopted the classical title in their **Al Arnab**, which degenerated into **Alarnebet**, **Elarneb**, and **Harneb**; and the Hebrews are said to have known it as **Arnebeth**; but the early Arabs designated the principal stars — α, β, γ, and δ — as **Al Kursiyy al Jabbār** and **Al ʿArsh al Jauzah**, the Chair of the Giant and the Throne of the Jauzah. Kazwini, repeating this, added, in Ideler's rendering, *Gott weiss wie sonst noch*, which Smyth assumed to be Ideler's comment thereon; but it was merely his translation of Kazwini's Arabic formula, *God is the Omniscient*, used when a writer did not wish to come to a decision. Smyth further wrote of it:

'Abdr rahmān Sūfī designates the throne — one of the many which the Arabs had in their heavens, although a squatting rather than a sitting people — **al-muakhkherah**, the succeeding, as following that formed by λ, β, ψ Eridani and τ Orionis.

Al Sufi also cited the occasional **Al Nihāl**, the Thirst-slaking Camels, for the four bright stars, in reference to the near-by celestial river, the Milky Way.

It is in the space occupied by Lepus, or perhaps by Monoceros, that Hommel locates the Euphratean **Udkagaba**, the Smiting Sun Face, although Brown assigns this to Sagittarius, "the original Sagittary being the sun."

Hewitt says that in earliest Egyptian astronomy Lepus was the **Boat of Osiris**, the great god of that country, identified with Orion. The Chinese knew it as **Tsih**, a Shed.

Caesius made the constellation represent one of the hares prohibited to the Jews; but Julius Schiller substituted for it **Gideon's Fleece**. The Denderah planisphere has in its place a **Serpent** apparently attacked by some bird of prey; and Persian zodiacs imitated this.

Gould catalogues in Lepus 103 stars down to the 7th magnitude.

Aelian, of our 2d century, in his Περί ζώων ἰδιότητος, referred to the early belief that the hare detested the voice of the raven,—a belief that has generally been put among the zoölogical fables of antiquity; but Thompson suggests for it an astronomical explanation, as "the constellation Lepus sets soon after the rising of Corvus"; and something similar may be said of Lepus in connection with Aquila, for the eagle in combat with the hare is frequent on gems, and on coins of Agrigentum, Messana, Elis, etc. . . . the wide occurrence of this subject . . . indicates a lost mythological significance, in which one is tempted to recognize a Solar or Stellar symbol.

Brown writes of the often discussed comparative location of Lepus and Orion:

The problem which perplexed the ancients, why the Mighty-hunter and his Dog should pursue the most timid of creatures, is solved when we recognize that Oriōn was originally a solar type, and that the Hare is almost universally a lunar type;

and mentions the very singular connection between this creature and the moon shown on Euphratean cylinders, Syrian agate seals, Chinese coins, the Moon-cakes of Central Asia, and in the legends of widely separated nations and savage tribes. Astronomical folk-lore has many allusions to this interesting association of animal with satellite, and indirectly with our constellation. The common idea that it is because all are nocturnal does not seem satisfactory; and there are others still less so, some being mentioned by Beaumont and Fletcher in the *Faithful Shepherd.*

A brief digression to some of these allusions may be allowed here. The Hindus called the moon Çaçin, or Ṣaṣānka, Marked with the Hare, from the story told of Sakya muni (Buddha). This holy man, in an early stage of his existence, was a hare, and, when in company with an ape and a fox, was applied to by the god Indra, disguised as a beggar, who, wishing to test their hospitality, asked for food. All went in search of it, the hare alone returning unsuccessful; but, that he might not fall short in duty to his guest, had a fire built and cast himself into it for the latter's supper. In return, Indra rewarded him by a place in the moon where we now see him. Other Sanskrit and Cingalese tales mention the palace of the king of the hares on the face of the moon; the Aztecs saw there the rabbit thrown by one of their gods; and the Japanese, the Jeweled Hare pounding *omochi*, their rice dough, in a mortar. Even the Khoikhoin, the Hottentots of South Africa, and the Bantus associated the hare and moon in their worship, and connected them in story, asserting that the hare, ill treated by the moon, scratched her face and we still see the scratches. Eskimos think the moon a girl fleeing from her brother, the sun, because he had disfigured her face by ashes thrown at her; but in Greenland the sex of these luminaries is interchanged, and the moon pursues his sister, the sun, who daubs her sooty hands over his face. The Khasias of the Himalayas say that every month the moon falls in love with his mother-in-law, who very properly repulses his affection by throwing ashes at him.

Other ideas to account for the lunar marks are current among many nations. One from our North American Indians appears in Longfellow's *Hiawatha:*

> Once a warrior very angry,
> Seized his grandmother, and threw her
> Up into the sky at midnight;
> Right against the moon he threw her;
> 'Tis her body that you see there.

The Incas knew them as a beautiful maiden who fell in love with the moon and joined herself forever to him; the New Zealanders, as a woman pulling *gnatuh;* the Hervey Islanders, as the lovely Ina, an earthly maiden carried away to be our satellite's wife, and still visible with her pile of *taro* leaves and tongs of a split cocoanut branch; and the Samoans, as a woman with her child and the mallet with which she is pounding out sheets of the native paper cloth. So that all these people long ago anticipated pretty Selene,[1] of whom Serviss tells us.

In southern Sweden a brewing-kettle is imagined on the moon's face; in northern Germany and Iceland, Hjuki and Bil with their mead burden, the originals of our Jack and Jill with their pail of water, the contents scattered or retained according to the lunar phases. In Frisia the marks were a man who had stolen cabbages, and whom, when discovered, his suffering neighbors wished in the moon, and so it turned out; or a sheep-stealer, with his dog, who enticed the animals to him by cabbages, and, when detected, was transported to the moon, where he is now seen, cabbages and all. But others said that he was caught with a bundle of osier willows that did not belong to him, and there he is on the moon's face with his plunder.

Danish folk-lore makes the moon a cheese formed from the milk that has run together out of the Milky Way; which recalls Rabelais' now familiar remark that some thought the moon made of green cheese.

Those biblically inclined saw here the **Magdalen** in tears; or **Judas Iscariot**; and, in the earlier record, the patriarch **Jacob; Isaac** with the wood for the sacrifice; the **Hebrew sinner** gathering sticks on the Sabbath; or **Cain** driven from the face of the earth to the face of the moon. This appeared even with Dante, Chaucer, and Shakespeare, for the first had in the *Paradiso:*

> But tell me what the dusky spots may be
> Upon this body, which below on earth
> Make people tell that fabulous tale of Cain;

[1] This may be seen on the western half of the moon after the ninth day of lunation, the face slightly upturned toward the east. It seems to have been first described some years ago by Doctor James Thompson; and an opera-glass of low power makes the phenomenon very distinct.

and in the *Inferno:*

> Touches the ocean wave Cain and the thorns.

In *A Midsummer Night's Dream* Quince says:

> Or else one must come in with a bush of thorns and a lanthorn and say, he comes to disfigure, or to present the person of moonshine;

and Chaucer described the figure as

> Bearing a bush of thorns on his back
> Whiche for his theft might clime so ner the heaven;

although Milton, from a higher plane of thought, wrote that the sinful wandered

> Not in the neighbouring moon as some have dreamed.

The Salish Indians of our northwest coast tell of a toad which, pursued by a wolf, jumped to the moon to escape his unwelcome attentions.

At the present day the handsome face of Selene shows itself in profile to the favored few; while the Old Man in the Moon is seen by all. It would be interesting to know who originated this, or, as in *Hudibras*,

> Who first found out the Man i' th' Moon,
> That to the ancients was unknown.

Yet Shakespeare knew him well, for we find in *The Tempest:*

> The man i' th' moon's too slow.

Ages before all this, however, the Egyptians had similar ideas; the Hindus called the moon Mriga, an Antelope; the Aethiopians saw that creature in it; while the Greeks knew it as the Gorgon's head, and Plutarch thought the phenomenon worthy a special treatise in his *De Facie in Orbe Lunae.* But perhaps too much attention has been paid to a probably very dead star; — let us return to those certainly alive, our more legitimate subject.

α, Double, 2.7 and 9.5, pale yellow and gray.

Arneb is from the Arabian name for the whole, but the *Century Dictionary* substitutes the early **Arsh.**

Other near-by stars, presumably in Lepus, were the Chinese **Kuen Tsing,** an Army Well, and **Ping Sing,** the Star Screen.

Arneb culminates on the 24th of January.

The components are 35″.4 apart, at a position angle of 156°; and 6′ away is Sir John Herschel's 3780, a sextuple star.

β, Double, 3.5 and 11, deep yellow and blue.

Nihal is from the collective title of α, β, γ, and δ,— **Nibal** with Burritt.

Holden says that the companion, nearly 3″ away, at a position angle of 292°, is suspected to be a planet; and Burnham has discovered other faint companions.

The variable R, 6th to 8.5 magnitudes, is **Hind's Crimson Star**, discovered by Mr. J. R. Hind in 1845,— "like a drop of blood on a black field." It lies in front of the Hare's head, on the border of Eridanus, but its discoverer announced it as in Orion. Its variability, in a very irregular period of about 438 days, was first recorded by Schmidt in 1855, but accurate observations of maxima and minima are difficult in high latitudes.

★

the scale of night
Silently with the stars ascended.

Longfellow's *Occultation of Orion.*

Libra, the Balance or Scales,

is the Italian **Libra** and **Bilancia,** the French **Balance,** the German **Wage,**— Bayer's **Wag** and Bode's **Waage,**— but the Anglo-Saxons said **Wæge** and **Pund,** and the Anglo-Normans, **Peise,** all meaning the Scales, or a Weight.

The early Greeks did not associate its stars with a Balance, so that many have thought it substituted in comparatively recent times for the **Chelae,** the **Claws of the Scorpion,** that previously had been known as a distinct portion of the double sign; Hyginus characterizing it as *dimidia pars Scorpionis*, and Ptolemy counting eight components in the two divisions of his Χηλαί,— *βόρειος* and *νότιος*,— with nine *ἀμόρφωτοι*. Aratos also knew it under that title, writing of it as a dim sign,— *φαέων ἐπιδυέες*,— though a great one,— *μεγάλας χηλάς*. Eratosthenes included the stars of the Claws with those of our Scorpio, and called the whole Σκορπίος, but alluded to the Χηλαί; as did Hipparchos, although with him the latter also were Ζυγόν, or Ζυγός, these words becoming common for our Libra, and turned by codices of the 9th century into **Zichos.** They were the equivalents of the Latin **Jugum,** the Yoke, or Beam, of the Balance, first used as a stellar title by Geminos, who, with Varro, mentioned it as the sign of the autumnal equinox. Ptolemy wrote these two Greek titles indiscriminately, and so did the Latin poets the three,— Chelae, Jugum, Libra,— although the scientific writers of Rome all adhered to Libra, and such has been its usual title from their day. The ancient name was persistent, however, for the *Latin Almagest* of 1551 gave a star as *in jugo sive chelis*, and Flamsteed used it in his description of Libra's stars.

The statement, often seen, that the constellation was invented when on the equinox, and so represented the equality of day and night, was current even with Manilius,—

> Then Day and Night are weigh'd in *Libra's* Scales
> Equal a while,—

repeated by James Thomson in the Autumn of his *Seasons*,—

> *Libra* weighs in equal scales the year,—

by Edward Young in his *Imperium Pelagi*, apostrophizing his king,—

> The Balance George! from thine
> Which weighs the nations, learns to weigh
> More accurate the night and day,—

and by Longfellow in his *Poet's Calendar* for September,—

> I bear the Scales, when hang in equipoise
> The night and day.

This idea gave rise to the occasional title **Noctipares**; yet Libra is rarely figured on an even balance, but as described by Milton where

> The fiend look'd up, and knew
> His mounted scale aloft.

The Romans claimed that it was added by them to the original eleven signs, which is doubtless correct in so far as they were concerned in its modern revival as a distinct constellation, for it first appears as Libra in classical times in the Julian calendar[1] which Caesar as *pontifex maximus*

[1] The much-vaunted Julian calendar was substantially the same in its method of intercalation as that formed 238 B. C. under Ptolemy III (Euergetes),— a fact discovered by Lepsius, in 1866, when he found the *Decree of Canopus* at Sanor Tanis.

took upon himself to form, 46 B. C., aided by Flavius, the Roman scribe, and Sosigenes, the astronomer from Alexandria.

Some have associated Andrew Marvell's line,

> Outshining Virgo or the Julian star,

with Libra, but this unquestionably referred to the comet of 43 B. C. that appeared soon after, and, as Augustus asserted, in consequence of, Caesar's assassination in September of that year, being utilized by the emperor and Caesar's friends to carry his soul to heaven. This comet, perhaps, was the same that has since appeared in 531, 1106, and 1680, and that may return in 2255.

Medals still in existence show Libra held by a figure that Spence thought represented Augustus as the dispenser of justice; thus recalling Vergil's beautiful allusion, in his 1st *Georgic*, to the constellation's place in the sky. Addressing the emperor, whose birthday coincided with the sun's entrance among the stars of the Claws, he suggested them as a proper resting-place for his soul when, after death, he should be inscribed on the roll of the gods:

> Anne novum tardis sidus te mensibus addas,
> Quā locus Erigonen inter Chelasque sequentes
> Panditur; ipse tibi jam brachia contrahit ardens
> Scorpius, et coeli justa plus parte relinquit;

so intimating that the place was then vacant, the Scorpion having contracted his claws to make room for his neighbor. But subsequently he wrote:

> Libra die somnique pares ubi fecerit horas;

and a few lines further on tells of twelve constellations,—*duodena astra.*

Milton has a reference in *Paradise Lost* to Libra's origin, where

> Th' Eternal, to prevent such horrid fray,
> Hung forth in heav'n his golden scales, yet seen
> Betwixt Astraea and the Scorpion sign;

and Homer's

> Th' Eternal Father hung
> His golden scales aloft,

is similar; but, although doubtless the original of Milton's verse, probably is not a reference to our Libra; for the Greek poet very likely antedated the knowledge of it in his country, and is supposed to have known but few of our stellar figures,—at all events, has alluded to but few in either the *Iliad* or the *Odyssey*.

Bayer said that the Greeks called it Σταθμός, a Weigh-beam, and Στάτηρ, a Weight; while Theon used for it the old Sicilian Λίτρα and Λίτραι, which, originally signifying a Weight, became the Roman Libra. Ampelius called it **Mochos**, after the inventor of the instrument; and Virgo's title, **Astraea**, the Starry Goddess, the Greek Δίκη, has sometimes been applied to these stars as the impersonation of Justice, whose symbol was the Scales. Addison devoted the 100th number of the *Tatler*—that of the 29th of November, 1709—to "that sign in the heavens which is called by the name of the Balance," and to his dream thereof in which he saw the Goddess of Justice descending from the constellation to regulate the affairs of men; the whole a very beautiful rendering of the ancient thought connecting the Virgin Astraea with Libra. He may have been thus inspired by recollections of his student days at Oxford, where he must often have seen this sign, as a Judge in full robes, sculptured on the front of Merton College.

Manilius, using the combined title, wrote of it in much the same way as of influence over the legal profession:

> This Rul'd at *Servius'* Birth, who first did give
> Our *Laws* a *Being*,—

a reference to Servius Sulpicius Rufus Lemonia, the great Roman lawyer, pupil, and friend of Cicero.

Cicero himself used Jugum as though it were well known; and, with evident intention of upsetting Caesar's claim to its invention, wrote:

> Romam in Jugo
> Cum esset Luna, natam esse dicebat.

The sacred books of India mention it as **Tulā**, the Tamil **Tulam** or **Tolam**, a Balance; and on the zodiac of that country it is a man bending on one knee and holding a pair of scales; but Varāha Mihira gave it as **Juga** or **Juka**, from ζυγόν, and so a reflex of Greek astronomy, which we know came into India early in our era; but he also called it **Fire**, perhaps a recollection of its early Altar form, mentioned further on.

In China it was **Show Sing**, the Star of Longevity, but later, copying our figure, it was **Tien Ching**, the Celestial Balance; and that country had a law for the annual regulation of weights supposed to have been enacted with some reference to this sign. In the early solar zodiac it was the **Crocodile**, or **Dragon**, the national emblem.

Manetho and Achilles Tatios said that Libra originated in Egypt; it plainly appears on the Denderah planisphere and elsewhere simply as a Scale-beam, a symbol of the Nilometer. Kircher gave its Coptic-Egyptian title as Λαμβαδία, *Statio Propitiationis.*

The Hebrews are said to have known it as **Moznayim,** a Scale-beam, Riccioli's **Miznaim,** inscribing it, some thought, on the banners of Asher, although others claimed Sagittarius for this tribe, asserting that Libra was unknown to the Jews and that its place was indicated by their letter **Tau,** while still others claimed Virgo for Asher, and Sagittarius for Joseph.

The Syrians called it **Masa'thā,** which Riccioli gave as **Masathre;** and the Persians, **Terāzū** or **Tarāzūk,** all signifying Libra; the Persian sphere showing a human figure lifting the Scales in one hand and grasping a lamb in the other, this being the usual form of a weight for a balance in the early East.

Arabian astronomers, following Ptolemy, knew these stars as **Al Zubānā,** the Claws, or, in the dual, **Al Zubānatain,** degenerating in Western use to the **Azubene** of the 1515 *Almagest;* but later on, when influenced by Rome, they became **Al Kiffatān,** the Trays of the Balance, and **Al Mīzān,** the Scale-beam, Bayer attributing the latter to the Hebrews. This appeared in the *Alfonsine Tables* and elsewhere as **Almisan, Almizen, Mizin;** Schickard writing it **Midsanon.** Kircher, however, said that **Wazn,** Weight, is the word that should be used instead of Zubānā; Riccioli adopting this in his **Vazneschemali** and **Vazneganubi,** or **Vaznegenubi,** respectively applied to the Northern and Southern Scale as well as to their *lucidae.*

Libra is stamped on the coins of Palmyra, as also on those of Pythodoris, queen of Pontus.

While it seems impossible to trace with any certainty the date of formation of our present figure and its place of origin, yet there was probably some figure here earlier than the Claws, and formed in Chaldaea in more shapes than one; indeed, Ptolemy asserted that it was from that country, while Ideler and modern critics say the same.

Brown thinks that its present symbol, ♎, generally considered a representation of the beam of the Balance, shows the top of the archaic Euphratean **Altar,** located in the zodiac next preceding Scorpio, and figured on gems, tablets, and boundary stones, alone or in a pair. Miss Clerke recalls the association of the 7th month, Tashrītu, with this 7th sign and with the Holy Mound, Tul Ku, designating the biblical Tower of Babel, surmounted by an altar,— the stars in this constellation, α, μ, ξ, δ, β, χ, ζ, and ν, well showing a circular altar. Sometimes this Euphratean figure was varied to that of a **Censer,** and frequently to a **Lamp;** Strassmaier confirming this by his translation of an inscription as **die Lampe als Nuru,** the Solar Lamp, synonymous with **Bir,** the Light, also found for the sky figure. In this connection it will be remembered that another of the names for our Ara, a reduplication of the zodiacal Altar, was Pharus, or Pharos, the Great Lamp, or Lighthouse, of Alexandria, one of the seven wonders of the world. This Lamp also has been found shown on boundary stones as held in the Scorpion's claws, and we see the same idea even as late as the Farnese globe and the *Hyginus* of 1488, where the Scales have taken the place of the Lamp. When the Altar, Censer, and Lamp were in the course of time forgotten, or removed to the South, the Claws were left behind, and perhaps extended, till they in turn were replaced by Libra. Miss Clerke additionally writes:

> The 8th sign is frequently doubled, and it is difficult to avoid seeing in the pair of zodiacal scorpions, carved on Assyrian cylinders, the prototype of the Greek Scorpion and Claws. Both Libra and the sign it eventually superseded thus owned a Chaldaean birthplace.

Brown also says that the Euphratean **Sugi,** the Chariot Yoke, which he identifies with α and β of this constellation, remind us by sound and signification of the Ζυγόν and Jugum of Greece and Rome respectively, and that astrology adds evidence in favor of a Chaldaean origin, for it has always claimed Libra — the Northern Scale at least — as a fruitful sign, taking this from the very foundations of astrology in the Chaldaean belief that "when the Sugi stars were clear the crops were good." In modern astrology, however, the reverse of this held in the case of the Southern Scale.

It seems not unreasonable to conclude that in Chaldaea the 7th sign had origin in all its forms.

In classical astrology the whole constituted the ancient **House of Venus,** for, according to Macrobius, this planet appeared here at the Creation; and, moreover, the goddess bound together human couples under the yoke of matrimony. From this came the title **Veneris Sidus,** although others asserted that Mars was its guardian; astrologers of the 14th century insisting that

> Whoso es born in yat syne sal be an ille doar and a traytor.

It was of influence, too, over commerce, as witness Ben Jonson in *The Alchemist:*

> His house of Life being Libra: which foreshc d
> He should be a merchant, and should trade with balance;

and governed the lumbar region of the human body. Its modern reign has been over Alsace, Antwerp, Austria, Aethiopia, Frankfürt, India, Lisbon, Livonia, Portugal, Savoy, Vienna, and our Charleston; but in classical times over Italy and, naturally enough from its history, especially over Rome, with Vulcan as its guardian. It thus became **Vulcani Sidus.**

To it was assigned control of the gentle west wind, Zephyrus,[1] personified as the son of Astraeus and Aurora.

Pious heathen called it **Pluto's Chariot,** in which that god carried off Proserpina, the adjacent Virgo; but early Christians said that it represented the **Apostle Philip**; and Caesius identified it with the **Balances** of the *Book of Daniel*, v, 27, in which Belshazzar had been weighed and "found wanting."

Argelander enumerated in it 28 stars down to 5.8 magnitude; and Heis, 53 down to 6.5; but its boundaries often have been confused with those of Scorpio. The central portion of the figure is marked by the trapezoid of stars α, ι, γ, and β.

The sun is in the constellation from the 29th of October to the 21st of November.

α^2 and α^1, Widely double, 3 and 6, pale yellow and light gray.

In Greek astronomy these were Χηλή νότιος, the Southern Claw, from the name of the whole division now our Southern Scale.

Our **Zubenelgenubi** is from **Al Zubān al Janūbiyyah,** the exact Arabian equivalent of Ptolemy's term; but **Zubenelgubi** and **Janib** are both wrong, and **Zubeneschamali** is worse, for it plainly belongs to β.

Chilmead's **Mizan Aliemin** is from an Arabian title for the constellation; yet that people also knew it as **Al Kiffah al Janūbiyyah,** the Southern Tray of the Scale, from which came the Arabo-Latin **Kiffa australis** of modern lists; and as **Al Wazn al Janūbiyyah,** the Southern Weight, distorted by Riccioli into **Vazneganubi.** The **Lanx meridionalis** of two centuries ago is synonymous with the first of these Arabian designations.

The *alphas* and β constituted the 14th *manzil*, **Al Zubānā,** although Al Bīrūnī said that this title should be **Zaban,** "to push," as though one of the stars were pushing away the other (!); while α marked the *nakshatra* **Viçakha,** Branched, under the rule of Indragni, the dual tutelar divinity Indra and Agni. This lunar station was figured as a decorated **Gateway,** and in later Hindu astronomy its borders were extended to include γ and ι, thus

[1] This was the same as Favonius,— Homer's Ζέφυρος, at first regarded as strongly blowing, but later as the genial Ζωηφόρος, the Life-bearing.

completing the resemblance to the object for which the asterism was named; ι was the junction star with Anuradha.

These same stars marked the *sieu* **Ti,** Bottom, anciently **Dsi,** and still earlier **I shi,** some Chinese authorities adding δ, μ, and ν.

The two *alphas* were the determinants of the 21st Babylonian ecliptic constellation **Nūru-sha-Shūtu,** the Southern Light; and some have included β and γ with them in the Euphratean **Entena-mas-luv,** the Star of the Tail-tip, as though they marked that part of the enormous, but undetermined, ancient **Hydra** of Chaldaea, the very early **Afr** of Arabia. Oppert considers them the **Idχu** that others apply to the star Altair.

They lie 10° southwest of β, close to the ecliptic and almost covered by the sun on the 5th of November, the components 230″ apart; but Bayer's map and text illustrate and mention only one star. They culminate on the 17th of June.

β, 2.7, pale emerald.

Zubeneschamali, sometimes **Zuben el Chamali,** is from **Al Zubān al Shamāliyyah,** the equivalent of Χηλή βόρειος, the Northern Claw; **Kiffa borealis** is Arabic and Latin for the Northern Scale Tray; Bayer's **Lanx septentrionalis** signifies the same thing; and **Vazneschemali,** the Southern Weight, was used by Riccioli. So that β, as well as α, seems always to have borne the name of that half of the constellation figure which it marked.

Miss Bouvier's and Burritt's **Zubenelgemabi** is entirely wrong, both in orthography and in application to this star.

Epping says that it marked the 22d ecliptic constellation of Babylonia, **Nuru sha-Iltānu,** the Northern Light; while Jensen assigns it and α to that country's lunar asterism **Zibanitu,** connecting this word with the similar Arabic Zubānā; but this is not generally accepted. Brown considers that, under the name of the **Sugi Stars,** they were associated with **Bilat,** the Lady, or **Beltis;** and that the Persians knew them as **Çrob,** the Horned; the Sogdians, as **Ghanwand,** the Claw-possessing, equivalent to the Khorasmian **Ighnuna,** and the Coptic **Prītithi,** the Two Claws,— all these being lunar stations. According to Ptolemy, an observation was made at Babylon on the 17th of January, 272 B. C.,— in the 476th year of Nabonassar, or Nabu-nazir,— of the very near approach of Mars[1] to β, one of the earliest records that we have of this planet. Hind, however, mentioned this approach as in connection with β of Scorpio.

[1] The Greeks knew it as Ἄρης and as Πυρόεις, the Fiery One; the Latins, as **Hercules,** in addition to its present title.

Professor Young states the opinion that β Librae formerly was brighter than Antares, now more than a full magnitude higher, for Eratosthenes distinctly called β "the brightest of all" in the combined Scorpion and Claws; and Ptolemy, 350 years later, gave to it and Antares the same brilliancy. Yet Antares may be the one that has increased.

The color is very unusual, perhaps unique, in conspicuous stars, for Webb says that in the heavens "deep green, like deep blue, is unknown to the naked eye."

Its spectrum is Sirian, and the star is approaching our system at the rate of six miles a second.

The globular cluster N. G. C. 5904, 5 M., discovered by Kirch in 1702, lies in Libra, above the beam of the Balance, not far from β and toward the 5th-magnitude 5 Serpentis. Messier could not resolve this, but Sir William Herschel, with his forty-foot reflector, counted in it more than two hundred 11th- to 15th-magnitude stars, besides those unresolved in the compressed nucleus. But it is chiefly noticeable from the recent photographic discovery by Bailey, at Arequipa, of at least forty-six, perhaps sixty, variables in the cluster,—a remarkable fact paralleled, so far as yet known, only in the cluster N. G. C. 5272, 3 M., of Canes Venatici. In 1890 Parker already had discovered two variables in 5904 by visual observation.

δ, Variable, 5 to 6.2, white,

seems to have been associated with μ Virginis in the Akkadian lunar asterism **Mulu-izi,** the Man of Fire, connected with the star-god Laterak; and in the Sogdian **Fasariva** and the Khorasmian **Sara-fasariva,** both titles signifying the One next to the Leader, *i. e.* the preceding moon station, ι, κ, and λ Virginis.

It is a variable of the Algol type, discovered by Schmidt in 1859, with a period of nearly two days and eight hours, the light oscillation occupying twelve hours.

η, 5.5,

lies between the Northern Scale and the northern arm of Scorpio.

Burritt called it **Zubenhakrabi,** a title properly belonging to γ Scorpii. His errors, however, as to the nomenclature of these stars in Libra have caused much confusion in our popular lists, sometimes none too clear at their best; yet the *Standard Dictionary* seems to have adopted all his titles, even to **Zubenelgubi** for γ Librae, which really is unnamed, as this word is merely a degenerate form of the name for the star a.

The Chinese asterism **Se Han,** named for a district of that country, lay around η, and included it with ε, ζ, θ, ξ, and *e*.

κ and λ, 5th-magnitude stars, bore the pretentious title **Jih,** the Sun.

ξ erroneously was called **Graffias** in Burritt's *Atlas* of 1835, but this title belongs to β Scorpii.

σ is the letter attached by Gould to the disputed γ Scorpii, as is more particularly noted at that star.

*

. . . another form
That men of other days have called the beast.
Poste's *Aratos.*

Lupus, the Wolf,

is the **Loup** of the French, **Lupo** with the Italians, and **Wolff** in Germany, an idea for the figure said to be from the astrologers' erroneous translation of **Al Fahd,** the Arabian title for this constellation, their Leopard, or Panther; although Suidas, the Greek lexicographer of 970, is reported to have called it Κνηκίας, a word for the wolf found in the fables of Babrias of the century before our era. The Greeks and Romans did not specially designate these stars, and thought of them merely as a Wild Animal, the Θηρίον of Aratos, Hipparchos, and Ptolemy; the **Bestia** of Vitruvius; **Fera** of Germanicus; **Quadrupes vasta** of Cicero; **Hostia,** the Victim, of Hyginus; **Hostiola,** cited by Bayer; **Bestia Centauri,** by Riccioli; and **Victima Centauri.**

The Wolf reappeared as Lupus in the *Alfonsine Tables*, and as **Fera Lupus** in the *Latin Almagests*, while Grotius said that **Panthera** was Capella's name for it.

Bayer also had **Equus masculus** and **Leaena**; and La Lande, **Leo marinus, Deferens leonem, Canis ululans, Leopardus, Lupa, Martius,**—the wolf being sacred to Mars,—and **Lycisca,** the Hybrid of the Wolf. **Belua,** the Monster, is found in early works.

The Arabians also called it **Al Asadah,** the Lioness,—found by Scaliger repeated on a Turkish planisphere and cited by Bayer as **Asida,**—and **Al Sabu',** the Wild Beast, Chilmead's **Al Subahh.** But the Desert astronomers seem to have mixed some of its smaller stars with a part of the Centaur as **Al Shamārīh,** the Palm Branches, and **Ḳaḍb al Karm,** the Vine Branch.

Zibu, the Beast, of Euphratean cylinders, may be for this constellation; and **Urbat,** the Beast of Death, or the Star of the Dead Fathers, is a title for it attributed to the Akkadians.

Caesius said that in Persia it was **Bridemif,** but Hyde, commenting on this from Albumasar, asserted that the word should be **Birdūn,** the Pack-horse, and was really intended for the Centaur.

Aratos wrote of it, "another creature very firmly clutched," and "the Wild-beast which the Centaur's right hand holds" as an offering to the gods upon the Altar, and so virtually a part of the Centaur; but Eratosthenes described it as a **Wine-skin** from which the Centaur was about to pour a libation; while others imagined both the Beast and the Wine-skin in the Centaur's grasp.

Mythologists thought it the animal into which Lycaon was changed; Caesius, that it was the Wolf to which Jacob likened Benjamin; but Julius Schiller saw in its stars **Benjamin** himself.

Although very ancient, Lupus is inconspicuous, lying partly in the Milky Way, south of Libra and Scorpio, east of the Centaur, with no star larger than 2.6 magnitude, while the few visible in the latitude of New York City — γ, δ, λ, and μ — are even smaller than this.

Gould enumerates 159 naked-eye stars, among which is an unusual proportion of doubles.

α, 2.6, seems to be unnamed except in China, where it was **Yang Mun** or **Men,** the South Gate.

On the Euphrates it probably was **Kakkab Su-gub Gud-Elim,** the Star Left Hand of the Horned Bull, said to have been a reference to the Centaur that was thus figured in that valley.

It culminates on the 14th of June, nearly due south from Arcturus and north of α Centauri.

β is the **Ke Kwan,** of the Reeves list of Chinese titles, a Cavalry Officer. This is a very close binary, of 3 and 3.5 magnitudes, both yellow, $0''.25$ apart, the position angle being 90°.

α and β are below the horizon of New York City.

Other Chinese asterisms appear within the boundaries of Lupus, all bearing titles pertaining to military affairs, and so of the second period of their star-naming.

*

Each after each, ungrouped, unnamed, revolve.

Brown's *Aratos.*

Lynx sive Tigris, the Lynx or Tiger,

is the Italian **Lince,** the German **Luchs** and **Linx,** the French **Lynx.**

Its stars may have been those intended by Aratos where he mentioned, in our motto, some in front of the Greater Bear; but for the modern figure we are indebted to Hevelius. He used in it nineteen stars, and in explaining the title said that those who would examine the Lynx ought to be lynx-eyed, in which he acknowledged the insignificance of the components. Of these Argelander has catalogued 42, and Heis 87; but the boundaries are not accurately determined.

The alternative name, now in disuse, came from the fancied resemblance of the many little stars to spots on the tiger; and the same word was applied by Bartschius in 1624, although as the river **Tigris,** to some stars that subsequently were made into the Polish Bull and the Little Fox with the Goose.

In the Lynx appeared in July, 1893, the much-discovered comet *b* of that year, the Rordame–Quenisset.

The constellation seems chiefly noticeable for the beauty of its numerous doubles, of which Espin mentions fifty in his edition of Webb's *Celestial Objects.* Of one of these Professor Young writes in his *Uranography:*

> 38, or ρ Lyncis; Mags. 4, 7.5; Pos. 240°; Dist. 2″.9; white and lilac. This is the northern one of a pair of stars which closely resembles the three pairs that mark the paws of Ursa Major. This pair makes nearly an isosceles triangle with the two pairs λ μ and ι κ Ursae Majoris.

It might well have been utilized by the modern constructor, whoever he was, of our Ursa Major to complete the quartette of feet.

Baily thought Fl. 44 Lyncis the original 18th of Ursa Major in early catalogues.

Fl. 31 Lyncis, of 4.4 magnitude, the 8th of Ptolemy's ἀμόρφωτοι of Ursa Major, is given by Assemani as the Arabic **Alsciaukat,** a Thorn (**Al-Shaukah**), and **Mabsuthat** (**Mabsūṭah**), Expanded.

The constellation comes to the meridian in February, due north from the star Castor.

✶

> Ariones harpe fyn.
>
> Chaucer's *Hous of Fame.*

Lyra, the Lyre or Harp,

is the **Leier** of Germany, **Lira** of Italy, and **Lyre** of France, and anciently represented the fabled instrument invented by Hermes and given to his half-brother Apollo, who in turn transferred it to his son Orpheus, the musician of the Argonauts, of whom Shakespeare wrote:

Everything that heard him play,
Even the billows of the sea,
Hung their heads, and then lay by.

While Manilius said that its service in its last owner's hands, in the release of Eurydice from Hades,

Gain'd it Heaven, and still its force appears,
As then the Rocks it now draws on the Stars.

From its ownership by these divinities came various adjectival titles: Ἑρμαίη and Κυλλεναίη, referring to Hermes and his birthplace; Cicero's **Clara Fides Cyllenea** and **Mercurialis,** that Varro also used; and the **Cithara,** or **Lyra, Apollinis, Orphei, Orphica, and Mercurii.** It also was **Lyra Arionis** and **Amphionis,** from those skilful players; but usually it was plain **Lyra** and, later on, **Cithara**; **Fides,**— the **Fidis** of Columella, who, with Pliny, also used **Fidicula**; **Decachordum**; and **Tympanum.** In this same connection we see **Fidicen,** the Lyrist; **Deferens Psalterium**; and **Canticum,** a Song.

The occasional early title **Aquilaris** was from the fact that the instrument was often shown hanging from the claws of the Eagle also imagined in its stars.

In Greece it was Κιθάρα; the ancient Φόρμιγξ, the first stringed instrument of the Greek bards; and Λύρα or Λύρη, and Λύρα κατοφερής, the Pendent Lyre.

Ovid mentioned its seven strings as equaling the number of the Pleiades; Longfellow confirming this number in his *Occultation of Orion:*

with its celestial keys,
Its chords of air, its frets of fire,
The Samian's great Aeolian Lyre,
Rising through all its sevenfold bars,
From earth unto the fixéd stars.

Still it has been shown with but six, and a vacant space for the seventh, which Spence, in the *Polymetis*, referred to the Lost Pleiad.

Manilius seems to have made two distinct constellations of this,— Lyra and Fides,— although we do not know their boundaries, and the subject is somewhat confused in his allusions to it.

The Persian Hafiz called it the **Lyre of Zurah,** and his countrymen translated Κιθάρα by **Ṣanj Rūmi**; the Arabians turning this into **Al Ṣanj,** from which Hyde and others derived **Asange, Asenger, Asanges, Asangue, Sangue,** and **Mesanguo,** all titles for Lyra in Europe centuries ago. But Assemani thought that these were from Schickard's **Azzango,** a Cymbal. The reproduced *Alfonsine Tables* of 1863–67 give **Alsanja**; while Ṣanj was again turned into **Arnig** and **Aznig** in the translation of Reduan's *Commentary*, and into the still more unlikely **Brinek**, as has been explained by Ideler.

In Bohemia our Lyra was **Hauslicky na Nebi**, the Fiddle in the Sky; but the Teutons knew it as **Harapha**, and the Anglo-Saxons as **Hearpe**, which Fortunatus of the 6th century, the poet-bishop of Poitiers, called the barbarians' **Harpa**. With the early Britons it was **Talyn Arthur**, that hero's Harp. Novidius said that it was **King David's Harp**; but Julius Schiller, that it was the **Manger of the Infant Saviour, Praesepe Salvatoris**.

Jugum has been wrongly applied to it, from the Ζυγόν of Homer, but this was for the Yoke, or Cross-bar, of the instrument, with no reference to the constellation, which Homer probably did not know; still the equivalent Ζύγωμα was in frequent use for it by Hipparchos.

Sundry other fancied figures have been current for these stars.

Acosta mentioned them as **Urcuchillay**, the parti-colored Ram in charge of the heavenly flocks of the ancient Peruvians; **Albegala** and **Albegalo** occur with Bayer and Riccioli, like the Arabic Al Baghl, a Mule, although their appropriateness is not obvious; and Nasr al Dīn wrote of α, ε, and ζ collectively as **Dik Paye** among the common people of Persia; this was the Χυτρό-πους, or Greek tripod, and the **Uthfiyyah** of the nomad Arabs.

Chirka, also attributed to Naṣr al Dīn, was, by some scribe's error for **Ḣazaf**, figured in this location on the Dresden globe as a circular vessel with a flat bottom and two handles; but on the Borgian it is a **Scroll**, commonly known, according to Assemani, as **Rabesco**.

The association of Lyra's stars with a bird perhaps originated from a conception of the figure current for millenniums in ancient India,— that of an **Eagle** or **Vulture**; and, in Akkadia, of the great storm-bird **Urakhga** before this was there identified with Corvus. But the Arabs' title, **Al Naṣr al Wāḳi'**,[1] — Chilmead's **Alvaka**,— referring to the swooping Stone Eagle of the Desert, generally has been attributed to the configuration of the group α, ε, ζ, which shows the bird with half-closed wings, in contrast to **Al Naṣr al Ṭā'ir**,[1] the Flying Eagle, our Aquila, whose smaller stars, β and γ, on either side of α, indicate the outspread wings. Scaliger cited the synonymous **Al Naṣr al Sākiṭ**, from which came the **Nessrusakat** of Bayer and **Nessrusakito** of Assemani.

Al Sufi, alone of extant Arabian authors, called it **Al Iwazz**, the Goose.

Chrysococca wrote of it as Γυψ καθήμενος, the Sitting Vulture, and it has been **Aquila marina**, the Osprey, and **Falco sylvestris**, the Wood Falcon.

[1] These are two of the few instances in Arab astronomy where more than one star were utilized to represent an animate object.

Its common title two centuries ago was **Aquila cadens,** or **Vultur cadens,** the Swooping Vulture, popularly translated the **Falling Grype,** and figured with upturned head bearing a lyre in its beak. Bartsch's map has the outline of a lyre on the front of an eagle or vulture.

Aratos called it Χέλυς ὀλίγη, the Little Tortoise or Shell, thus going back to the legendary origin of the instrument from the empty covering of the creature cast upon the shore with the dried tendons stretched across it. Lowell thus described its discovery and use by Hermes:

> So there it lay through wet and dry,
> As empty as the last new sonnet,
> Till by and by came Mercury,
> And, having mused upon it,
> "Why, here," cried he, "the thing of things
> In shape, material and dimension!
> Give it but strings and, lo! it sings —
> A wonderful invention."

The equivalent Latin word *Chelys* does not seem to have been often applied to the constellation, but the occasional adjectival titles **Lutaria,** Mud-inhabiting, and **Marina** were, and are, appropriate, while **Testudo** has been known from classical times. Horace thus alluded to it:

> Decus Phoebi, et dapibus supremi
> Grata testudo Jovis; O laborum
> Dulce lenimen;

the poet doubtless having in mind the current story that the Tortoise-Lyre was placed in the sky near Hercules for the alleviation of his toil. The Alfonsine illustration is of a Turtle, **Galapago** in the original Spanish, which Caesius turned into the indefinite **Belua aquatica,** and La Lande into **Mus** and **Musculus,** some marine creature, not the little rodent.

Other names were **Testa,** the creature's Upper Shell; and **Pupilla,** which, by a roundabout process of continued blundering explained by Ideler, was derived from Testa, or, as seems more likely, from Aquila. Bayer's Βάσανος is probably a mistranslation of Testa that also signified a Test.

Smyth said that another **Testudo** was at one time proposed as a constellation title for some of the outside stars of Cetus, between the latter's tail and the cord of Pisces.

When the influence of Greek astronomy made itself felt in Arabia, many of the foregoing designations, or adaptations thereof, became current; among them **Nablon,** from Νάβλα, or **Nablium,** the Phoenician Harp; **Al Lurā,** which degenerated into **Allore, Alloure, Alohore, Alchoro,** etc., found in the *Alfonsine Tables* and other bygone lists; **Shalyāķ** and **Sulaḥfāt,** words for the Tortoise, Ulug Beg's translator having the former as **Shelyāk,** which Piazzi repeated in his catalogue; **Salibāķ,** which heads Kazwini's chapter on the Lyre;—Ideler tracing these Arabic words to Χέλυς. They were turned into **Azulafe** and **Zuliaca** in the original *Alfonsine Tables*, and **Schaliaf** in Chilmead's *Treatise*. The *Almagest* of 1515 combines all these figures for Lyra's stars in its *Allore: et est Vultur cadens: et est Testudo;* while that of 1551 says **Lyrae Testudo.**

But, notwithstanding the singularly diverse conceptions as to its character, the name generally has been Lyra, and the figure so shown. Roman coins still in existence bear it thus, as does one from Delos, Apollo's birthplace in the Cyclades; and Cilician money had this same design with the head of Aratos on the obverse. The *Leyden Manuscript* has the conventional instrument, with side bars of splendid horns issuing from the tortoise-shell base; the Venetian *Hyginus* of 1488, with a similar figure, calls it **Lura** as well as Lyra; but the drawing of Hevelius shows "an instrument which neither in ancient nor in modern times ever had existence." Dürer's illustration, as well as others, places it with the base towards the north.

Lyra is on the western edge of the Milky Way, next to Hercules, with the neck of Cygnus on the east, and contains 48 stars according to Argelander, 69 according to Heis. Its location is noted as one of the various regions of concentration of stars with banded spectra, Secchi's 3d type, showing a stage of development probably in advance of that of our sun.

From near its κ, 5° southwest of Wega, radiate the swiftly moving **Lyraids,** the meteors which are at their maximum of appearance on the 19th and 20th of April, but visible in lesser degree from the 5th of that month to the 10th of May. These have been identified as followers of the comet I of 1861.

> . . . azure Lyra, like a woman's eye,
> Burning with soft blue lustre.
>
> Willis' *The Scholar of Thebet ben Khorat.*

α, 0.3, pale sapphire.

Wega, less correctly **Vega,** originated in the *Alfonsine Tables* from the Wāķiʿ of the Arabs, Bayer having both titles; Scaliger, **Waghi;** Riccioli, **Vuega** *vel* **Vagieh;** and Assemani, **Veka.**

The Greeks called it Λύρα, which, in the 16th-century *Almagests* and *Tables*, was turned into **Allore, Alahore,** and **Alohore.**

Among Latin writers it was **Lyra,** in classical days as in later, seen in the *Almagest* of 1551 as *Fulgens quae in testa est & vocatur Lyra ;* and in Flamsteed's *Testa fulgida dicta Lyra ;* but Cicero also used **Fidis** specially for the star, as did Columella and Pliny **Fides** and **Fidicula,** its preëminent brightness fully accounting for the usurpation of so many of its constellation's titles, indeed undoubtedly originating them. In Holland's translation of Pliny it is the **Harp-star.**

The Romans made much of it, for the beginning of their autumn was indicated by its morning setting. It was this star that, when the hour of its rising was alluded to, called forth Cicero's remark, "Yes, if the edict allows it,"— a contemptuous reference to Caesar's arbitrary, yet sensible, interference with the course of ancient time in the reformation of the calendar, an interference that occasioned as much dissatisfaction in his day as did Pope Gregory's reform [1] in the 16th century.

Sayce identifies Wega, in Babylonian astronomy, with **Dilgan,** the Messenger of Light, a name also applied to other stars; and Brown writes of it:

> At one time Vega was the Pole-star called in Akkadian **Tir-anna** ("Life of Heaven"), and in Assyrian **Dayan-same** ("Judge of Heaven"), as having the highest seat therein;

but fourteen millenniums have passed since Wega occupied that position!

The Chinese included it with ε and ζ in their **Chih Neu,** the Spinning Damsel, or the Weaving Sister, at one end of the Magpies' Bridge over the Milky Way,— Aquila, their Cow Herdsman, being at the other; but the story, although a popular one not only in China, but also in Korea and Japan, is told with many variations, parts of Cygnus sometimes being introduced.

These same three stars were the 20th *nakshatra*, **Abhijit,** Victorious, the most northern of these stellar divisions and far out of the moon's path, but apparently utilized to bring in this splendid object; or, as Mueller says, because it was of specially good omen, for under its influence the gods had vanquished the Asuras; these last being the Hindu divinities of evil, similar to the Titans of Greece. It was the doubtful one of that country's lunar stations, included in some, but omitted in others of their lists in all ages of their astronomy, and entirely different from the corresponding *manzil* and *sieu*, which lay in Capricorn. The Hindus figured it as a

[1] The English refused to adopt this reform till 1752, when they abandoned the Old Style on the 2d of September, and made the succeeding day September 14th, New Style: a change, however, that "was made under very great opposition, and there were violent riots in consequence in different parts of the country, especially at Bristol, where several persons were killed. The cry of the populace was 'Give us back our fortnight,' for they supposed they had been robbed of eleven days."

Triangle, or as the three-cornered nut of the aquatic plant Cringata, Wega marking its junction with the adjoining Çravana.

Hewitt says that in Egypt it was **Ma'at,** the Vulture-star, when it marked the pole,— this was 12000 to 11000 B. C. (!),— and Lockyer, that it was the orientation point of some of the temples at Denderah long antecedent to the time when γ Draconis and α Ursae Majoris were so used,— probably 7000 B. C.,— one of the oldest dates claimed by him in connection with Egyptian temple worship.

Owing to precession, it will be the Polaris of about 11500 years hence, by far the brightest in the whole circle of successive pole-stars, and then 4½° from the exact point, as it was about 14300 years ago. In 1880 it was 51° 20′ distant. Professor Lewis Boss and Herr Stumpe place near it the Apex of the Sun's Way.

Picard failed in his efforts to obtain its parallax in the 17th century, but Struve thought that he had succeeded in this by his observations previous to 1840; still much discrepancy exists in the recent determinations. Elkin, in 1892, gave it as 0″.092; or, to put it in popular language, if the distance from the earth to the sun be regarded as one foot, that from Wega would be 158 miles. The 10th-magnitude companion, about 48″ away, used for some of these determinations, is entirely independent of it, although difficult to be seen owing to the great brilliancy of Wega. At least two other still fainter companions also have been found.

This was the first star submitted to the camera, by the daguerreotype process, at the Harvard Observatory on the 17th of July, 1850.

It lies on the western edge of the constellation figure, and, after Sirius, is the most prominent of the stars showing spectra of the Sirian type; yet, with all its splendor, affords but $\frac{1}{9}$ of the latter's light. Still it is supposed to be enormously larger than our sun, and proportionately very much hotter. It is moving toward our system at the rate of about 9½ miles a second, and makes "the nearest approach in the northern hemisphere to an independently blue star"; while its flashing brilliancy justifies its being called the **Arc-light** of the sky. Miss Mitchell strangely called it pale yellow.

Wega rises at sunset far toward the north on the 1st of May, and, being visible at some hour of every clear night throughout the year, is an easy and favorite object of observation. It culminates on the 12th of August.

With ϵ and ζ it formed one of the Arabs' several **Athāfiyy,** this one being "of the people," while the others, fainter, in Aries, Draco, Musca, and Orion, were "of the astronomers"; for sky objects are often very plain to them that are invisible to the ordinary observer.

β, Variable and binary, 3.4 to 4.5, very white.

Sheliak, Shelyak, and **Shiliak** are from **Al Shilyāk,** one of the Arabian names for Lyra. The star lies about 8° southeast from Wega and 2½° west from γ.

With δ and ι it was **Tsan Tae** in China.

The changes in its brilliancy, detected by Goodricke in 1784, were fully investigated by Argelander from 1840 to 1859, and showed a regularly increasing period of variability which now is 12 days, 21¾ hours, with several fluctuations of a somewhat complex nature.

Like γ Cassiopeiae and other variables of the Sirian type, it shows in its spectrum,— perhaps the best specimen of Pickering's 4th class,— not only the usual dark lines, but also the bright lines of glowing gases, hydrogen and helium being especially conspicuous. Pickering concluded, from the singular character and behavior in the shifting of these lines, that the chief star must consist of at least two luminous bodies rotating around a common centre of gravity at a very great rate of speed, perhaps three hundred miles a second, the period of revolution equaling the period of variability. Scheiner says of it, "There is great probability that more than two bodies are concerned in the case of β Lyrae"; and yet it may not be impossible, in view of the recent discoveries at the Johns Hopkins Laboratory, that variations of *pressure* may be concerned in this remarkable shifting of lines.[1]

γ, 3.3, bright yellow,

2½° east of β is **Sulafat,** from another of the titles of the whole constellation.

Jugum, formerly seen for it, may have come from a misunderstanding of Bayer's text, where it probably is used merely to designate the star's position on the frame of the Lyre, his words being *ad dextrum cornu*, Ζυγόν, *Iugum*, — a fair example of the indefiniteness of much of his stellar nomenclature.

At a point ⅓ of the distance from β to γ is the wonderful **Ring Nebula,** N. G. C. 6720, 57 M., discovered in 1772 by Darquier from Toulouse, although its apparent annular form was not revealed till later by Sir William Herschel's observations. In our day high-powers show its oval form somewhat undefined at the edges, with a dark opening in the centre containing a few very faint stars, among which, visible only in the largest telescopes, but prominent in photographs, is a central condensation of light like a star.

[1] A full and interesting discussion of this appears in *Popular Astronomy* for July, 1898.

The spectrum of nebula and central "star" is purely gaseous. Although appearing oval to us, it is supposed to be nearly circular, but seen obliquely. It is the only annular nebula visible through small telescopes, although there are six others now known.

ε^1, or Fl. 4, Binary, 4.6 and 6.3, yellow and ruddy;

ε^2, or Fl. 5, Binary, 4.9 and 5.2, both white.

These are the celebrated **Double Double,** each pair probably separately revolving in a period of over two hundred years, and both pairs perhaps revolving around their common centre of gravity; but if so, the period is to be reckoned only by millenniums, for the measures of the last fifty years show no sensible orbital motion. This is by far the finest object of the kind in all the heavens.

They are 207″ apart, and, to the ordinary eye, form an elongated star; but exceptionally sharp sight will resolve them without aid. The pairs are 3″.2 and 2″.45 apart respectively, and a good 2¼-inch glass with a power of 140 will separate each pair. The position angle of the components of ε^1 is 12°; and of those of ε^2, 132°; while that of ε^1 and ε^2 is 173°. Their "double-double" character was first published by the Jesuit father Christian Mayer in 1779, although its discovery has generally been attributed to Sir William Herschel.

The distance between ε^1 and ε^2, small as it is, is nearly twice that noticed by astronomers, in 1846,— 128″—between the actual and the computed positions of the planet Uranus, a discrepancy which convinced them of the existence of a still more remote planet and led to the discovery of Neptune. Such is the marvelous nicety of modern astronomical measurements!

Between these stars lie three very much fainter, two of which, of the 13th magnitude, are the **Debilissima,** Excessively Minute, of Sir John Herschel, discovered by him in 1823.

ε and ζ form an equilateral triangle with Wega, the sides about 2° long; ε being at the northern angle. These three stars were one of the **Athāfiyy** of the early Arabs.

η, a 4.4-magnitude, is **Aladfar** in the *Century Atlas*, by some confusion with the star μ; and with θ, of the same brilliancy, was, in China, **Lëen Taou,** Paths within the Palace Grounds.

μ, of the 5th magnitude, was Kazwini's **Al Aṭhfār,** the Talons (of the Falling Eagle), which he described as a fainter star in front of the bright one, *i. e.* west of Wega.

Machina Electrica,

one of Bode's constellations of 1800, lies south of the central portion of Cetus. With him it was the **Elektrisir Machine** and **Machine Electrique**; the Italians call it **Machina Elettrica.**

It is now generally omitted from the maps and catalogues.

★

Microscopium,

formed by La Caille south of Capricornus and west of Piscis Australis, although small and unimportant, contains sixty-nine stars, varying in magnitude from 4.8 to 7, the *lucida* being θ^1. The constellation comes to the meridian in September, nearly due south of β Aquarii.

In its vicinity, perhaps including it, was an early figure referred to, in a German astronomical work of 1564 from Frankfürt, as **Neper,** the Auger, Ideler's **Bohrer,** which he thus described:

> It is situated at the tail of Sagittarius and Capricornus, and has many stars. At the head of the Neper two, and on the iron three.

Brown alludes to it as an unknown object, and illustrates it in the 47th volume of *Archaeologia* as from a German astronomical manuscript of the 15th century; but Flammarion, in *les Étoiles*, probably referring to this same manuscript, thus mentions Neper, as the predecessor of Monoceros:

> Il est question de la constellation du Neper ou Foret, qui n'est autre que la Licorne.

★

Monoceros, the Unicorn,

das Einhorn in Germany, **la Licorne** in France, and **il Unicorno** or **Liocorno** in Italy, lies in the large but comparatively vacant field between the two Dogs, Orion, and the Hydra, the celestial equator passing through it lengthwise from the Belt of Orion to the tail of the animal, just below the head of Hydra. Proctor assigned to it the alternative title **Cervus**.

Its 4.6-magnitude S, or Fl. 15, marks the head of the figure, facing towards the west.

This is a modern constellation, generally supposed to have been first charted by Bartschius as **Unicornu**; but Olbers and Ideler say that it was of much earlier formation, the latter quoting allusions to it, in the work of 1564, as "the other Horse south of the Twins and the Crab"; and Scaliger found it on a Persian sphere.

Flammarion's identification of it with the still earlier Neper has already been mentioned under Microscopium.

Monoceros seems to have no star individually named, but the Chinese asterisms **Sze Fūh**, the Four Great Canals; **Kwan Kew**; and **Wae Choo**, the Outer Kitchen, all lay within its boundaries.

It contains 66 naked-eye stars according to Argelander,— Heis says 112,— and is interesting chiefly from its many telescopic clusters, and as being located in the Milky Way.

It comes to the meridian in February, due south from Procyon.

a, the *lucida*, is Fl. 30, of 3.6 magnitude.

*

Mons Maenalus,

at the feet of Boötes, was formed by Hevelius, and published in his *Firmamentum Sobiescianum*; this title coinciding with those of neighboring stellar groups bearing Arcadian names. It is sometimes, although incorrectly, given as **Mons Menelaus**,— perhaps, as Smyth suggested, after the Alexandrian astronomer referred to by Ptolemy and Plutarch.

The Germans know it as the **Berg Menalus**; and the Italians, as **Menalo**.

Landseer has a striking representation of the Husbandman, as he styles Boötes, with sickle and staff, standing on this constellation figure. A possible explanation of its origin may be found in what Hewitt writes in his *Essays on the Ruling Races of Prehistoric Times*:

> The Sun-god thence climbed up the mother-mountain of the Kuṣhika race as the constellation Hercules, who is depicted in the old traditional pictorial astronomy as climbing painfully up the hill to reach the constellation of the Tortoise, now called Lyra, and thus attain the polar star Vega, which was the polar star from 10000 to 8000 B. C.

May not this modern companion constellation, Mons Maenalus, be from a recollection of this early Hindu conception of our Hercules transferred to the adjacent Boötes?

It culminates in June, due south from β Boötis and north of β Librae.

★

Mons Mensae, the Table Mountain,

now abbreviated by astronomers to **Mensa,** is translated by the French as **Montagne de la Table**; by the Italians, as **Monte Tavola**; and by the Germans, as **Tafelberg.**

La Caille, who did so much for our knowledge of the southern heavens, formed the figure from stars under the Greater Cloud, between the poles of the equator and the ecliptic, just north of the polar Octans; the title being suggested by the fact that the Table Mountain, back of Cape Town, "which had witnessed his nightly vigils and daily toils," also was frequently capped by a cloud.

Gould found in the constellation 44 naked-eye stars, the brightest being of 5.3 magnitude; but within its borders is a portion of the Nubecula Major.

★

Musca Australis vel Indica, the Southern, or Indian, Fly,

the French **Mouche Australe ou Indienne,** the German **Südliche Fliege,** and the Italian **Mosca Australe,** lies partly in the Milky Way, south of the Cross, and east of the Chamaeleon.

This title generally is supposed to have been substituted by La Caille, about 1752, for Bayer's **Apis,** the Bee; but Halley, in 1679, had called it **Musca Apis**; and even previous to him, Riccioli catalogued it as **Apis** *seu* **Musca.** Even in our day the idea of a Bee prevails, for Stieler's *Planisphere* of 1872 has **Biene,** and an alternative title in France is **Abeille.**

The modern Chinese translate Bayer's title as **Meih Fung,** and have so known it since the 16th century.

Julius Schiller united it with the Bird of Paradise and the Chamaeleon as mother **Eve.**

Gould assigned to it 75 stars, of magnitudes from 2.9 to 7; these culminating, with the Cross, about the middle of May.

Musca Borealis, the Northern Fly,

the small group of 3½- to 5th-magnitude stars over the back of the Ram, is the Italian **Mosca,** the French **Mouche,** and the German **Fliege.**

Houzeau attributed its formation to Habrecht, but others to Bartschius, who called it **Vespa,** the Wasp, although also **Apis,** the Bee; and, still further changing the figure, wrote that it represented **Beel-zebul,** the god of flies, the Phoenician Baal-zebub; this insect being the ideograph of that heathen divinity, varied at times by the Scarabaeus. La Lande's **Apes** probably is a typographical error. To whom we owe its present title I cannot learn; but it is thus given in the Flamsteed *Atlas* of 1781.

The constellation has been retained in some popular astronomical works, although not figured by the scientific Argelander, Heis, nor Klein, nor recognized in the *British Association Catalogue.*

Ptolemy included its stars in the five ἀμόρφωτοι of his Κριός, the Ram.

Its chief components, Fl. 41, 33, 35, and 39 of Aries, were common to the 28th *nakshatra*, **Barani,** Bearer, or **Apha Barani,**—Yama, the ruler of the spirit world, being the presiding divinity; Fl. 35 being the junction star towards the *nakshatra* **Krittikā.** They also formed the *sieu* **Oei** or **Wei,** anciently **Vij**; and the *manzil* **Buṭain.** But as these Chinese and Arabic titles, signifying Belly, *i. e.* of the Ram, do not coincide with the present location of the stars, we may infer a change from the earlier drawings of Aries. Al Tizini's **Nā'ir al Butain,** the Bright One of the Little Belly, probably was 41, a 3.6-magnitude. These same stars, μ being added, were the Persian lunar station **Pish Parvis,** the Sogdian **Barv,** the Khorasmian **Farankhand,** the Forerunners, and the Coptic **Koleōn,** the Belly, or Scabbard. Flamsteed's 41, 35, and 39 formed another of the Arabs' **Athāfiyy.**

Musca comes to the meridian on the 17th of December.

Instead of the Fly, Royer figured here, in 1679, the Lily, **le Lis** or **le Fleur de Lis,** with the French coat of arms, but this has entirely passed out of the books and maps.

★

Noctua, the Night Owl,

has been added by some modern to the already overweighted Hydra. It is shown by Burritt perched upon the extreme tail-tip of that figure, but encroaching on the boundary of the Southern Scale.

Its location formerly was occupied by Le Monnier's **Solitaire,** but neither of these asterisms is now recognized.

★

Norma et Regula, the Level and Square,

originally was composed of some unformed stars of Ara and Lupus, within the branches of the Milky Way, just north of Apus; but later it became the **Southern Triangle** of Theodor and Bayer. According to Ideler, it was altered by La Caille to its present form, and associated with a Pair of Compasses, the constellation Circinus, next to it on the north, adjoining the fore feet of the Centaur. Modern astronomers, however, call it simply Norma, and locate it as an entirely distinct constellation to the north of and adjoining the Triangle.

It is sometimes given as **Quadra Euclidis**, Euclid's Square, not Quadrant as it often is incorrectly translated.

The French edition of Flamsteed's *Atlas* of 1776 has it as **Niveau,** the Level; and Houzeau cites **Libella** of the same meaning; but in France it now is **l'Équerre et la Règle**; in Italy, **Riga e Squadra**; and in Germany, **Lineal** or **Winkelmass.**

Norma contains 64 naked-eye stars, from 4.6 to 7th magnitudes, but none seem to be named. They culminate about the 4th of July, their northern limit 15° south from the star Antares, and so are visible only in low latitudes.

La Caille's α Normae lies within the present limits of our Scorpio.

In Norma appeared in 1893 a 7th-magnitude *nova* detected by Mrs. Margaret Fleming on a photograph taken on the 1st of July at the Harvard Observatory's station near Arequipa, although it never was visually observed. Special interest attaches to it from the identity of its spectrum with that of the *nova* Aurigae of the preceding year, the first two of their kind discovered.

The appearance of two new stars at such a short interval is also noticeable, as Miss Clerke says that only about eighteen had been recorded since the days of Hipparchos; Professor Young reducing this to eleven as certainly known down to 1892; but observers have greatly increased in recent years, the heavens are better known than formerly, and the camera shows what the eye, aided even by the best telescope, cannot,— all factors in the problem of the detection of these strangers. The photographs retain impressions of thousands of stars, while the visual observer practically is limited to a few hundred.

★

Nubeculae Magellani, the Magellanic Clouds,

were the **Cape Clouds** of the earliest navigators, being the prominent heavenly objects seen as they neared the Cape of Good Hope; but after Magellan became noted and fully described them, they took and have retained his name. The Latin word is the diminutive of *nubes*, and literally signifies "the Little Clouds."

Miss Mitchell alluded to them as the **Magellan Patches;** and Smyth, as the **Sacks of Coals** of English navigators; but the latter term generally has been applied to the darkly vacant spaces in the Milky Way near the Northern and the Southern Cross, and to one near the Robur Carolinum.

Although Bayer seems to have been the first to figure them, they were thus mentioned by Peter Martyr in Eden's *Decades:*

Coompasinge abowte the poynt thereof, they myght see throughowte al the heaven about the same, certeyne shynynge whyte cloudes here and there amonge the starres, like unto theym whiche are seene in the tracte of heaven cauled Lactea via, that is the mylke whyte waye:

and by Corsali:

[We] sawe manifestly twoo clowdes of reasonable bygnesse movynge abowt the place of the pole continually now rysynge and now faulynge, so keepynge theyr continuall course in circular movynge, with a starre ever in the myddest which is turned abowt with them abowte .xi. degrees frome the pole.

This star is γ Hydri, a 3.2-magnitude red, now 15° from the pole.

According to Ellis, the Polynesian Islanders called the clouds **Mahu,** Mist, distinguishing them as Upper and Lower; and Gill, in his stories of the natives of the Hervey group, cited their somewhat similar **Nga Maū.**

Russell's photographs, taken at Sydney in 1890, show them to be spiral in formation, each with two centres of condensation, and, as Doctor William Whewell wrote in his *Plurality of Worlds*, composed of "masses of stars, clusters of stars, nebulae regular and irregular, and nebulous streaks and patches." The space around them is very blank, especially in the case of the Minor, "as if the cosmical material in the neighborhood had been swept up and garnered in these mighty groups."

Together they serve to show the location of the pole, marking two angles of a nearly equilateral triangle, of which the polar point is the third.

★

Nubecula Major, the Greater Cloud,

Nubes Major with Royer, is the Italian **Nube Maggiore,** the French **Grand Nuage,** and the German **Grosse Wolke.**

It lies in the constellations Dorado and Mons Mensae, 20° from the south pole, covering an irregular space in the sky of about forty-two square degrees; but the intensity of its light is inferior to that of the Lesser Cloud and is obliterated by the full moon. According to Flammarion, it contains 291 distinct nebulae, 46 clusters, and 582 stars.

Al Sufi mentioned it as **Al Bakr,** the White Ox, of the southern Arabs, and invisible from Baghdad, or northern Arabia, but visible from the parallel of the Strait of Babd al Mandab, in 12° 15′ of north latitude. Ideler translated this as the **Oxen of Tehama,**— Tehama being a province on the Red Sea; this title probably includes the companion cloud.

Julius Schiller combined it with Dorado and Piscis Volans in his biblical figure **Abel the Just.**

★

Nubecula Minor, the Lesser Cloud,

Nubes Minor with Royer, is the **Nube Minore** of the Italians, the **Petit Nuage** of the French, and the **Kleine Wolke** of the Germans. It lies within the borders of Hydrus and Tucana, with which Julius Schiller fashioned it into the archangel **Raphael.**

According to Flammarion, it contains 37 nebulae, 7 clusters, and 200 stars, and covers about ten square degrees, the immediately surrounding space being almost devoid of stars, or, as Sir John Herschel wrote, "most oppressively desolate," and access to it on all sides "is through a desert."

Close to it, between η Hydri and κ Tucanae, is the centre of the constellational vacancy of 2400 to 2000 B. C., marking the place of the south pole of that date.

Hic vertex nobis semper sublimis; at illum
Sub pedibus Styx atra videt, Manesque profundi.

Vergil's 1st *Georgic.*

Octans Hadleianus,

now known simply as **Octans,** was formed and published by La Caille in 1752 in recognition of the octant invented in 1730 by John Hadley. It is the French **Octant,** the German **Oktant,** and the Italian **Ottante.** The French edition of Flamsteed's *Atlas* has it as **l'Octans Réflexion.**

Gould assigns to it 88 naked-eye stars down to the 7th magnitude; the brightest, ν, being only of 3.8; but the constellation is noteworthy as marking the south pole, its 5.8-magnitude σ being about ¾ of a degree away. A straight line from α Crucis to β Hydri almost touches the pole at ⅓ of the distance from the latter star.

Ancient references to a south pole are of course infrequent; Ovid, however, makes Phoebus allude to it in his instructions to Phaëthon; Vergil mentions it as in our motto; Creech thus renders from Manilius:

the *lower pole* resemblance bears
To this *above,* and shines with equal Stars;

and Pliny tells us that the Hindus had given it a name, **Dramasa,**—

Austrinum Polum Indi Dramasa vocant.

The heathen Arabs, too, seem to have had some knowledge of it, for they imagined that, like its northern counterpart, it exercised a healing power on all afflicted persons who would attentively observe it.

The early navigators commented more or less correctly on the blankness of the heavens in this region, and Peter Martyr wrote:

They knewe no starre there lyke unto this pole, that myght be decerned aboute the poynte;

Pigafetta, in his description of the Magellanic Clouds:

Betweene these, are two starres not very bigge, nor much shyninge, which move a little: and these two are the pole Antartike,—

probably the colored stars β and γ Hydri of about the 3d magnitude; and Camões:

Vimos a parte menos rutilante,
E por falta d'estrellas menos bella
Do polo fixo,

which probably refers to the same thing, but which his translator Aubertin claims as an allusion to the **Coal-sack,** or **Soot-bag.** Vespucci, on the other hand, strangely stated, in his *Lettera* of 1505, that "the stars of the pole of the south . . . are numerous, and much larger and more brilliant than those of our pole"; and that he saw in the southern sky about twenty stars as bright as Venus and Jupiter. Ideler's comment on Vespucci, in this connection, is "the greater part of his news is of this reliable character!" Even now it is the popular opinion that the South is richer in stars than is the North; Tennyson expressing this in *Locksley Hall:*

> Larger constellations burning.

★

Officina Typographica, the Printing Office,

was formed by Bode — at all events, first published by him — from stars immediately east of Sirius; but it is seldom found on the maps of our day, nor recognized by astronomers, although Father Secchi inserted it on his planisphere of 1878.

Italian lists have it as **Tipografia,** and the German as **Buchdrucker Presse,** or **Buchdrucker Werkstadt.**

★

> . . . the length of Ophiuchus huge
> In th' arctic sky.
>
> Milton's *Paradise Lost.*

Ophiuchus vel Serpentarius, the Serpent-holder,

not **Ophiuchus Serpentarius,** is **Ofiuco** with the Italians, **Schlangenträger** with the Germans, and **Serpentaire** with the French.

It stretches from just east of the head of Hercules to Scorpio; partly in the Milky Way, divided nearly equally by the celestial equator; but, although always shown with the Serpent, the catalogues have its stars entirely distinct from the latter. The classical Hyginus, however, united the two figures into a single constellation, and some early nations, especially the Sogdians and Khorasmians, did the same, the stars being intermingled in their nomenclature.

The original title, Ὀφιοῦχος, appeared in the earliest Greek astronomy; μογερός, "toiling," being an adjectival appellation in the *Phainomena*.

Transliterated as in our title it was best known to the Latins, but also as **Ophiulchus, Ophiulcus, Ophiultus,** and, in the diminutive, **Ophiuculus** and **Ophiulculus**; while the classical word plainly shows itself in the **Afeichus, Afeichius,** and **Alpheichius** of the 16th and 17th centuries.

Serpentarius first appeared with the scholiast on Germanicus, while **Serpentiger, Serpentis Lator, Serpentis Praeses,** and **Serpentinarius** are seen for it; as also the **Anguifer** of Columella, which was **Anguiger** elsewhere. Cicero and Manilius had the peculiar **Anguitenens.** Golius insisted that this sky figure represents a **Serpent-charmer,** one of the Psylli of Libya, noted for their skill in curing the bites of poisonous serpents; and this would seem to be confirmed by the constellation's title **le Psylle** in Schjellerup's edition of Al Sufi's work.

But the Serpent-holder generally was identified with Ἀσκληπιός,[1] **Asclepios,** or **Aesculapius,** whom King James I described as "a mediciner after made a god," with whose worship serpents were always associated as symbols of prudence, renovation, wisdom, and the power of discovering healing herbs. Educated by his father Apollo, or by the Centaur Chiron, Aesculapius was the earliest of his profession and the ship's surgeon of the Argo. When the famous voyage was over he became so skilled in practice that he even restored the dead to life, among these being Hippolytus, of whom King James wrote:

> Hippolyte. After his members were drawin in sunder by foure horses, Esculapius at Neptun's request glewed them together and revived him.

But several such successful operations and numerous remarkable cures, and especially the attempt to revive the dead Orion, led Pluto, who feared for the continuance of his kingdom, to induce Jove to strike Aesculapius with a thunderbolt and put him among the constellations.

The figure also was associated with **Caecius,** the Blinding One, slain by Hercules and celebrated by Dante in the *Inferno;* indeed, it is said that the

[1] According to Greek tradition, he was a lineal ancestor of the great physician Hippocrates; and Doctor Francis Adams, in his *Genuine Works of Hippocrates*, writes:

> A genealogical table, professing to give a list of names of his forefathers, up to Aesculapius, has been transmitted to us from remote antiquity.

This list, from the *Chiliads* of Tzetzes of our 12th century, makes Hippocrates the 15th in descent from Aesculapius through his son Podalirius, who, with his brother Machaon, was an army surgeon, as well as a valiant fighter before the walls of Troy.

The name and the profession were continued in the Asclepiadae, an order of priest-physicians long noted in Greece.

Hero himself was assigned to these stars by Hyginus, and gave them his name: a confusion that may have arisen because the boundaries between the two stellar groups were at first ill defined, or from the similarity of their original myths to that of Izhdubar and the dragon Tiāmat. It also represented **Triopas,** king of the Perrhaebians; **Carnabon, Carnabas,** and **Carnabus,** the slayer of Triopas; **Phorbas,** his Thessalian son, who freed Rhodes from snakes; **Cadmus** changed to a serpent; **Jason** pursuing the golden-fleeced Aries; **Aesacus,** from the story of Hesperia; **Aristaeus,** from the story of Eurydice; **Laocoön** struggling with the serpent; and **Caesius,** or **Glaucus,** the sea-god, although this latter title, identified by some with that of **Androgēus,** may have come from that namesake who was restored to life by Aesculapius.

The Arabians translated the Greek name into **Al Ḥawwāʿ,** which Assemani repeated as **Alhava,** *Collector serpentum;* but it appeared on the globes as **Al Haur,** turned by the Moors into **Al Hague,** and by early astronomical writers into **Alangue, Hasalangue,** and **Alange;** the Turks having the similar **Yilange.** It has been suggested, however, that these may have come from the Latin *Anguis*, a word that the astronomical Arabians and Moors well knew.

Euphratean astronomers knew it, or a part of it with Serpens, as **Nutsir-da;** and Brown associates it with **Sa-gi-mu,** the God of Invocation.

Pliny said that these stars were dangerous to mankind, occasioning much mortality by poisoning; while Milton compared Satan to the burning comet that "fires" this constellation,— a comparison perhaps suggested by the fact that noticeable comets appeared here in the years 1495, 1523, 1537, and 1569, which might well have been known to Milton, for Lord Bacon wrote in his *Astronomy:*

> Comets have more than once appeared in our time; first in Cassiopeia, and again in Ophiuchus.

Novidius changed the figure to that of **Saint Paul with the Maltese Viper;** Caesius gave it as **Aaron,** whose staff became a serpent, or as **Moses,** who lifted up the Brazen Serpent in the Wilderness; but Julius Schiller, far more appropriately, made of it **Saint Benedict** in the midst of the thorns, for it was this founder of the order of the Benedictine monks who, with his followers in the 6th century, inspired and carried on all the learning of the times, as Aesculapius-Ophiuchus had done in his day.

The constellation generally has been shown as an elderly man, probably copied from the celebrated statue at Epidaurus; but the *Leyden Manuscript* and the planisphere of the monk Geruvigus represent it as an unclad boy standing on the Scorpion and holding the Serpent in his hands; and the *Hyginus* of 1488 has a somewhat similar representation.

Bayer added to his titles for Ophiuchus **Grus** *aut* **Ciconia** *Serpenti cum inscriptione*, *Elhague*, *insistens*, which he said was from the Moors, but Ideler asserted was from a drawing of a Crane, or Stork, on a Turkish planisphere instead of the customary figure; and the *Almagest* of 1551 alludes to Ciconia as if it were a well-known title. All this, perhaps, may be traced to ancient Iṇdia, whose mythology was largely astronomical, and the Adjutant-bird, *Ciconia argala*, prominent in worship as typifying the moon-god Soma, so that its devotees would only be following custom in locating it among the stars.

Although this is not one of the zodiac twelve, Mr. Royal Hill writes:

> Out of the twenty-five days, from the 21st of November to the 16th of December, which the sun spends in passing from *Libra* to *Sagittarius*, only nine are spent in the *Scorpion*, the other sixteen being occupied in passing through *Ophiuchus*.

Thus, according to his idea of the boundaries, this actually is more of a zodiacal constellation than is the Scorpion. But the boundaries are very variously given by uranographers.

Argelander enumerates in it 73 naked-eye stars, and Heis 113.

It was in Ophiuchus that appeared, A. D. 123, the second *nova* of which we have reliable record, the first having been that of Hipparchos, 134 B. C., in Scorpio. At least three other such have appeared in Ophiuchus: one in 1230; another, the so-called **Kepler's Star,** discovered by Kepler's pupil Brunowski, on the 10th of October, 1604, in the eastern foot near θ, which gave Galileo opportunity for his "onslaught upon the Aristotelian axiom of the incorruptibility of the heavens"; and a third, discovered on the 28th of April, 1848, by Hind as of the 4th magnitude, and still visible as of the 11th or 12th.

Citing Firmicus as authority, La Lande wrote:

> Il met le **Renard** au nord du Scorpion avec Ophiuchus;

but I do not find this Fox elsewhere alluded to.

α, 2.2, sapphire.

Ras alhague, or **Rasalague,** is from **Rās al Ḥawwā'**, the Head of the Serpent-charmer, the Moorish **El Hauwe,** the first being its only title with Bayer. The *Alfonsine Tables* of 1521 have **Rasalauge,** and the original has been variously altered into **Ras Alhagas, Ras Alhagus, Rasalange, Ras al Hangue, Rasalangue, Ras Alaghue, Rasalhagh, Alhague,** and **Alangue.** The occasional **Azalange** has been traced to the Turkish title for the constellation; but "a universal star-name from that nation does not seem probable," and it is more likely that the Turks adopted and altered the Arabic. **Ras al Hayro** also has been seen for the star; and the *Century Cyclopedia* mentions **Hawwa** as rarely used.

Kazwini cited Al Rā'i, the Shepherd, from the early Arabs, which, although now a title for γ Cephei, may have come here from the adjacent Rauḍah, or Pasture; the near-by α Herculis, 6° to the west, being Kalb al Rā'i, the Shepherd's Dog; while neighboring stars, the present Club of Hercules, marked the Flock.

In China α was **How,** the Duke; and the small surrounding stars, **Hwan Chay,** a title duplicated at those in the hand.

Its spectrum is Sirian, and the star is receding from us about twelve miles a second. It culminates on the 28th of July.

β, 3.3, yellow.

Cebalrai, Celbalrai, and **Cheleb** are from **Kalb al Rā'i.** "The Heart of the Shepherd," which Brown gives as the meaning of his **Celabrai,** is erroneous, doubtless from confusion of the Arabic Ḳalb, Heart, and Kalb, Dog.

The star is 9° southeast of α, and 5° west of Taurus Poniatovii, the Polish Bull, now included in Ophiuchus.

γ, 4.3,

has been called **Muliphen,** but I cannot trace it here, although this title is famous in other parts of the sky.

β and γ were **Tsung Ching** in China.

70 Ophiuchi, east of β and γ in the stars of the Polish Bull, now discarded, is a most interesting binary system, with a period of about eighty-eight years. The component stars are of 4.1 and 6.1 magnitudes, yellow and purple in color, their distance varying from 1″.7 to 6″.7; in 1898 it was 2″.05, and the position angle 280°. Its parallax, 0″.16, indicates a distance of twenty light years, and certain irregularities in motion show that there may be an invisible companion.

δ, 2.8, deep yellow,

is **Yed Prior,** the Former of the two stars in the Hand,— the Arabic **Yad,**— originating with Bayer, adopted by Flamsteed, and now common. It is sometimes written **Jed.**

It was **Leang,** a Mast, in China.

ϵ, 3.8, red.

Yed Posterior, the star Behind, or Following, δ, is found on our modern lists, but was not given by Bayer.

In China it was **Tsoo,** the name of one of the feudal states; and, with ι and some other stars, is said to have formed **Hwan Chay.**

The two stars Yed, with ζ and η Ophiuchi and α, δ, and ϵ of Serpens, constituted the **Nasaḳ al Yamaniyy,** the Southern Boundary Line of the Rauḍah, or Pasture, which here occupied a large portion of the heavens; other stars in Ophiuchus and Hercules forming the **Nasaḳ al Shāmiyyah,** or Northern Boundary. The stars between these two Nasaḳ marked the Rauḍah itself and **Al Aghnām,** the Sheep within it, now the **Club of Hercules.** These sheep were guarded by the Shepherd and his Dog, the two *lucidae* marking the heads of Ophiuchus and Hercules.

ϵ was the Euphratean **Nitaχ-bat,** the Man of Death. Coincidently, "in modern astrology, which contains some singular survivals, the Hand of Ophiuchus is said to be a star 'of evil influence.'"

δ and ϵ point out the left hand grasping the body of the Serpent; τ and ν, the other hand, holding the tail.

ζ, 2.8, near the left knee, was the Chinese **Han,** an old feudal state.

It sometimes shared with η the title **Sābiḳ,** or Preceding One, attached to the latter star in Al Tizini's catalogue.

Brown thinks that, with ϵ, it marked the Akkadian lunar asterism **Mulubat,** the Man of Death; with η, θ, and ξ, the Persian **Garafsa,** or Serpent-tamer; with η, the Sogdian **Bastham,** Bound, "*i. e.* Ophiuchus enveloped in the coils of Ophis"; and the Khorasmian **Sardhiwa,** the Head of the Evil One.

η, 2.6, pale yellow,

is **Sābiḳ** with Al Tizini, ζ often being included; but Beigel thought that the name should be **Sāīḳ,** the Driver.

Brown combines η, θ, and ξ in the Akkadian **Tsir,** or **Sir,** the Snake.

In China it was **Sung,** another of the early feudal states.

θ, 3.4,

lies on the right foot, only a little to the southwest of the place of the noted Kepler's Star, the *nova* of 1604.

Epping says that the 25th ecliptic constellation of Babylonia was marked by it as **Kash-shud Sha-ka-tar-pa,** of undetermined signification.

With ξ it was the Sogdian **Wajrik,** the Magician; the Khorasmian **Markhashik,** the Serpent-bitten; and the Coptic **Tshiō,** the Snake, and **Aggia,** the Magician; η being included in the last two.

With adjacent stars it was the Chinese **Tien Kiang,** the Heavenly River.

ι, a 4½-magnitude, was **Ho,** one of the dry measures of China, but this title included κ and two other near-by stars of Hercules.

Gould thinks that it may be variable.

λ, Binary, 4 and 6, yellowish white and smalt blue.

Marfic, or **Marfik,** is from the similar Arabic **Al Marfiḳ,** the Elbow, which it marks. Bayer, Burritt, and probably others have it **Marsic,** doubtless from confounding the antique forms of the letters *f* and *s*. This same title appears for κ Herculis.

With neighboring stars the Chinese knew it as **Lee Sze,** a Series of Shops.

The components are 1″.6 apart, with a position angle of 53° in 1897, and an estimated period of revolution of 234 years.

υ, a 4½-magnitude, was **She Low,** a Market Tower; and the 5th-magnitudes ϕ, χ, ψ, and ω were **Tung Han,** the name of a district in China.

*

While far Orion o'er the waves did walk
That flow among the isles.
Shelley's *The Revolt of Islam.*

Orion with his glittering belt and sword
Gilded since time has been, while time shall be.
. . . .
Thou splendid soulless warrior! What to thee,
Marching along the bloodless fields, are we!
Lucy Larcom's *Orion.*

Orion, the Giant, Hunter, and Warrior,

admired in all historic ages as the most strikingly brilliant of the stellar groups, lies partly within the Milky Way, extending on both sides of the celestial equator entirely south of the ecliptic, and so is visible from every part of the globe.

With Theban Greeks of Corinna's time, about the year 490 before our era, it was Ὠαρίων, the initial letter having taken the place of the ancient digamma, Ϝ, which, pronounced somewhat like the letter *W*, rendered the early word akin to our Warrior. Corinna's pupil Pindar followed in Ὠαριώνειος, but by the time of Euripides the present Ὠρίων prevailed, and we see it thus in Polymestor's words in the Ἑκάβη of 425 B.C.:

> through the ether to the lofty ceiling,
> Where Orion and Seirios dart from their eyes
> The flaming rays of fire.

Catullus transcribed **Oarion** from Pindar, shortened to **Arion**, and sometimes changed to **Aorion**; but the much later **Argion**, attributed to Firmicus, was for Procyon, probably from Ἄργος, the faithful dog of Ulixes.

The derivation of the word has been in doubt, but Brown refers it to the Akkadian **Uru-anna**,[1] the Light of Heaven, originally applied to the sun, as Uru-ki, the Light of Earth, was to the moon; so that our title may have come into Greek mythology and astronomy from the Euphrates. The Οὐρίον, Οὖρον, or Ὑριών of the Hyriean, or Byrsaean, story, the **Urion** of the original *Alfonsine Tables*, graphically explained by Minsheu, is in no sense an acceptable title, although Hyginus and Ovid vouched for it, thus showing its currency in their day. Caesius' derivation from Ὧρα, as if marking the Seasons, seems fanciful.

At one time it was Ἀλετροπόδιον, found in the *Uranologia* of Petavius of the 16th century, which Ideler said should be Ἀλεκτροπόδιον, Cock's Foot, likening the constellation to a Strutting Cock; but Brown goes back to Ἄλη, Roaming, and so reads it Ἀλητροπόδιον, the Foot-turning Wanderer, mythologically recorded as roaming in his blindness till miraculously restored to sight by viewing the rising sun.

The Boeotians, according to Strabo, fellow-countrymen of the earthly Orion, called his stars Κανδάων, their alternative title for Ἄρης, the god of war, well agreeing with, perhaps originating, the Greek conception of the Warrior.

Ovid said that the constellation was **Comesque Boötae**; and some authors asserted that Orion never set, an idea possibly coming from the early confusion in name with Boötes already alluded to; although even as to that constellation the assertion would not have been strictly correct. Matthew Arnold similarly wrote in his *Sohrab and Rustum:*

[1] This divinity was the later Chaldaeo-Assyrian sun-god Dumu-zi, the Son of Life, or Tammuz, widely known in classical times as Adonis. Aries also represented him in the sky.

> the northern Bear,
> Who from her frozen height with jealous eye
> Confronts the Dog and Hunter in the South.

Dianae Comes, and **Amasius**, Companion, and Lover, of Diana, were other titles, the Hero, after his death from the Scorpion's sting inflicted for his boastfulness, having been located by Jove in his present position, at the request of the goddess, that he might escape in the west when his slayer, the Scorpion, rose in the east,— as Aratos said:

> When the Scorpion comes
> Orion flies to utmost end of earth.

Thompson sees in this alternate rising and setting of these two sky figures an astronomic explanation of the symbolism in classic ornithology of the mutual pursuit and flight of Haliaëtos and Keiris, the Sea Eagle and Kingfisher, compared in the poem *Ciris* to these opposed constellations.

In Horace's *Odes* the constellation is termed *pronus;* and Tennyson had

> Great Orion sloping slowly to the west,

which, with the rest of the beautiful opening passage, adds much to the charm of his *Locksley Hall.*

Homer, who made but a single allusion in the *Iliad* to this constellation, followed by a parallel passage in the *Odyssey*, wrote of "the might of huge Orion," and described the earthly hero as the "Illustrious Orion, the tallest and most beautiful of men,—even than the Aloidae," adjectives all well applied to our stellar figure; Hesiod said:

> When strong Orion chaces to the deep the Virgin stars;

Pindar, that he was of monstrous size; as did Manilius in his *Magna pars maxima coeli;* and nearly all authors, as well as illustrators, have thus described Orion, and as an armed warrior. In the Ἑκάβη we read:

> with his glittering sword Orion arm'd;

in Ovid's works, of *ensiger Orion;* in Lucan's, of *ensifer;* and Vergil has a fine passage in the *Aeneid* quaintly translated in 1513 by the "Scottis" Gavin Douglas, where Palinurus

> Of every sterne the twynkling notis he
> That in the still hevin move cours we se,
> Arthurys house, and Hyades betaikning rane,
> Watling strete, the Horne and the Charlewane,
> The fiers Orion with his goldin glave;

these last a very liberal translation of the much quoted *armatumque auro.* But later on in the voyage, when the fleet was off Capreae, the old pilot, in his astronomical enthusiasm *dum sidera servat,* lost his balance, and tumbled overboard.

The constellation's stormy character appeared in early Hindu, and perhaps even in earlier Euphratean days, and is seen everywhere among classical writers with allusions to its direful influence. Vergil termed it *aquosus, nimbosus,* and *saevus;* Horace, *tristis* and *nautis infestus;* Pliny, *horridus sideribus;* and the Latin sailors had a favorite saying, *Fallit saepissime nautas Orion.* Polybios, the Greek historian of the second century before Christ, attributed the loss of the Roman squadron in the first Punic war to its having sailed just after "the rising of Orion"; Hesiod long before wrote of this same rising:

> then the winds war aloud,
> And veil the ocean with a sable cloud:
> Then round the bark, already haul'd on shore,
> Lay stones, to fix her when the tempests roar;

and Milton, in *Paradise Lost:*

> when with fierce winds Orion arm'd
> Hath vex'd the Red-sea coast, whose waves o'erthrew
> Busiris and his Memphian chivalry.

Many classical authors variously alluded to it as a calendar sign, for its morning rising indicated the beginning of summer, when, as we find in the *Works and Days,* the husbandman was instructed to

> Forget not, when Orion first appears,
> To make your servants thresh the sacred ears;

his midnight rising marked the season of grape-gathering; and his evening appearance the approach of winter and its attendant storms: an opinion that prevailed as late as the 17th century, for in the *Geneva Bible,* familiarly known as the *Breeches Bible,* the marginal reading in the *Book of Job,* xxxviii, 31, is "which starre bringeth in winter." Plautus, Varro, and others called the constellation **Jugula** and **Jugulae,** the Joined, referring to the *umeri,* the two bright stars in the shoulders, as if connected by the *jugulum,* or collar-bone. Such, at least, is the generally received derivation, but Buttmann claimed it as from *jugulare,* and hence the **Slayer,** a fitting title for the Warrior.

The Syrians knew it as **Gabbārā;** the Arabians, as **Al Jabbār,** both signifying "the Giant," Γίγας with Ptolemy,— and in Latin days occasionally **Gigas;**

the Arabian word gradually being turned into **Algebra, Algebaro,** and, especially in poetry, **Algebar,** which Chilmead gave as **Algibbar.**

In early Arabia Orion was **Al Jauzah,** a word also used for stars in Gemini, and much, but not very satisfactorily, discussed as to its derivation and meaning in its stellar connection. It is often translated **Giant,** but erroneously, for it, at first, had no personal signification. Originally it was the term used for a black sheep with a white spot on the middle of the body, and thus may have become the designation for the middle figure of the heavens, which from its preëminent brilliancy always has been a centre of attraction. Some think that the Belt stars, δ, ε, ζ, known to the Arabs as the **Golden Nuts,** first bore the name Jauzah, either from another meaning of that word,— Walnut,— or because they lay in the centre of the splendid quadrangle formed by α, β, γ, and κ; or from their position on the equator, the great central circle; the title subsequently passing to the whole figure. Grotius adopted the first of these derivations, quoting from Festus the passage *quasi nux juglans*, that a lesser light, Robert Hues, thus enlarged upon:

> Now *Geuze* signifieth a *Wall-nut;* and perhaps they allude herein to the Latine word *Jugula*, by which name *Festus* calleth *Orion;* because he is greater than any of the other Constellations, as a Wall-nut is bigger then any other kinde of nut.

In mediaeval as well as in later astronomy, the original appears in degenerate forms, such as **Elgeuze, Geuze, Jeuze,** and the **Geuzazguar** of Grotius.

Al Sufi's story of the feminine Jauzah has been noticed at the star Canopus and under Canis Minor.

Hyde quoted from an Arabian astronomer, **Al Babādur,** the Strong One, as a popular term for the constellation. **Sugia** and **Asugia** were thought by Scaliger to be corruptions of the Arabs' **Al Shujāʽ,** the Snake, applied to Orion in the sense of **Audax, Bellator** and **Bellatrix, Fortis** and **Fortissimus, Furiosus** and **Sublimatus,** and all proper names for it in Bayer's and other early astronomical works, Chilmead translating Asugia as "the Madman." Similar titles at one time obtained for Hydra.

Al Firuzabadi's **Al Nusuḳ** may be equivalent to the Nasaḳ, a Line, or Row, applied to the Belt stars, but there signifying a String of Pearls.

Niphla, attributed to Chaldaea, has not been confirmed by modern scholars.

In Egypt, as everywhere, Orion was of course prominent, especially so in the square zodiac of Denderah, as **Horus** in a boat surmounted by stars, followed by Sirius, shown as a cow, also in a boat; and nearly three thousand years previously had been sculptured on the walls of the recently discovered step-temple of Saḳkara, and in the great Ramesseum of Thebes about 3285 B. C. as **Sahu.** This twice appears in the *Book of the Dead:*

> The shoulders of the constellation Sahu ;

and:

> I see the motion of the holy constellation Sahu.

A similar title, but of Akkad origin, appeared for Capricornus. Egyptian mythology laid to rest in this constellation the soul of **Osiris,** as it did in the star Sirius that of Isis; and, again, in the *Book of the Dead* we read:

> The Osiris N is the constellation Orion ;

in this connection, Orion was known as **Smati-Osiris,** the Barley God.

The Giant generally has been represented with back turned toward us and face in profile, armed with club, or sword, and protected by his shield, or, as Longfellow wrote,

> on his arm the lion's hide
> Scatters across the midnight air
> The golden radiance of its hair.

Dürer drew him facing the Bull, whose attack he is warding off; but the *Leyden Manuscript* has a lightly clad youth with a short, curved staff in the right hand, and the Hare in the background.

The head is marked by λ, ϕ^1, and ϕ^2, the stars α and γ pointing out the shoulders, β and κ the left foot and right knee. But Sir John Herschel observed from southern latitudes that the inverted view of the constellation well represents a human figure; the stars that we imagine the shoulders appearing for the knees, Rigel forming the head, and Cursa of Eridanus, one of the shoulders.

In astrology the constellation was **Hyreides,** Bayer's **Hyriades,** from Ovid's allusion to it as **Hyriea proles,** thus recalling the fabled origin from the bull's hide still marked out in the sky. This, formerly depicted as a shield of rawhide, is now figured as a lion's skin; and it perhaps was this Hyriean story that gave the stellar Orion the astrological reputation, recorded by Thomas Hood, of being "the verie cutthrote of cattle"; at all events, it certainly gave rise to the τρίπατρος and **Tripater,** applied to him.

Saturnus has been another title, but its connection here I cannot learn, although I hazard the guess that as this divinity was the sun-god of the Phoenicians, his name might naturally be used for Uruanna-Orion, the sun-god of the Akkadians.

Anterior to much of this, we find in the various versions of the *Book of Job* and *Amos* the word Orion for the original Hebrew word **Kᵉsîl,** literally signifying "Foolish," "Impious," "Inconstant," or "Self-confident."

This perhaps is etymologically connected with Kislev, the name for the ninth month of the Hebrew calendar, the tempestuous November–December. Julius Fürst considered this **Kislev** an early title for Orion. The epithet "Inconstant" has fancifully been referred to the storms usual at his rising.

The Kesīlīm of *Isaiah* xiii, 10, rendered "constellations" in some versions, is also thought to refer to it and other prominent sky figures; in fact, Cheyne translates the word as "the Orions" in the *Polychrome Bible;* while **Rahab,** in the Revised Version of the *Book of Job*, ix, 13,— the "proud helpers" in the Authorized,— is referred by Ewald, Renan, and others to this,— possibly to some other group of stars,— with the same significations as those of Kesīl, or perhaps "Arrogance," "Rebellion," "Strength," or "Violence."

Later on the Jews called Orion **Gibbōr,** the Giant, considered as **Nimrod** bound to the sky for rebellion against Jehovah, whence perhaps came the Bands, or Bonds, of Orion, which some say should be Cords, or a Girdle; but the conception of Nimrod as "the mighty Hunter before the Lord," at least in the ordinary sense of that word, is erroneous, for the original, according to universal Eastern tradition, signifies a Lurking Enemy, or a Hunter of men rather than of beasts. This idea may have led to a Latin title, **Venator,** for the stellar Orion.

But, relative to the renderings of biblical words supposed to refer to sky groups, the Reverend Doctor Adam Clarke wrote in his *Commentary*

> that 'Aish has been generally understood to signify the Great Bear; Kesil Orion; and Kimah the Pleiades, may be seen everywhere; but that they do signify these constellations is perfectly uncertain. We have only conjectures concerning their meaning.
>
> . . . . . .
>
> As to the Hebrew words, they might as well have been applied to any of the other con stellations of heaven; indeed, it does not appear that constellations at all are meant.

The discordance between the various renderings would indicate the probable correctness of these comments, and that we are in no respect assured as to the identification of Bible star-names. Yet it is worth noting that the three constellations adopted by the translators of the *Book of Job* and of *Amos* in the Revised Version fitly represent the cardinal points of the sky: the Bear in the north, Orion in the south, and the Pleiades rising and setting in the east and west.

In the Hindu *Brahmanas* Orion is personified as **Praja-pāti,**[1] under the form of a stag, **Mriga,** in pursuit of his own daughter, the beautiful roe Rohini, our Aldebaran. In his unnatural chase he was transfixed by the

[1] He was also, and differently, represented in the sky by Hindu astronomers as an immense figure stretching from Boötes through Virgo, Corvus, and Libra into Scorpio.

three-jointed arrow — the Belt stars — shot by the avenging Hunter, Sirius, which even now is seen sticking in his body. This hero was the father of twenty-seven daughters, the wives of King Soma, the Moon, with whom the latter equally divided his time, thus referring to the *nakshatras*.

The Chinese made up their 4th *sieu* from the seven conspicuous stars in the shoulders, belt, and knees of Orion, with the title **Shen,** or **Tsan,** Three Side by Side, anciently **Sal,** which may have originated from the Belt having at first alone formed the *sieu*. Indeed, the lunar asterism was mentioned in the *She King* as the **Three Stars.** δ was its determinant; but it overlapped the corresponding *nakshatra*, although entirely distinct from the 4th *manzil* in the feet of the Twins. Orion was worshiped in China during the thousand years before our era as **Shen,** or **Shï Ch'en,** from the moon station; but it also was known as the **White Tiger,** a title taken from the adjacent Taurus.

The Khorasmians adopted Orion's stars as a figure of their zodiac in place of Gemini.

The early Irish called it **Caomai,** the Armed King; the Norsemen, **Orwandil;** and the Old Saxons, **Ebuŏrung,** or **Ebiŏring,**— words that Grimm thought connected with Iringe, or Iuwaring, of the Milky Way.

Caesius cited the singular title **Ragulon,** perhaps from Al Rijl, the Arabic designation for the star β, but he made this the equivalent of the Latin *Vir*, the Man *par excellence*, the Hero; and suggested that Orion represented **Jacob** wrestling with the angel; or **Joshua,** the Hebrew warrior; but Julius Schiller, that it was **Saint Joseph,** the husband of the Blessed Virgin. Weigel figured it as the Roman **Two-headed Eagle;** and De Rheita, of 1643, found somewhere among its stars **Christ's Seamless Coat** and a **Chalice;** but he was addicted to such discoveries.

Argelander has 115 stars here; Heis, 136; and Gould, 186; while the whole is as rich in wonderful telescopic objects as it is glorious to the casual observer. Flammarion calls it the **California of the sky.**

α, Irregularly variable, 0.7, orange.

Betelgeuze is from **Ibṭ al Jauzah,** the Armpit of the Central One; degenerated into **Bed Elgueze, Beit Algueze, Bet El-geuze, Beteigeuze,** etc., down to the present title, which itself also is written **Betelgeuse, Betelguese, Betelgueze, Betelgeux,** etc. The *Alfonsine Tables* had **Beldengenze,** and Riccioli, **Bectelgeuze** and **Bedalgeuze.**

The star also was designated by various Arabian authors as **Al Mankib,** the Shoulder; **Al Dhirā',** the Arm; and **Al Yad al Yamnā',** the Right Hand,

—all of the Giant; but Chilmead wrote "**Ied Algeuze,**— that is, Orion's Hand," quoted from Christmannus.

The title **Mirzam,** from **Al Murzim,** the Roarer, or perhaps the Announcer, originally used for γ, also is applied to this as heralding the rising of its companions. La Lande, borrowing the full name of that star for this, quoted it as **Almerzamo nnagied.**

Sayce and Bosanquet identify α with the Euphratean **Gula,** other stars possibly being included under this title; and Brown says that **Kakkab Sar,** the Constellation of the King, or **Ungal,** refers to α with γ and λ. We can see in this signification the origin of the astrologer's idea that Betelgeuze portended fortune, martial honors, wealth, and other kingly attributes.

α alone constituted the 4th *nakshatra*, **Ārdrā,** Moist, depicted as a Gem, with Rudra, the storm-god, for its presiding divinity, and so, perhaps, the origin of the long established stormy character of Orion. This lunar station, therefore, formed but a part of the 4th *sieu*, and differed entirely from the 4th *manzil.* Individually the star was the Sanskrit **Bāhu,** Arm, probably from the Hindu conception of the whole figure as a running Stag, or Antelope, of which α, β, γ, and κ marked the legs and feet, with α on the left forearm; the adjacent Sirius being the hunter Mrigavyādha.

Brown mentions its equivalent Persian title, **Besn,** the Arm, and the Coptic **Klaria,** an Armlet.

Bayer quoted *γλήνεα* from Aratos, but it is not in the original; and Chrysococca had *Ὦμος διδύμων*, the Shoulder of—*i.e.* next to—the Twins.

Among the many queerly worded descriptions in the 1515 *Almagest*, perhaps none is more so than that of this star, reading in part thus: *ipsa tendit ad rapinam quae appropinquat ad terram.* This *tendit ad rapinam*, also used for the star Antares, apparently has been an unsolved puzzle; and as I have never seen any explanation, my own suggestion may not be amiss. The 1515 *Almagest* followed Ulug Beg's *Tables*, and these followed Ptolemy, who characterized the color of α as *ὑπόκιρρος*, which Ulug Beg's translator turned into *rubedinem*, "ruddiness," and the *Almagest* into the not very different word of the quotation, expressing ideas of war and carnage, astrology's attributes of red stars. The *appropinquat ad terram* doubtless refers to the comparatively low elevation of the star above the horizon.

Professor Young says that at times, when near a minimum, it closely matches Aldebaran in color and brightness, and Lassell described it as a rich topaz. Secchi makes it the typical star of his third class with a banded spectrum, suggesting that it may be approaching the point of extinction. Elkin finds its parallax insensible; according to Vogel, it is receding from the earth at the rate of 10½ miles a second.

It was first seen to be variable by Sir John Herschel in 1836, from which time till 1840 "its variations were most marked and striking." A similar period began in 1849, and on the 5th of December, 1852, "it was actually the largest star in the northern hemisphere." It was especially brilliant in 1894. Argelander found a period of 196 days, but Schoenfeld thought periodicity questionable.

Its position is less than 3° west of the solstitial colure; it rises at sunset on the 30th of December, and culminates on the 29th of January. It has an 8th-magnitude companion 20′ away, first observed by Wilhelm Struve as double, 18″.5 apart, and the great glasses of the present day reveal other members in the combination still nearer and smaller than the original companion; while Barnard has discovered about it large and diffused nebulosity.

β, Double, 0.3 and 8, both bluish white.

Algebar and **Elgebar** are seen in poetry for this star, but it universally is known as **Rigel,** from **Rijl Jauzah al Yusrā'**, the Left Leg of the Jauzah, by which extended title the Arabians knew it after the word Jauzah had become a personal title; the modern name first appearing in the *Alfonsine Tables* of 1521. These say of it, in connection with Eridanus:

Lucida que est in pede sinistro: et est communis ei et aquae: et dicitur Algebar nominatur etiam Rigel.

Riccioli had **Regel**; Schickard, **Riglon**; and Chilmead, **Rigel Algeuze,** or **Algibbar**.

Al Sufi gave the earlier popular name **Rā'i al Jauzah,** the Herdsman of the Jauzah, whose camels were the stars α, γ, δ, and κ; and **Al Najīd,** the Conqueror, which also was given to α and γ.

Chrysococca termed it Πούς δίδυμων, the Foot of — *i.e.* next to — the Twins; and Bayer, the Hebrew **Kesil,** of the constellation.

Smyth wrote that

independent of the "*nautis infestus Orion*" character of the constellation, Rigel had one of his own; for it was to the astronomical rising of this "*marinus aster,*" in March, that *St. Marinus* and *St. Aster* owe their births in the Romish calendar.

He gave, however, no explanation of this, and these saints certainly are not familiar in any stellar connection. Possibly its "marine" character came from its location at the end of the River, and from its being given in the various editions of the *Syntaxis* and in the *Alfonsine Tables* as common to both constellations; although the supposed stormy character of the whole group in affecting navigation may have induced the epithet for Orion's greatest star.

Astrologers said that splendor and honors fell to the lot of those who were born under it.

In the Norsemen's astronomy Rigel marked one of the great toes of Orwandil, the other toe having been broken off by the god Thor when frost-bitten, and thrown to the northern sky, where it became the little Alcor of the Greater Bear.

Although lettered below Betelgeuze, it is usually superior to it in brightness, being estimated in the *Harvard Photometry* as exactly equal to Arcturus, Capella, and Wega. Its spectrum is like that of Sirius, and it is receding from our system about 10¼ miles a second.

The smaller star, at a position angle of 200°, is 9″.1 away, but not easily seen owing to the brightness of the principal. It is strongly suspected that this smaller star itself is closely double.

Another minute companion is 44″.5 away.

γ, Slightly variable, 2, pale yellow.

Bellatrix, the Female Warrior, the **Amazon Star,** is from the translation, rather freely made in the *Alfonsine Tables*, of its Arabic title, **Al Najīd,** the Conqueror. Kazwini had this last, but Ulug Beg said **Al Murzim al Najīd,** the Roaring Conqueror, or, according to Hyde, the Conquering Lion heralding his presence by his roar, as if this star were announcing the immediate rising of the still more brilliant Rigel, or of the whole constellation. This Murzim occasionally appears in our day as **Mirzam,** which is also applied to both of the stars β in the two Dogs as heralds of Sirius and Procyon.

Al Sufi had **Al Ruzam,** which Hyde said was another of the very many Arabic words for the lion, but Beigel thought it also a reference to the camel, another roarer. Still it is well to remember in this connection Ideler's remark that "etymology has full play with a word which has not traveled beyond astronomical language,"—a statement equally applicable to very many other star-names.

Caesius cited **Algauza** from the name for the whole.

γ marks the left shoulder of Orion, and naturally shared the Arabs' **Mankib,** and the Hindus' **Bahū,** titles of the star α on the right shoulder of Orion and forearm of the Stag.

In Amazon River myth Bellatrix is a **Young Boy in a Canoe** with an old man, the star Betelgeuze, chasing the Peixie Boi, a dark spot in the sky near Orion.

In astrology it was the natal star of all destined to great civil or military honors, and rendered all women born under its influence lucky and loquacious; or, as old Thomas Hood said, "women born under this constellation shall have mighty tongues."

Its spectrum is Sirian in character, and indicates that it is receding from our system at the rate of about 5¾ miles a second.

δ, Double and slightly variable, 2.4 and 6.8, brilliant white and pale violet.

Mintaka, from **Al Minṭakah,** the Belt, is the first star seen in that portion of the rising constellation. Burritt has it **Mintika.**

Astrologers considered it of importance as portending good fortune.

It is about 23′ of arc south of the celestial equator, the components 53″ apart, at a position angle of 0°. The spectrum is Sirian, and the star seems to have very little motion either of approach or recession.

Burnham has discovered still another companion of the 13th to 14th magnitudes, one of the faintest ever seen near a brilliant star.

ε, 1.8, bright white.

Alnilam, Anilam, Ainilam, and **Alnihan** are from **Al Niẓhām,** or **Al Naẓhm,** the String of Pearls, or, as Recorde said, the Bullions set in the middle of Orion's Belt.

It portended fleeting public honors to those born under its influence.

The spectrum is Sirian, and the star recedes from us at the rate of about 16½ miles a second.

It is the central one of the Belt, culminating on the 25th of January.

ζ, Triple, 2.5, 6.5, and 9, topaz yellow, light purple, and gray.

Alnitak, or **Alnitah,** for this, the lowest star in the Belt, is from **Al Niṭāk,** the Girdle.

The spectrum is Sirian, and the star recedes from us about nine miles a second.

One of its components, 2″.4 distant from the largest, at a position angle of 155°, was singularly missed by Sir William Herschel, but discovered by Kunowski in 1819, and seems of some nondescript hue about which observers do not agree. The elder Struve called it, in one specially manufactured word, *olivaceasubrubicunda*, "slightly reddish olive."

Orion's studded belt.
Scott's *Lay of the Last Minstrel.*

These Arabian titles of δ, ε, and ζ, although now applied to them individually, were at first indiscriminately used for the three together; but they had other names also,— **Al Nijād,** the Belt; **Al Nasaḳ,** the Line; **Al Alḳāṭ,** the Golden Grains, Nuts, or Spangles; and **Faḳār al Jauzah,** the Vertebrae in the Jauzah's back. Niebuhr cited the modern Arabic **Al Mīzān al H·aḳḳ,** the Accurate Scale-beam, so distinguishing them from the curved line of the fainter *c*, θ, ι, *d*, and κ, **Al Mīzān al Baṭil,** the False Scale-beam. The Chinese similarly knew them as a Weighing-beam, with the stars of the sword as a weight at one end.

They were the **Jugula** and **Jugulae** of Plautus, Varro, and others in Roman literature; the **Balteus,** or Belt, and the **Vagina,** or Scabbard, of Germanicus. The **Zona** of Ovid may have been taken from the Ζώνη of Aristotle.

The early Hindus called them **Iṣus Trikāṇḍā,** the Three-jointed Arrow; but the later transferred to it the *nakshatra* title, **Mrigaçiras.**

The Sogdian **Rashnawand** and the Khorasmian **Khawiya** have significations akin to our word "Rectitude," which this straight line of stars personified. The Rabbi Isaac Israel said that it was the **Mazzārōth, Mazẓālōth,** or **Mazlātha** that most of his nation applied to the zodiac.

Riccioli cited **Baculus Jacobi,** which became in popular English speech **Jacob's Rod** or **Staff,**—the German **Jakob Stab,**—from the tradition given by Eusebius that Israel was an astrologer, as, indeed, he doubtless was; and some had it **Peter's Staff.** Similarly, it was the Norse **Fiskikallar,** or Staff; the Scandinavian **Frigge Rok,** Frigg's, or Freya's, Distaff,— in West Gothland **Frigge Rakken,**— and **Maria Rok,** Mary's Distaff; in Schleswig, **Peri-pik.** In Lapland it was altered to **Kalevan Miekka,** Kaleva's Sword, or still more changed to **Niallar,** a Tavern; while the Greenlanders had a very different figure here,— **Siktut,** the Seal-hunters, bewildered when lost at sea, and transferred together to the sky.

The native Australians knew the stars as **Young Men** dancing a *corroboree*, the Pleiades being the Maidens playing for them; and the Poignave Indians of the Orinoco, according to Von Humboldt, as **Fuebot,** a word that he said resembled the Phoenician.

The University of Leipsic, in 1807, gave to the Belt and the stars in the Sword the new title **Napoleon,** which a retaliating Englishman offset by **Nelson**; but neither of these has been recognized on star-maps or -globes.

Seamen have called it the **Golden Yard-arm**; tradesmen, the **L,** or **Ell,** the **Ell and Yard,** the **Yard-stick,** and the **Yard-wand,** as occupying 3° between the outer stars,— the **Elwand** of Gavin Douglas; Catholics, **Our Lady's Wand**; and the husbandmen of France and along the Rhine, **Râteau,** the Rake. In Upper Germany it has been the **Three Mowers**; and it is often the **Magi,** the **Three Kings,** the **Three Marys,** or simply the **Three Stars,** that Tennyson had in his *Princess*,—

> those three stars of the airy Giants' zone
> That glitter burnished by the frosty dark.

The celestial equator now passes through the Belt, but was 12° below it 4000 years ago.

η, Triple, 3.5, 5, and 5,

occasionally and very appropriately has been designated **Saiph,** from **Saif al Jabbār,** the Sword of the Giant; but this title included other adjacent stars in the same line of sight,— the **Ensis** of Cicero,— and all supposed to have been a separate constellation with Pliny.

Al Sufi called them **Al Alkāṭ,** which we have seen applied to the Belt; and Burritt, the **Ell,** because this line of stars "is once and a quarter the length of the yard."

θ^1, 4.6, pale white,

although not individually named, marks the Fish-mouth of the **Great Nebula,** N. G. C. 1976, 42 M., in the sword scabbard of the figure, with the celebrated **Trapezium** in its midst. De Quincey gave a characteristic description of it in one of his *Essays in Philosophy*.

This nebula, faintly visible to the naked eye, was not even mentioned by Galileo, and is generally thought to have been accidentally discovered by Christian Huygens in 1656, and described in his *Systema Saturnium* half a century after Galileo's adaptation of the principle of the telescope to astronomical use; but Cysatus of Lucerne had already known it in 1618. This was the first[1] object to which Sir William Herschel directed, on the 4th of March, 1774, the first serviceable telescope of his own construction after two hundred failures; and the first nebula to be successfully photographed, as it was by Professor Henry Draper, at Hastings-upon-Hudson, on the 30th of September, 1880.

[1] Similarly, too, it was the last object viewed by Sir William through his forty-foot reflector, on the 19th of January, 1811, when the great glass was laid aside forever.

Its spectrum is purely gaseous, and spectroscopic investigations by Sir William and Lady Huggins seem to show "a unity of composition of the [trapezium] stars and nebulae which surround them and link them together." Keeler finds from spectroscopic observations that it and our system are separating at the rate of ten miles a second. Holden thinks it of fluctuating brightness.

The nebula proper covers a space equal to the apparent size of the moon, but nebulosity extends over a very much larger area, for recent observations by Swift, by William H. Pickering in 1889 from Wilson's Peak, and by Barnard with the camera on Mount Hamilton in October, 1894, reveal nebulous matter, 14° to 15° in diameter, that includes the Belt and much of the body of Orion. Barnard says of it: "Compared with this enormous nebula, the old θ, or so-called Great Nebula, is but a pigmy." A million of globes, each equal in diameter to that of the earth's orbit, would not equal this in extent. One of the Harvard photographs of 1889 showed a certain amount of spiral structure in the Great Nebula.

The adjacent nebula, N. G. C. 1982, catalogued separately by Messier as 43, is shown on a photograph of the 30th of November, 1886, by Roberts, to be connected with it by threads of nebulosity.

At least six stars are found in the Trapezium, the four largest being of the 5th, 6th, 7th, and 8th magnitudes, easily visible in a 2¼-inch glass with a power of 140. They may form a system. Huygens noted the triplicity of θ^1 when he discovered the nebula; the 4th component was first seen in 1684; the 5th was "discovered by Robert Hooke in 1664, but forgotten and rediscovered by Struve in 1826"; and the 6th was first seen by Sir John Herschel, on the 13th of February, 1830. More are claimed by some recent observers, but Burnham disputes their existence.

In 3.36 square degrees of the θ^1 nebula Bond catalogued nearly 1000 stars.

ι, Triple and nebulous, 3.5, 8.5, and 11, white, pale blue, and grape red.

Al Tizini designated this as **Nā'ir al Saif,** the Bright One in the Sword, but it is practically unnamed with us, although far more deserving of the title Saiph than is the succeeding star κ.

In China it was **Fa,** a Middle-man, v and intermediate stars being included under this name; but Edkins translates the word "Punishment," and gives another title for it,—**Tui,** or **Jui,** the Sharp Edge, analogous to the Arabian Saif and perhaps taken from it.

It lies just south of θ, inclosed in faint nebulosity. The two larger stars are 11''.5 apart, with a position angle of 142°; the 11th-magnitude companion is 49'' away, at a position angle of 103°.

κ, 2.4,

located near the right knee, was appropriately described by the Arabic astronomers as **Rijl Jauzah al Yamnā',** the Right Leg of the Jauzah, but we now know it as **Saiph,** from Al Saif, the Sword, although it is at some distance from that weapon, and the name really belongs to η, ι, and stars near by.

> In his vast Head immerst in boundless spheres
> Three Stars less bright, but yet as great, he bears.
> But further off remov'd, their Splendor's lost.
> Creech's *Manilius.*

λ, Double, 3.8 and 6, pale white and violet.

Al Maisān, the title of γ Geminorum, by some error of Firuzabadi was applied to this star as **Meissa,** and is now common for it. Al Sufi called it **Al Taḥāyī;** but Al Ferghani and Al Tizini knew it as **Rās al Jauzah,** the Head of the Jauzah, which it marks.

The original Arabic name, **Al Haḳ'ah,** a White Spot, was from the added faint light of the smaller ϕ^1 and ϕ^2 in the background, and has descended to us as **Heka** and **Hika.** These three stars were another of the Athāfiyy of the Arabs; and everywhere in early astrology were thought, like all similar groups, to be of unfortunate influence in human affairs.

They constituted the Euphratean lunar station **Mas-tab-ba-tur-tur,** the Little Twins, a title also found for γ and η Geminorum; and individually were important stars among the Babylonians, rising to them with the sun at the summer solstice, and, with α and γ, were known as **Kakkab Sar,** the Constellation of the King. In other lunar zodiacs they were the Sogdian **Marezānā,** and the Khorasmian **Ikhma,** the Twins; the Persian **Aveçr,** the Coronet; and the Coptic **Klusos,** Watery. They also were the 3d *manzil,* **Al Haḳ'ah;** the *sieu* **Tsee,** or **Tsuy He,** the Beak, or Pouting Lips, anciently **Tsok,** which Reeves gave as **Keo;** and the *nakshatra* **Mrigaçiras,** or **Mrigaçirshā,** the Head of the Stag,— Soma, the Moon, being its presiding divinity, and λ the junction star towards Ārdrā, and its determinant. As to this lunar station Professor Whitney very reasonably wrote:

> It is not a little strange that the framers of the system should have chosen for marking the 3d station this faint group, to the neglect of the brilliant and conspicuous pair β and

ζ Tauri, the tips of the Bull's horns. There is hardly another case where we have so much reason to find fault with their selection.

But they were possibly influenced by recollection of the fact that the vernal equinox lay here 4500 B. C. In addition to the customary Hindu title, Weber mentioned **Andhakā,** Blind, apparently from its dimness; **Āryikā,** Honorable, or Worthy; and **Invakā,** of doubtful meaning, sometimes read **Invalā**.

In China these stars were **Sï ma ts'ien,** the Head of the Tiger.

Ulug Beg, as well as Naṣr al Dīn, likened the group to the letter of the Persian alphabet that was similar in form to the Greek Λ. La Lande wrote of them:

qui ressemblent à un jeu de trois noix, ce qui a fait appeller cette constellation Nux, ou Juglans, Stella jugula.

Hipparchos did not allude to them, but Ptolemy called them *ὁ νεφελοειδής*, the Nebulous One, for such is their appearance to the casual observer, and has been their designation in all early catalogues, even to Flamsteed's in his *in capite Orionis nebulosa.*

Although called double, λ has a second faint companion 149″ above it, visible by a 3½-inch glass; and another, of the 12th magnitude, 27″ distant. The two largest stars are 4″.2 apart, at a position angle of 40°.3.

λ and the two stars *phi* furnish an easy refutation of the popular error as to the apparent magnitude of the moon's disc, Colas writing of this in the *Celestial Handbook* of 1892:

In looking at this triangle nobody would think that the moon could be inserted in it; but as the distance from λ to ϕ^1 is 27′, and the distance from ϕ^1 to ϕ^2 is 33′, it is a positive fact;

the moon's mean apparent diameter being 31′ 7″. This illusion, prevalent in all ages, has attracted the attention of many great men; Ptolemy, Roger Bacon, Kepler, and others having treated of it. The lunar disc, seen by the naked eye of an uninstructed observer, appears, as it is frequently expressed, "about the size of a dinner-plate," but should be seen as only equal to a peppercorn, or as a circle a half-inch in diameter fifty-seven inches away; or, to write it astronomically, equal to the planet Jupiter viewed at opposition through a telescope magnifying forty diameters; or equal to Mars magnified seventy-four times when at his nearest approach to the earth and distant thirty-four millions of miles. To still better illustrate this, Professor Young tells us that the planet Venus,

when about midway between greatest elongation and inferior conjunction, has an apparent diameter of 40″, so that, with a magnifying power of only 45, she looks exactly like the moon four days old, and of precisely the same apparent size.

ν, 4.7, and ξ, 4.6,

were the Chinese **Shwuy Foo,** a Water-depot.

They mark Orion's right hand, ξ being the radiant point of the fine meteor stream, the **Orionids,** of the 18th of October.

o^1, o^2, π^1, π^2, π^3, π^4, π^5, π^6, and *g*,

all of the 4th to the 5th magnitudes, in a vertical line at the right of the figure, indicate the lion's skin; but Al Tizini said that they were the Persians' **Al Tāj,** the Crown, or Tiara, of their kings; and the Arabians' **Al Kumm,** the Sleeve of the garment in which they dressed the Giant, the skin being omitted.

Ulug Beg called them **Al Dhawāib,** Anything Pendent; and the Borgian globe had the same, perhaps originated it; but Al Sufi's title was **Manica,** a Latin term for a protecting Gauntlet; and Grotius gave a lengthy dissertation on the **Mantile** which some anonymous person applied to them, figured as a cloth thrown over the Giant's arm.

With Pliny these stars in the lion's skin are supposed to have been a separate constellation known as the **Shield,** made from the bull's hide of the Hyriean legend.

They were the Chinese **Tsan Ke,** the Three Flags.

τ, 3.6, lies just north of Rigel, and was known in China as **Yuh Tsing,** the Golden Well.

υ, 4.7.

Thabit is Burritt's name for an unlettered star on his *Atlas*, the *v* of Heis. It lies on the lower edge of the tunic, but I cannot learn the derivation or history of the title, although the Arabic Al Thābit signifies the " Endurer."

★

Junonis volucrem, quae caudā sidera portat.
Ovid's *Metamorphoses.*

Pavo, the Peacock,

lying south of Sagittarius and the Southern Crown, is one of Bayer's twelve constellations, and the Italian **Pavone,** the French **Paon,** the German **Pfau.**

The title is an appropriate one for enduring stars, as this bird has long been a symbol of immortality, fancifully said to be from the annual renewing of its feathers; but this is common to all birds, and the symbolism probably is from the fact that its starry tail rendered the peacock sacred to Juno, the immortal queen of the heavens, and thus in classical times, as in the days of chivalry, an object of adjuration. This bird was still further astronomical in originally having been Argos, the builder of the ship Argo, who was changed by Juno to a peacock when his vessel was transferred to the sky, where he has since rejoined her.

In China the constellation was **Joo Tseo,** their translation of our word.

Julius Schiller united it with Indus in his biblical figure "**S. Job.**"

Gould catalogued 129 component stars, from the 2d to the 7th magnitudes, but none seem to be individually named, as is the case among all the new southern figures.

✶

That poetic steed,
With beamy mane, whose hoof struck out from earth
The fount of Hippocrene.

Bryant's *The Constellations.*

Pegasus,

called thus in Germany, but **Pégase** in France and **Pegaso** in Italy, lies north of the Urn of Aquarius and the easternmost Fish, the stars of the Great Square inclosing the body of the Horse.

Mythologically he was the son of Neptune and Medusa, sprung by his father's command from the blood of the latter which dropped into the sea after her head had been severed by Perseus; and he was named either from *Πηγαί*, the Springs of the Ocean, the place of his birth, or from *Πηγός*, Strong. He was snowy white in color, and the favorite of the Muses, for he had caused to flow their fountain Pirene on Helicon,— or Hippocrene on the Acrocorinthus,— whence came one of the constellation titles, **Fontis Musarum Inventor.** Longfellow prettily reproduced in modern dress this portion of the story, in his *Pegasus in Pound*, where "this wondrous winged steed with mane of gold," straying into a quiet country village, was put in pound; but, finding his quarters uncomfortable, made his escape, and

> To those stars he soared again.
> . . . .
> But they found upon the greensward
> Where his struggling hoofs had trod,
> Pure and bright a fountain flowing
> From the hoofmarks in the sod.

He seems, however, to have come back to earth again, for he was subsequently caught by Bellerophon at the waters of his fountain, and ridden by him when he slew the Chimaera, helping in the latter's destruction. By this time classical legend had given him wings, and Bellerophon sought by their aid to ascend to heaven; but Jupiter, incensed by his boldness, caused an insect to sting the steed, which threw his rider, and, as Wordsworth wrote:

> Bold Bellerophon (so Jove decreed
> In wrath) fell headlong from the fields of air.

Pegasus then rose alone to his permanent place among the stars, becoming the Thundering Horse of Jove that carried the divine lightning.

Ptolemy mentioned the wings as well recognized in his day; and this has continued till ours, for the sky-figure is now known as the **Winged Horse**,— a recurrence to Etruscan, Euphratean, and Hittite ideas, for the wings are clearly represented on a horse's figure on tablets, vases, etc., of those countries, where this constellation may have been known in pre-classical times. Indeed, it is said to have been placed in the heavens by the early Aryans to represent Asva, the Sun.

Early classical mythology did not associate the Horse with Perseus, although artists and authors do not seem to have remembered this, for the celebrated picture by Rubens in the Berlin Gallery shows the winged Pegasus held by a Cupid, while Perseus in full armor is unbinding Andromeda from the rocks, Cetus raging in the waters close by; and the late Lord Leighton left unfinished his *Perseus on Pegasus* at the cliffs of Joppa, with the Gorgoneion in his hand; while in *Troilus and Cressida* Shakespeare mentioned "Perseus' horse."

The Greeks called the constellation simply Ἵππος, although Aratos added ἱερός, "divine," and Eratosthenes alluded to it as Πήγασος, but distinctly asserted that it was without wings, and until after middle classical times it generally was so drawn, although loose plumes at the shoulders occasionally were added. The figure was considered incomplete, a possible reason for this being given under Aries. Thus it was characterized as ἡμιτελής and ἡμίτομος, "cut in two," or as if partly hidden in the clouds; while Nonnus had Ἡμιφανής Λίβυς ἵππος, the Half-visible Libyan Horse.

Thus the Equi Sectio used by Tycho and others for Equuleus would seem equally appropriate for this.

Euripides is said to have called it **Melanippe,** after a daughter of Chiron, also known as Euippe, changed by the goddess Artemis into a Black Mare and placed in the sky; but Bayer quoted from some later writer **Menalippe.** The *Θειανα*, or **Theano,** of Nonnus does not seem intelligible.

Translated from Greece by the Romans, it was **Equus,** and later on **Equus Ales,** qualified at times by the adjectives *alter*, *major*, *Gorgoneus*, and *Medusaeus;* but Isidorus and Lampridius degraded it to **Sagmarius Caballus,** a Pack-horse; La Lande cited **Ephippiatus,** Caparisoned; and elsewhere it was **Cornipes,** Horn-footed; **Sonipes,** Noisy-footed; and **Sonipes Ales.** Germanicus was apparently the first of Latin authors to style it Pegasus.

In the *Alfonsine Tables* it was **Alatus,** Winged, Secundus sometimes being added to distinguish it from Equuleus, which preceded it on the sphere; the *Almagest* of 1551 had **Equus Pegasus,** which the 17th-century astronomers extended to **Pegasus Equus alatus.** Caesius cited **Pegasides,** and Bayer quoted **Equus posterior, volans, aëreus,** and **dimidiatus, Bellerophon,** and **Bellerophontes.**

Jewish legends made it the mighty **Nimrod's Horse;** Caesius, one of those of *Jeremiah* iv, 13, that "are swifter than eagles"; other pious people, the **Ass** on which Christ made his triumphal entry into Jerusalem; but Julius Schiller exalted it into the **Archangel Gabriel.** Weigel drew it as the heraldic **Lüneburg Horse.**

Pegasus appears on coins of Corinth from 500 to 430 B. C., and from 350 to 338 B. C., and 200 years thereafter, on the *decadrachma*, complete and with wings; as well as on coins of Lampsacus, Scepsis, and Carthage,—on these last with the asterisk of the sun, or with the winged disc, and the hooded snakes over its back. It is also shown on a coin of Narbonne as a sectional winged figure, and as a winged horse on a Euphratean gem, with a bull's head, a crescent moon, and three stars in the field. A coin of Panormus, the modern Palermo, has the Horse's head with what was probably intended for a dorsal plume.

Bochart said that the word is a compound of the Phoenician Pag, or Pega, and Sūs, the Bridled Horse, used for the figurehead on a ship, which would account for the constellation being shown with only the head and fore quarters; but others have considered it of Egyptian origin, from Pag, "to cease," and Sūs, "a vessel," thus symbolizing the cessation of navigation at the change of the Nile flow. From this, Pegasus seems to have been regarded, in those countries at least, as the sky emblem of a ship. In the old work the *Destruction of Troye*, we read of "a ship built by Perseus, and named Pegasus, which was likened to a flying horse."

Brugsch mentions as in its location an Egyptian constellation, the **Servant**; and some of its stars would seem to be shown on the Denderah planisphere as a **Jackal.**

The Arabs knew the familiar quadrangle as **Al Dalw,** the Water-bucket, the **Amphora** of some Latin imitator, which generally was used for the Urn in Aquarius; and the Arabian astronomers followed Ptolemy in **Al Faras al Thānī,** the Second Horse, which Bayer turned into **Alpheras**; Chilmead, into **Alfaras Alathem**; and La Lande, into **Alpharès.**

Argelander catalogued 108 stars here, down to the 6th magnitude; and Heis, 178, to the 6½.

The starless region toward Pisces was Al Bīrūnī's **Al Baldah,** the Fox's Kennel, a term for whose stellar connection I find no explanation.

Before leaving this constellation, it is worth while to note that an asterism, now virtually lost to us and seldom mentioned except in the lists of Al Sufi, Al Amasch, and Kazwini, is described by the last-named under the title **Al Faras al Tāmm,** the Complete Horse. Although somewhat indefinitely marked out, it is said to have occupied the space between the eastern wing of the Swan, the chest of Pegasus, Equuleus, and the tail of Lacerta, drawing for its components from the last three; but Beigel held that it could have existed only with the grammarians,—the Tāmm in its title being easily confused, in transcription, with the Thānī in the Arabians' name for Pegasus. Ideler's *Sternnamen* is the sole modern work in which I find any reference to this Complete Horse, and even that author, in one passage, seems to regard Monoceros as the modern representative of this somewhat mythical constellation; but this is impossible if Kazwini's description be accepted. Indeed, Ideler himself, later on in his book, changed his opinion to agree with that of Beigel.

α, 2.5, white.

Markab — Flamsteed's **Marchab** — is the Arabs' word for a Saddle, Ship, or Vehicle,— anything ridden upon,— that was early applied to this star; but they also designated it as **Matn al Faras,** the Horse's Withers or Shoulder, and Bayer cited **Yed Alpheras,** the Horse's Hand, or, more properly, Forearm,— the Arabian **Yad.** Kazwini knew it and β as **Al ʽArḳuwah,** the Cross-bar of the well in which Al Dalw, the Bucket, was used.

In India it was noted as the junction star of the Bhādra-pàdā *nakshatras*, detailed under β.

In China it was **Shih**, a title borrowed from the *sieu* that it marked.

Brown thinks that, with γ and ζ, it was the Euphratean asterism **Lik-bar-ra**, the Hyaena,—perhaps **Ur-bar-ra.**

Among astrologers it portended danger to life from cuts, or stabs, and fire. It culminates on the 3d of November, and when on the meridian forms, with γ, the southern side of the Great Square, β and δ forming the northern, and all 15° to 18° apart.

Markab's spectrum is Sirian, and it is receding from us at the rate of three quarters of a mile a second.

It is one of the so-called lunar stars, much observed in navigation.

β, Irregularly variable, 2.2 to 2.7, deep yellow.

This is the **Scheat** of Tycho, the *Palermo Catalogue*, and modern lists generally, either from **Al Sā'id,** the Upper Part of the Arm, or, as Hyde suggested, from the early **Sa'd,** appearing in the subsequent three pairs of stars. Bayer had **Seat Alpheras**; Chilmead, **Seat Alfaras**; Riccioli, **Scheat Alpheraz**; and Schickard, **Saidol-pharazi.**

Arabian astronomers knew it as **Mankib al Faras,** the Horse's Shoulder, mentioned by Ulug Beg and still occasionally seen as **Menkib.** Chilmead had **Almenkeb.**

The Great Square, of which β formed one corner, constituted the double asterism, the 24th and 25th *nakshatras*, **Pūrva,** Former, and **Uttara,** Latter, **Bhādra-padā,** Beautiful, Auspicious, or Happy Feet, sometimes also called **Proshtha-padā,** *Proshtha* meaning a Carp or Ox; but Professor Whitney translated it "Footstool Feet," and said that the authorities do not agree as to the figures by which they are represented, for by some the one, by others the other, is called a Couch or Bed, the alternate one, in either case, being pronounced a Bifaced Figure, or Twins. This Couch is a not inapt representation of the group if both asterisms are taken together, the four stars well marking the feet. Weber calls them **Pratishṭhana,** a Stand or Support, as Whitney wrote,

an evident allusion to the disposition of the four bright stars which compose it, like the four feet of a stand, table, bedstead, or the like;

the regents of these *nakshatras* being Aja Ekapāt, the One-footed Goat, and Ahi Budhya, the Bottom Snake, "two mythical figures, of obscure significance, from the Vedic Pantheon." The 24th *manzil*, formed by α and β, was **Al Fargh al Muḳdim,** the Fore Spout, *i. e.* of the water-bucket,—Al

Bīrūnī's **Al Fargh al Awwal,** the First, or the Upper, Spout; and the 24th *sieu* was these same stars known as **Ying She,** or **Shih,** a House, anciently **Sal** and **Shat;** but it also comprised parts of Aquarius and Capricornus. They also were the Persian **Vaht,** the Sogdian and Khorasmian **Farshat Bath,** and the Coptic **Artulos,** all signifying something pertaining to Water; while in astrology β indicated danger to mankind from that element.

Within the area of this Square Argelander counted only about 30 naked-eye stars, but in the clearer sky of Athens Schmidt saw 102.

It was in the 24th *sieu* that the Chinese record a conjunction of the planets Mercury, Mars, Jupiter, and Saturn, on the 28th of February, 2449 B. C., according to Bailly's computations; but we sometimes see this statement made as to five planets, Venus being added, and as having taken place on the 29th of February, that year being bissextile. Smyth indefinitely mentions this conjunction as at some point between *α* Arietis and the Pleiades; Flammarion states that it was in Capricorn; and Steele alludes to it as of 2246 B. C., and between the tenth and eighteenth degrees of Pisces. At that date the signs and constellations were about coincident.

The variability of β was discovered by Schmidt in 1847, and Argelander found a period of forty-one days; but Schoenfeld thinks that irregular oscillations, in a period of thirty to fifty days, are more probable.

The spectrum of Scheat is of the third type of Secchi's classification, which includes the red and orange stars and most of the variables: "*α* Orionis, *α* Herculis, Antares, and *o* Ceti (Mira) are good examples."

The star is receding from us about four miles a second.

γ, 3, white,

erroneously placed by Tycho in Pisces, marks the extreme tip of the Horse's wing, so that its name **Algenib** has been considered as derived from **Al Janāḥ,** the Wing, but it probably is from **Al Janb,** the Side. It has sometimes been written **Algemo.** Al Bīrūnī quoted it, with δ (*α* Andromedae), as **Al Fargh al Thānī,** the Second, or Lower, Spout, *i. e.* of the Bucket. This also is the title of the 25th *manzil,* but appears in Professor Whitney's list as **Al Fargh al Mu'hir,** the Rear Spout, and in Smyth's as **Al Fargu.**

Chrysococca called it Πήγασος from the constellation.

Reeves said that it is the Chinese **Peih,** a Wall or Partition, thus taking the title of the 25th *sieu,* which it marked and, with δ, constituted. It lies at the junction of the *nakshatras* Bhādrapadā and Revatī; and, with δ, was included in the corresponding lunar station of several other nations.

With the same star and β Cassiopeiae it makes up the **Three Guides,** all these being almost exactly on the prime meridian, the vernal equinox lying in a starless region of Pisces about 15° south of γ Pegasi. Two 11th-magnitude stars are close by.

δ, 2.2, white.

This, as already noted, is the same as Alpheratz (α Andromedae), and recognized by astronomers of every age as in either constellation; or, as Aratos wrote, ξυνός ἀστήρ, "a common star." It seems to be unnamed as a member of Pegasus.

Al Achsasi included it with γ in the Fargh al Mu'ĥir.

ϵ, Triple, 2.5, 11.5, and 8.8, yellow, ——, and blue.

Enif, Enf, and **Enir,** all titles for this, are from **Al Anf,** the Nose, by which the Arabians designated it. Scaliger had **Enf Alpheras,** and Schickard **Aniphol Pharasi.** It was also **Fum al Faras,** the Horse's Mouth; and **Al Jaḥfalah,** the Lip, this last being found on one of their globes.

Bayer quoted from "the interpreters of the *Almagest*" **Grumium** and **Muscida,** respectively Jaw and Muzzle, so describing its position; but these have become proper names for ξ Draconis and π Ursae Majoris. Flamsteed knew it as **Os Pegasi.**

With θ, and the star α Aquarii, it was the 23d *sieu*, **Goei,** or **Wei,** Steep or Danger, anciently **Gui.**

Enif's spectrum is Solar, and it is receding from us about five miles a second. Gould thinks it probably variable.

ζ, 3.7, light yellow.

Homam seems to have been first given to this in the *Palermo Catalogue*, from **Sa'd**[1] **al Humām,** the Lucky Star of the Hero, in which Ulug Beg included ξ; other lists have **Homan.** But Hyde said that the original was **Al Hammām,** the Whisperer. Al Tizini mentioned it as **Sa'd al**

[1] This Arabic Sa'd is our "Good Luck" and a component word of many titles in the Desert sky, all of which seem to have been applied to stars rising in the morning twilight at the commencement of the pleasant season of spring. **Al Sa'dain,** the dual form, was the title for Jupiter and Venus, the Two Fortunate Planets; **Al Nahsān,** the Unlucky, referring to Mars and Saturn.

Na'amah, the Lucky Star of the Ostriches; and Al Achsasi, as **Nā'ir Sa'd al Bahāim,** the Bright Fortunate One of the Two Beasts, which Al Sufi had said were θ and ν. Thus ξ was one of the general group **Al Su'ūd al Nujūm,** the Fortunate Stars.

The Chinese called it **Luy Tien,** Thunder.

7° to the north of ζ is the point assigned by Denning as the radiant of the first stream of **Pegasids,** the meteors visible about the 28th of June; although Espin locates it near δ Cygni.

η, Double, 3.2,

on the left forearm, is the **Matar** of Whitall's *Planisphere*, from **Al Sa'd al Maṭar,** the Fortunate Rain; as such, however, *o* was included with it.

θ, 3.8, and ν, 4.8,

were Al Sufi's **Sa'd al Bahāim,** the Good Luck of the Two Beasts; Al Achsasi adding to the group the still brighter ζ. θ alone is **Baham** in some modern lists; but Ulug Beg had **Bihām,** the Young of domestic animals.

It appears on the Dresden globe as **Al Ḥawā'im,** the Thirsty Camels.

κ, Triple and binary, 4.8, 5.3, and 10.8, yellowish and orange,

marking the right forearm, is unnamed except in China, where it is **Jih,** the Sun, a title also for κ and λ Librae.

The two largest stars were divided by Burnham in 1880 and found to be 0″.2 apart, this decreasing to 0″.1 in 1891. Their orbital period of revolution is 11½ years, and, with that of δ Equulei, the most rapid known to astronomers until See discovered the binary character of Ll. 9091 in Orion. The first and third stars are 11″ apart, at a position angle of 308°.5.

λ, 4.1, and μ, 3.4,

were **Sa'd al Bāri'**, the Good Luck of the Excelling One; but Kazwini designated it as **Sa'd al Nāzi'**, the Good Luck of the Camel Striving to Get to Pasture.

ν was **Fum al Faras** and **Al Jaḥfalah,** but both titles are more correctly applied to ε.

π was the Chinese **Woo,** a Pestle.

τ, 4.5,

with υ, was Al Sufi's **Sa'd al Na'amah,** which Knobel thinks should be **Al Na'āim,** the Cross-bars over a well; but they also were known as **Al Karab,** the Bucket-rope.

The usual titles for τ — **Markab** and **Sagma** or **Salma** — are from Bayer, but the last two should be **Salm,** a Leathern Bucket.

λ μ, η ο, and υ τ, forming a group of three pairs, were a noted asterism in China, under the title **Li Kung.**

This long list of names for rather inconspicuous stars shows unusual early interest in the constellation.

*

There was the knight of fair-hair'd Danaë born,
Perseus.

Elton's translation of the *Shield of Hercules.*

Perseus, even amid the stars, must take
Andromeda in chains aetherial!

Mrs. Browning's *Paraphrases on Nonnus.*

Perseus, the Champion,

the French **Persée,** the Italian **Perseo,** and the German **Perseus,** formerly was catalogued as **Perseus et Caput Medusae.**

He is shown in early illustrations[1] as a nude youth wearing the *talaria,* or winged sandals, with a light scarf thrown around his body, holding in his left hand the Gorgoneion, or head of Medusa-Guberna, the mortal one of the Gorgons, and in his right the ἅρπη, or *falx,* which he had received from Mercury. Dürer drew him thus, but added a flowing robe, a figuring that Bayer, Argelander, and Heis have followed, as they have, in the main, all of that great artist's constellation figures.

A title popular at one time, and still seen, was the **Rescuer,** for, according to the story, Perseus, when under obligations to furnish a Gorgon's head to Polydectes, found the Sisters asleep at the Ocean; and, using the shield of

[1] Tintoretto's celebrated painting of the hero's exploit now hangs in the Hermitage Gallery of Saint Petersburg.

Minerva as a mirror, that he might not be petrified by Medusa's glance, cut off her head, which he then utilized in the rescue of Andromeda. Some one has written about this:

> In the mirror of his polished shield
> Reflected, saw Medusa slumbers take,
> And not one serpent by good chance awake;
> Then backward an unerring blow he sped,
> And from her body lopped at once her head.

Aratos characterized the stellar hero as "stirring up a dust in heaven," either from the fact that his feet are in the celestial road, the Milky Way, or from the haste with which he is going to the rescue of Andromeda; and Manilius, describing his place in the sky, wrote:

> Her Perseus joyns, her Foot his Shoulder bears
> Proud of the weight, and mixes with her Stars.

His story probably was well known in Greece anterior to the 5th century B. C., for Euripides and Sophocles each wrote a drama based on Andromeda's history; and with them, as with the subsequent Greeks, he was Περσεύς, a word that may be derived from the Hebrew Pārāsh, a Horseman, although Ctesias, in his Περσικά of about 400 B. C., had **Parsondas** as a stellar name from Babylonia that may be this. **Parasiea,** current in late Indian astronomy, is only another form of the Greek original.

Ἱππότης, the Horseman, and **Profugus,** the Flying One, also are titles for these stars.

Classical poets called it **Pinnipes,** referring to the *talaria*; **Cyllenius,** the Hero having been aided by Mercury; **Abantiades** and **Acrisioniades,** from his grandfather and father; **Inachides,** from a still earlier ancestor, the first king of Argos; and **Deferens caput Algol, Victor Gorgonei monstri, Gorgonifer, Gorgonisue,** and **Deferens cathenam,** from the association of Perseus with Medusa and the chain of Andromeda.

Alove probably came, by some error in transcription, from Al Ghūl, more correctly applied to the star β; while **Bershawish, Fersaus,** and **Siaush** are plainly the Arabians' orthography of the Greek title, the letter *P* not being found in their alphabet. They, however, commonly called it **Ḥāmil Rā's al Ghūl,** the Bearer of the Demon's Head, which became **Almirazgual** in Moorish Spain, and was translated from Ulug Beg as **Portans caput larvae,** the same being still seen in the German **Träger des Medusen Kopf.**

The **Celeub, Cheleub,** and **Chelub** of the 1515 *Almagest*, *Alfonsine Tables*, and Bayer's *Uranometria* probably are from the Arabic Kullāb, the Hero's weapon, although Grotius and others have referred them to **Kalb,** a Dog, which would render intelligible the occasional title **Canis.**

La Lande identified the figure with the Egyptian **Khem,** and with **Mithras** of Persia, Herodotus having asserted that Perseus, through his and Andromeda's son Perses, gave name to that country and her people, who previously were the Chephenes, as descended from Chepheus, the son of Belus, identified by some with the Cepheus of the sky. The kings of Cappadocia and of Pontus, similarly descended, represented the Hero on their coins.

Cacodaemon was the astrologers' name for this constellation, with special reference to Algol as marking the demon's head; while Schickard, Novidius, and the biblical school generally said that it was **David with the head of Goliath;** but others of the same kind made of it the **Apostle Paul with his Sword and Book.** Mrs. Jameson thought that the legend of Perseus and Cetus was the foundation of that of Saint George and the Dragon, one version making this saint to have been born at Lydda, only nine miles from Joppa, the scene of Perseus' exploit.

The constellation is 28° in length,— one of the most extended in the heavens,— stretching from the upraised hand of Cassiopeia nearly to the Pleiades, and well justifying the epithet *περιμήκετος*, "very tall," applied to it by Aratos. It offers a field of especial interest to possessors of small telescopes, while even an opera-glass reveals much that is worthy of observation. Argelander gives a list of 81 naked-eye stars, and Heis 136.

The former has suggested that within its boundaries may lie the possible central point of the universe, which Mädler located in the Pleiades and Maxwell Hall in Pisces,— all probably unwarranted conclusions.

δ, ψ, σ, α, γ, η, and others on the figure's right side, form a slight curve, open towards the northeast, that has been called the **Segment of Perseus.**

α, 2.1, brilliant lilac and ashy.

Algenib, with the early variations of **Algeneb, Elgenab, Ģenib, Chenib,** and **Alchemb,** is from **Al Janb,** the Side, its present position on the maps; Chrysococca similarly called it Πλευρά Περσάους.

Another name, **Marfak** or **Mirfak,** the Elbow, sometimes written **Mirzac,** comes from the Arabians' **Marfiḳ al Thurayya,** thus qualified as being next to the Pleiades to distinguish it from the other elbow. But this may indicate a different representation of Perseus in their day,— a suspicion stengthened by the nomenclature of others of his stars, especially of ξ and o.

Assemani alluded to a title on the Borgian globe,—**Mughammid,** or **Muḥammir, al Thurayya,** the Concealer of the Pleiades,— which, from its location, may be for this star.

With γ, δ, and others it was the Chinese **Tien Yuen,** the Heavenly Enclosure.

Algenib never sets in the latitude of New York City, but just touches the horizon at its lower culmination. Its spectrum is of Secchi's second, or Solar, type, and the Potsdam observations indicate that the star is approaching our system at the rate of 6½ miles a second.

the Gorgon's head, a ghastly sight,
Deformed and dreadful, and a sign of woe.
Bryant's translation of the *Iliad.*

β, Spectroscopic binary and variable, 2.3 to 3.5, white.

Algol, the **Demon,** the **Demon Star,** and the **Blinking Demon,** from the Arabians' **Rā's al Ghūl,** the Demon's Head, is said to have been thus called from its rapid and wonderful variations; but I find no evidence of this, and that people probably took the title from Ptolemy. Al Ghūl literally signifies a Mischief-maker, and the name still appears in the Ghoul of the *Arabian Nights* and of our day. It degenerated into the **Alove** often used some centuries ago for this star.

Ptolemy catalogued it as τῶν ἐν γοργονίω ὁ λαμπρός, "the bright one of those in the Gorgon's head," which Al Tizini followed in his **Nā'ir,** for, with π, ρ, and ω, it made up that well-known group, itself being the **Gorgonea prima**; the Γοργόνιον of Chrysococca, **Gorgoneum Caput** of Vitruvius, **Caput Gorgonis** of Hyginus, and the **Gorgonis Ora** of Manilius.

With astronomical writers of three centuries ago Algol was **Caput Larvae,** the Spectre's Head.

Hipparchos and Pliny made a separate constellation of the Gorgon stars as the **Head of Medusa,** this descending almost to our own day, although always connected with Perseus.

The Hebrews knew Algol as **Rōsh ha Sāṭān,** Satan's Head, Chilmead's **Rosch hassatan, the Divels head**; but also as **Lilīth,** Adam's legendary first wife,[1] the nocturnal vampyre from the lower world that reappeared in the demonology of the Middle Ages as the witch Lilis, one of the characters in Goethe's *Walpurgis Nacht.*

The Chinese gave it the gruesome title **Tseih She,** the Piled-up Corpses.

[1] We are indebted to the Talmudists for this story, which probably originated in Babylonia; and they added that, after Adam had separated from Lilīth and their demon children, Eve was created for him. Our Authorized Version renders the original word, in *Isaiah* xxxiv, 14, by "screech owl"; the Revised Version, by "night-monster"; Cheyne adopts the Hebrew Lilīth in the *Polychrome Bible;* and *Luther's Bible* had Kobold, but this corresponded to the Scottish Brownie and the English "Robin Goodfellow,"—Shakespeare's "Puck." Saint Jerome's *Vulgate* translated it "Lamia," the Greek and Roman title for the fabled woman, beautiful above, but a serpent below, that Keats reproduced in his *Lamia.*

Astrologers of course said that it was the most unfortunate, violent, and dangerous star in the heavens, and it certainly has been one of the best observed, as the most noteworthy variable in the northern sky. It "continues sensibly constant at 2.3 magnitude during 2½ days, then decreases, at first gradually, and afterward with increasing rapidity, to 3.5 magnitude"; its light oscillations occupying about nine hours; its total period being stated as 2 days 20 hours 48 minutes 55 seconds. Al Sufi, a good observer for his day, yet strangely making no allusion to its variability, called it a 2d-magnitude; and the phenomenon was first scientifically noted by Montanari during several years preceding 1672. This was confirmed by Maraldi's observations of 1694, and, later, by those of the Saxon farmer Palitsch,[1] but its approximate period seems to have been first announced by Goodricke in 1782, who even then advanced the theory of a dark companion revolving around it with immense velocity, which periodically cut off its light. This, reaffirmed by Pickering in 1880, was made certain by the spectroscope in the hands of Vogel of Potsdam in 1889. Chandler thinks that there must exist another invisible body larger than either Algol or its companion, around which both revolve in a period of 130 years; but Tisserand has shown that the phenomenon on which Chandler bases this opinion can be explained in a different and simpler way. Its name is used for the type indicating short-period variables whose changes may be explained by this theory of "eclipses." Of these seventeen are now known.

Although classed among the white stars with a Sirian spectrum, Al Sufi wrote of it as red, which Schmidt confirmed as seen by him at Athens for a short time in 1841. It seems to be approaching us at the rate of about a mile a second; and is estimated as a little more than a million miles in diameter.

When on the meridian Algol is almost exactly in the zenith of New York City. This is at nine o'clock in the evening of the 23d of December.

ε, Double, 3.5 and 9, greenish white and lilac.

In China this, with the 4th-magnitude ν and some others, was **Keuen She.** It has been suspected of variation in color as well as in light. The components are about 9″ apart, at a position angle of 10°, and form an interesting object for a four-inch telescope.

[1] Palitsch also was famous for his discovery of Halley's comet on Christmas night, 1758.

η, Double, 5 and 8.5, orange and smalt blue,

is unnamed except in China, where, with γ, it was **Tien Chuen,** Heaven's Ship. But it is noticeable in having three small stars on one side nearly in line, and one on the other, forming a miniature representation of Jupiter and his satellites. The components are 28″ apart, at a position angle of 300°.

λ and μ, 4th- to 5th-magnitude stars, were **Tseih Shwuy,** Piled-up Waters.

ξ, a 4½-magnitude, is the **Menkib** of Burritt, from **Mankib al Thurayya,** the Shoulder of—*i. e.* next to—the Pleiades in the Arabian figure, although on modern charts it marks the left ankle.

o, a double star of 4th and 9th magnitudes, is **Ati** and **Atik,** from the word **Al ʽĀtiḳ** found on the Borgian globe, at the space between the shoulders, and applied to it by Ulug Beg; but it is now located near the left foot.

π, a 4½-magnitude, was **Gorgonea secunda**; and ρ, a variable from 3.4 to 4.2, orange in color, was **Gorgonea tertia.**

τ, a 4½-magnitude, with others in the constellation, was known by the Chinese as **Ta Ling,** the Great Mound.

υ, 3.8,

marking the tip of the weapon in Perseus' hand, bears many titles with Bayer, all referring to its location; but none of these—indeed, no name at all—is seen in modern lists. Bayer wrote of them:

In falce adamanthinā trium praecedens. Falx dicitur & curvus Harpes, Gladius falcatus, & incurvus, Arab. Nembus, Maroni Ensis falcatus, & curvus Saturni dens.

The "*Arab.*" would seem erroneous, for **Nembus** is neither Arabic nor Latin, and if intended for *Nimbus*, is equally wrong, as there is no suspicion of nebulosity about the star. *Curvus Saturni dens* was Vergil's designation in the *Georgics* for a "pruning-hook," and the equivalent of *Falx* and Ἅρπη, so well known in connection with Perseus.

χ, a multiple star, and the little h mark two clusters noticeable with the naked eye, Nos. 884 and 869 of the *New General Catalogue*, 30′ and 15′ in diameter, almost connected, and apparently a protuberant part of the Milky Way. They were the Arabians' **Miʽṣam al Thurayya,** the Wrist of —*i. e.* next to—the Pleiades.

Hipparchos seems to have been the first to record them, which he did as νεφελοειδής, a "cloudy spot"; Ptolemy, as συστροφή, a "dense mass"; and subsequent astronomers down to Galileo's day similarly considered them nebulous. The *Alfonsine Tables* said, *revolutio nebulosa*, and the *Almagest* of 1551, *girus ille in capulo ensis*, this *girus*—correctly *gyrus*—signifying a circle. They seem strangely to have escaped the notice of astrologers, who, as a rule, devoted much attention to clusters as harmful objects which portended accidents to sight and blindness.

In China they were **Foo Shay.**

These stars and clusters are now known as the **Sword Hand of Perseus,** i, g, ϕ, and v marking the outstretched sword. In small telescopes the twin clusters form one of the most beautiful objects within their reach.

Between χ and η lies the diverging point of the **Perseids,** the prominent meteor stream visible from the 19th of July to the 17th of August, its maximum occurring about the 10th of the latter month and continuing several days. These appear in the early part of the night, at an elevation of from fifty-six to seventy miles, moving with moderate speed and leaving streaks of yellow light; the radiant advancing nearly 30° eastward during their period of visibility. Schiaparelli found their orbit coincident with that of Tuttle's comet, III of 1862. The Perseids were recorded as far back as 811, seven appearances being mentioned down to 841, and they are supposed to have been members of the solar system for thousands of years, although now, perhaps, steadily decreasing in number. Dante may have made reference to them in the *Purgatorio:*

> Vapors enkindled saw I ne'er so swiftly
> At early nightfall cleave the air serene,
> Nor, at the set of sun, the clouds of August;

and in the later Middle Ages they were known as the **Larmes de Saint Laurent, Saint Laurence's**[1] **Tears,** his martyrdom upon the red-hot gridiron having taken place on the 10th of August, 258.

ω, of the 5th magnitude, was **Gorgonea quarta.**

★

Phoenix,

the French **Phénix,** the German **Phönix,** and the Italian **Fenice,** is one of Bayer's new figures, between Eridanus and Grus, south of Fornax and Sculptor,—its α, κ, μ, β, ν, and γ in a line curving toward the south like that of a primitive **Boat,** by which figure, as **Al Zauraḳ,** the Arabs knew them. Al Sufi cited another name,—**Al Riʽāl,** the Young Ostriches,—which Hyde wrongly read **Al Zibal,** perhaps a synonymous title; and Kazwini used Al Sufi's term in describing some stars of Al Nahr, the River, in which our Phoenix was then included by Arabian astronomers.

[1] It is in the church of this Saint Laurence at Upton that the remains of Sir William Herschel lie buried, and over them is the fitting inscription:

Coelorum perrupit claustra.

Others changed the figure to that of a **Griffin,** or **Eagle,** so that the introduction of a Phoenix into modern astronomy was, in a measure, by adoption rather than by invention.

But, whether Bayer knew it or not, his title is an appropriate one, for with various early nations — at all events, in China, Egypt, India, and Persia — this bird has been "an astronomical symbol of cyclic period," some versions of the well-known fable making its life coincident with the Great Year of the ancients beginning at noon of the day when the sun entered among the stars of Aries; and, in Egypt, with the Sothic Period when the sun and Sirius rose together on the 20th of July. Thompson further writes of this:

> A new Phoenix-period is said to have commenced A. D. 139, in the reign of Antoninus Pius; and a recrudescence of astronomical symbolism associated therewith is manifested on the coins of that Emperor.

Coincidently, Ptolemy adopted as the epoch of his catalogue the year 138, the first of Antoninus. With the Egyptians, who knew this bird as Bennu and showed it on their coins, it was an emblem of immortality; indeed it generally has been such in pagan as well as in Christian times.

In China the constellation was **Ho Neaou,** the Fire Bird, showing its derivation there from the Jesuits.

Julius Schiller combined it with Grus in his **Aaron the High Priest.**

Gould catalogues 139 naked-eye stars here, from 2.4 to 7.

a, of 2.2 magnitude, was Al Tizini's **Nā'ir al Zauraḳ,** the Bright One in the Boat, rendered in Hyde's translation *lucida Cymbae.* It culminates just above the horizon of New York City on the 17th of November, and is quite conspicuous from its solitary position southeast from Fomalhaut.

A 14th-magnitude companion, purple in tint, has recently been discovered by See, 9″ away, with a position angle of 280°.

★

> And here fantastic fishes duskly float,
> Using the calm for waters, while their fires
> Throb out quick rhythms along the shallow air.
>
> Mrs. Browning's *A Drama of Exile.*

Pisces, the Fishes,

are the German **Fische,** the Italian **Pesci,** the French **Poissons,** the Anglo-Norman **Peisun,** and the Anglo-Saxon **Fixas.** The *Alfonsine Tables* of

1521 had **Pesces,** and the *Almagest* of 1515 **Echiguen,** Bayer's **Ichiguen,** a word that has defied commentators unless Caesius has explained it as being a corruption of **Ichthues.**

The figures are widely separated in the sky, the northeastern one lying just south of β Andromedae, headed towards it, and the southwestern one east from and headed towards Aquarius and Pegasus, the *lucida* marking the knot of the connecting bands. Both are north of the ecliptic, the first culminating on the 28th of November, and the second about three weeks earlier. In early days they were shown close together, one above the other, but in reversed directions, although united as now.

By reason of precession this constellation is now the first of the zodiac, but entirely within its boundaries lies the sign Aries; the vernal equinox being located in a comparatively starless region south of ω in the tail of the southwestern Fish, and about 2° west of "a line from α Andromedae through γ Pegasi continued as far again." This equinoctial point is known as the First of Aries, and the Greenwich of the Sky; and from their containing it, the Fishes are called the **Leaders of the Celestial Host.**

The Greeks knew them as Ἰχθύε and Ἰχθύες, in the dual and plural; the Romans as we do, often designating them as **Imbrifer Duo Pisces, Gemini Pisces,** and **Piscis Gemellus.** Classic authors said **Aquilonius,** sometimes **Aquilonaris;** and very appropriately, for the Aquilo of the Romans, perhaps derived from *aqua*, or *aquilus*, signified a rain-bringing wind from the north, and well represented the supposed watery character of the constellation, as also its northerly position. Ampelius, however, ascribed Aquilo to Gemini, and Eurus, or Vulturnus, the Southeast Wind, to Pisces.

Miss Clerke thinks that the dual form of this constellation recalls the additional month which every six years was inserted into the Babylonian calendar of 360 days; and Sayce, agreeing in this opinion, translates the early title for these stars as the **Fishes of Hea** or **Ia.** It has also been found on Euphratean remains as **Nuni,** the Fishes, a supposed equivalent of its other title, **Zib,** of the later Graeco-Babylonian astronomy; although this last word may mean "Boundary" as being at the end of the zodiac. Another signification is the **Water,** which we have already seen with Aratos for this part of the sky; this also is the meaning of the word **Atl,** the Aztecs' name for Pisces.

It was the Babylonian **Nūnu,** the Syriac **Nūno,** the Persian **Mahīk,** and the Turkish **Balīk,** all translated "Fish"; while Kircher cited, from Coptic Egypt, Πικοτώριων, *Piscis Hori*, which Brown translates "Protection," but claims for a Coptic lunar asterism formed by β and γ Arietis.

In earliest Chinese astronomy, with Aquarius, Capricornus, and a part of

Sagittarius, it was the northern one of the four quarters of the zodiac, the **Dark Warrior,** or the residence of the **Dark,** or **Northern, Emperor;** but later, in their zodiac of twelve figures, it was the Pig, **Tseu Tsze;** and, after the Jesuits, **Shwang Yu,** the Two Fishes.

With the Arabians it was **Al Samakah,**— Chilmead's **Alsemcha,**— or, in the dual, **Al Samakatain;** and **Al Hūt,** the Fish, referring to the southern one, the **Vernal Fish,** as marking that equinox; the northern being confounded with Andromeda's stars and so not associated with the zodiac. From these came **Sameh, Haut, El Haut,** and **Elhautine** in Bayer's *Uranometria.*

Dante combined the two in his **Celeste Lasca,** the Celestial Roach or Mullet, saying that here and in Aquarius geomancers saw their Fortuna Major; and thus described **I Pesci:**

> quivering are the Fishes on the horizon,
> And the Wain wholly over Caurus lies.

This was on a Saturday morning, and the positions of the constellations indicate that the time was just before sunrise in the month of April; Caurus, or Corus, the Northwest Wind, symbolizing that quarter of the heavens.

Varāha Mihira mentioned the constellation as **Ittha,** in which the Greek word appears; but before his day it was **Anta, Jitu,** and **Mina** or **Minam** in the Tamil dialect.

The 26th *nakshatra,* **Revatī,** Abundant or Wealthy, lay here in the thirty-two stars from ζ northwards, figured as a **Drum** or **Tabor.** But the *manzil,* **Batn al Hūt,** the Fish's Belly, or **Al Rishā',** the Cord, and the corresponding *sieu,* **Koei,** or **Kwei,** Striding Legs, were formed by sixteen stars in a figure 8 from ψ Piscium to ν Andromedae, and mainly lay in this constellation, although β and ζ in Andromeda seem to have been their determinant points. All of these stations, however, may have been even more extended, for there certainly is "a perplexing disagreement in detail among the three systems."

Al Bīrūnī asserted that "the name of the sign in all languages signifies only one fish," and it is probable that the original asterism was such, for, according to Eratosthenes, it symbolized the great Syrian goddess **Derke** or **Derketo,** and so, later, was named **Dea Syria, Dercis, Dercetis, Dercete, Proles Dercia,** and **Phacetis.** The Greeks called this Ἀτάργατις;[1] and from a supposed derivation of this word from Adïr and Dag (Great and Fish) it was drawn with a woman's head upon a huge fish's body. In this manner it was connected with the Syrian **Dāgōn** and the Jews' **Dagaïm,** their

[1] Allusion was made to this Atargatis in the apocryphal *2d Book of Maccabees,* xii, 26; and gems now in the British Museum show the fish-god with a star or other astronomical symbol.

title for the Two Fishes,— Riccioli's **Dagiotho.** Avienus called the constellation **Bombycii Hierapolitani**; Grotius correcting the error in orthography to **Bambycii,** as Derke was worshiped at Bambyce,— the Mabog of Mesopotamia, or Hierapolis,— on the borders of Syria. Thus, too, it was **Dii Syrii.**

But the Greeks confounded this divinity with another Syrian goddess, Astarte, identified with *'Αφροδίτη* (Venus), who precipitated herself, with her son *Ἔρως* (Cupid), into the Euphrates when frightened by the attack of the monster Typhon; these becoming two fishes that afterwards were placed in the zodiac. Latin classical authors, with the same groundwork of the story, made Pisces the fishes that carried Venus and her boy out of danger, so that, as Manilius said,

> Venus ow'd her Safety to their Shape.

The constellation was thus known as **Venus et Cupido, Venus Syria cum Cupidine, Venus cum Adone, Dione,** and **Veneris Mater**; and it has been *'Ουρανία* and **Urania,** the Sarmatian Aphrodite. All this, perhaps, was the foundation of the Syrians' idea that fish were divine, so that they abstained from them as an article of food; Ovid repeating this in the *Fasti*, in Gower's rendering:

> Hence Syrians hate to eat that kind of fishes;
> Nor is it fit to make their gods their dishes.

But Xenophon limited this restriction to the fish of the river Chalos.

A scholiast on Aratos, commented on by Grotius, said that the "Chaldaeans" called the northernmost Fish *Χελιδόνιας ἰχθύς*, shown with the head of a swallow, a representation that Scaliger attributed to the appearance of the bird in the spring, when the sun is in this region of the sky. Dupuis had much to say about this changed figure, calling it **l'Hirondelle,** but as of the Arabs; and this idea has led to confusion in the Piscine titles already noticed under Apus. The Greek word, however, was common for a Tunny, so that there is reason enough for its application to either of the Pisces in their normal shape. This northern Fish has sometimes been considered as representing the monster sent to devour Andromeda, and its proximity to the latter would render this more appropriate than the comparatively distant Cetus; in fact, Κῆτος was as often used by the Greeks for the Tunny as it was for the Whale.

Some of the Jews ascribed the joint constellation to the joint tribes of Simeon and Levi, whose sanguinary character Jacob on his death-bed so vividly portrayed; others, to Gad the Marauder. Perhaps it was from this that Pisces was considered of such malignant influence in human affairs,—"a dull, treacherous, and phlegmatic sign"; yet this opinion, doubtless, was anterior to the patriarch's time, for the Egyptians, the instructors of the Hebrews in astrology, are said to have abstained from eating sea-fish out of dread and abhorrence; and when they would express anything odious, represented a fish in their hieroglyphics. Pliny, too, asserted that the appearance of a comet here indicated great trouble from religious differences besides war and pestilence; but this became the common reputation of comets wherever they showed themselves.

In early astrology the constellation appropriately was under the care of the sea-god Neptune, and so the **Neptuni Sidus** of Manilius; and it was the **Exaltation of Venus,** as Chaucer said in the *Wyf of Bathes Tale,*—

In Pisces where Venus is exaltat,—

which Sir Thomas Browne, the author-physician of the 17th century, thus commented upon:

> Who will not commend the wit of astrology? Venus, born out of the sea, hath her exaltation in Pisces.

Thus it naturally ruled the Euphrates, Tigris, and the Red Sea, and Parthia; but in later days was assigned to the guardianship of Jupiter, whose **House** it was, reigning over Egypt, Calabria, Galicia, Normandy, Portugal, Spain, and Ratisbon. It was predominant in influence with mariners, and had charge of the human feet; the designated color being a glistening white, as of fish just out of the water; and it was fruitful, like its namesakes, for, according to Manilius:

Pisces fill the Flood.

Ptolemy distinguished the members of the constellation as ἑπόμενος, "the rear or eastern," and ἡγουμένος, "the front or western"; the Southern Fish being his νότιος; a precaution rendered necessary by the frequent confounding of these three by classical writers. A notable instance of this is seen in the *Poeticon Astronomicon*, where our Pisces are made to receive the water from the Urn. In Humboldt's *Cosmos* they are **Pisces boreales.**

The constellation is popularly thought to have taken its name from its coincidence with the sun during the rainy season; and the symbol for the sign, ♓, to represent the two Fishes joined; but Sayce thinks it the Hittite determinative affix of plurality.

Postellus asserted that the Fishes represented those with which Christ fed "about five thousand men, beside women and children"; and Caesius, that they were the ΙΧΘΥΣ of Ἰησοῦς Χριστός Θεοῦ Υἱός Σωτήρ, a fish always being the symbol of the early Christians' faith; but when the old twelve figures were turned into those of the apostles, these became **Saint Matthias,** successor to the traitor Judas.

The Fishes were changed to a **Dolphin** in the zodiac sculptured on the wall of Merton College, taken from the armorial bearings of Fitz James, bishop of London, and warden of the college from 1482 to 1508; a dolphin being of as sacred significance among pagans as a fish was among Christians.

Within their boundaries took place the three distinct conjunctions of Jupiter and Saturn in the year 747 of Rome,— the year to which for a long time was assigned Christ's birth; these phenomena strikingly agreeing in some of their details with Saint Matthew's account of the Star of Bethlehem. The opinion that these appearances guided the Magi in their visit to Judaea was first advanced and advocated by the celebrated Kepler, and worked out in 1826 by Ideler, and in 1831 by Encke.[1] It is noticeable that the Rabbis held the tradition, recorded by Abrabanel in the 15th century, that a similar conjunction took place in Pisces three years previous to the birth of Moses, and they anticipated another at their Messiah's advent. Thus the Fishes were considered the national constellation of the Jews, as well as a tribal symbol. Jupiter and Saturn were again together here in February, 1881, Venus being added to the group,— a well remembered and most beautiful sight.

Here, too, was the seat of the predicted conjunction of three planets that Stoffler said would cause another Deluge in 1524,— an announcement that created universal consternation; but, unfortunately for the prophet's reputation, the season was unusually dry.

It was in Pisces, on the 2d of September, 1804, that Harding, of Lilienthal in Hanover, discovered the minor planet Juno.

In his *Shepheard's Kalendar* for November, Edmund Spenser thus described the constellation's place in the sky:

> But nowe sadde Winter welked hath the day,
> And Phoebus, weary of his yerely taske,
> Ystabled hath his steedes in lowly laye,
> And taken up his ynne in Fishes haske.

La Lande, quoting indirectly from Firmicus, mentioned as from the Egyptian sphere of Petosiris:

> au nord des Poissons, il place le Cerf, & une autre constellation du Lièvre;

[1] More recent determinations, by the late Reverend Mr. Charles Pritchard of Oxford, have somewhat altered the previous conclusions, while our chronologists, meanwhile, have changed the date of the Nativity, so that the time-honored identification of the Star of the Magi with these planetary conjunctions now seems to be discarded.

but this second **Hare** I cannot trace, although Bayer had Cerva as a title for Cassiopeia "north of the Fishes."

There is a sprinkling of indistinct stars between the Fishes and the Whale that Vitruvius called Ἑρμεδόνη, explained by Hesychios as the Stream of Faint Stars, but by some French commentator as *les délices de Mercure*, whatever that may be. Riccioli, calling it **Hermidone,** said that it was *effusio Aquarii*, the classical designation for the Stream from the Urn; but Baldus, with Scaliger, said that the word was Ἁρπεδόνη, the Cord, although this seems equally inapplicable here. These stars may be the proposed new **Testudo** noted under β Ceti.

Argelander gives 75 components visible to the naked eye, and Heis 128; but the *lucida* is only of the 4th magnitude.

α, Double and probably binary, 4 and 5.5, pale green and blue.

Al Rescha, or **Al Rischa,** derived from the Arabians' **Al Rishā',** the Cord, is 20° south from the head of Aries, 2°.7 north of the celestial equator, and marks the knot in the united cords of the Fishes; the same title being applied to β Andromedae. This word originally may have come from the Babylonian Riksu, Cord.

Hipparchos and Ptolemy designated it as Σύνδεσμος τῶν Ἰχθύων, or τῶν Λίνων, the Knot of the Fishes, or of the Threads, varied by Aratos and Geminos in Δεσμός; these words being transcribed by Germanicus and the scholiasts as **Sundesmos** and **Desmos.** They were rendered by Cicero and others as **Nodus, Nodus coelestis,** and **Nodus Piscium**; by Pliny as **Commissura Piscium**; and in the 1515 *Almagest* as **Nodus duorum filorum.**

The Arabians translated these by **'Ukd al Ḥaiṭain,** which, as **Okda** and **Kaitain,** are not unusual titles now.

The uniting cords, branching from α through ο, π, η, and ρ to the tail of the northernmost Fish, and through ξ, ν, μ, *f*, *e*, ζ, ε, and δ to ω that marks the tail of the one to the south, were Ptolemy's λίνον, "thread," the λίνοι of other authors. Cicero called them **Vincla,** the Bonds; and the scholiast on Germanicus, **Alligamentum linteum** or **luteum,** divided by Hevelius into **Linum boreum** and **austrinum.** Some of these terms also were applied to the star δ as marking one of the cords.

The Arabians knew these cords as **Al Ḥaiṭ al Kattāniyy,** the Flaxen Thread; and Al Aṣma'ī, about the year 800, mentioned them in his celebrated romance *Antarah* as a distinct constellation; but Pliny had done the same long before him.

Al Rischa, although lettered first, is somewhat fainter than γ and η. It culminates on the 7th of December.

The component stars are 3″ apart, at a position angle of 324°.

β, a 4½-magnitude, is given by Al Achsasi as **Fum al Samakah,** the Fish's Mouth, descriptive of its position near that feature in the westernmost of the two. With γ, θ, ι, and ω it was the Chinese **Peih Leih**, Lightning.

δ, 4.1,

has in Bayer's *Uranometria* many of the titles already noted under *a*, but they would seem to be words merely indicative of the star's position on the Cord, although some have used them as proper names. δ, *a*, ε, ζ, μ, ν, and ξ made up the Chinese figure **Wae Ping,** a Rolled Screen.

ζ, a double 5th- and 6.3-magnitude, apparently unnamed, was prominent in Hindu astronomy as marking the initial point of the celestial sphere about the year 572, when it coincided within 10′ of longitude with the vernal equinox. It formed part of the Khorasmian lunar station **Zidadh,** the Sogdian[1] **Riwand,** and of the 26th *nakshatra*, **Revatī,** Rich, being the junction star between Revatī and Açvini. With ε it was the Persian lunar station **Kaht** and the Coptic **Kuton,** Cord.

η, Double, 4 and 11.

Epping asserts that this marked the 1st ecliptic constellation of the Babylonians, **Kullat Nūnu,** the Cord of the Fish, which, if correct, would show the origin of the Greek title, and the probable great antiquity of the present figure. Another signification may be the Dwelling of the Fish.

In China, with ο, ρ, and χ, it was **Yew Kang,** the Right-hand Watch.

The components of η are 1″ apart, at a position angle of 12°.9.

κ and λ, 4th-magnitude stars just above the ventral fin of the western Fish, were the Chinese **Yun Yu,** the Cloud and Rain.

ο, 4.6, appeared in the 1515 *Almagest* as **Torcularis septentrionalis,** a translation of ληνός, erroneously written for λίνος, this star being on the Thread northeast from *a*. But the Latin word should read **Torcular.**

Fl. 65, a 6th-magnitude double, has been regarded by Maxwell Hall as the Central Sun of the Universe.

[1] The Arabs considered Sogdiana one of the four fairest lands on earth; its capital, Samarkhand, was the home of the great Tamerlane and of Ulug Beg, his grandson.

> Next swims the *Southern Fish* which bears a Name
> From the South wind, and spreads a feeble Flame.
> To him the *Flouds* in spacious windings turn.
>
> Creech's *Manilius.*

Piscis Australis, the Southern Fish,

is the Italian **Pesce Australe**; the French **Poisson Australe**; and the German **Südliche Fisch.** It lies immediately south of Capricorn and Aquarius, in that part of the sky early known as the Water, Aratos describing the figure as " on his back the Fish," and

> The Fish reversed still shows his belly's stars;

but modern representations give it in a normal attitude. In either case, however, it is very unnaturally drinking the whole outflow from the Urn.

This idea of the Fish drinking the Stream is an ancient one, and may have given rise to the title **Piscis aquosus,** found with Ovid and in the 4th *Georgic*, which has commonly been referred to this constellation; Vergil mentioning it in his directions as to the time for gathering the honey harvest; but the proper application of this adjectival title is uncertain, for Professors Ridgeway and Wilkins, in their admirable article on *Astronomia* in Doctor Smith's *Dictionary of Greek and Roman Antiquities*, write:

> The Piscis in question has been variously supposed to be one of the Fishes in the Zodiac — the Southern Fish — Hydra — the Dolphin — or even the Scorpion.

Smyth said that

> In the early Venetian editions of Hyginus, there is a smaller fish close under it, *remora* fashion, interfering with the *Solitarius* by which that astronomer, from its insulated position, designated Piscis Notius.

Accordingly the edition of 1488, with this representation, had it **Pisces,** and the German manuscript of the 15th century showed it with a still larger companion.

The figure is strangely omitted from the Farnese globe, the stream from the Urn of Aquarius ending at the tail of Cetus.

In early legend our *australis* was the parent of the zodiacal two, and has always been known under this specific title, varied by the other adjectives of equivalent signification, *austrinus*, *meridionalis*, and *notius.*

La Lande asserted that Dupuis had proved this to be the sky symbol of the god **Dāgōn** of the Syrians, the **Phagre** and **Oxyrinque** adored in Egypt; and it even has been associated with the still greater **Oannes.**

It also was Ἰχθύς and Ἰχθύς νότιος; Ἰχθύς μέγας and **Piscis magnus;** Ἰχθύς μονάζων and **Piscis solitarius; Piscis Capricorni,** from its position; and it is specially mentioned by Avienus as the **Greater Fish.** Longfellow, in the notes to his translation of the *Divine Comedy*, called it the **Golden Fish,** probably as being so much more conspicuous than those in the north.

When the Arabians adopted the Greek constellations and names this became **Al Ḥūt al Janūbiyy,** the Large Southern Fish, distorted in late mediaeval days into **Haut elgenubi,** and given by Chilmead as **Ahaut Algenubi;** but their figure was extended further to the south than ours, and so included stars of the modern Grus. Smyth wrote of it:

> The Mosaicists held the asterism to represent the **Barrel of Meal** belonging to Sarephtha's widow; but Schickard pronounces it to be the **Fish** taken by St. Peter with a piece of money in its mouth.

Bayer said that it partook of the astrological character of the planet Saturn.

Gould assigns to it 75 naked-eye components.

α, 1.3, reddish.

Fomalhaut, from the Arabic **Fum al Ḥūt**, the Fish's Mouth, has long been the common name for this star, Smyth saying that **Fom Alhout Algenubi** appears, with its translation *Os Piscis Meridiani*, in a still existing manuscript almanac of 1340.

Aratos distinctly mentioned it as

> One large and bright by both the *Pourer's* feet,

which is its location in the maps of to-day, although sometimes it has marked the eye of the Fish, and formerly was still differently placed, as is noted at β.

In addition to putting it in its own constellation, Ptolemy inserted it in his Ὑδροχόος, and Flamsteed followed him in making it his 24 of Piscis Australis and 79 of Aquarius, calling it *Aquae Ultima Fomalhaut.*

No other star seems to have had so varied an orthography.

The *Alfonsine Tables* of 1521 locate it in Aquarius as **Fomahant** and of the 1st magnitude, but they describe it in **Piscis Meridionalis** as *in ore*, omitting its title and calling it a 4th-magnitude. The other editions of these *Tables*, and Kazwini, do not mention it at all in this constellation, but in Aquarius; nor does Bullialdus in his edition of the *Rudolphine Tables*, although in his reproduction of the *Persian Tables* of Chrysococca he calls it **Os Piscis notii** and **Fumahaud.** The *Astronomica Danica* of Longomontanus includes it in Aquarius as *ultima in effusione* **Fomahant,** giving no Piscis at all; Tycho's *Rudolphine Tables*, in Kepler's edition of 1627, have the same, and Hevelius also puts it there as **Fomahandt.** Bayer cites it, in Piscis Notius, as **Fumahant, Fumahaut** *rectius* **Fumalhaut**; Chilmead, **Phom Ahut**; Caesius has **Fomahand** and **Fontabant**; Riccioli's names for it are **Fomauth, Phomaut, Phomault, Phomant, Phomaant, Phomhaut, Phomelhaut**; La Caille's, **Phomalhaut**; La Lande's are **Fumalhant, Fomahaut,** and **Phomahant**; and Schickard's, **Fomalcuti.** Costard gives it as **Fomahout**; and Sir William Herschel had it **Fomalhout,** writing to his sister:

> Lina,— Last night I "popt" upon a comet . . . between Fomalhout and β Ceti.

More correctly than all these, Hyde wrote it **Pham Al Ḥūt.** Burritt's *Atlas* has the present form Fomalhaut, but his *Planisphere*, **Fomalhani.** It generally, but wrongly, is pronounced **Fomalo,** as though from the French.

The *Harleian Manuscript* of Cicero's *Aratos* has the words *Stella Canopus* at the Fish's mouth, which is either an erroneous title, or another use of the word for any very bright star, as is noted under *α* Argūs,— Canopus.

Among early Arabs Fomalhaut was **Al Difdiʿ al Awwal,** the First Frog; and in its location on the Borgian globe is the word **Ṭhalīm,** the Ostrich, evidently another individual title.

Flammarion says that it was **Hastorang** in Persia 3000 B. C., when near the winter solstice, and a Royal Star, one of the four Guardians of Heaven, sentinels watching over other stars; while about 500 B. C. it was the object of sunrise worship in the temple of Demeter at Eleusis; and still later on, with astrologers, portended eminence, fortune, and power.

The Chinese knew it as **Pi Lo Sze Mun.**

With Achernar and Canopus it made up Dante's **Tre Facelle**; and sixty years ago, Boguslawski thought that it might be the Central Sun of the Universe.

It lies in about 30° 15′ of south declination, and so is the most southerly of all the prominent stars visible in the latitude of New York City, but it is in the zenith of Chile, the Cape of Good Hope, and South Australia. To the uninstructed observer it seems a full 1st-magnitude, perhaps from the absence of near-by stars. It culminates on the 25th of October. As one of the so-called lunar stars it is of importance in navigation, and appears in the *Ephemerides* of all modern sea-going nations.

See calls its color white, and has discovered a 14.8 bluish companion 30″ away, at a position angle of 36°.2.

β, Double, 4.3 and 8.

Al Tizini knew this, instead of α, as **Fum al Ḥūt,**—evidence either of a different figuring of the constellation from that of Ptolemy, which we follow, or of its extension towards the northeast by the Arabian astronomers. This may account for the location of Fomalhaut in Aquarius by some early authors.

With δ and ζ it was the Chinese **Tien Kang,** the Heavenly Rope.

Al Tizini mentioned the stars, now γ, α, and β of Grus, as the Tail, the Bright One, and the Rear One of the Fish,—additional proof that our *lucida* of Piscis Australis was not his *nā'ir* of Al Ḥūt al Janūbiyy.

η, θ, ι, and μ were **Tien Tsien,** Heavenly Cash.

Bayer's lettering extended only to μ, and there seems to be no star lettered κ in the constellation.

★

Piscis Volans, the Flying Fish,

now known by astronomers as **Volans,** is the **Poisson Volant** of the French and the **Fliegende Fisch** of the Germans. The *Rudolphine Tables* have it **Passer,** the Sparrow, and, as such, it is translated **Fe Yu** by the Chinese. This is another of the new southern constellations formally introduced by Bayer, comprising forty-six stars south of Canopus and Miaplacidus,—α and β Argūs.

Julius Schiller included it with Dorado and the Nubecula Major in his biblical figure of **Abel the Just.**

The *lucida* is β, a colored 3.9-magnitude, culminating on the 12th of March.

★

Psalterium Georgii or Georgianum,

sometimes **Harpa Georgii,** was formed in 1781 by the Abbé Maximilian Hell, and named in honor of King George II of England. On the Stieler *Planisphere* it is **Georg's Harfe,** from Bode's **Georgs Harffe.**

It lies between the fore feet of Taurus and the River Eridanus, its stars all very inconspicuous, unless it be the 4½-magnitude o^2 Eridani, which was borrowed for its formation. But the loan has been returned, for Psalterium is not now recognized by astronomers.

★

Pyxis Nautica, the Mariner's Compass.

Pyxis was formed by La Caille from stars in the Mast of Argo, and so associated with the Ship, although there, of course, it is an anachronism.

Baily reannexed it to Argo, since four of its members had been placed by Ptolemy where La Caille found them, so that for a time it fell into disuse; but Gould inserted it in his *Uranometria Argentina* of 1879, with sixty-six stars from 3.8 to 7th magnitudes.

★

Quadrans Muralis, the Mural Quadrant,

between the right foot of Hercules, the left hand of Boötes, and the constellation Draco, was formed by La Lande in 1795, as a souvenir of the instrument with which he and his nephew, Michel Le Français, observed the stars subsequently incorporated under this title into the latter's *Histoire Celeste Française*.

It is the **Mauer Quadrant** of Stieler's *Planisphere*, and the **Quadrante** of the Italians, but is not figured by Argelander or Heis, nor recognized by modern astronomers.

It comes to the meridian with β Ursae Minoris on the 19th of June.

A rich meteor stream, the **Quadrantids**, radiates from this group on the 2d and 3d of January.

★

Reticulum Rhomboidalis, the Rhomboidal Net,

is generally supposed to be of La Caille's formation as a memorial of the reticle which he used in making his celebrated southern observations; but it was first drawn by Isaak Habrecht, of Strassburg, as the **Rhombus**, and so probably only adopted by its reputed inventor. It lies north of Hydrus and the Greater Cloud, containing thirty-four stars from 3.3 to 7th magnitudes.

It is the French **Reticule** or **Rhombe**, the German **Rhomboidische Netz**, and the Italian **Reticolo.**

*

Robur Carolinum, Charles' Oak,

the **Quercia** of Italy and the **Karlseiche** of Germany, was formally published by Halley in 1679 in commemoration of the Royal Oak of his patron, Charles II, in which the king had lain hidden for twenty-four hours after his defeat by Cromwell in the battle of Worcester, on the 3d of September, 1651. This invention secured for Halley his master's degree from Oxford, in 1678, by the king's express command. But La Caille complained that the construction of the figure, from some of the finest stars in the Ship, ruined that already incomplete constellation, "and the Oak ceases to flourish after half a century of possession," although Bode sought to restore it, and Burritt incorporated it into his maps, assigning to it twenty-five stars. Halley's 2d-magnitude α Roburis was changed to β Argūs, now in Carina.

Reeves' list of Chinese star-titles has only one entry under Robur —

Nan Chuen, the Southern Ship, θ, etc., but doubtful, incorrectly laid down.

*

There is in front another Arrow cast
Without a bow; and by it flies the Bird
Nearer the north.

Brown's *Aratos.*

Sagitta, the Arrow,

the French **Flèche,** the German **Pfeil,** and the Italian **Saetta,** lies in the Milky Way, directly north of Aquila and south of Cygnus, pointing eastward; and, although ancient, is insignificant, for it has no star larger than the 4th magnitude, and none that is named.

It has occasionally been drawn as held in the Eagle's talons, for the bird was armor-bearer to Jove; but Eratosthenes described it separately, as Aratos had done, and as it now is on our maps. The common belief that the latter included it with his Αἰετός was based, Grotius said, on an error in the version of Germanicus. And it has been regarded as the traditional weapon that slew the eagle of Jove, or the one shot by Hercules towards the adjacent Stymphalian birds, and still lying between them, whence the title **Herculea;** but Eratosthenes claimed it as the arrow with which Apollo exterminated the Cyclopes; and it sometimes was the **Arrow of Cupid.** The *Hyginus* of 1488 showed it overlying a bow; indeed, Eratosthenes called it Τόξον, a Bow, signifying Arrows in its plural form; Aratos mentioned it as the Feathered Arrow and the Well-shaped Dart, the ἄλλος ὀϊστός of our motto, "another arrow," in distinction from that of Sagittarius. Still, it has often been thought of as the latter's weapon strayed from its owner. Hipparchos and Ptolemy had plain 'Οϊστός.

Latin authors of classical times and since knew it as **Canna, Calamus,** and **Harundo,** all signifying the Reed from which the arrow-shafts were formed; and as **Missile, Jaculum,** and **Telum,** the Weapon, Javelin, and Dart; Telum descending even to Kepler's day. But Sagitta was its common title with all the Romans who mentioned its stars; Cicero characterizing it as *clara* and *fulgens*, which, however, it is not.

Bayer, who ascribed to it the astrological nature of Mars and Venus, picked up several strange names: **Daemon, Feluco,** and **Fossorium,** apparently unintelligible here; **Obelus,** one of the σεμεῖαι, or *notae*, of ancient grammarians, or, possibly, an Obelisk, which it may resemble; **Orfercalim,** cited by Riccioli and Beigel from Albumasar for the Turkish **Otysys Kalem,** a Smooth Arrow; **Temo meridianus,** the Southern Beam; **Vectis,** a Pole; **Virga** and **Virgula jacens,** a Falling Wand. The **Missore** attributed to Cicero is erroneous, and was never used by the latter as a star-name, but for the one who shot the arrow; while the **Musator** of Aben Ezra is either a barbarism for Missore, or may be from the Arabic Saṭar, a Straight Line.

The Hebrews called it **Ḥēṣ** or **Ḥēts;** the Armenians and Persians, **Tigris;** and the Arabians, **Al Sahm,** all meaning an Arrow; this last, given on the Dresden globe, being turned by Chilmead into **Alsoham,** by Riccioli into **Schaham,** and by Piazzi into **Sham.**

In some of the *Alfonsine Tables* appeared **Istusc,** repeated in the *Almagest* of 1515 as **Istiusc,** both probably disfigured forms of ὀϊστός; and the *Alfonsine Tables* of 1521 had **Alahance,** perhaps from the Arabic **Al Ḥams** or **Ḥamsah,** the Five (Stars), its noticeable feature. The same *Almagest* also had **Albanere,** adding *est nun*, all unintelligible except from Scaliger's note:

legendum Alhance, id est Sagitta, hebraicae originis, converso Dages in Nun, ut saepe accidit in Arabismo et Syriasmo.

Schickard wrote it **Alchanzato.**

Sagitta is not noticed in the Reeves list of Chinese asterisms.

Caesius imagined it the **Arrow** shot by Joash at Elisha's command, or one of those sent by Jonathan towards David at the stone Ezel; and Julius Schiller, the **Spear,** or the **Nail, of the Crucifixion.**

Originally only 4° in length, modern astronomy has stretched the constellation to more than 10°; Argelander assigning to it 16 naked-eye stars, and Heis 18. Eratosthenes gave it only 4.

It comes to the meridian on the 1st of September.

None of Sagitta's stars seem to have been named, but its triple ζ is an interesting system. It has long been known as double, but the larger star was discovered by the late Alvan G. Clark to be itself an extremely close double and rapid binary.

The components are of 6, 6, and 9 magnitudes; the two larger 0″.1 apart in 1891, at a position angle of 182°.8. The smallest star is 8″.5 distant. The colors are greenish, white, and blue.

★

. . . glorious in his Cretian Bow,
Centaur follows with an aiming Eye,
His Bow full drawn and ready to let fly.
Creech's *Manilius.*

Sagittarius, the Archer,

the French **Sagittaire,** the Italian **Sagittario,** and the German **Schütze,**— Bayer's **Schütz,**— next to the eastward from Scorpio, was Τοξευτής, the Archer, and Ρύτωρ τόξου, the Bow-stretcher, with Aratos; Τοξευτήρ with other Greeks; and Τοξοτής with Eratosthenes, Hipparchos, Plutarch, and Ptolemy. The Βελοκράτωρ cited by Hyde, though not a lexicon word, probably signifies the Drawer of the Arrow.

These were translated by Lucian and the Romans into our title, although Manilius had **Sagittifer;** Avienus, **Sagittiger;** and Cicero, **Sagittipotens,** a term peculiar to him. His equivalent **Arquitenens,** the ancient form of **Arcitenens,**— reappearing with Ausonius and with Al Bīrūnī in Sachau's translation,— was also used by early classic writers for this constellation; although where the word is seen with Vergil it is for the god Apollo.

Flamsteed's *Atlas* has **Sagittary,** common for centuries before him; Shakespeare calling Othello's house — probably the Arsenal in Venice — the Sagittary,[1] *i. e.* bearing the zodiac sign. The word was early written **Sagitary;** and **Sagittarie** and **Saagittare** in Chaucer's *Astrolabe*, from his Anglo-Norman predecessor, De Thaun. The Anglo-Saxons had **Scytta.**

Columella called it **Crotos,** and Hyginus, **Croton,** the Herdsman; but how these names are applicable does not appear.

Others have been Ἱππότης, On Horseback; **Semivir,** the Half Man; **Taurus** and **Minotaurus,** from his fabled early shape, although now figured in equine form; while Cicero's **Antepes** and **Antepedes** may be for this, or for our Centaur. **Cornipedes,** Horn-Footed, also has been applied to it.

Sometimes the whole was personified by its parts, as with Aratos, where we see Τόξον, the Bow, the **Arcus** of Cicero and Germanicus; and the **Haemonios Arcus** of Ovid; in Egypt, where it is said to have been known as an **Arrow** held in a human hand; and with Ovid again in **Thessalicae Sagitta,** Thessaly being the birthplace of the Centaurs. This induced Longfellow's lines in his *Poets' Calendar* for November:

> With sounding hoofs across the earth I fly,
> A steed Thessalian with a human face.

And it has been **Sagitta arcui applicata;** or plain **Telum** with Capella of Carthage. Bayer cited **Pharetra,** the Quiver, and, recurring to the Bow, **Elkausu** or **Elkusu,** Schickard's **Alkauuso,** from the Arabic **Al Ḳaus.** The translator of Ulug Beg added to its modern name *quem etiam Arcum vocant*, which the *Almagest* of 1515 confirmed in its *et est Arcus.* It was the Persian **Kamān** and **Nimasp;** the Turkish **Yai;** the Syriac **Ḳeshtā** and the Hebrew **Ḳesheth;** Riccioli's **Kertko,** "from the Chaldaeans"; all signifying a Bow, whence some early maps illustrated Sagittarius simply as a **Bow and Arrow.** This was an idea especially prevalent in Asiatic astronomy.

Among the Jews it was the tribal symbol of Ephraim and Manasseh, from Jacob's last words to their father Joseph, "his bow abode in strength."

Novidius claimed it as **Joash, the King of Israel,** shooting arrows out of "the window eastward," at the command of the dying Elisha; but the

[1] In *Troilus and Cressida*, where Agamemnon says:

> The dreadful Sagittary appals our numbers,

the reference is not a stellar one, but to the famous imaginary monster introduced into the armies of the Trojans by the fabling writer Guido delle Colonne, whose work was translated and versified in the *Troye Book* by Lydgate, the great poet of the 15th century.

biblical set generally identified it with **Saint Matthew the Apostle,** although Caesius claimed that Sagittarius was **Ishmael.**

The formation of this constellation on the Euphrates undoubtedly preceded that of the larger figure, the Centaur Chiron; but the first recorded classic figuring was in Eratosthenes' description of it as a **Satyr,** probably derived from the characteristics of the original Centaur, Hea-bani, and it so appeared on the more recent Farnese globe. But Manilius mentioned it, as in our modern style, *mixtus equo*, and with threatening look, very different from the mild aspect of the educated Chiron, the Centaur of the South; while it sometimes is given in later manuscripts and maps with flowing robes; but his crown always appears near his fore feet, and his arrow is always aimed at the Scorpion's heart.

Dupuis said that it was shown in Egypt as an **Ibis** or **Swan;** but the Denderah zodiac has the customary Archer with the face of a lion added, so making it bifaced. Kircher gave its title from the Copts as Πιμάηρε, *Statio amoenitatis.*

The illustrated manuscript partly reproduced in the 47th volume of *Archaeologia* has a centaur-like figure, **Astronochus,** which, perhaps, is our Archer; but the title is of unexplained derivation, unless it be the Star-holder, as Ophiuchus is the Serpent-holder, and Heniochus, the Rein-holder.

It is in this same manuscript that is illustrated a sky group, **Joculator,**[1] usually rendered the "Jester," and representing the Court Fool of mediaeval days; but I find no trace of this elsewhere.

We have already noticed the confusion in the myths and titles of this zodiacal Centaur with those of the southern Centaur, some thinking Sagittarius the Χείρων of the Greeks,—**Chiron** with Hyginus and the Romans; although Eratosthenes and others, as did the modern Ideler, understood this name to refer to the Centaur proper. Ovid's **Centaurus,** however, and Milton's **Centaur** are the zodiac figure, as has been the case with some later poets; James Thomson writing in the Winter of his *Seasons:*

> Now when the chearless empire of the sky
> To Capricorn the Centaur Archer yields.

Early tradition made the earthly Chiron the inventor of the Archer constellation to guide the Argonauts in their expedition to Colchis; although, and about as reasonably, Pliny said that Cleostratos originated it, with Aries, during the 6th or 5th century B. C. As to this we may consider

[1] The Latin word, the equivalent of the early French *Jongleur*, is seen with old Bishop Thomas Percy for a Minstrel, applied to King Alfred.

that, while Cleostratos, possibly, was the first to write on it, certainly none of the Greeks gave it form or title, for we see abundant evidence of its much greater antiquity on the Euphrates.

Cuneiform inscriptions designate Sagittarius as the **Strong One,** the **Giant King of War,** and as the **Illuminator of the Great City,** personifying the archer god of war, **Nĕrgal** or **Nĕrigal,**[1] or under his guardianship, as the Great Lord. This divinity is mentioned in the *Second Book of Kings*, xvii, 30. An inscription, on a fragment of a planisphere, transcribed by Sayce as **Utucagaba,** the Light of the White Face, and by Pinches as **Udgudua,** the Flowing (?) Day, or the Smiting Sun Face, is supposed to be an allusion to this constellation; while on this fragment also appear the words **Nibat Anu,** which accord with an astrolabe of Sennacherib, and were considered by George Smith as the name of its chief star. Another inscribed tablet, although somewhat imperfect, is thought to read **Kakkab Kastu,** the Constellation, or Star, of the Bow,— in Akkadian **Ban,**— indicating one or more of the bow stars of the Archer. This will account for the Τόξον of Aratos and the Arcus of the Latins, Sayce agreeing with this in his rendering **Mulban,** the Star of the Bow. **Pa** and **χut,** Dayspring, also seem to have been titles, the latter because our Archer was a type of the rising sun. Upon some of the boundary stones of Sippara (Sepharvaim of the *Old Testament*), a solar city, Sagittarius "appears sculptured in full glory." In Assyria it always was associated with the ninth month, Kislivu, corresponding to our November–December, with which we have already seen Orion associated. From all the foregoing it would seem safe to assume the Archer to be of Euphratean origin.

India also claimed Sagittarius for its zodiac of 3000 years ago, figured as a **Horse, Horse's head,** or **Horseman,— Açvini,**— a word that appeared in Hindu stellar nomenclature in different parts of the sky. Al Bīrūnī said that the constellation was the Sanskrit **Dhanu,** or **Dhanasu,** the Tamil **Dhamsu,** given by Professor Whitney as **Dhanus**; while we have a very early statement that the stars of the bow and human part of the Archer represented the fan of lions' tails twirled by Mula, the wife of Chandra Gupta, the Sandrokottos of 300 B. C., ruler over the Indian kingdom Maurya and the Gangaridae and Prasii along the Ganges. But in later Indian astronomy it became **Taukshika,** derived from the Greek Τοξότης.

The Hindus located here another of their double *nakshatras*, the 18th and 19th, the **Former** and the **Latter Ashādhā,** Unconquered, which, in the main, were coincident with the *manazil* and *sieu* of the same numbering. These were under the protection of the divinities Āpas, Waters, and Viçve

[1] This may be seen in the Mandaeans' name to-day — Nerig — for the planet Mars.

Devās, the Combined Gods; each being figured as an **Elephant's Tusk,** and both together as a **Bed.**

In ancient Arabia the two small groups of stars now marking the head and the vane of the Archer's arrow were of much note as relics of still earlier asterisms, as well as a lunar station. The westernmost of these,— γ, δ, ϵ, and η,— were **Al Na'ām al Wārid,** the Going Ostriches; and the easternmost,— σ, ζ, ϕ, χ, and τ,—**Al Na'ām al Ṣādirah,** the Returning Ostriches, passing to and from the celestial river, the Milky Way, with the star λ for their **Keeper.** Ideler thought it inexplicable that these non-drinking creatures should be found here in connection with water, and Al Jauharī compared the figures to an **Overturned Chair,** which these stars may represent. But Al Bīrūnī said that Al Zajjāj had a word that signifies the Beam over the mouth of a well to which the pulleys are attached; while another authority said that pasturing **Camels,** or **Cattle,** were intended. There evidently is much uncertainty as to the true reading and signification of this title. All of the foregoing stars, with μ^1 and μ^2, were included in the 18th *manzil,* **Al Na'ām.**

The 19th *manzil* lay in the vacant space from the upper part of the figure toward the horns of the Sea-Goat, and was known as **Al Baldah,** the City, or District, for this region is comparatively untenanted. It was marked by one scarcely distinguishable star, probably π, and was bounded by six others in the form of a Bow, the Arabs' **Ḳaus,** which, however, was not our Bow of Sagittarius. It also was **Al Kilādah,** the Necklace; and **Al Udḥiyy,** the Ostrich's Nest, marked by our τ, ν, ψ, ω, A, and ζ; while the space between this and the preceding mansions was designated by Al Bīrūnī as "the head of Sagittarius and his two locks." In his discussion of this subject, quoting, as he often did, from Arab poets, he compared this 19th *manzil* to "the interstice between the two eyebrows which are not connected with each other,"— a condition described by the word *'Ablād,* somewhat similar to the Baldah generally applied to it.

The 18th *sieu,* **Ki,** a Sieve, anciently **Kit,** was the first of these groups; and the 19th, **Tew, Tow,** or **Nan Tow,** a Ladle or Measure, anciently **Dew,** was the second; both being alluded to in the *She King:*

> In the south is the **Sieve**
> Idly showing its mouth
> . . . . .
> But it is of no use to sift;

the commentator explaining that the two stars widest apart were the **Mouth,** and the two closer together the **Heels;** but he does not give the connection of these with the Sieve. And of the second group:

> In the north is the **Ladle**
> Raising its handle to the west
> . . . . .
> But it lades out no liquor;

so that our **Milk Dipper,** ζ, τ, σ, ϕ, and λ, in the same spot, is not a modern conceit after all. The stars of this Ladle were objects of special worship in China for at least a thousand years before our era; indeed, also were known as a **Temple.**

The whole constellation was the Chinese **Tiger,** Williams giving, as another early name, **Seih Muh,** the Cleft Tree, or Branches cut for fire-wood, and the later name, from the Jesuits, **Jin Ma,** the Man-Horse. A part of it was included with Scorpio, Libra, and some of Virgo's stars in the large zodiacal division the **Azure Dragon.** The astrologers incorporated it with Capricornus in their **Sing Ki.**

Astrologically the constellation was the **House of Jupiter,** that planet having appeared here at the Creation, a manuscript of 1386 calling it the **Schoter** "ye principal howce of Jupit"; although this honor was shared by Aquarius and Leo. Nor did Jupiter monopolize its possession, for it also was the domicile of Diana, one of whose temples was at Stymphalus, the home of the Stymphalian birds. These last, when slain by Hercules, were transferred to the sky as Aquila, Cygnus, and Vultur Cadens, and are all paranatellons of Sagittarius, as has been explained under Aquila. Thus the constellation was known as **Dianae Sidus.** It inclined to fruitfulness, a character assigned to it as far back as the Babylonian inscriptions; and was a fortunate sign, reigning over Arabia Felix, Hungary, Liguria, Moravia, and Spain, and the cities of Avignon, Cologne, and Narbonne; while Manilius said that it ruled Crete, Latium, and Trinacria. Ampelius associated it with the south wind, Auster, and the southwest wind, Africus; Aries and Scorpio being also associated with the latter. Yellow was the color attributed to it, or the peculiar green sanguine; and Arcandum in 1542 wrote that a man born under this sign would be thrice wedded, very fond of vegetables, would become a matchless tailor, and have three special illnesses, the last at eighty years of age. Such was much of the science of his day!

Sagittarius is shown on a coin of Gallienus of about A. D. 260, with the legend *Apollini Conservatori;* and on those of King Stephen emblematic of his having landed in England in 1135 when the sun was here.

La Caille took the star η out of this constellation for the β of his new Telescopium. This was the 25th of Ptolemy's list in the *σφυρόν*, or pastern, which would indicate that with him the feet had a very different situation from that on the present maps.

The symbol of the sign, ♐, shows the arrow with part of the bow.

Sagittarius contains 54 naked-eye stars according to Argelander, and 90 according to Heis, although none is above the 2d magnitude.

The sun passes through the constellation from the 16th of December to the 18th of January, reaching the winter solstice[1] near the stars μ on the 21st of December, but then of course in the sign Capricorn.

A noticeable feature in the heavens lies within the boundaries of Sagittarius, an almost circular black void near the stars γ and δ, showing but one faint telescopic star; and to the east of this empty spot is another of narrow crescent form.

An extraordinarily brilliant *nova* is said to have appeared low down in the constellation in 1011 or 1012, visible for three months. This was recorded in the Chinese annals of Ma Touan Lin.

α, 4.

This is **Rukbat,** but variously written **Rucba, Rucbah, Rukbah,** and **Rucbar,** from Ulug Beg's **Rukbat al Rāmī,** the Archer's Knee; in some early books it is **Al Rāmī,** the Archer himself. The *Standard Dictionary* has **Ruchbar ur Ranich.**

The Euphratean **Nibat Anu,** already alluded to, may be for this, or for some other of the chief components of the constellation; perhaps for ε if, in early days, that star was comparatively as bright as now.

β^1, Double, 3.8 and 8, and β^2, 4.4.

Arkab and **Urkab** are from **Al 'Urḳūb,** translated by Ideler as the Tendon uniting the calf of the leg to the heel, and this coincides with their location in the figure on modern maps, as well as with their Euphratean title **Ur-nergub,** the Sole of the Left Foot; but Al Sufi and the engraver of the Borgian globe assigned these stars to the rear of the horse's body.

Kazwini knew *a* and the two *betas* as **Al Ṣuradain,** the two Surad, desert birds differently described,— by some as "larger than sparrows" and variegated black and white (magpies?); by others as yellow and larger than doves.

γ, 3.1, yellow.

Al Naṣl, the Point, is Al Tizini's word designating this as marking the head of the Arrow; but Hyde cited **Zujj al Nushshābah** of similar meaning.

[1] The solstices are first mentioned by Hesiod in three different passages of his *Works and Days.*

The Borgian globe termed it **Al Wazl,** the Junction, indicating the spot where the arrow, bow, and hand of the Archer meet.

This star, with δ and ε and with β of the Telescope, was the *sieu* **Ki,** but in the worship of China the three were **Feng Shï,** the General of Wind.

δ, Double, 3 and 14.5, orange yellow and bluish.

Kaus Meridionalis, or **Media,** is Arabic and Latin for the Middle (of the) Bow. It marked the junction of the two Ashādhā; and, with γ and ε, was the Akkadian **Sin-nun-tu,** or **Si-nu-nu-tum,** the Swallow.

The companion was 26″ away in 1896, at a position angle of 276°.4.

ε, Double, 2 and 14.3, orange and bluish,

is **Kaus Australis,** the Southern (part of the) Bow.

In Euphratean days it may have been **Nibat Anu.**

ε comes to the meridian on the 8th of August.

The companion is 32″.5 away, at a position angle, in 1896, of 295°.

A comparison of the magnitudes of α, β, γ, δ, and ε in Sagittarius, each one being brighter than the preceding, goes far to show that Bayer was not guided in his star-lettering by any such rule of alphabetical arrangement in order of brilliancy as has been attributed to him.

ζ, Binary, 3.9 and 4.4.

The *Latin Almagest* of 1515 gives this as **Ascella,** *i. e. Axilla,* the Armpit of the figure, still its location on the maps.

The two components have the rapid orbital revolution of 18½ years.

With σ, τ, and ϕ it formed a portion of the 18th *manzil*, **Al Ṅa'ām,** or **Al Na'āïm al Ṣādirah,** and the whole of that *nakshatra;* but the corresponding *sieu* included λ and μ, with ϕ as the determinant.

λ, 3.1, yellow.

Kaus Borealis, the Northern (part of the) Bow, was Al Tizini's **Rā'i al Na'āïm,** the Keeper of the Na'ams, the uncertainty as to the meaning of which has already been noticed; but Kazwini evidently understood by it Ostriches, for in his list it is, with the stars μ, **Al Ṭhalimain,** plainly meaning these desert birds.

With the same stars it may have been the Akkadian **Anu-ni-tum,** said to have been associated with the great goddess **Istar.**

Near λ appeared in A. D. 386 a bright *nova*, the fourth on record; and 7° northeasterly the cluster 25 M. is visible to the naked eye.

μ^1, Triple, 3.5, 9.5, and 10, and μ^2, 5.8,

form a wide naked-eye double on the upper part of the bow, and are named in Akkadia and Arabia with the preceding star.

They mark the point of the winter solstice two thirds of the way southward towards, and in line with, the cluster N. G. C. 6523, 8 M., visible to the naked eye, with other noticeable clusters and nebulae close by. One of these, N. G. C. 6603, 24 M., towards the northeast, is Secchi's **Delle Caustiche,** from its peculiar arrangement of curves, while the celebrated **Trifid Nebula,** N. G. C. 6514, 20 M., lies not far off to the southwest. This was discovered in 1764, and so named from its three dark rifts; it is now specially noted from a suspected recent change in its position with regard to a star in one of these rifts. Spectroscopic observations of this object show considerable discordance in their results.

Brown says that the stars in the bow were the Persian **Gau** and the Sogdian and Khorasmian **Yaugh,** but by these nations were imagined as a **Bull;** the Copts knew them as **Polis,** a Foal.

ν^1 and ν^2, red stars of the 5th magnitude, 12′ apart, and both double, were **'Ain al Rāmī,** the Archer's Eye. Ptolemy catalogued them as a nebulous double star,— νεφελοειδής καί διπλοῦς,— among the first to be so designated.

With ξ and o they were the Chinese **Kien Sing,** a Flag-staff.

π, a 3d-magnitude on the back of the head, was Al Tizini's **Al Baldah,** from the 19th *manzil*, which it marked; Al Achsasi considering it as **Al Nā'ir,** the Bright One, of that lunar station.

σ, 2.3.

This has been identified with **Nunki** of the Euphratean *Tablet of the Thirty Stars*, the Star of the Proclamation of the Sea, this **Sea** being the quarter occupied by Aquarius, Capricornus, Delphinus, Pisces, and Piscis Australis. It is the same space in the sky that Aratos designated as the **Water;** perhaps another proof of the Euphratean origin of much of Greek astronomy.

In India it marked the junction of the *nakshatra* Ashādhā with Abhijit.

It lies on the vane of the arrow at the Archer's hand.

σ, with ζ and π, may have been the Akkadian **Gu-shi-rab-ba,** the Yoke of the Sea.

The 5th-magnitude stars ψ^1, χ^1, and χ^2 were the Chinese asterism **Kow,** the Dog.

ω, 4.8; A, 5; *b*, 4.7; and *c*, 4,

forming a small quadrangle on the hind quarter of the horse, were the *τετράπλευρον* of Ptolemy, which Bayer repeated in the Low Latin **Terebellum,** still often seen for these stars. The *Standard Dictionary* gives it thus, but mentions the components as ω, or a^1, *b* and *e*.

The Chinese knew this little figure as **Kow Kwo,** the Dog's Country.

★

Sceptrum Brandenburgicum, the Brandenburg Sceptre,

was charted in 1688 by Gottfried Kirch, the first astronomer of the Prussian Royal Society of Sciences, and, more than a century thereafter, was published by Bode, who thus rescued it for a time from the oblivion into which, however, it seems to have lapsed again. It contains but four stars, of the 4th and 5th magnitudes, standing in a straight line north and south, below the first bend in the River, west from Lepus.

The Chinese here had an asterism, **Kew Yew,** the nine Scallops of a Pennon, but in this they included μ, ω, and *b* of Eridanus.

There was, in the sky, still another **Sceptre** held by the **Hand of Justice,** introduced by Royer in 1679 in honor of King Louis XIV, in the place of Lacerta; but this also has been forgotten.

★

. . . that cold animal
Which with its tail doth smite amain the nations.

Longfellow's translation of Dante's *Purgatorio.*

Scorpio, or Scorpius, the Scorpion,

was the reputed slayer of the Giant, exalted to the skies and now rising from the horizon as Orion, still in fear of the Scorpion, sinks below it; although the latter itself was in danger,— Sackville writing in his Induction to the *Mirror of Magistrates*, about 1565:

> Whiles Scorpio, dreading Sagittarius' dart
> Whose bow prest bent in flight the string had slipped,
> Down slid into the ocean flood apart.

Classical authors saw in it the monster that caused the disastrous runaway of the steeds of Phoebus Apollo when in the inexperienced hands of Phaëthon.

For some centuries before the Christian era it was the largest of the zodiac figures, forming with the Χηλαὶ, its Claws,— the *prosectae chelae* of Cicero, now our Libra,— a double constellation, as Ovid wrote:

> Porrigit in spatium signorum membra duorum;

and this figuring has been adduced as the strongest proof of Scorpio's great antiquity, from the belief that only six constellations made up the earliest zodiac, of which this extended sign was one.

With the Greeks it universally was Σκορπίος; Aratos, singularly making but slight allusion to it, added Μεγαθηρίον, the Great Beast, changed in the 1720 edition of Bayer to Μελαθυρίον; while another very appropriate term with Aratos was Τέρας μέγα, the Great Sign. This reputed magnitude perhaps was due to the mythological necessity of greater size for the slayer of great Orion, in reference to which that author characterized it as πλειότερος προφανείς, "appearing huger still."

The Latins occasionally wrote the word **Scorpios,** but usually **Scorpius,** or **Scorpio**; while Cicero, Ennius, Manilius, and perhaps Columella gave the kindred African title **Nepa,** or **Nepas,** the first of which the *Alfonsine Tables* copy, as did Manilius the Greek adjective 'Οπισθο-βάμων, Walking Backward. Astronomical writers and commentators, down to comparatively modern times, occasionally mentioned its two divisions under the combined title **Scorpius cum Chelis**; while some representations even showed the Scales in the creature's Claws.

Grotius said that the Barbarians called the Claws **Graffias,** and the Latins, according to Pliny, **Forficulae.**

In early China it was an important part of the figure of the mighty but genial **Azure Dragon** of the East and of spring, in later days the residence of the heavenly **Blue Emperor**; but in the time of Confucius it was **Ta Who,** the Great Fire, a primeval name for its star Antares; and **Shing Kung,** a Divine Temple, was applied to the stars of the tail. As a member of the early zodiac it was the **Hare,** for which, in the 16th century, was substituted, from Jesuit teaching, **Tien He,** the Celestial Scorpion.

Sir William Drummond asserted that in the zodiac which the patriarch Abraham knew it was an **Eagle;** and some commentators have located here the biblical **Chambers of the South,** Scorpio being directly opposite the Pleiades on the sphere, both thought to be mentioned in the same passage of the *Book of Job* with two other opposed constellations, the Bear and Orion; but the original usually is considered a reference to the southern heavens in general. Aben Ezra identified Scorpio, or Antares, with the **K^esil** of the Hebrews; although that people generally considered these stars as a Scorpion, their **'Akrabh,** and, it is claimed, inscribed it on the banners of Dan as the emblem of the tribe whose founder was "a serpent by the way." When thus shown it was as a **crowned Snake** or **Basilisk.** A similar figure appeared for it at one period of Egyptian astronomy; indeed it is thus met with in modern times, for Chatterton, that precocious poet of the last century, plainly wrote of the Scorpion in his line,

> The slimy Serpent swelters in his course;

and long before him Spenser had, in the *Faerie Queen:*

> and now in Ocean deepe
> Orion flying fast from hissing snake,
> His flaming head did hasten for to steepe.

But the Denderah zodiac shows the typical form.

Kircher called the whole constellation Ἰσιας, *Statio Isidis,* the bright Antares having been at one time a symbol of Isis.

The Arabians knew it as **Al 'Akrab,** the Scorpion, from which have degenerated **Alacrab, Alatrab, Alatrap, Hacrab,**— Riccioli's **Aakrab** and **Hacerab;** and similarly it was the Syrians' **Akrevā.** Riccioli gave us **Acrobo** *Chaldaeis,* which may be true, but in this Latin word he probably had reference to the astrologers.

The Persians had a Scorpion in their **Ghezhdūm** or **Kazhdūm,** and the Turks, in their **Koirūghi,** Tailed, and **Uzun Koirūghi,** Long-tailed.

The Akkadians called it **Girtab,** the Seizer, or Stinger, and the **Place where One Bows Down,** titles indicative of the creature's dangerous character; although some early translators of the cuneiform text rendered it the **Double Sword.** With later dwellers on the Euphrates it was the symbol of darkness, showing the decline of the sun's power after the autumnal equinox, then located in it. Always prominent in that astronomy, Jensen thinks that it was formed there 5000 B. C., and pictured much as it now is;

perhaps also in the semi-human form of two Scorpion-men, the early circular Altar, or Lamp, sometimes being shown grasped in the Claws, as the Scales were in illustrations of the 15th century. In Babylonia this calendar sign was identified with the eighth month, Arakh Savna, our October–November.

Early India knew it as **Āli, Viçrika,** or **Vrouchicam,**— in Tamil, **Vrishaman**; but later on Varāha Mihira said **Kaurpya,** and Al Bīrūnī, **Kaurba,** both from the Greek Scorpios. On the Cingalese zodiac it was **Ussika.**

Dante designated it as **Un Secchione,**

> Formed like a bucket that is all ablaze;

and in the *Purgatorio* as **Il Friddo Animal** of our motto, not a mistaken reference to the creature's nature, but to its rising in the cold hours of the dawn when he was gazing upon it. Dante's translator Longfellow has something similar in his own *Poets' Calendar* for October:

> On the frigid Scorpion I ride.

Chaucer wrote of it, in the *Hous of Fame*, as the **Scorpioun**; his Anglo-Norman predecessors, **Escorpiun**; and the Anglo-Saxons, **Throwend.**

Caesius mistakenly considered it one of the **Scorpions of Rehoboam**; but Novidius said that it was

> the scorpion or serpent whereby Pharaoh, King of Egypt, was enforced to let the children of Israel depart out of his country;

of which Hood said "there is no such thing in history." Other Christians of their day changed its figure to that of the **Apostle Bartholomew**; and Weigel, to a **Cardinal's Hat.**

In some popular books of the present day it is the **Kite,** which it as much resembles as it does a Scorpion.

Its symbol is now given as ♏, but in earlier times the sting of the creature was added, perhaps so showing the feet, tail, and dart; but the similarity in their symbols may indicate that there has been some intimate connection, now forgotten, between Scorpio and the formerly adjacent Virgo (♍).

Ampelius assigned to it the care of Africus, the Southwest Wind, a duty which, he said, Aries and Sagittarius shared; and the weather-wise of antiquity thought that its setting exerted a malignant influence, and was accompanied by storms; but the alchemists held it in high regard, for only when the sun was in this sign could the transmutation of iron into gold be performed. Astrologers, on the other hand, although they considered it a fruitful sign, "active and eminent," knew it as the accursed constellation, the baleful source of war and discord, the birthplace of the planet Mars, and so the **House of Mars,** the **Martis Sidus** of Manilius. But this was located in the sting and tail; the claws, as Ζυγός, Jugum, or the Yoke of the Balance, being devoted to Venus, because this goddess united persons under the yoke of matrimony. It was supposed to govern the region of the groin in the human body, and to reign over Judaea, Mauritania, Catalonia, Norway, West Silesia, Upper Batavia, Barbary, Morocco, Valencia, and Messina; the earlier Manilius claiming it as the tutelary sign of Carthage, Libya, Egypt, Sardinia, and other islands of the Italian coast. Brown was its assigned color, and Pliny asserted that the appearance of a comet here portended a plague of reptiles and insects, especially of locusts.

Although nominally in the zodiac, the sun actually occupies but nine days in passing through the two portions that project upwards into Ophiuchus, so far south of the ecliptic is it; indeed, except for these projections, it could not be claimed as a member of the zodiac.

Scorpio is famous as the region of the sky where have appeared many of the brilliant temporary stars, chief among them, perhaps, that of 134 B. C., the first in astronomical annals, and the occasion, Pliny said, of the catalogue of Hipparchos, about 125 B. C. The Chinese *She Ke* confirmed this appearance by its record of "the strange star" in June of that year, in the *sieu* Fang, marked by β, δ, π, ρ, and others in Scorpio. Serviss thinks it conceivable that the strange outbursts of these *novae* in and near Scorpio may have had some effect in causing this constellation to be regarded by the ancients as malign in its influence. But this character may, with at least equal probability, have come from the fiery color of its *lucida*, as well as from the history of the constellation in connection with Orion, and the poisonous attributes of its earthly namesake.

In southern latitudes Scorpio is magnificently seen in its entirety,—nearly 45°,— Gould cataloguing in it 184 naked-eye stars.

Along its northern border, perhaps in Ophiuchus, there was, in very early days, a constellation, the **Fox**, taken from the Egyptian sphere of Petosiris, but we know nothing as to its details.

> . . . capricious Antares
> Flushing and paling in the Southern arch.
>
> Willis' *The Scholar of Thebet Ben Khorat.*

α, Binary, 0.7 and 7, fiery red and emerald green.

Antares, the well-nigh universal title for this splendid star, is transcribed from Ptolemy's ἀντάρης in the *Syntaxis*, and generally thought to be from ἀντί Ἄρης, "similar to," or the "rival of," Mars, in reference to its color, — the Latin *Tetrabiblos* had *Marti comparatur;* or, in the Homeric signification of the words, the "equivalent of Mars," either from the color-resemblance of the star to the latter, or because the astrologers considered the Scorpion the **House** of that planet and that god its guardian. Thus it naturally followed the character of its constellation,—perhaps originated it,— and was always associated with eminence and activity in mankind.

Grotius, however, said that the word signifies a Bat, which, as **Vespertilio,** Sophocles perhaps called it; but Bayer erroneously quoted from Hesychios Ἀνταρτης, a Rebel, and **Tyrannus.** Caesius appropriately styled the constellation **Insidiata,** the Lurking One.

Others say that it was **Antar's Star,**— but they forget Ptolemy,— the celebrated Antar or Antarah who, just previous to the time of Muḥammād, was the mulatto warrior-hero of one of the *Golden Mu'allaḳāt.*[1]

Our word, however, is sometimes written **Antar,** which Beigel said is the Arabic equivalent of "Shone"; but the Latin translator of the 1515 *Almagest* connected it with *Natar*, Rapine, and so possibly explaining the generally unintelligible expression *tendit ad rapinam* applied to Antares in that work and in the *Alfonsine Tables* of 1521; or the expression here may refer to the character of Ἄρης, the god of war. The *Rudolphine Tables* designated it as *rutilans*, Pliny's word for "glowing redly."

The Arabians' **Ḳalb al 'Aḳrab,** the Scorpion's Heart, which probably preceded the Καρδία Σκορπίου and **Cor Scorpii** of Greece and Rome respectively, became, in early English and Continental lists, **Kelbalacrab, Calbalacrab, Calbolacrabi, Calbalatrab,** and **Cabalatrab;** Riccioli having the unique **Alcantub,** although he generally wrote **Kalb Aakrab.** Antares alone constituted the 16th *manzil*, **Al Ḳalb,** the Heart, one of the fortunate stations; but the Chinese included σ and τ, on either side, for their *sieu*, the synonymous **Sin,** anciently **Sam,** σ being the determinant; although Brown says that this Heart refers to that of **Tsing Lung,** the Azure Dragon, one of the four great divisions of their zodiac. They also have a record of a comet 531 B. C., "to the left of **Ta Shin,**" which last Williams identified with Antares; while, as the Fire Star, **Who Sing,** it seems to have been invoked in worship centuries before our era for protection against fire. With some adjacent it was one of the **Ming t'ang,** or Emperor's Council-hall; his sons and courtiers, other stars, standing close by, to whom Antares, as **Ta Who,** announced the principles of his government.

[1] These were the famous seven selected poems of Arabia, said to have been inscribed in letters of gold on silk, or Egyptian linen, and suspended, as their title signifies, in the Ka'bah at Mecca.

The Hindus used α, σ, and τ for their *nakshatra* **Jyesthā,** Oldest, also known as **Rohinī,** Ruddy, from the color of Antares,—Indra, the sky-goddess, being regent of the asterism that was figured as a pendent **Ear Jewel.**

It was one of the four Royal Stars of Persia, 3000 B. C., and probably the Guardian of the Heavens that Dupuis mentioned as **Satevis**; but, as their lunar asterism, it was **Gel,** the Red; the Sogdians changing this to **Maghan sadwis,** the Great One saffron-colored. The Khorasmians called it **Dharind,** the Seizer; and the Copts, **Kharthian,** the Heart.

It pointed out to the Babylonians their 24th ecliptic constellation, **Hurru,** of uncertain meaning, itself being **Urbat** according to an astrolabe discovered in the palace of Sennacherib and interpreted by the late George Smith; Brown, however, assigns this title to stars in Lupus. Other Euphratean names were **Bilu-sha-ziri,** the Lord of the Seed; **Kak-shisa,** the Creator of Prosperity, according to Jensen, although this is generally ascribed to Sirius; and, in the lunar zodiac, **Dar Lugal,** the King, identified with the god of lightning, **Lugal Tudda,** the Lusty King. Naturally the inscriptions make much of it in connection with the planet Mars, their Ul Suru, showing that its Arean association evidently had very early origin; and from them we read **Masu** (?) **Sar,** the Hero and the King, and **Kakkab Bir,** the Vermilion Star. Brown identifies it with the seventh antediluvian king, 'Ευεδώραυχος, or Udda-an-χu, the Day-heaven-bird.

From his Assyrian researches Cheyne translates the 36th verse from the 38th chapter of the *Book of Job*:

> Who hath put wisdom into the Lance-star?
> Or given understanding to the Bow-star?

Jensen referring this **Lance-star** to Antares. Hommel, however, identifies it with Procyon of Canis Minor.

In Egyptian astronomy it represented the goddess **Selkit, Selk-t,** or **Serk-t,** heralding the sunrise through her temples at the autumnal equinox about 3700–3500 B. C., and was the symbol of **Isis** in the pyramid ceremonials. Renouf included it with Arcturus in the immense figure **Menat.**

Penrose mentions the following early Grecian temples as oriented towards the rising or setting of Antares at the vernal equinox: the Heraeum at Argos, in the year 1760, perhaps the oldest temple in the cradle of Greek civilization; the first Erechtheum at Athens, 1070; one at Corinth, 770; an early temple to Apollo at Delphi, rebuilt with this orientation in 630; and one of the same date to Zeus at Aegina; — all of these before our era.

It rises at sunset on the 1st of June, culminating on the 11th of July, and is one of the so-called lunar stars; and some have asserted that it was the first star observed through the telescope in the daytime, although Smyth made this claim for Arcturus. Ptolemy lettered it as of the 2d magnitude, so that in his day it may have been inferior in brilliancy to the now very much fainter β Librae.

Antares belongs to Secchi's third type of suns, which Lockyer says are "in the last visible stage of cooling," and nearly extinct as self-luminous bodies; although this is a theory by no means universally accepted.

The companion is 3″.5 away, and suspected of revolution around its principal; their present position angle is 270°.

A photograph by Barnard in 1895 first showed the vast and intricate **Cloud Nebula** stretching to a great distance around Antares and the star σ. It was here, two or three degrees north of Antares, that was discovered, on the 9th of June, Coddington's comet, *c* of 1898, the third comet made known by the camera.

β, Triple, 2, 10, and 4, pale white, ——, and lilac.

Graffias generally is said to be of unknown derivation; but since Γραψαῖος signifies "Crab," it may be that here lies the origin of the title, for it is well known that the ideas and words for crab and scorpion were almost interchangeable in early days, from the belief that the latter creature was generated from the former.[1] It was thought by Grotius to be a "Barbarian" designation for the Claws of the double constellation; and Bayer said the same, although he used the word for ξ Scorpii in the modern northern claw. In Burritt's *Atlas* of 1835 it appears for ξ of the northern Scale, the ancient northern Claw; but in the edition of 1856 he applied it to our β Scorpii, and in both editions he has a second β at the base of the tail, west of ε. The *Century Dictionary* prints it **Grassias,** probably from erroneously reading the early type for the letter *f*. β is near the junction of the left claw with the body, or in the arch of the Kite bow, 8° or 9° northwest of Antares. In some modern lists it is **Acrab,**— Riccioli's **Aakrab schemali.**

It was included in the 15th *manzil*, **Iklīl al Jabhah,** the Crown of the Forehead, just north of which feature it lies, taking in with this, however, the other stars to δ and π; some authorities occasionally adding ν and ρ. This was one of the fortunate stations, and from this *manzil* title comes the occasional **Iclil.** The Hindus knew the group as their 15th *nakshatra*, **Anurādhā,** Propitious or Successful,— Mitra, the Friend, one of the Adityas, being the presiding divinity; and they figured it as a **Row** or **Ridge,** which

[1] This was held even by the learned Saints Augustine and Basil of the 4th century, and confidently expressed by Saint Isidore in his *Origines et Etymologiae*.

the line of component stars well indicates. The corresponding *sieu*, **Fang,** a Room or House, anciently **Fong,** consisted of β with δ, π, and ρ, although Professor Whitney thought it limited to the determinant π, the faintest of the group and farthest to the south. It shared with Antares the title **Ta Who,** and was the central one of the seven lunar asterisms making up the Azure Dragon, **Tsing Lung.** But individually β seems to have been known as **Tien Sze,** the Four-horse Chariot of Heaven, and was worshiped by all horsemen. It probably also was **Fu Kwang,** the Basket with Handles, and highly regarded as presiding over the rearing of silkworms, and as indicating the commencement of the season of that great industry of China.

Timochares saw β occulted by the moon in the year 295 B. C.; and Hind repeats a statement by Ptolemy, from Chaldaean records, that the planet Mars almost occulted it on the 17th of January, 272 B. C.; Smyth, however, substituted β Librae in this phenomenon and 271 B. C. as the date.

The two largest components are 14″ apart, at a position angle of 25°; the third being 0″.9 from the first, with a position angle of 89°.

Half-way from β to Antares lies the fine cluster N. G. C. 6093, 80 M., on the western edge of a starless opening 4° broad. It was this that called forth Sir William Herschel's exclamation:

> Hier ist wahrhaftig ein Loch im Himmel!

although powerful telescopes reveal in it many minute stars. His son afterwards described forty-nine such spots in various parts of the sky. This cluster, that Sir William thought might perhaps have been formed by stars drawn from that vacancy, "was lit up in 1860 for a short time by the outburst of a temporary star."

γ, 3.25, red,

lies, in Bayer's map, on the tip of the southern claw, and is the same star as Flamsteed's 20 Librae; but Smyth strangely alluded to it as being at the end of the sting and nebulous; and Burritt placed Bayer's letter at the object mentioned by Smyth. Indeed for at least three hundred years there has been disagreement among astronomers as to this star; for although Argelander and Heis follow Bayer, Gould writes:

> Since it appears out of the question that it should ever again be regarded as belonging to Scorpius, I have ventured to designate it by the letter σ [Librae].

Bayer cited for it **Brachium,** the Arm, as from Vergil, but this was erroneous in so far as being a title for this star, the original *brachia* in the *Georgics* simply signifying the "claws" that it marks; Bayer added **Cornu,** the Horn, as from some anonymous writer.

In Arabia it was **Zubān al 'Aḳrab,** the Scorpion's Claw, which has become **Zuban al Kravi, Zuben Acrabi**; and Bayer said **Zuben Hakrabi** and **Zuben el Genubi,** contracted from **Al Zubān al Janūbiyyah,** the Southern Claw. Similar titles also appear for stars in Libra, the early Claws.

In China it was **Chin Chay,** the Camp Carriage.

Brown included it, with others near by in Hydra's tail, in the Akkadian **Entena-mas-luv,** or **Ente-mas-mur,** the Assyrian **Etsen-tsiri,** the Tail-tip.

δ, 2.5.

Dschubba is found in the Whitall *Planisphere*, probably from **Al Jabhah,** the Front, or Forehead, where it lies.

In the *Palermo Catalogue* the title **Iclarkrav** is applied to a star whose assigned position for the year 1800 would indicate our δ. If this be the case, it may have been a specially coined word from the Arabs' **Iklīl al 'Aḳrab,** the Crown of the Scorpion; and this conjecture would seem justified by our previous experience of that catalogue's star nomenclature as seen in its remarkable efforts with α and β Delphini. Riccioli had **Aakràb genubi.**

δ was of importance in early times, for with β and π, on either side in a bending line, it is claimed for the Euphratean **Gis-gan-gu-sur,** the Light of the Hero, or the Tree of the Garden of Light, "placed in the midst of the abyss," and so reminding us of that other tree, the Tree of Life, in the midst of the Garden of Eden. It was selected by the Babylonian astronomers, with β, to point out their 23d ecliptic constellation, which Epping calls **Qablu (und qābu) sha rīshu aqrabi,** the Middle of the Head of the Scorpion. The earliest record that we have of the planet Mercury is in connection with these same two stars seen from that country 265 B. C. In the lunar zodiac δ, β, and π were the Persian **Nūr,** Bright; the Sogdian and Khorasmian **Bighanwand,** Clawless; and the Coptic **Stephani,** the Crown.

In China the 2d-magnitude ε, with μ, ζ, η, θ, ι, κ, v, and λ, formed the 17th *sieu*, **Wei,** the Tail, anciently known as **Mi** and as **Vi,** μ being the determinant; but, although this Tail coincided with that part of our Scorpion, Brown thinks that reference is rather made to the tail of the Azure Dragon, one of the quadripartite divisions of the Chinese zodiac which lay here.

θ, a 2d-magnitude red star, was the Euphratean **Sargas,** lying in the Milky Way just south of λ and v, with which it formed one of the seven pairs of Twin Stars; as such it was **Ma-a-su.** And it may have been, with ι, κ, λ, and v, the **Girtab** of the lunar zodiac of that valley, the **Vanant** of

Persia and **Vanand** of Sogdiana, all meaning the "Seizer," "Smiter," or "Stinger"; but the Persian and Sogdian words generally are used for our Regulus. In Khorasmia these stars were **Khachman,** the Curved. θ has a 14th-magnitude greenish companion that may be in revolution around it, 6″.77 away in 1897, at a position angle of 316°.9. See writes of this:

a magnificent system of surpassing interest; one of the most difficult of known double stars.

λ, 1.7.

Shaula probably is from **Al Shaulah,** the Sting, where it lies; but, according to Al Bīrūnī, from **Mushālah,** Raised, referring to the position of the sting ready to strike. These words have been confused with the names for the adjoining v, and in the course of time corrupted to **Shauka, Alascha, Mosclek,** and **Shomlek;** Chilmead writing of these last:

It is also called **Schomlek,** which Scaliger thinkes is read by transposition of the letters for Mosclek, which signifieth the bending of the taile.

Naturally it was an unlucky star with astrologers.

λ and v were the 17th *manzil*, **Al Shaulah,** and the *nakshatra* **Vicritāu,** the Two Releasers, perhaps from the Vedic opinion that they brought relief from lingering disease.

Some Hindu authorities, taking in all the stars from ε to v, called the whole **Mūlā,** the Root, with the divine Nirrity, Calamity, as regent of the asterism, which was represented as a **Lion's Tail;** this title appearing also for stars of Sagittarius. In Coptic Egypt λ and v were **Minamref,** the Sting; and, on the Euphrates, **Sarur.**

An imaginary line extended from v through Shaula serves to point out the near-by clusters 6 M. and N. G. C. 6475, 7 M., visible together in the field of an opera-glass. These probably were the ancient termination of the sting to which Smyth alluded in his comments on λ and v, although he is not quite clear about the matter; they certainly were the νεφελοειδής of Ptolemy, among his ἀμόρφωτοι of Σκορπίος; and *Girus ille nebulosus* in the *Latin Almagest* of 1551. Ulug Beg's translator had *Stella nebulosa quae sequitur aculeum Scorpionis*,— **Tālī' al Shaulah,** That which follows the Sting.

In the legends of the Polynesian Islanders, notably those of the Hervey group, the stars in the Scorpion, from the two lettered μ to λ and v, were the **Fish-hook of Maui,** with which that god drew up from the depths the great island Tongareva; and the names and legend that Ellis, in his *Polynesian Researches*, applied to Castor and Pollux in Gemini, the Reverend

Mr. W. W. Gill asserts, in his *Myths and Songs of the South Pacific*, belong here, and are the favorites among the story-tellers of the Hervey Islands. They make the star μ^1 a little girl, **Piri-ere-ua**, the Inseparable, with her smaller brother, μ^2, fleeing from home to the sky when ill treated by their parents, the stars λ and v, who followed them and are still in pursuit.

This μ^1 has recently been discovered to be a spectroscopic binary, with a period of about 35 hours. It is a 3.3-magnitude, and of Secchi's 1st class.

μ^2 is of 3.7 magnitude.

ν, Quadruple, 4, 5, 7.2, and 8.3,

is **Jabbah** in the *Century Cyclopedia*, perhaps from its being one of the *manzil* **Iklīl al Jabhah.**

It lies 2° east of β, and is another **Double Double** like ε Lyrae, although less readily resolved, the larger pair being only 0″.89 apart, and the smaller about 1″.9. Espin-Webb says: "Probably a quadruple system." Burnham finds it surrounded by a remarkable winglike nebula some 2° in diameter.

ξ, Triple, 5, 5.2, and 7.5, bright white, pale yellow, and gray.

Bayer wrote that the "Barbarians" called this **Graffias,** a title that Burritt assigned in 1835 to ξ of Libra; but he transferred this in his *Atlas* of 1856 to β Scorpii, 8½° to the north, leaving this star nameless. On the Heis map ξ is near the tip of the northern claw, so close to the northern scale that Flamsteed made it the 51 Librae of his catalogue.

The components are 1″.4 and 7″.3 apart, and may form a triple system with a possible period of about 105 years.

σ, Double, 3 and 9, creamy white, and τ, 2.9,

were **Al Niyāṭ**, the *Praecordia*, or Outworks of the Heart, on either side of, and, as it were, protecting, Antares, the Heart of the Scorpion. Knobel, in his translation of Al Achsasi's work, explains the word as "the vein which suspends the heart"!

υ, 2.8.

Lesath, or **Lesuth,** is from **Al Las'ah,** the Sting, which, with λ, it marks; yet Smyth, who treats of these two stars at considerable length, says that the word is formed by Scaliger's conjecture from Alascha, which is a corruption of *al-shaúlah*. Lesath, therefore, is not a term used by the Arabs, who designate all these bumps, which form the tail, *Al-fikrah*, vertebrated twirls; they are formed by ε, μ, ζ, η, θ, ι, κ, λ, and υ, and it is supposed that the sting, *punctura scorpionis*, was formerly carried to the following star, γ, marked nebulous by Ptolemy.

But this γ is surely wrong; that letter really applying to a star in the right claw very far to the west of the sting,— as far as the make-up of the creature will allow. Still Burritt located it as Smyth did. Al Bīrūnī wrote that λ and υ were in the **H·arazāh,** the Joints of the Vertebrae. Riccioli mentioned υ as **Lesath** *vel potius* **Lessaa Elaakrab Morsum Scorp.** *vel* **Denneb Elaakrab**; and Bayer, **Leschat** *recté* **Lesath, Moschleck, Alascha,** which we have seen for λ; but the proximity of these stars renders this duplication not unnatural.

The Chinese knew them as **Keen Pi,** the Two Parts of a Lock.

Ideler thought υ the γ of Telescopium, but this does not agree with Bode's drawing of the latter.

ω^1, 4.1, and ω^2, 4.6, red.

The Arabians called these **Jabhat al ʽAḳrab,** the Forehead, or Front, of the Scorpion; and the Chinese, **Kow Kin,** a Hook and Latch.

They are an interesting naked-eye pair, 14½′ apart, lying just south of β; but Bayer mentions and shows only a single star.

★

Sculptor,

as it is now generally known, was formed by La Caille from stars between Cetus and Phoenix. He called it **l'Atelier du Sculpteur,** the Sculptor's Studio or Workshop, which Burritt and others have changed to **Officina Sculptoria,** or occasionally **Apparatus Sculptoris.** The Italians say **Scultore,** and the Germans **Bildhauerwerkstätte,**—Bode's **Bildhauer Werkstadt.**

It is an inconspicuous figure, but contains the intensely scarlet variable R, one of the most brilliantly colored stars in the heavens, with a period of variability from 5.8 to about 7.7 in 207 days.

The constellation culminates with the bright star of the Phoenix on the 17th of November, and is visible from the latitude of New York City.

Gould catalogues 131 stars, from 4.2 to 7th magnitudes.

Scutum Sobiescianum, Sobieski's Shield,

the French **Écu,** or **Bouchiere, de Sobieski,** the Italian **Scudo di Sobieski,** and the German **Sobieskischer Schild,** was formed by Hevelius from the seven unfigured 4th-magnitude stars in the Milky Way west of the feet of Antinoüs, between the tail of the Serpent and the head of Sagittarius. Heis increased this number to eleven. The title is often seen as **Scutum Sobieskii** or **Sobiesii,** sometimes as **Clypeus Sobieskii,** more correctly written **Clipeus;** but our astronomers follow Flamsteed in his plain **Scutum.**

It is pictured as the **Coat of Arms** of the third John Sobieski, king of Poland, who so distinguished himself in the defensive wars of his native land, as well as in his successful resistance of the Turks in their march on Vienna when turned back at the Kalenberg on the 12th of September, 1683. It was just after this, when he had made his triumphal entry into the city, that at the cathedral service of thanksgiving the officiating priest read the passage:

> There was a man sent from God, whose name was John.

Seven years subsequently this new constellation was named for him by Hewel, with a glowing tribute to his merit and heroic deeds; the sign of the Cross for which he fought being emblazoned on his Shield as we have it to-day. Some identify this **Cross,** however, with that of the fighting Franciscan friar, Saint John Capistrano, famous at Belgrade in 1456, and now honored by a colossal statue on the exterior of the Vienna cathedral. The four stars on the border of the Shield are for the four sons of the king.

Although Scutum is a recent creation with us, it has long been known in China as **Tien Pien,** the Heavenly Casque, but in this are included some components of Antinoüs.

It comes to the meridian about the 10th of August.

It has no named star,— indeed the figure itself does not appear upon some modern maps,— and is chiefly noticeable from the peculiar brightness of the surrounding Galaxy; for within its boundaries, in five square degrees of space, Sir William Herschel estimated that there are 331,000 stars; and it is very rich in nebulae. Of these the notable cluster N. G. C. 6705, 11 M., discovered by Kirch in 1681 and likened by Smyth to a flight of wild ducks, lies on the dexter chief of the Shield. This is just visible to the naked eye, and Sir John Herschel called it "a glorious object."

Just below the constellation is the celebrated **Horseshoe,** or **Ω, Nebula,** N. G. C. 6618, 17 M., one of the most interesting in the heavens, although in small glasses it bears more resemblance to a swan seen on the water, whence comes another title, the **Swan Nebula.**

*

> The starry Serpent . . .
> Southward winding from the Northern Wain,
> Shoots to remoter spheres its glittering train.
>
> *Statius.*

Serpens,

le Serpent in France, **il Serpente** in Italy, and **die Schlange** in Germany, probably is very ancient, and always has been shown as grasped by the hands of Ophiuchus at its pair of stars δ, ϵ, and at ν, τ Ophiuchi. The head is marked by the noticeable group ι, κ, γ, ϕ, υ, ρ, and the eight little stars all lettered τ, and consecutively numbered, 10° south from the Crown and 20° due east from Arcturus; the figure line thence winding southwards 15° to Libra, and turning to the southeast and northeast along the western edge of the Milky Way, terminating at its star θ, 8° south of the tail of the Eagle and west of that constellation's δ.

Of the four stellar Snakes this preëminently is the **Serpent,** its stars originally being combined with those of Ophiuchus, although Manilius wrote

> Serpentem Graiis Ophiuchus nomine dictus dividit;

but it now is catalogued separately, and occasionally divided into **Caput** and **Cauda** on either side of the Serpent-holder.

The Greeks knew it as Ὄφις Ὀφιούχου, or simply as Ὄφις, and familiarly as Ἑρπετόν and Ἐγχέλυς, respectively the Serpent and the Eel; the Latins, occasionally as **Anguilla, Anguis,** and **Coluber;** but universally as **Serpens,** often qualified as the **Serpent of Aesculapius, Caesius, Glaucus, Laocoön,** and **of Ophiuchus;** and as **Serpens Herculeus, Lernaeus,** and **Sagarinus.** The 1515 *Almagest* and the *Alfonsine Tables* of 1521 had **Serpens Alangue,** thus combining their corrupted Latin with their equally corrupted Arabic, as often is the case with those works. It also was **Draco Lesbius** and **Tiberinus,** and, perhaps, Ovid's and Vergil's **Lucidus Anguis.**

In the astronomy of Arabia it was **Al Ḥayyah,** the Snake,— Chilmead's **Alhafa;** but before that country was influenced by Greece there was a very different constellation here, **Al Rauḍah,** the Pasture; the stars β and γ, with γ and β Herculis, forming the **Nasak Shāmiyy,** the Northern Boundary; while δ, α, and ϵ Serpentis, with δ, ϵ, ζ, and η Ophiuchi, were the **Nasak Yamāniyy,** the Southern Boundary. The enclosed sheep were shown by the stars now in the **Club of Hercules,** guarded on the west by the **Shepherd and his Dog,** the stars α in Ophiuchus and Hercules.

To the Hebrews, as to most nations, this was a Serpent from the earliest times, and, Renan said, may have been the one referred to in the *Book of Job*, xxvi, 13; but Delitzsch, who renders the original words as the "Fugitive Dragon," and others with him, consider our Draco to be the constellation intended, as probably more ancient and widely known from its ever visible circumpolar position. The biblical school made it the serpent seducer of Eve, while in our day imaginative observers find another heavenly **Cross** in the stars of the head, one that belongs to Saint Andrew or Saint Patrick.

Serpens shared with Ophiuchus the Euphratean title of **Nu-tsir-da,** the Image of the Serpent; and is supposed to have been one of the representatives of divinity to the Ophites, the Hivites of Old Testament times.

The comparatively void space between ν and ϵ was the Chinese **Tien Shï Yuen,** the Enclosure of the Heavenly Market.

Argelander counts 51 stars within the constellation boundaries, and Heis 82. In its cluster N. G. C. 5904, 5 M., Bailey has discovered 85 variables.

α, 3, pale yellow.

Unuk[1] **al Hay,**— or **Unukalhai,**— is from **'Unk al Hayyah,** the Neck of the Snake, the later Arabic name for this star; the **Uunk al Hay** of the *Standard Dictionary* is erroneous,— a type error perhaps for Unuk. It was also **Alioth, Alyah,** and **Alyat,** often considered as terms for the broad and fat tail of the Eastern sheep that may have been at some early day figured here in the Orientals' sky; but we know nothing of this, and these are not Arabic words, so that their origin in Al Hayyah of the constellation is more probable. Smyth somewhat indefinitely states that **Alangue** and **Ras Alaugue** appear in the *Alfonsine Tables*, presumably for this star.

α may have been the *lucidus anguis* of Ovid and Vergil, as it certainly was the **Cor Serpentis** of astrology.

With λ it was known as **Shuh,** the title of certain territory in China; and Edkins rather unsatisfactorily writes:

> The twenty-two stars in the Serpent are named after the states into which China was formerly divided.

[1] Although errors in the adoption of Arabic star-names into our popular lists are common, indeed almost universal, this **Unuk** is peculiarly wrong, for 'Unūk is the plural of 'Unk.

As their radiant point it has given name to the **Alpha Serpentids** of the 15th of February.

It is of Secchi's 2d type of spectra, and receding from us about 14 miles a second. It culminates on the 28th of July; and a 12th-magnitude blue companion is 58″ distant.

β, Double, 3 and 9.2, both pale blue.

This was **Chow** with the Chinese, the title of one of their imperial dynasties; but it does not seem to have been named by any other nation. The components are 30″.6 apart, at a position angle of 265°.

Near it is the radiant point of the **Beta Serpentids**, a minor stream of meteors visible from the 18th to the 20th of April.

γ, a 4th-magnitude, was **Ching**, and δ, **Tsin**, in Chinese lists.

This last, a white and bluish 4th- and 5th-magnitude double, was first noted as a binary by Sir William Herschel. The components are 3″.6 apart, with a position angle at present of about 185°.

ε, of 3.7 magnitude, was **Pa**, the name of a certain territory in China.

ζ, a 4½-magnitude, and η were **Tung Hae**, the heavenly Eastern Sea of that country; the latter star being a golden-yellow 3.3-magnitude with a small, pale lilac companion.

θ, Binary and perhaps slightly variable, 4 and 4.5, pale yellow and gold yellow.

Alya, of the *Palermo Catalogue* and others (sometimes, but erroneously, **Alga**), probably is from the same source as the similar title of the *lucida*.

The Chinese knew it as **Sen**, one of their districts.

It is the terminal star in the Serpent; and lies southwest of Aquila, in a comparatively starless region between the two branches of the Milky Way. The components are 21″ apart, at a position angle of 104°.

ξ, 3.7, on the lower part of the body, was **Nan Hae**, the Southern Sea; and υ, 5.3, on the back of the head, was **Cha Sze**, a Carriage-shop.

*

Sextans Uraniae

was formed by Hevelius to commemorate the **Sextant** so successfully used by him in stellar measurements at Dantzig from 1658 to 1679. The original figure comprised the twelve unclaimed stars between Leo and Hydra, west of Crater; and Smyth writes:

With more zeal than taste, he fixed the machine upon the Serpent's back, under the plea that the said Sextant was not in the most convenient situation, but that he placed it between Leo and Hydra because these animals were of a fiery nature, to speak with astrologers, and formed a sort of commemoration of the destruction of his instruments when his house at Dantzic was burnt in September, 1679; or, as he expresses it, when Vulcan overcame Urania.

Its inventor's great name has kept it in the sky till now, and it is still generally recognized by astronomers as **Sextans.**

Here, on the frame of the instrument, 9° south by east from the star Regulus, De Rheita thought that he had found a representation of the **Sudarium Veronicae,** the sacred handkerchief of Saint Veronica. Commenting upon this discovery, Sir John Herschel said that "many strange things were seen among the stars before the use of powerful telescopes became common."

The *lucida*, a 4th-magnitude, is 12° south from Regulus.

One of the Sextant stars, which Reeves gives as *q*, Bode's 2306, a 6th-magnitude, was the Chinese **Tien Seang,** the Heavenly Minister of State.

Argelander catalogues 17 naked-eye stars, and Heis 48.

★

Solarium, the Sun-dial,

lies east from Horologium, between the head of Hydrus and the tail of Dorado; but I can nowhere find anything as to the origin of the figure, although Miss Bouvier included it in her list, and Burritt drew it on his *Atlas*. It seems to be ignored by our astronomers, its stars being combined with those of the neighboring constellations.

✶

Tarandus vel Rangifer, the Reindeer,

a small and faint asterism between Cassiopeia and Camelopardalis, was formed by Pierre Charles Le Monnier, under the title **Renne,** as a memento of his stay in Lapland when engaged in geodetic work in 1736. The Germans know it as **Rennthier,** and Bode so inserted it in *Die Gestirne.*

It has seldom been figured, and now is never mentioned.

★

> Ere the heels of flying Capricorn
> Have touched the western mountain's darkening rim,
> I mark, stern Taurus, through the twilight gray,
> The glinting of thy horn,
> And sullen front, uprising large and dim,
> Bent to the starry Hunter's sword at bay.
>
> Bayard Taylor's *Hymn to Taurus.*

Taurus, the Bull,

le Taureau of France, **il Toro** of Italy, and **der Stier** of Germany, everywhere was one of the earliest and most noted constellations, perhaps the first established, because it marked the vernal equinox from about 4000 to 1700 B. C., in the golden age of archaic astronomy; in all ancient zodiacs preserved to us it began the year. It is to this that Vergil alluded in the much quoted lines from the 1st *Georgic*, which May rendered:

> When with his golden hornes bright Taurus opes,
> The yeare; and downward the crosse Dog-starre stoopes;

and the poet's description well agrees with mythology's idea of Europa's bull, for he always was thus described, and snowy white in color. This descended to Chaucer's **Whyte Bole,** in *Troilus and Criseyde*, from the *candidus Taurus* of the original. The *averso*, "crosse," in the second line of this passage:

> . . . averso cedens Canis occidit astro,—

adversus with Ovid, and *aversaque Tauri sidera* with Manilius,— generally has, however, been translated "backward," as a supposed allusion to the constellation rising in reversed position; but quite as probably it is from the mutual hostility of the earthly animals.

Ταῦρος, its universal title in Greek literature, was more specifically given as Τομή and Προτομή, the Bust, the Bull generally being drawn with only his forward parts, Cicero following this in his *prosecto corpore Taurus*, and Ovid in his

> Pars prior apparet Posteriora latent,

which the mythologists accounted for by saying that, as Taurus personified the animal that swam away with Europa, his flanks were immersed in the waves. This association with Europa led to the constellation titles **Portitor,** or **Proditor, Europae; Agenoreus,** used by Ovid, referring to her father; and **Tyrius,** by Martial, to her country. This incomplete figuring of Taurus induced the frequent designation, in early catalogues, **Sectio Tauri,** which the Arabians adopted, dividing the figure at the star *o*, but retaining the hind quarters as a sub-constellation, **Al Ḣaṭṭ,** recognized by Ulug Beg, and, in its translation, as **Sectio,** by Tycho, the line being marked by *o*, *ξ*, *s*, and *f*. Ancient drawings generally showed the figure as we do, although some gave the entire shape, Pliny and Vitruvius writing of the Pleiades as *cauda Tauri*, so implying a complete animal.

Aratos qualified his Ταῦρος by πεπτηώς, "crouching"; Cicero, by *inflexoque genu*, "on bended knee"; Manilius, by *nixus*, "striving"; and further, in Creech's translation:

> The mighty Bull is lame; His leg turns under;

and

> Taurus bends as wearied by the Plough;

this crouching position also being shown in almost all Euphratean figuring, as are the horns in immense proportions. The last descended to Aratos, who styled the constellation Κεραόν, and is seen in the **Cornus** of Ovid.

The latter author wrote again of the sky figure:

> Vacca sit an taurus non est cognoscere promptum,

from the conflicting legends of Io and Europa; for some of the poets, changing the sex, had called these stars **Io,** the Wanderer, another object of Jupiter's attentions, whom Juno's jealousy had changed to a cow. They also varied the title by the equivalent **Juvenca Inachia** and **Inachis,** from her father Inachus. She afterwards became the ancestress of our Cepheus and Andromeda. Still another version, from the myth of early spring, made Taurus **Amasius Pasiphaes,** the Lover of Pasiphaë; but La Lande's **Chironis Filia** seems unintelligible.

The story that the Bull was one of the two with brazen feet tamed by the Argonaut Jason, perhaps, has deeper astronomical meaning, for Thompson writes:

> The sign Taurus may have been the Cretan Bull; and a transit through that sign may have been the celestial Βόσπορος of the Argonautic voyage.

It bore synonymous titles in various languages: in Arabia, **Al Thaur,** which degenerated to **El Taur, Altor, Ataur, Altauro,** by Schickard; **Tur,** by Riccioli; and even now **Taur,** in our *Standard Dictionary.* In Syria it was **Taurā**; in Persia, **Tora, Ghav,** or **Gāu**; in Turkey, **Ughuz**; and in Judaea, **Shōr,** although also known there as **Rᵉ'ēm,** a word that zoölogically appears in the Authorized Version of our *Bible* as the "unicorn," but better in the Revised as the "wild ox."

Latin writers mentioned it under its present name, to which Germanicus added **Bos** from the country people, although it also was **Princeps armenti,** the Leader of the herd, and **Bubulcus,** the peasant Driver of the Oxen, a title more usual and more correct, however, for Boötes; La Lande quoting it as **Bubulum Caput.**

Manilius characterized Taurus as *dives puellis,* "rich in maidens," referring to its seven Hyades and seven Pleiades, all daughters of Atlas, and the chief attraction in a constellation not otherwise specially noticeable. An early Grecian gem shows three nude figures, hand in hand, standing on the head of the Bull, one pointing to seven stars in line over the back, which Landseer referred to the Hyades; but as six of the stars are strongly cut, and one but faintly so, and the letter *P* is superscribed, Doctor Charles Anthon is undoubtedly correct in claiming them for the Pleiades, and the three figures for the Graces, or Charites. These were originally the Vedic Harits, associated with the sun, stars, and seasons; and this astronomical character adhered to the Charites, for their symbols in their ancient temple in Boeotia were stones reputed to have fallen from the sky.

A coin, struck 43 B. C. by P. Clodius Turrinus, bore the Pleiades in evident allusion to the consular surname; while earlier still — 312–64 B. C.— the Seleucidae of Syria placed the humped bull in a position of attack on their coins as symbol of this constellation. The gold *muhrs,* or mohurs, and the zodiacal rupees, attributed to Jehangir Shah, of 1618, show Taurus as a complete, although spiritless, creature, with the gibbous hump peculiar to Indian cattle. This is always drawn in the Euphratean stellar figure, and was described as Κυρτός by an early commentator on the *Syntaxis.* But the silver rupees of the same monarch have the customary half animal in bold, butting attitude exactly as it is now, and as it was described by Manilius in his *flexus* and *nisus,* and by Lucan in his *curvatus.* A very ancient coin of Samos, perhaps of the 6th century before Christ, bears a half-kneeling, sectional figure of a bull, with a lion's head on the obverse; and one of Thurii, in Lucania, of the 4th century B. C., has the complete animal in position to charge. Another of this same city bears the Bull with a bird on its back, perhaps symbolizing the Peleiad Doves.

Plutarch wrote, in his *De Facie Orbe Lunae*, that when the planet Saturn was in Taurus, *i. e.* every thirty years, there took place the legendary migration from the external continent beyond the Cronian, or Saturnian, Sea to the Homeric Orgyia, or to one of its sister islands.

South American savage tribes held ideas similar to our own about Taurus, for La Condamine, the celebrated French scientist of the last century, said that the Amazon Indians saw in the > of the Hyades the head of a bull; while Goguet more definitely stated that, at the time of the discovery of that river, by Yañez Pinzon in 1500, the natives along its banks called the group **Tapüra Rayoaba,** the Jaw of an Ox; and even in civilized countries it has been fancifully thought that its shape, with the horns extending to β and ζ, gave title to the constellation.

In China it formed part of the **White Tiger,** and also was known as **Ta Leang,** the Great Bridge, from a very early designation of the Hyades and Pleiades; but as a zodiac constellation it was the **Cock,** or **Hen,** recalling the modern Hen and Chickens of the Pleiades. When the Jesuits introduced their Western nomenclature it became **Kin Neu,** the Golden Ox.

After Egyptian worship of the bull-god **Osiris** had spread to other Mediterranean countries, our Taurus naturally became his sky representative, as also of his wife and sister **Isis,** and even assumed her name; but the starry **Bull** of the Nile country was not ours, at least till late in that astronomy. Still this constellation is said to have begun the zodiacal series on the walls of a sepulchral chamber in the Ramesseum; and, whatever may have been its title, its stars certainly were made much of throughout all Egyptian history and religion, not only from its then containing the vernal equinox, but from the belief that the human race was created when the sun was here. In Coptic Egypt it, or the Pleiades, was Ὥριας, the Good Season, Kircher's *Statio Hori*, although it was better known as **Apis,** the modern form of the ancient **Hapi,** whose worship as god of the Nile may have preceded even the building of the pyramids.

As first in the early Hebrew zodiac it was designated by **A** or **Āleph,** the first letter of that alphabet, coincidently a crude figure of the Bull's face and horns; some of the Targums assigning it to the tribes of Manasseh and Ephraim, from Moses' allusion to their father Joseph in the 33d chapter of *Deuteronomy*,— "his horns are the horns of the wild ox"; but others said that it appeared only on the banners of Ephraim; or referred it to Simeon and Levi jointly, from Jacob's death-bed description of their character,—"they houghed an ox"; or to Issachar, the "strong ass" which shared with the ox the burdens of toil and carriage.

It has been associated with the animal that Adam first offered in sacrifice, or with the later victims in the Jewish temple; and the Christian school of which Novidius was spokesman recognized in Taurus the **Ox** that stood with the ass by the manger at the blessed Nativity. Hood said of this: "But whether there were any ox there or no, I know not how he will prove it." In the "apostolic zodiac" it became **Saint Andrew**; but Caesius said that long before him it was **Joseph the Patriarch.**

Representations of the **Mithraic Bull**, on gems of four or five centuries before Christ, reproduced in Lajarde's *Culte de Mithra*, prove that Taurus was at that time still prominent in Persico-Babylonian astronomy as well as in its religion. One of these representations, showing the front of the Bull's head, may very well be the origin of our present symbol of this sign, ♉, although it also has been considered a combination of the full and crescent moon, associated with this constellation as a nocturnal sign; and some assert that Taurus was drawn as a demi-bull from his representing the crescent moon. This appears on a Babylonian cylinder seal of about 2150 B.C. Still earlier in Akkadia it seems to have been known as the **Bull of Light,** its double title, **Te Te,** referring to its two groups, the Hyades and Pleiades, which in every age have been of so much interest to mankind; and a cylinder has **Gut-an-na,** the Heavenly Bull, mentioned in connection with rain, so recalling the rainy Hyades. Epping says that it was the Babylonians' **Shūr,** and that four of their ecliptic constellations were marked by its stars; while Jensen mentions it as symbolic of **Mardūk,** the Spring Sun, son of Ia, whose worship seems to have been general 2200 B.C.,—probably long before,—and that it was originally complete and extended as far as the Fish of Ia, the northern of the two Fishes. This high authority carries the formation of Taurus still farther back, to about 5000 B.C., even before the equinox lay here. The name of the second of the antediluvian Babylonian kings, the mythical Alaparos, seems connected with this constellation or with the *lucida*, Aldebaran; and its stars certainly were associated with the second month of the Assyrian year, A-aru, the Directing Bull, our April–May, as they were in the *Epic of Creation* with the conquest of the Centaur.

Taurus was the Cingalese **Urusaba,** the early Hindu **Vrisha, Vrishan,** or **Vrouchabam,**—in the Tamil tongue, **Rishabam**; but subsequently Varāha Mihira gave it as **Taouri,** his rendering of Taurus, and Al Bīrūnī, in his *India*, as **Tāmbiru.**

With the Druids it was an important object of worship, their great religious festival, the Tauric, being held when the sun entered its boundaries; and it has, perhaps fancifully, been claimed that the tors of England were the old sites of their Taurine cult, as our cross-buns are the present representatives of the early bull cakes with the same stellar association, tracing back through the ages to Egypt and Phoenicia. And the Scotch have a story that on New Year's eve the **Candlemas Bull** is seen rising in the twilight and sailing across the sky,—a matter-of-fact statement, after all.

The Anglo-Saxon *Manual of Astronomy* four centuries ago gave it as **Fearr.**

Astrologers made this sign the lord of man's neck, throat, and shoulders; Shakespeare having an amusing passage in *Twelfth Night*, in the dialogue between Sirs Toby Belch and Andrew Aguecheek, when both blunder as to this character of Taurus. And it was considered under the guardianship of Venus, sharing this distinction with the body of Scorpio,—some said with Libra,—whence it was known as **Veneris Sidus, Domus Veneris nocturna,** and **Gaudium Veneris**: an idea also perhaps influenced by its containing the Πελειάδες, the Doves, the favorite birds of that goddess. It ruled over Ireland, Greater Poland, part of Russia, Holland, Persia, Asia Minor, the Archipelago, Mantua, and Leipzig in modern astrology, as it did over Arabia, Asia, and Scythia in ancient; Ampelius assigned to it the care of the much dreaded west-northwest wind, Pliny's Argestes. White and lemon were the colors allotted to it. On the whole, it was an unfortunate constellation, although a manuscript almanac of 1386 had "whoso is born in yat syne schal have grace in bestis"; and thunder, when the sun was here, "brought a plentiful supply of victuals."

The extent and density of the stars in Taurus are shown by the fact that, according to Argelander, it contains 121 visible to the naked eye; 188, according to Heis.

> . . . go forth at night,
> And talk with Aldebaran, where he flames
> In the cold forehead of the wintry sky.
>
> Mrs. Sigourney's *The Stars.*

α, 1.2, pale rose.

Aldebaran is from **Al Dabarān,** the Follower, *i. e.* of the Pleiades, or, as Professor Whitney suggested, because it marked the 2d *manzil* that followed the first.

The name, now monopolized by this star, originally was given to the entire group of the Hyades and the lunar mansion which, as **Nā'ir al Dabarān,** the Bright One of the Follower, our star marked; yet there was diversity of opinion as to this, for the first edition of the *Alfonsine Tables* applied it solely to α, while that of 1483, and Al Sufi, did not recognize α as included in the title. Riccioli usually wrote it **Aldebara,** occasionally

Aldebaram, adopted in the French edition of Flamsteed's *Atlas* of 1776; Spenser, in the *Faerie Queen* wrote **Aldeboran,** which occasionally still appears; Chaucer, in the *Hous of Fame*, and even the modern La Lande, had **Aldeberan;** Schickard gave the word as **Addebiris** and **Debiron;** and Costard, in his *History of Astronomy*, cited **Aldebaron.**

Al Bīrūnī quoted, as titles indigenous to Arabia, **Al Fanīḳ,** the Stallion Camel; **Al Fatīḳ,** the Fat Camel; and **Al Muḥdij,** the Female Camel,— the smaller adjacent stars of the Hyades being the **Little Camels;** and it was **Tāli al Najm** and **Hādī al Najm,** equivalents of the **Stella Dominatrix** of classical ages, as if driving the Pleiades before it. Indeed in the last century Niebuhr heard the synonymous **Sāïḳ al Thurayya** on the Arabian shores of the Persian Gulf. A later name was **ʽAin al Thaur,**— which Western astronomers corrupted to **Atin** and **Hain Altor,**— identical with Ὄμμα Βοός, **Oculus Tauri,** and the early English **Bull's Eye,** even now a common title. Riccioli gave this more definitely as **Oculus australis,** and Aben Ezra as the **Left Eye.**

The *Alfonsine Tables*, however, said **Cor Tauri,** the Bull's Heart, which is far out of the way; and it has borne the constellation's Arabic title, changed to **El Taur.**

Aldebaran was the divine star in the worship of the tribe Misām, who thought that it brought rain, and that its heliacal rising unattended by showers portended a barren year.

The Hindu **Rohinī,** a Red Deer, used also for the *nakshatra* in Scorpio marked by Antares, was unquestionably from the star's ruddy hue, Leonard Digges writing, in his *Prognostication* for 1555, that it is "ever a meate rodde [red]"; and the *Alfonsine Tables* had *quae trahit ad aerem clarum valde — est ut cerea.*

Palilicium,[1] in various orthography, but correctly **Parilicium,** used for the whole group of the Hyades, descended as a special designation for Aldebaran through all the catalogues to Flamsteed's, where it is exclusively used. Columella called it **Sucula** as chief of the peasants' Suculae. Ptolemy's Λαμπαδίας, Torch-bearer, was Λαμπαύρας in Proclus' *Paraphrase.*

The 1603 and 1720 editions of Bayer's *Uranometria* distinctly terminate their lists of Aldebaran's titles with the words **Subruffa** and **Aben Ezra;** but Bayer's star-names are often by no means clear, and here incorrect. The latter of these is merely the name of the famous Jewish commentator to whom he often refers; and the former a designation of the light red color (*Subrufa*)

1 This word is from Palilia, or Parilia, the feast of Pales,— the Latin shepherds' divinity and their feminine form of Pan,— which marked the birthday of Rome the 21st of April, when this star vanished in the twilight.

of the star which we all recognize. Some poet has written "red Aldebáran[1] burns"; and William Roscoe Thayer, in his *Halid:*

> I saw on a minaret's tip
> Aldébaran[1] like a ruby aflame, then leisurely slip
> Into the black horizon's bowl.

In all astrology it has been thought eminently fortunate, portending riches and honor; and was one of the four Royal Stars, or Guardians of the Sky, of Persia, 5000 years ago, when it marked the vernal equinox. As such Flammarion quoted its title **Taschter,** which Lenormant said signified the Creator Spirit that caused rain and deluge; but a different conception of these Guardian Stars among the Hindus is noted under Argo, and still another is given by Edkins, who makes Aldebaran **Sataves,** the leader of the western stars.

Flammarion has assigned to it the Hebrew **Āleph** that we have seen for Taurus, rendering it **God's Eye;** and Aben Ezra identified it with the biblical **Kīmāh,** probably in connection with all the Hyades and as being directly opposed on the sphere to **K^e^sīl** which he claimed for Antares.

Sharing everywhere in the prominence given to its constellation, this was especially the case in Babylonian astronomy, where it marked the 5th ecliptic asterism **Pidnu-sha-Shame,** the Furrow of Heaven, perhaps representing the whole zodiac, and analogous to the Hebrew and Arabic Padan and Fadan, the Furrow. So that, before the Ram had taken the Bull's place as Leader of the Signs, Aldebaran was **Ku, I-ku,** or **I-ku-u,** the Leading Star of Stars. Still more anciently it was the Akkadian **Gis-da,** also rendered the "Furrow of Heaven"; and **Dil-gan,** the Messenger of Light,— this, as we have seen, being applied to Hamal, Capella, Wega, and perhaps to other bright stars, as their positions changed with respect to the equinox. In the same way the Syriac word **'Iyūthā,** which we have seen for the star Capella, seems to have been used also for Aldebaran.

As marking the lunar station it was the Persian **Paha** and the Khorasmian-Sogdian **Baharu,** signifying the Follower.

Riccioli cited, from Coptic Egypt, Πιώριων, *Statio Hori;* and Renouf identified Aldebaran with the indigenous Nile figure **Sarit.**

An old Bohemian title is **Hrusa.**

The Hervey Islanders associated it, as **Aumea,** with Sirius in their legend of the Pleiades.

Al Bīrūnī quoted strange Arabic titles for the comparatively vacant space

[1] Thus the pronunciation of the word seems to be in doubt, although the best usage follows the original Arabic in Aldeb'aran.

westward towards the Pleiades,— **Al Ḍaiḳā,** Growing Small, *i. e.* from its rapid setting, and **Kalb al Dabarān,** the Dog of Aldebaran,— asserting that it was considered a place of evil omen. But there seems to have been dispute as to its location, for he added that those authors were wrong who marked this Dog by the 21st and 22d stars of Taurus,— κ and v.

Aldebaran is but slightly south of the ecliptic, and, lying in the moon's path, is frequently occulted, thus often showing the optical illusion of projection. As one of the lunar stars it is much used in navigation. It is the only star in the *Harvard Photometry* which is exactly of the 1st magnitude, although by the Estimates of that catalogue it is 1.2. It thus has three times the brilliancy of Polaris.

The parallax is given by Elkin as 0″.101, showing a distance from us of twenty-eight light years; or, if the interval between the earth and the sun, the astronomers' unit of stellar measurement, be considered as one inch, that between the sun and this star would be twenty-seven miles. It is receding from our system at the rate of thirty miles a second, and, next to ζ Herculis, seems to have the greatest velocity in the line of sight of any of the bright stars yet determined. The spectrum is Solar, and a beautiful example of the type.

Aldebaran comes to the meridian on the 10th of January. It has a 10th-magnitude companion, 109″ away, which has long been known, but Burnham recently divided this into 11 and 13.5, 1″.8 apart, at a position angle of 279°; and, in 1888, discovered a 14th-magnitude companion 31″.4 distant, at a position angle of 109°.

The **Taurids** of the 20th of November radiate from a point north of, and preceding, this star. These meteors "are slow, and fireballs occasionally appear among them."

*

The Hyades marked by the sailor.
Potter's translation of Euripides' Ἴων.

As when the seaman sees the Hyades
Gather an army of Cimmerian clouds,
Auster and Aquilon with winged steeds.
Christopher Marlowe's *History of Doctor Faustus.*

The Hyades,

α, θ^1, θ^2, γ, δ, and ϵ Tauri, 10° southeast of the Pleiades,

Whitening all the Bull's broad forehead,

form one of the most beautiful objects in the sky, and have been famous for ages, especially with the classical authors.

Mythologically they were daughters of Atlas and Aethra, and hence half-sisters of the Pleiades, with whom they made up the fourteen Atlantides; or the Dodonides, the nymphs of Dodona, to whom Jupiter entrusted the nurture of the infant Bacchus, and raised them to the sky when driven into the sea by Lycurgus. Similarly they were said to be the Nysiades, the nymphs of Nysa, and teachers of Bacchus in India.

Anciently supposed to be seven in number, we moderns count but six, and Hesiod named only five,— Kleea, Eudora, Koronis, Phaeo, and Phaesula; but Pherecydes gave a complete list of them, although one of his names has been lost, and the rest, preserved by Hyginus, vary from those given by Hesiod, and doubtless are somewhat corrupted in form. These were Aesula or Pedile, Ambrosia, Dione, Thyene or Thyone, Eudora, Koronis, and Polyxo or Phyto.[1] Pherecydes probably took in β and ζ at the tips of the horns, omitting some of the fainter stars now included in the group; Thales, however, is said to have acknowledged but two,— α and ε in the eyes,— "one in the Northern Hemisphere, and the other in the South"; Hipparchos and Ptolemy named only α and γ as Ὑάδων; Euripides, in the *Phaëthon*, counted three; and Achaeus, four. Ovid used **Thyone** for the whole, but none of the sisters' names have been applied to the individual stars as in the case of the Pleiades.

They are among the few stellar objects mentioned by Homer,— and by him, Hesiod, Manilius, Pliny, and doubtless others, given separately from Taurus. Pliny called them **Parilicium,** from their *lucida*, Aldebaran.

The Greeks knew them as Ὑάδες, which became "Hyades" with the cultured Latins, supposed by some to be from ὕειν, "to rain," referring to the wet period attending their morning and evening setting in the latter parts of May and November; and this is their universal character in the literature of all ages. Thus we have *Hyades Graiis ab imbre vocat* of Ovid's *Fasti; pluviasque Hyadas* of the *Aeneid* and of Ovid again; and *pluviae* generally, which Manilius expressed in his

> Sad Companions of the turning Year.

While far back of all these, in the *She King:*

> The Moon wades through Hyads bright,
> Foretelling heavier rain.

Pliny wrote of them as being "a violent and troublesome star causing stormes and tempests raging both on land and sea"; in later times Edmund Spenser called them the **Moist Daughters**; Tennyson, in his *Ulysses*, said:

> Thro' scudding drifts the rainy Hyades vext the dim sea;

[1] Grotius has much information as to their titles in his *Syntagma Arateorum*.

and Owen Meredith has "the watery Hyades" in *The Earl's Return.* The queer old *Guide into Tongues* of John Minsheu, calling them the **Seven Stars,**— the only instance of this title that I have met for this group,— makes still more intimate their connection with the showers; for at its word *Hyades* the reader is referred to the word *Raine*, where we see:

Hyades, ὑάδες, dictae *stellae quaedam* in cornibus Tauri; *quae ortu occasuq.* sus pluvias largosque imbres *concitant.*

And in Doctor Johnson's *Dictionary* the word is defined as "a watery constellation." Thus they have always been considered most noteworthy by husbandmen, mariners, and all who were dependent upon the weather, even to the last two or three centuries.

Ovid called them **Sidus Hyantis,** after their earthly brother, Hyas, whose name, after all, would seem to be the most natural derivation of the title; and it was their grief at his death which gave additional point to Horace's *tristes Hyadas*, and, in one version of their story, induced Jove to put them in the sky.

But their colloquial title among the Roman country-people was **Suculae,** the Little Pigs, as if from *Sus*, Sow, the Greek Ὑς, Homer's Σῦς, which indeed might as well be the derivation of Ὑάδες as ὕειν. This name constantly occurs in astronomical literature from the time of Columella and Pliny to Kepler, Hevelius, and Flamsteed; Pliny accounting for it by the fact that the continual rains of the season of their setting made the roads so miry that these stars seemed to delight in dirt, like swine! And this idea, trivial though it seems, was sufficiently prevalent for Cicero, a century before Pliny, to think worthy of contradiction in his *De Natura Deorum.* Smyth said that the title might come from the resemblance of the group to a pig's jaws; or because Aldebaran and its companion stars were like a sow with her litter. Peck suggests, in his *Dictionary of Classical Literature and Antiquities*, that Suculae was the oldest Roman name, given before the Greek appellation was known, and to be compared with our popular stellar titles such as the Dipper, Charles' Wain, etc. Isidorus traced it to *sucus*, "moisture," a pleasanter derivation, and possibly more correct, than that held in ancient Italy. This will account for Bayer's **Succidae.**

Bassus and others knew the group as ὑ-ψιλόν, the symbol with Pythagoras for human life; and the **Roman V,** as it resembles those letters,— *a* and *ε* being the extremes, *γ* at the vertex. But Ulug Beg's translator wrote:

Quinque stellae quae sunt in facie, in forma Lambdae Graecorum et formā τοῦ Dāl.

In the *Alfonsine Tables* we find **Lampadas,** the accusative plural of Lampada, a Torch.

Occasional Arabic titles were **Al Mijdaḥ,** a Triangular Spoon, and **Al Ḳilāṣ,** the Little She Camels, referring to the smaller stars in distinction from Aldebaran, the Large Camel; Al Ferghani wrote the word **Ḳalā'iṣ.** These Little Camels appeared in one Arabic story as driven before the personified Aldebaran, in evidence of his riches, when he went again to woo Al Thurayya, the Pleiades, who previously had spurned him on account of his poverty. Another author made the word **Al Ḳallāṣ,** the Boiling Sea, so continuing in Arabia the Greek and Roman ideas of its stormy and watery character. Generally, however, in that country, the Hyades were **Al Dabarān,** which was adopted in the 1515 *Almagest*, as well as in the *Alfonsine Tables* of 1521, where we read *sunt stellae aldebaran*, specially referring to the star γ "of those in the face." The Arabic title, therefore, was identical with that of the 2d *manzil*, which these stars constituted, as they also did the 2d *nakshatra*, **Rohinī,** Aldebaran marking the junction with the adjacent Mrigaçīrsha.

The Hindus figured this asterism as a **Temple,** or **Wagon;** and there are many astrological allusions to it in the *Siddhāntas*, the collective term for the various standard astronomical books of that people.

The Chinese utilized it for their 2d *sieu*, **Pi,** or **Peih,** anciently **Pal,** a Hand-net, or a Rabbit-net, but included λ and σ; although some limited this station to ϵ, the farthest to the north. The *She King* thus described it:

> Long and curved is the Rabbit Net of the sky;

but with that people generally it was the **Star of the Hunter,** and, with the astrologers, the **Drought Car.** This title, however, was inappropriate, for the Hyades seem to have been as closely identified with rain in China as in Greece or Rome,—indeed were worshiped as **Yü Shī,** the General, or Ruler, of Rain, from at least 1100 B.C. Still this character was not native, but must have been derived from western Asia, where the early rains coincided with the heliacal rising of these stars, which was not the case in China by nearly two months. The adjacent small stars, with ξ, were **Tien Lin,** the Celestial Public Granary; and the whole group was known as the **Announcer of Invasion on the Border.**

The Hyades have been identified with the scriptural **Mazzārōth,** but there is little foundation for this; even less than for their identification, by Saint Jerome and by Riccioli, with the **Kīmāh** of the *Book of Job*, ix, 9.

Anglo-Saxon titles are **Raedgastran, Raedgasnan,** and **Redgaesrum,** whatever these may mean; and the **Boar-Throng** which that people saw in the sky may have been this group rather than Orion as generally is supposed.

It is thought that the Hyades have a united proper motion towards the west. They are rich in doubles and full of interest to the owners of even small glasses.

β, Double, 2.1 and 10, brilliant pure white and pale gray.

El Nath is from **Al Nāṭiḥ,** the Butting One, because located on the tip of the northern horn, 5° from ζ, similarly placed on the southern. This title also appears for Aries and its star Hamal.

Bayer said that many included it and ζ in the Hyades group, but this seems improbable, although Pherecydes had it thus.

β Tauri is identical with γ Aurigae, and has been considered as belonging to either constellation; Burritt's *Atlas* calling it **Aurigae** or **El Nath.** As a member of Auriga it lies on the left ankle, and was the Arabians' **Ḳabḍ al ʽInān,** usually translated the Heel of the Rein-holder.

Smyth, who is often humorous amid his exact science, referring to the position of this star at the greatest possible distance from the hoof, says: "Can this have given rise to the otherwise pointless sarcasm of 'not knowing B from a bull's foot'?"

With Capella and other stars in Auriga it was the Chinese **Woo Chay,** a Fire-carriage.

In Babylonia it was **Shur-narkabti-sha-iltanu,** the Star in the Bull towards the North, or the Northern Star towards the Chariot,— not our Wain, but the Chariot of Auriga,— and marked the 6th ecliptic constellation. The sun stood near this star at the commencement of spring 6000 years ago.

Among the Hindus it represented **Agni,** the god of fire, and commonly bore that title; as also the similar **Hutabhuj,** the Devourer of the Sacrifice.

Astrologers said that El Nath portended eminence and fortune to all who could claim it as their natal star.

It has a Sirian spectrum, and is receding from us at the rate of about five miles a second.

Between it and ψ Aurigae was discovered on the 24th of January, 1892, the now celebrated *nova* Aurigae that has occasioned so much interest in the astronomical world.

γ, 4.2, yellow.

Hyadum I is generally seen for this, and, synonymously, **Primus Hyadum,** or, more correctly, as with Flamsteed, **Prima Hyadum;** but this was not original with him, for long before it evidently was an Arabic designation, as Al Achsasi had **Awwal al Dabarān,** the First of the Dabarān.

Hipparchos described it as ἐν τῷ ῥύγχει, "in the muzzle," still its location at the vertex of the triangle.

With others adjacent it was **Choo Wan,** the Many Princes, of China.

δ, 4.2, is **Hyadum II.**

ϵ, 3.6, one of the Hyades, according to Whitall, is **Ain,** from the Arabic ʻAin, the Eye, near which it lies, Flamsteed calling it **Oculus boreus,** the Northern Eye.

Some think that it alone constituted the 2d *sieu*, Pi.

Close by is a small nebula, N. G. C. 1555, one of the few known to be variable in light.

ζ, 3.5,

was the determinant of the 7th ecliptic constellation of Babylonia, **Shur-narkabti-sha-shūtū,** the Star in the Bull towards the South, or the Southern Star towards the Chariot.

Reeves gave it, with others near by, as **Tien Kwan,** the Heavenly Gate.

In astrology ζ has been considered of mischievous influence.

It marks the tip of the southern horn and the singular **Crab Nebula,** a little to the northwest, the first in Messier's catalogues,[1] and now known as N. G. C. 1952, 1 M. Although Bevis had seen this in 1731, it was accidentally rediscovered by Messier on the 12th of September, 1758, while observing ζ and a neighboring comet, and led to his two catalogues of 103 nebulae and clusters, published from 1771 to 1782, the first attempt at a complete list of these objects. The return of Halley's comet was first observed in August, 1835, close to this star, when the nebula was a perfect mare's-nest to astronomical tyros.

★

The seven sweet Pleiades above.

Owen Meredith's *The Wanderer*.

The group of sister stars, which mothers love
To show their wondering babes, the gentle Seven.

Bryant's *The Constellations*.

The Pleiades,

the **Narrow Cloudy Train of Female Stars** of Manilius, and the **Starry Seven, Old Atlas' Children,** of Keats' *Endymion*, have everywhere been

[1] The work of Messier, shared by La Caille and Mechain, was supposed to have brought together all objects of that class in the heavens; but twenty years afterwards Sir William Herschel had added 2500 to their lists, and his son's *General Catalogue* of 1864 has 5079 nebulae and clusters. This was enlarged by Dreyer, in his *New General Catalogue*, to 9416 discovered up to December, 1887; and since then at least 1000 more have been added by Swift and the observers at Marseilles. Halley, in 1716, knew only six, and of these four are clusters.

among the most noted objects in the history, poetry, and mythology of the heavens; though, as Aratos wrote,

> not a mighty space
> Holds all, and they themselves are dim to see.

All literature contains frequent allusions to them, and in late years they probably have been more attentively and scientifically studied than any other group.

They generally have been located on the shoulder of the Bull as we have them, but Hyginus, considering the animal figure complete, placed them on the hind quarter; Nicander, Columella, Vitruvius, and Pliny, on the tail,

> In cauda Tauri septem quas appellavere Vergilias;—

although Pliny also is supposed to have made a distinct constellation of them. Proclus and Geminos said that they were on the back; and others, on the neck, which Bayard Taylor followed in his *Hymn to Taurus*, where they

> Cluster like golden bees upon thy mane.

Eratosthenes, describing them as over the animal, imitated Homer and Hesiod in his Πλειάς; while Aratos, calling them, in the Attic dialect, Πληϊάδης, placed them near the knees of Perseus; thus, as in most of his poem, following Eudoxos, whose sphere, it is said, clearly showed them in that spot. Hipparchos in the main coincided with this, giving them as Πλειάς and Πλειάδες; but Ptolemy used the word in the singular for four of the stars, and did not separate them from Taurus. The Arabians and Jews put them on the rump of Aries; and the Hindu astronomers, on the head of the Bull, where we now see the Hyades.

The Pleiades seem to be among the first stars mentioned in astronomical literature, appearing in Chinese annals of 2357 B. C., Alcyone, the *lucida*, then being near the vernal equinox, although now 24° north of the celestial equator; and in the Hindu lunar zodiac as the 1st *nakshatra*, **Krittikā**,[1] **Karteek**, or **Kartiguey**, the General of the Celestial Armies, probably long before 1730 B. C., when precession carried the equinoctial point into Aries. Al Bīrūnī, referring to this early position of the equinox in the Pleiades, which he found noticed "in some books of Hermes,"[2] wrote:

[1] The Krittikās were the six nurses of Skanda, the infant god of war, represented by the planet Mars, literally motherless, who took to himself six heads for his better nourishment, and his nurses' name in Karttikeya, Son of the Krittikās.

[2] These Hermetic Books were the sacred canon of Egypt, in forty-two volumes, treating of religion and the arts and sciences, their authorship being ascribed to the god Thoth, whom the Greeks knew as Hermes Trismegistos, Thrice Great Hermes.

This statement must have been made about 3000 years and more before Alexander.

And their beginning the astronomical year gave rise to the title "the Great Year of the Pleiades" for the cycle of precession of about 25,900 years.

The Hindus pictured these stars as a **Flame** typical of Agni, the god of fire and regent of the asterism, and it may have been in allusion to this figuring that the western Hindus held in the Pleiad month Kartik (October–November) their great star-festival Dībalī, the Feast of Lamps, which gave origin to the present Feast of Lanterns of Japan. But they also drew them, and not incorrectly, as a **Razor** with a short handle, the radical word in their title, *kart*, signifying "to cut."

The Santals of Bengal called them **Sar en**; and the Turks, **Ulgher.**

As a Persian lunar station they were **Perv, Perven, Pervis, Parvig,** or **Parviz,** although a popular title was **Peren,** and a poetical one, **Parur.** In the *Rubá'ís*, or *Rubá'iyát*, of the poet-astronomer Omar Khayyám, the tent-maker of Naishápúr in 1123, "who stitched the tents of science," they were **Parwin,** the **Parven** of that country to-day; and, similarly, with the Khorasmians and Sogdians, **Parvi** and **Parur**; — all these from Peru, the Begetters, as beginning all things, probably with reference to their beginning the year.

In China they were worshiped by girls and young women as the **Seven Sisters of Industry,** while as the 1st *sieu* they were **Mao, Mau,** or **Maou,** anciently **Mol,** The Constellation, and **Gang,** of unknown signification, Alcyone being the determinant.

On the Euphrates, with the Hyades, they seem to have been **Mas-tab-ba-gal-gal-la,** the Great Twins of the ecliptic, Castor and Pollux being the same in the zodiac.

In the 5th century before Christ Euripides mentioned them with 'Αετός, our Altair, as nocturnal timekeepers; and Sappho, a century previously, marked the middle of the night by their setting. Centuries still earlier Hesiod and Homer brought them into their most beautiful verse; the former calling them 'Ατλάγενης, Atlas-born. The patriarch Job is thought to refer to them twice in his word **Kīmāh,** a Cluster, or Heap, which the Hebrew herdsman-prophet Amos, probably contemporary with Hesiod, also used; the prophet's term being translated "the seven stars" in our Authorized Version, but "Pleiades" in the Revised. The similar Babylonian-Assyrian **Kimtu,** or **Kimmatu,** signifies a "Family Group," for which the Syrians had **Kīmā,** quoted in Humboldt's *Cosmos* as **Gemat**; this most natural simile is repeated in Seneca's *Medea* as *densos Pleiadum greges.* Manilius had **Glomerabile Sidus,** the Rounded Asterism, equivalent to the

Globus Pleiadum of Valerius Flaccus; while Brown translates the Πληϊάδης of Aratos as the **Flock of Clusterers.**

In Milton's description of the Creation it is said of the sun that

> the gray
> Dawn and the Pleiades before him danc'd,
> Shedding sweet influence,—

the original of these last words being taken by the poet from the *Book of Job*, xxxviii, 31, in the Authorized Version, that some have thought an astrological reference to the Pleiades as influencing the fortunes of mankind, or to their presumed influential position as the early leaders of the Lunar Mansions. The Revised Version, however, renders them "cluster," and the *Septuagint* by the Greek word for "band," as if uniting the members of the group into a fillet; others translate it as "girdle," a conception of their figure seen in Amr al Ḳais' contribution to the *Mu'allaḳāt*, translated by Sir William Jones:

> It was the hour when the Pleiades appeared in the firmament like the folds of a silken sash variously decked with gems.

Von Herder gave Job's verse as:

> Canst thou bind together the brilliant Pleiades?

Beigel as:

> Canst thou not arrange together the rosette of diamonds of the Pleiades?

and Hafiz wrote to a friend:

> To thy poems Heaven affixes the Pearl Rosette of the Pleiades as a seal of immortality.

An opening rose also was a frequent Eastern simile; while in Sadi's *Gulistan*, the Rose-garden, we read:

> The ground was as if strewn with pieces of enamel, and rows of Pleiades seemed to hang on the branches of the trees;

or, in Graf's translation:

> as though the tops of the trees were encircled by the necklace of the Pleiades.

William Roscoe Thayer repeated the Persian thought in his *Halid:*

> slowly the Pleiades
> Dropt like dew from bough to bough of the cinnamon trees.

That all these wrote better than they knew is graphically shown by Miss Clerke where, alluding to recent photographs of the cluster by the Messrs. Henry of Paris, she says:

> The most curious of these was the threading together of stars by filmy processes. In one case seven aligned stars appeared strung on a nebulous filament "like beads on a rosary." The "rows of stars," so often noticed in the sky, may therefore be concluded to have more than an imaginary existence.

The title, written also **Pliades** and, in the singular, **Plias,** has commonly been derived from πλεῖν, "to sail," for the heliacal rising of the group in May marked the opening of navigation to the Greeks, as its setting in the late autumn did the close. But this probably was an afterthought, and a better derivation is from πλεῖος, the Epic form of πλέως, "full," or, in the plural, "many," a very early astronomical treatise by an unknown Christian writer having *Plyades ā pluralitate.* This coincides with the biblical Kīmāh and the Arabic word for them — **Al Thurayya.** But as Pleione was the mother of the seven sisters, it would seem still more probable that from her name our title originated.

Some of the poets, among them Athenaeus, Hesiod, Pindar, and Simonides, likening the stars to Rock-pigeons flying from the Hunter Orion, wrote the word Πελειάδες, which, although perhaps done partly for metrical reasons, again shows the intimate connection in early legend of this group with a flock of birds. When these had left the earth they were turned into the Pleiad stars. Aeschylus assigned the daughters' pious grief at their father's labor in bearing the world as the cause of their transformation and subsequent transfer to the heavens; but he thought these **Peleiades** ἄπτεροι, "wingless." Other versións made them the Seven Doves that carried ambrosia to the infant Zeus, one of the flock being crushed when passing between the Symplegades, although the god filled up the number again. This story probably originated in that of the dove which helped Argo through; Homer telling us in the *Odyssey* that

> No bird of air, no dove of swiftest wing,
> That bears ambrosia to the ethereal king,
> Shuns the dire rocks; in vain she cuts the skies,
> The dire rocks meet and crush her as she flies;

and the doves on Nestor's cup described in the *Iliad* have been supposed to refer to the Pleiades. Yet some have prosaically asserted that this columbine title is merely from the loosing of pigeons in the auspices customary at the opening of navigation. These stories may have given rise to the Sicilians' **Seven Dovelets,** the **Sette Palommielle** of the *Pentameron.*

Another title analogous to the foregoing is **Butrum** from Isidorus,— Caesius wrongly writing it **Brutum,**— in the mediaeval Latin for Βότρυς, a Bunch of Grapes, to which the younger Theon likened them. It is a happy simile, although Thompson [1] considers it merely another avian association like that seen in the poetical Peleiades and the Alcyone of the *lucida.*

Vergiliae and **Sidus Vergiliarum** have always been common for the cluster as rising after Ver, the Spring,— the *Breeches Bible* having this marginal note at its word " Pleiades " in the *Book of Job*, xxxviii, 31 :

which starres arise when the sunne is in Taurus which is the spring time and bring flowers.

And these names obtained from the times of the Latin poets to the 18th century, but often erroneously written **Virgiliae.** Pliny, describing the glow-worms, designated them as *stellae* and likened them to the Pleiades:

Behold here before your very feet are your Vergiliae; of that constellation are they the offspring.

And the much quoted lines in *Locksley Hall* are similar:

Many a night I saw the Pleiads, rising thro' the mellow shade,
Glitter like a swarm of fire-flies tangled in a silver braid.

Bayer cited **Signatricia Lumina.**

Hesiod called them the **Seven Virgins** and the **Virgin Stars;** Vergil, the **Eoae Atlantides;** Milton, the **Seven Atlantic Sisters;** and **Hesperides,** the title for another batch of Atlas' daughters from Hesperis, has been applied to them. Chaucer, in the *Hous of Fame*, had **Atlantes doughtres sevene;** but his " Sterres sevene " refer to the planets. As the **Seven Sisters** they are familiar to all; and as the **Seven Stars** they occur in various early Bible versions; in the **Sifunsterri** of the Anglo-Saxons, though they also wrote **Pliade;** in the *Septistellium vestis institoris*, cited by Bayer; and in the modern German **Siebengestirn.** This numerical title also frequently has been applied to the brightest stars of the Greater Bear, as in early days it was to the " seven planets,"—the Sun, Moon, Mercury, Venus, Mars, Jupiter, and Saturn. Minsheu had the words " Seven Starres " indiscriminately for

[1] He traces the word back as equivalent to Οἰνάς, a Dove, probably *Columba oenas* of Old World ornithology, and so named from its purple-red breast like wine,—οἶνος,— and naturally referred to a bunch of grapes; or perhaps because the bird appeared in migration at the time of the vintage. This is strikingly confirmed by the fact that coins of Mallos in Cilicia bore doves with bodies formed by bunches of grapes; these coins being succeeded by others bearing grapes alone; and we often see the bird and fruit still associated in early Christian symbolism.

the Pleiades, Hyades, and Ursa Major, saying, as to the first, "that appear in a cluster about midheaven."

As the group outline is not unlike that of the Dipper in Ursa Major, many think that they much more deserve the name **Little Dipper** than do the seven stars in Ursa Minor; indeed that name is not uncommon for them. And even in our 6th century, with Hesychios, they were *Σάτιλλα*, a Chariot, or Wagon, another well-known figure for Ursa Major.

Ideler mentioned a popular designation by his countrymen,—**Schiffahrts Gestirn,** the Sailors' Stars,—peculiarly appropriate from the generally supposed derivation of their Greek title and meteorological character of 2000 years ago; but the *Tables of some Obscure Wordis* of King James I anticipated this in "**Seamens Starres** — the seaven starres."

The Teutons had **Seulainer;** the Gaels, **Griglean, Grioglachan,** and **Meanmnach;** the Hungarians, who, Grimm says, have originated 280 native names for stars, called the Pleiades **Fiastik** and **Heteveny,**—this last in Finland **Het′e wā‘ne;** the Lapps of Norway knew them as **Niedgierreg;** while the same people in Sweden had the strange **Suttjenēs Rauko,** Fur in Frost, these seven stars covering a servant turned out into the cold by his master. The Finns and Lithuanians likened them to a **Sieve** with holes in it; and some of the French peasantry to a **Mosquito Net, Cousinière,**—in the Languedoc tongue **Cousigneiros.** The Russians called them **Baba,** the Old Wife; and the Poles, **Baby,** the Old Wives.

As we have seen the Hyades likened to a Boar Throng, so we find with Hans Egede, the first Norse missionary to Greenland, 1721–34, that this sister group was the **Killukturset** of that country, Dogs baiting a bear; and similarly in Wales, **Y twr tewdws,** the Close Pack.

Weigel included them among his heraldic constellations as the **Multiplication Table,** a coat of arms for the merchants.

Sancho Panza visited them, in his aërial voyage on Clavileño Aligero, as **las Siete Cabrillas,** the Seven Little Nanny Goats; and **la Racchetta,** the Battledore, is a familiar and happy simile in Italy; but the astronomers of that country now know them as **Plejadi,** and those of Germany as **Plejaden.**

The Rabbis are said to have called them **Sukkōth R^εnōth,** usually translated "the Booths of the Maidens" or "the Tents of the Daughters," and the *Standard Dictionary* still cites this supposed Hebrew title; but Riccioli reversed it as **Filiae Tabernaculi.** All this, however, seems to be erroneous, as is well explained in the *Speaker's Commentary* on the *2d Book of the Kings* xvii, 30, where the words are shown to be intended for the Babylonian goddess Zarbanit, Zirat-banit, or Zir-pa-nit, the wife of Bēl Mardūk.

The *Alfonsine Tables* say that the "Babylonians," by whom were probably meant the astrologers, knew them as **Atorage,** evidently their word for the *manzil* **Al Thurayya,** the Many Little Ones, a diminutive form of Tharwān, Abundance, which Al Bīrūnī assumed to be either from their appearance, or from the plenty produced in the pastures and crops by the attendant rains. We see this title in Bayer's **Athoraie;** in Chilmead's **Atauria** *quasi Taurinae;* and otherwise distorted in every late mediaeval work on astronomy. Riccioli, commenting on these in his *Almagestum Novum*, wrote *Arabicē non* **Athoraiae** *vel* **Atarage** *sed* **Altorieh** *seu* **Benat Elnasch,** *hoc est filiae congregationis;* the first half of which may be correct enough, but the **Benat,** etc., singularly confounded the Pleiad stars with those of Ursa Major. In his *Astronomia Reformata* he cited **Athorace** and **Altorich** from Aben Ragel. **Turanyā** is another form, which Hewitt says is from southern Arabia, where they were likened to a **Herd of Camels** with the star Capella as the driver.

A special Arabic name for them was **Al Najm,** the Constellation *par excellence*, and they may be **the Star,** or the **Star of piercing brightness,** referred to by Muḥammād in the 53d and 86th *Suras* of the *Ḳur'ān*, and versified from the latter by Sir Edwin Arnold in his *Al Hafiz, the Preserver:*

> By the sky and the night star!
> By Al Tārik the white star!
> To proclaim dawn near;
> Shining clear —
> When darkness covers man and beast —

the planet Venus being intended by Al Ṭāriḳ. Grimm cited the similar **Syryän Voykodzyun,** the Night Star.

They shared the watery character always ascribed to the Hyades, as is shown in Statius' *Pliadum nivosum sidus;* and Valerius Flaccus distinctly used the word "Pliada" for the showers, as perhaps did Statius in his *Pliada movere;* while Josephus states, among his very few stellar allusions, that during the investment of Jerusalem by Antiochus Epiphanes, 170 B. C., the besieged suffered from want of water, but were finally relieved "by a large shower of rain which fell at the setting of the Pleiades." In the same way they are intimately connected with traditions of the Flood found among so many and widely separated nations, and especially in the Deluge-myth of Chaldaea. Yet with all this well established reputation, we read in the *Works and Days:*

> When with their domes the slow-pac'd snails retreat,
> Beneath some foliage, from the burning heat
> Of the Pleiades, your tools prepare.

They were a marked object on the Nile, at one time probably called **Chu** or **Chow**, and supposed to represent the goddess **Nit** or **Neith,** the Shuttle, one of the principal divinities of Lower Egypt, identified by the Greeks with Athene, the Roman Minerva. Hewitt gives another title from that country, **Athur-ai,** the Stars of Athyr (Hathor), very similar to the Arabic word for them; and Professor Charles Piazzi Smyth suggests that the seven chambers of the Great Pyramid commemorate these seven stars.

Grecian temples were oriented to them, or to their *lucida;* those of Athene on the Acropolis, of different dates, to their correspondingly different positions when rising. These were the temple of 1530 B. C.; the Hecatompedon of 1150 B. C.; and the great Parthenon, finished on the same site 438 B. C. The temple of Bacchus at Athens, 1030 B. C., looked toward their setting, as did the Asclepieion at Epidaurus, 1275 B. C., and the temple at Sunium of 845 B. C. While at some unknown date, perhaps contemporaneous with these Grecian structures, they were pictured in the New World on the walls of a Palenque temple upon a blue background; and certainly were a well-known object in other parts of Mexico, for Cortez heard there, in 1519, a very ancient tradition of the destruction of the world in some past age at their midnight culmination.

A common figure for these stars, everywhere popular for many centuries, is that of a **Hen with her Chickens,**— another instance of the constant association of the Pleiades with flocking birds, and here especially appropriate from their compact grouping. Aben Ragel and other Hebrew writers thus mentioned them, sometimes with the **Coop** that held them,— the **Massa Gallinae** of the Middle Ages; these also appearing in Arabic folk-lore, and still current among the English peasantry. In modern Greece, as the **Hen-coop,** they are Πούλια or Πούλεια, not unlike the word of ancient Greece. Miles Coverdale, the translator in 1535 of the first complete English *Bible*, had as a marginal note to the passage in the *Book of Job:*

> these vii starres, the **clock henne with her chickens;**

and Riccioli, in his *Almagestum Novum:*

> Germanicē **Bruthean:** Anglicē **Butrio** id est gallina fovens pullos.

We see in the foregoing the **Butrum** of Isidorus, Riccioli's great predecessor in the Church. The German farm laborers call them **Gluck Henne;** the Russian, **Nasēdha,** the Sitting Hen; the Danes, **Aften Hoehne,** the Eve Hen; while in Wallachia they are the **Golden Cluck Hen and her five Chicks.** In Servia a **Girl** is added in charge of the brood, probably the star Alcyone, Maia appropriately taking her place as the Mother. The French and

Italians designate them, in somewhat the same way, as **Pulsiniere, Poussinière,** and **Gallinelle,** the Pullets, Riccioli's **Gallinella.** Aborigines of Africa and Borneo had similar ideas about them. Pliny's translator Holland called them the **Brood-hen star Vergiliae.**

Savage tribes knew the Pleiades familiarly, as well as did the people of ancient and modern civilization; and Ellis wrote of the natives of the Society and Tonga Islands, who called these stars **Matarii,** the Little Eyes:

The two seasons of the year were divided by the Pleiades; the first, Matarii i nia, the Pleiades Above, commenced when, in the evening, those stars appeared on the horizon, and continued while, after sunset, they were above. The other season, Matarii i raro, the Pleides Below, began when, at sunset, they ceased to be visible, and continued till, in the evening, they appeared again above the horizon.

Gill gives a similar story from the Hervey group, where the Little Eyes are **Matariki,** and at one time but a single star, so bright that their god Tane in envy got hold of Aumea, our Aldebaran, and, accompanied by Mere, our Sirius, chased the offender, who took refuge in a stream. Mere, however, drained off the water, and Tane hurled Aumea at the fugitive, breaking him into the six pieces that we now see, whence the native name for the fragments, **Tauono,** the Six, quoted by Flammarion as **Tau,** both titles singularly like the Latin Taurus. They were the favorite one of the various *avelas,* or guides at sea in night voyages from one island to another; and, as opening the year, objects of worship down to 1857, when Christianity prevailed throughout these islands. The Australians thought of them as **Young Girls** playing to Young Men dancing,—the Belt stars of Orion; some of our Indians, as **Dancers**; and the Solomon Islanders as **Togo ni samu,** a Company of Maidens. The Abipones of the Paraguay River country consider them their great Spirit **Groaperikie,** or Grandfather; and

in the month of May, on the reappearance of the constellation, they welcome their Grandfather back with joyful shouts, as if he had recovered from sickness, with the hymn, "What thanks do we owe thee! And art thou returned at last? Ah! thou hast happily recovered!" and then proceed with their festivities in honor of the Pleiades' reappearance.

Among other South American tribes they were **Cajupal,** the Six Stars.

The pagan Arabs, according to Hafiz, fixed here the seat of immortality; as did the Berbers, or Kabyles, of northern Africa, and, widely separated from them, the Dyaks of Borneo; all thinking them the central point of the universe, and long anticipating Wright in 1750 and Mädler in 1846, and, perhaps, Lucretius in the century before Christ.

Miss Clerke, in a charming and instructive chapter in her *System of the Stars* which should be read by every star-lover, tells us that:

With November, the "Pleiad-month," many primitive people began their year; and on the day of the midnight culmination of the Pleiades, November 17, no petition was presented in vain to the ancient Kings of Persia; the same event gave the signal at Busiris for the commencement of the feast of Isis, and regulated less immediately the celebration connected with the fifty-two-year cycle of the Mexicans. Savage Australian tribes to this day dance in honor of the "Seven Stars," because "they are very good to the black fellows." The Abipones of Brazil regard them with pride as their ancestors. Elsewhere, the origin of fire and the knowledge of rice-culture are traced to them. They are the "**hoeing-stars**" of South Africa, take the place of a farming-calendar to the Solomon Islanders, and their last visible rising after sunset is, or has been, celebrated with rejoicings all over the southern hemisphere as betokening the "waking-up time" to agricultural activity.

They also were a sign to ancient husbandmen as to the seeding-time; Vergil alluding to this in his 1st *Georgic*, thus rendered by May:

> Some that before the fall 'oth' Pleiades
> Began to sowe, deceaved in the increase,
> Have reapt wilde oates for wheate.

And, many centuries before him, Hesiod said that their appearance from the sun indicated the approach of harvest, and their setting in autumn the time for the new sowing; while Aristotle wrote that honey was never gathered before their rising. Nearly all classical poets and prose writers made like reference to them.

Mommsen found in their rising, from the 21st to the 25th of the Attic month Θαργηλιών, May–June, the occasion for the prehistoric festival Πλυντήρια, Athene's Clothes-washing, at the beginning of the corn harvest, and the date for the annual election of the Achaeans; while Drach surmised that their midnight culmination in the time of Moses, ten days after the autumnal equinox, may have fixed the day of atonement on the 10th of Tishri. Their rising in November marked the time for worship of deceased friends by many of the original races of the South,—a custom also seen with more civilized peoples, notably among the Parsis and Sabaeans, as also in the Druids' midnight rites of the 1st of November; while a recollection of it is found in the three holy days of our time, All Hallow Eve, All Saints' Day, and All Souls' Day.

Hippocrates made much of the Pleiades, dividing the year into four seasons, all connected with their positions in relation to the sun; his winter beginning with their setting and ending with the spring equinox; spring lasting till their rising; the summer, from their appearing to the rising of Arcturus; and the autumn, till their setting again. And Caesar made their heliacal rising begin the Julian summer, and their cosmical setting the commencement of winter. In classic lore the Pleiades were the heavenly group chosen with the sun by Jove to manifest his power in favor of Atreus by causing them to move from east to west.

Notwithstanding, however, all that we read so favorable to the high regard in which these stars were held, they were considered by the astrologers as portending blindness and accidents to sight, a reputation shared with all other clusters. The Arabs, especially, thought their forty days' disappearance in the sun's rays was the occasion of great harm to mankind, and Muḥammād wrote that "when the star rises all harm rises from the earth." But Hippocrates had differently written in his *Epidemics*, a thousand years before, of the connection of the Pleiades with the weather, and of their influence on diseases of autumn:

until the season of the Pleiades, and at the approach of winter, many ardent fevers set in;

and:

in autumn, and under the Pleiades, again there died great numbers.

Although the many legends of their origin are chiefly from Mediterranean countries, yet the Teutonic nations have a very singular one associated with our Saviour. It says that once, when passing by a baker's shop, and attracted by the odor of newly baked bread, He asked for a loaf; but being refused by the baker, was secretly supplied by the wife and six daughters standing by. In reward they were placed in the sky as the Seven Stars, while the baker became a cuckoo;[1] and so long as he sings in the spring, from Saint Tiburtius' Day, April 14th, to Saint John's Day, June 24th, his wife and daughters are visible. Following this story, the Pleiades are the Gaelic **Crannarain,** the Baker's Peel, or Shovel, a title shared with Ursa Major.

Another, still homelier, but appropriately feminine, name is hinted at in Holland's translation from the *Historia Naturalis*, where Pliny treats of "the star Vergiliae":

So evident in the heaven, and easiest to be known of all others, it is called by the name of a garment hanging out at a Broker's shop.

Those who have traced out the origin of the title Petticoat Lane for the well-known London street will recognize what Pliny had in mind.

In various ages their title has been taken for noteworthy groups of seven in philosophy or literature. This we see first in the Philosophical Pleiad of 620 to 550 B. C., otherwise known as the Seven Wise Men of Greece, or the Seven Sages, generally given as Bias, Chilo, Cleobūlus, Epimenides or

[1] May it not be from this that comes the English term "Cuckoo Bread," that we find in Mrs. Dana's and Miss Satterlee's delightful book, *How to Know the Wild Flowers*, for the June-flowering *Oxalis*, the dainty Wood Sorrel of our northern groves?

Periander, Pittacus, Solon, and the astronomer Thales; again in the Alexandrian Literary Pleiad, or the Tragic Pleiades, instituted in the 3d century B. C. by Ptolemy Philadelphus, and composed of the seven contemporary poets, variously given, but often as Apollonius of Rhodes, Callimachus or Philiscus, Homer the Younger of Hierapolis in Caria, Lycophron, Nicander, Theocritus, and our Aratos; in the Literary Pleiad of Charlemagne, himself one of the Seven; in the Great Pléiade of France, of the 16th century, brought together in the reign of Henri III, some say by Ronsard, the "Prince of Poets," others by d'Aurat, or Dorat, the "Modern Pindar," called "Auratus," either in punning allusion to his name or from the brilliancy of his genius, and the "Dark Star," from his silence among his companions; and in the Lesser Pléiade, of inferior lights, in the subsequent reign of Louis XIII. Lastly appear the Pleiades of Connecticut, the popular, perhaps ironical, designation for the seven patriotic poets after our Revolutionary War: Richard Alsop, Joel Barlow, Theodore Dwight, Timothy Dwight, Lemuel Hopkins, David Humphreys, and John Trumbull,— all good men of Yale.

I have not been able to learn when, and by whom, the titles of the seven sisters were applied to the individual stars as we have them; but now they are catalogued nine in all, the parents being included. These last, however, seem to be a comparatively modern addition, the first mention of them that I find — in Riccioli's *Almagestum Novum* of 1651 — reading:

> Michaël Florentius Langrenius [1] illarum exactam figuram observavit, & ad me misit, in qua additae sunt duae Stellae aliis innominatae, quas ipse vocat Atlantem, & Pleionem; nescio an sint illae, quas Vendelinus ait observari tanquam novas, quia modō apparent, modō latent.

> . . . the great and burning star,
> Immeasurably old, immeasurably far,
> Surging forth its silver flame
> Through eternity, . . . Alcyone!
>
> Archibald Lampman's *Alcyone*.

η, or Fl. 25, 3, greenish yellow.

Alcyone represents in the sky the Atlantid nymph who became the mother of Hyrieus by Poseidon; but, though now the **Light of the Pleiades,** its mythological original was by no means considered the most beautiful. Riccioli wrote the word **Alcione** and **Alcinoe,** and some early manuscripts have **Altione.**

The early Arabs called it **Al Jauz,** the Walnut; **Al Jauzah** or **Al Wasaṭ,** the Central One; and **Al Na'ir,** the Bright One;—all of **Al Thurayya.** The

[1] This Michel Florent van Langren was of Antwerp, a contemporary and friend of Riccioli, and associated with him in giving names to the various features of the moon's surface.

later Al Achsasi added to this list **Thaur al Thurayya,** which, literally the Bull of the Pleiades, *i. e.* the Leading One, probably was a current title in his day, for his Italian contemporary Riccioli said, in his *Astronomia Reformata*, that the *lucida* "Alcinoe" was **Altorich** *non* **Athorric.** Hipparchos has been supposed to allude to it in his ὀξύς, and ὀξύτατος, τῆς Πλειάδος, the Bright One, and the Brightest One, of the Pleiad. Yet, in the face of these epithets, Ptolemy apparently did not mention it in the *Syntaxis;* while Baily, in his edition of Hyde's translation of Ulug Beg's *Tables*, affixed Flamsteed's 25 and Bayer's η to the 32d star of Taurus, which is described as *stella externa minuta vergiliarum*, *quae est ad latus boreale*,—our Atlas.

In Babylonia it determined the 4th ecliptic constellation, **Temennu,** the Foundation Stone.

In India it was the junction star of the *nakshatras* Krittikā and Rohinī, and individually **Amba,** the Mother; while Hewitt says that in earlier Hindu literature it was **Arundhati,** wedded to Vashishṭha, the chief of the Seven Sages, as her sisters were to the six other Rishis of Ursa Major; and that every newly married couple worshiped them on first entering their future home before they worshiped the pole-star. He thinks this a symbol of the prehistoric union of the northern and southern tribes of India.

We often see the assertion that our title is in no way connected with Ἀλκυών, the Halcyon, that "symbolic or mystical bird, early identified with the Kingfisher," the ornithological *Alcedo* or *Ceryle;* so that although the myth of the Halcyon Days, that "clement and temperate time, the nurse of the beautiful Halcyon,"

> When birds of calm sit brooding on the charmed wave,

is not yet understood, some of Thompson's conjectures as to its stellar aspect will be found interesting. He writes that

> the story originally referred to some astronomical phenomenon, probably in connexion with the Pleiades, of which constellation Alcyone is the principal star. In what appears to have been the most vigorous period of ancient astronomy (not later than 2000 B. C., but continuing long afterwards to influence legend and nomenclature) the sun rose at the vernal equinox, in conjunction with the Pleiad, in the sign Taurus: the Pleiad is in many languages associated with bird-names . . . and I am inclined to take the bird on the bull's back in coins of Eretria, Dicaea, and Thurii for the associated constellation of the Pleiad. . . . Suidas definitely asserts that the Pleiades were called Ἀλκυόνες. At the winter solstice, in the same ancient epoch, the Pleiad culminated at nightfall in mid-heaven. . . . This culmination, between three and four months after the heliacal rising of the Pleiad in Autumn, was, I conjecture, symbolized as the nesting of the Halcyon. Owing to the antiquity and corruption of the legend, it is impossible to hazard more than a conjecture; but that the phenomenon was in some form an astronomic one I have no doubt.

Mädler located in Alcyone the centre of the universe, but his theory has been shown to be fallacious. There is no satisfactory reason for his conclusion, and not much more for Miss Clerke's remarks as to the probable size and distance of Alcyone,— that it shines to its sister stars with eighty-three times the lustre of Sirius in terrestrial skies, while its intrinsic brilliancy, as compared with that of the sun, is 1000 times greater. All this rests upon the extremely doubtful assumption of a parallax of $0''.013$ deduced from the star's proper motion.

It culminates on the 31st of December.

The three little companions, easily visible with a low-power, form a beautiful triangle $3'$ away from Alcyone.

Multi ante occasum Maiae coepere.

Vergil's 1st *Georgic.*

Fl. 20, or Bessel's *c*, 4.

Maia appears in the motto as personifying all the Pleiad stars, and the poet cautions the farmer against sowing his grain before the time of its setting.

She was the first-born and most beautiful of the sisters, and some have said that her star was the most luminous of the group; in fact, Riccioli, in his *Almagestum Novum*, distinctly wrote of Maia: *dicta lucida Pleiadum & tertii honoris, quae mater Mercurii perhibetur*, although in the *Astronomia Reformata* his "Alcinoe" is the *lucida;* so that we are uncertain which of these stars was the **Pleias** that he used for some one of the group. But the mythological importance of the goddess whose name Maia bears would indicate that Riccioli may have been correct as to the first of these identifications, and that the titles of the two stars perhaps should be interchanged.

The name also is written **Mea** and **Maja,** the feminine form of *majus*, an older form of *magnus.* Cicero had the word **Majja,** calling the Pleiad *sanctissima*, for in his day Maia was only another figure for the great and much named Rhea-Cybele, Fauna, Faula, Fatua, Ops, familiarly known as Ma, or Maia Maiestas, the Bona Dea, or Great and Fruitful Mother, who gave name to the Roman month, our May.

Ovid added to her title **Pleias uda,** the Moist Pleiad, as another symbol for the group; and Dante used her title for the planet Mercury, as the Atlantid was mother of that god.

The equivalent **Maou,** for the Pleiades in China, is singularly like the Latin word.

The nebula attached to this star, a part of the general nebulosity that envelops the group, was first noticed in 1882 on photographs by Pickering and the Messrs. Henry.

> . . . the lost Pleiad seen no more below.
>
> Byron's *Beppo*.

Fl. 17, or *b*, 4.6.

Electra, although for at least two or three centuries the title of a clearly visible star, has been regarded as the Lost Pleiad, from the legend that she withdrew her light in sorrow at witnessing the destruction of Ilium, which was founded by her son Dardanos,—as witness Ovid in the *Fasti:*

> Electra Trojae spectare ruinas
> Non tulit ante oculos, opposuitque manum;

or, as Hyginus wrote, left her place to be present at its fall, thence wandering off as a hair-star, or comet; or, reduced in brilliancy, settled down close to Mizar as Ἀλώπηξ, the Fox, the Arabs' Al Suhā, and our Alcor. In the *Harleian Manuscript* the word is written **Electa.**

Ovid called her **Atlantis**, personifying the family.

The Pirt-Kopan-noot tribe of Australia have a legend of a Lost Pleiad, making this the queen of the other six, beloved by their heavenly Crow, our Canopus, and who, carried away by him, never returned to her home.

> Thy beauty shrouded by the heavy veil
> Thy wedlock won.
>
> Elizabeth Worthington Fiske.

Fl. 23, or *d*, 5, silvery white.

Merope often is considered the Lost Pleiad, because, having married a mortal, the crafty Sisyphus, she hid her face in shame when she thought of her sisters' alliances with the gods, and realized that she had thrown herself away. She seems, however, to have recovered her equanimity, being now much brighter than some of the others. The name itself signifies "Mortal."

This star is enveloped in a faintly extended, triangular, nebulous haze, visually discovered by Tempel in October, 1859; and there is a small, distinct nebula, discovered by Barnard in November, 1890, close by Merope, almost hidden in its radiance, although intrinsically very bright.

Taygete simul os terris ostendit honestum Pleias.

Vergil's 4th *Georgic.*

Fl. 19, or *e*, Double, 5.1 and 10, lucid white and violet.

Taygete, or **Taygeta,** a name famous in Spartan story for the mother of Lacedaemon by Zeus, was mentioned by Ovid and Vergil as another representative of this stellar family; the former calling it **Soror Pleiadum,** and the latter using it to fix the two seasons of the honey harvest, as in Davidson's translation of the passage beginning with our motto:

as soon as the Pleiad Taygete has displayed her comely face to the earth, and spurns with her foot the despised waters of the ocean; or when the same star, flying the constellation of the watery Fish, descends in sadness from the sky into the wintery waves.

Ulug Beg applied to it **Al Wasaṭ,** the Central One, usually and more appropriately given to Alcyone.

Bayer lettered it *q*, describing it as *Pleiadum minima;* but the *Century Cyclopedia's* ε is a misprint for *e*.

And is there glory from the heavens departed?
—Oh! void unmarkéd!—thy sisters of the sky
Still hold their place on high,
Though from its rank thine orb so long hath started,
Thou, that no more art seen of mortal eye.

Mrs. Hemans' *The Lost Pleiad.*

Fl. 16, or *g*, 6.5, silvery white.

Celaeno, or **Celeno,** has been called the Lost Pleiad, which Theon the Younger said was struck by lightning!

It gives but one half the light of Taygete; still it can be seen with the naked eye, if a good one, and is so given in the Heis *Verzeichniss.*

The Sister Stars that once were seven
Mourn for their missing mate in Heaven.

Alfred Austin.

Fl. 21 and Fl. 22, or *k* and *l*, 6.5 and 7.

Sterope I and **Sterope II,** less correctly **Asterope,** are a widely double star at the upper edge of the rising cluster, and faintly visible only by reason of the combined light; so that Al Sufi's 5th magnitude seems large.

Ovid made use of **Steropes sidus** to symbolize the whole, but the present magnitudes would show that his star—if, indeed, he referred to any special star at all, as is improbable — was not ours, or else that a change in brilliancy has taken place. In fact, this also, and not without reason, has been called the Lost Pleiad.

> Atlas, that on his brazen shoulders rolls
> Yon heaven, the ancient mansion of the gods.
>
> Potter's translation of Euripides' Ἴων.

Fl. 27, or *f*, Double, 4.5, intense white.

Atlas was **Pater Atlas** with Riccioli, apparently having been added in his day to the original group of the seven daughters. It was of him that Ovid wrote:

> Pleïades incipiunt umeros relevare paternos;

for their setting relieved the father of some of his burden as bearer of the heavens.

With Pleione it marks the end of the handle of the Pleiad Dipper, and probably has a very minute, close companion, said to have been discovered by Struve in 1827, and again revealed, at an occultation by the moon, on the 6th of January, 1876.

> Hinc sata Pleïone cum caelifero Atlante
> Jungitur, ut fama est, Pleïadasque parit.
>
> Ovid's *Fasti*.

Fl. 28, or *h*, 6.5.

Pleione, Riccioli's **Mater Pleione,** and **Plione,** were equally modern additions, although Valerius Flaccus used the word to personify the whole.

As the spectrum of this star shows the bright lines of hydrogen like that of P Cygni, Pickering suggests that it may similarly have had a temporary brilliancy and thus be the Lost Pleiad: a scientific and — if there ever has been in historic time a star in the cluster that is now missing — the most probable solution of this much discussed question; so that the mother seems to have been lost, as well as many of the daughters!

The *Harleian Manuscript* of Cicero's *Aratos* represents the Sisters by plain female heads under the title *VII Pliades et Athlantides*, and individually as Merope, Alcyone, Celaeno, Electra, Ta Ygete, Sterope, and Maia.[1] Grotius has them in the same way, but in far more attractive style, from

[1] Other names, too, were assigned to the mythological septette; the scholiast on Theocritus giving them as Coccymo, Plancia, Protis, Parthemia, Lampatho, Stonychia, and the familiar Maia.

the old *Leyden Manuscript*, where we find the orthography Asterope and Mea, the former of which, appearing with Germanicus, has become common in our day. The German manuscript, dating from the 15th century, shows seven full-length figures, the Dark Sister smaller than the others, and wearing a dark-blue head-dress, the rest brighter in color, with faces of true German type.

While this list includes all the named Pleiad stars, some practically invisible without optical aid, yet every increase of power reveals a larger number. Riccioli wrote about this in 1651:

> Telescopio autem spectatae visae sunt Galileo plus quam 40. ut narratur in Nuncio Sidereo;

a first-rate field-glass, taking in 3¼° and magnifying seven diameters, shows 57; Hooke, in 1664, saw 78 with the best telescope of his day; Swift sees 300 with his 4½-inch, and 600 with his 16-inch; and Wolf catalogued, at the Paris Observatory in 1876, 625 in a space of 90′ by 135′. But with the camera the Messrs. Henry photographed 1421 in 1885, and two years later, by a four-hours' exposure, 2326 down to the 16th magnitude within three square degrees,— more than are visible at any one time by the naked eye in the whole sky. And a recent photograph by Bailey, with the Bruce telescope, reveals 3972 stars in the region 2° square around Alcyone; although there is no certainty that all of these belong to the Pleiades group. Statements as to their magnitudes and distances make many of them exceed Sirius in size, and to be 250 light years away; but these are based upon an assumption of parallax as yet only hypothetical. But, if correct, how appropriate are Young's verses in his *Night Thoughts:*

> How distant some of these nocturnal Suns!
> So distant (says the Sage) 'twere not absurd
> To doubt, if Beams set out at Nature's Birth,
> Are yet arrived at this so foreign World
> Tho' nothing half so rapid as their Flight;

and Longfellow's stanza in his *Ode to Charles Sumner:*

> Were a star quenched on high,
> For ages would its light,
> Still travelling downward from the sky,
> Shine on our mortal sight.

While some of these undoubtedly are only optically connected with the true Pleiades, yet the larger part seem to form a more or less united group, which the spectroscope shows to be of the same general type; this fact being first brought out by Harvard observers in 1886, from comparisons of the spectra of forty of its stars. They are supposed to be drifting together toward the south-southwest, and so may be called a natural constellation.

Nicander wrote of them as *ὀλίζωνας*, "the smaller ones"; Manilius, as *tertia forma*, "the third-sized"; and many think that the light of some has decreased, not only from the legends of the Lost Pleiad and the fact that some of the sisters' names are applied to stars which could not possibly have been seen by the unaided eye, but also because only six are now visible to the average observer, and whoever can see seven can as readily see at least two more. Miss Airy counted twelve; Mr. Dawes, thirteen; and Kepler said that his scholar Michel Möstlin could distinguish fourteen, and had correctly mapped eleven before the invention of the telescope, while others have done about as well; indeed Carl von Littrow has seen sixteen. In the clear air of the tropic highlands more of the group are visible than to us in northern latitudes,—from the Harvard observing station at Arequipa, Peru, eleven being readily seen; so that Willis was unconsciously right in his verses:

> the linkéd Pleiades
> Undimm'd are there, though from the sister band
> The fairest has gone down; and South away!

Smyth wrote:

> If we admit the influence of variability at long periods, the seven in number may have been more distinct, so that while Homer and Attalus speak of six, Hipparchus and Aratus may properly mention seven.

Yet we find Humboldt, in *Cosmos*, saying that Hipparchos refuted the assertion of Aratos that only six are to be seen with the naked eye, and that

> One star escaped his attention, for when the eye is attentively fixed on this constellation, on a serene and moonless night, seven stars are visible.

But Aratos' words do not justify this statement as to his opinion. He wrote:

> seven paths aloft men say they take,
> Yet six alone are viewed by mortal eyes.
> From Zeus' abode no star unknown is lost
> Since first from birth we heard, but thus the tale is told;

this "seven paths," ἑπτάποροι, being first found in the Ῥῆσος attributed to Euripides. Eratosthenes called it Πλειάς ἑπτάστερος, the Seven-starred Pleiad, although he described one as Παναφανής, All-invisible; Ovid repeated from the *Phainomena* the now trite

> Quae septem dici, sex tamen esse solent;

and again:

> Six only are visible, but the seventh is beneath the dark clouds.

Cicero thought of them in the same way; and Galileo wrote *Dico autem sex, quando quidem septima ferē nunquam apparet.* But the early Copts knew them as Ἕξαστρον, the Six-starred Asterism, and many Hindu legends mention only six.

Discarding, of course, all the mythical explanations of the Lost Pleiad, I would notice some of the modern and serious attempts at an elucidation of the supposed phenomenon. Doctor Charles Anthon considered it founded solely upon the imagination, and not upon any accurate observation in antiquity. Jensen thinks that, as a favorite object in Babylonia, the astronomers of that country attached to it, with no regard to exactitude, their number of perfection or completeness, 7 playing with them a more important part even than it did among the Jews; thence it descended to Greece, where, its origin being lost sight of, was caused the discrepancy which we cannot now explain, as well as the legends and folk-lore on the subject. Lamb asserted that the astronomers of Assyria could see in their sky seven stars in the group, and so described them; but the Greeks, less favorably situated, finding only six, invented the story of the missing sister. Riccioli propounded a theory — which I have nowhere found adopted by any later writer — that the seventh and missing Pleiad may have been a *nova* appearing before that number was recorded by observers, but extinguished about the date of the Trojan war; this last idea accounting, too, for the association of Electra with the lost one. Still another explanation is hinted at by Thompson under Coma Berenices; and the really scientific theories of Smyth and Pickering have already been noticed. It is in these last two, I think, that the solution of this interesting question will be found, if at all; and with the astronomers I would leave it, as perhaps I ought to have done before.

Ptolemy mentioned Πλειάς for only four stars in Ταῦρος that Baily said were Flamsteed's 18, 19, 23, and 27, our Alcyone singularly being disregarded, as well as four others of our named stars; and Al Sufi, who revised Ptolemy's observations, stated that this "Alexandrian Quartette" also were the brightest in his day — the 10th century. But Ulug Beg, although he is supposed to have followed Ptolemy, applied "Al Thurayya" to the five that Baily said were Fl. 19, 23, 21, 22, and 25 (Alcyone). Baily himself, editing Hyde's translation of Ulug Beg, gave only Fl. 19 and 23 as of "**Al Thuraja.**"

Recent photographic observations have revealed other nebulous matter, in different degrees of condensation, scattered throughout the cluster, connecting its various members; while Barnard in 1894 found vast nebulosity extendin g almost as far as ζ Persei.

The Pleiades afford so convincing a proof of the popular misapprehension as to the moon's apparent magnitude that I am tempted to introduce another illustration drawn from these stars. The angular distance between Alcyone and Electra and between Merope and Taygeta is greater by several minutes than the mean angular diameter of the moon's disc,— 31′ 7″,— so that the latter could be inserted within the quadrangle formed by those four stars with plenty of room to spare; although in looking at the cluster the impression is that our satellite would cover the whole. An occultation of the Pleiades by the moon gives a vivid realization of this fact; and as this is a not infrequent phenomenon, I commend its observation to any unbeliever.

θ^1 and θ^2, 4.1 and 3.6, pearly white and yellowish,

form a naked-eye double in the Hyades to which Mr. William Peck applies the name **Alya**; but, as this is inappropriate and found with no other author for these stars, may we not suspect error in transcription? — this title belonging by universal recognition to another θ^1,— that of Serpens.

Although 337″ apart, our *thetas* may be in physical relation to each other.

ι, with *k*, *l*, *n*, and *o*, between the horns, all of about the 5th magnitude, were the Chinese **Choo Wang**, the Many Princes.

κ^1 and κ^2, 4.4 and 6.5, and υ, 4.3;

φ, Double, 5.1 and 8, and χ, Double, 5.6 and 8,

stretching from the left eye to the left ear of the Bull, were the Arabs' **Al Kalbain,** the Two Dogs, *i. e.* of Al Dabarān, who, as the Driver of the Pleiades, would naturally have his dogs as near-by attendants.

Reeves included ϕ, χ, and ψ in the Chinese **Li Shih,** a Coarse Sandstone; χ and v in **Tien Keae,** the Heavenly Street; and π and ρ, of the 5th magnitude, with other small stars near the Hyades, in **Tien Tsze,** Heaven's Festival,

A pair of 11th-magnitude stars, 4″.9 apart, lies between the *kappas;* the *phi* stars, yellow and orange in color, are 53″.6 apart; and the components of χ, white and bluish white, are 19″.3 apart.

*

Taurus Poniatovii, Poniatowski's Bull,

the **Taurus Regalis** of Houzeau, is the **Taureau Royal** of the French; **Toro di Poniatowski** of the Italians; **Poniatowsky's Stier** of the Germans; and, on the Stieler *Planisphere*, **Poln Stier,** the **Polish Bull.**

It was made up from unformed stars of Ophiuchus, Smyth writes,

> in 1777 by the Abbé Poczobut, of Wilna, in honour of Stanislaus Poniatowski, King of Poland; a formal permission to that effect having been obtained from the French Academy. It is between the shoulder of Ophiuchus and the Eagle, where some stars form the letter V, and from a fancied resemblance to the zodiac-bull and the Hyades, became another Taurus. Poczobut was content with seven component stars, but Bode has scraped together no fewer than eighty,—

of course chiefly telescopic, for only 20 to 25 are visible to the unaided eye; but as a distinct constellation it is not generally recognized by astronomers, and its stars have been returned to Ophiuchus.

We have no individual names for any of these, but sundry small ones in the head were the Chinese **Tsung Ting, or Tsung Jin,** a Relative.

A century and a half before Poczobut's time these stars, with those of our Vulpecula, had been introduced by Bartsch into his plates as the **River Tigris,** although this probably had previously been a recognized constellation. Its course was from β and γ, in the right shoulder of Ophiuchus, onwards between Aquila and the left hand of Hercules; thence between Albireo (β Cygni) and Sagitta to Equuleus and the front parts of Pegasus, ending at the latter's neck. This Tigris continued until as late as 1679 with Royer, but has long since disappeared from the maps, and indeed from the memory of most observers; while the Royal Bull itself seems to be lapsing into similar obscurity.

Three or four centuries before all this the Arabian engraver of the Borgian globe appropriately represented the stars of this constellation by a **triangular figure.**

It comes to the meridian on the 10th of August.

Although it has no named star, its "70 Ophiuchi," the middle one in the eastern leg of the V, is a celebrated binary, with a period of about ninety years, the components 2″ apart, at a position angle, in the year 1897, of 276°.58. A third invisible companion is suspected.

★

Telescopium, or Tubus Astronomicus,

was formed by La Caille between Ara and Sagittarius on the edge of the Milky Way, but in such irregular form that it encroached upon four of the old constellations; η Sagittarii having been taken as β to mark the Telescope's stand; *d* Ophiuchi for its θ; σ was in Corona Australis; and γ was the v of Scorpio. Bode had it in his *Gestirne* of 1805 as the **Astronomische Fernrohr,** crowding it in between Sagittarius and Scorpio; but Baily and Gould restricted it to the south of Scorpio, Sagittarius, and Corona Australis.

Gould assigned to it 87 naked-eye stars, the brightest a 3½-magnitude.

Small as these are, two bore individual titles in Chinese astronomy; α being known as **We,** Danger; and γ as the mythological **Chuen Shwo.**

The constellation culminates on the 13th of August, at the same time as Wega of the Lyre.

★

Telescopium Herschelii,

formed by the Abbé Hell in 1781, in honor of Sir William Herschel, was first published by Bode in 1800. It lay between the Lynx and Gemini and appears on Burritt's *Atlas*; but since his day has passed away from the maps and catalogues.

The star π of Gemini marks its former location, the western end having been among the ψ stars of Auriga, not far from the latter's β.

★

Five splendid Stars in its *unequal* Frame
Deltoton bears, and from the shape a Name;
But those that grace the sides dim Light display
And yield unto the *Basis* brighter Ray.
Creech's *Manilius.*

Triangulum,

the German **Dreieck,** the French and English **Triangle,** and the Italian **Triangolo,** appeared as **Triangulus** in the *Rudolphine Tables*, always qualified as *major* till the Lesser Triangle was discarded. It lies just south from γ Andromedae on the edge of the Milky Way, and although small and faint notwithstanding our poet's description, is one of the old constellations evidently more noticed by the ancients than by us. They drew it as equilateral, but now it is a scalene figure, β, δ, γ at the base and α at the vertex.

Hood strangely said that it was placed in the heavens only that the head of Aries might be better known, which recalls the blunder of Aratos as to the faintness of Aries' stars.

It was Δελτωτόν with the earlier Greeks, from their similarly shaped letter Δ, to which Ovid in his *Nux* likened it; as did Aratos in his lines that Brown renders, more literally than rhythmically:

> Below Andromeda, in three sides measured
> Like-to-a-Delta; equal two of them
> As it has, less the third, yet good to find
> The sign, than many better stored with stars.

Transcribed by Cicero and Hyginus as **Deltoton,** it became **Deltotum** with the Romans, as well as with astronomers to the 17th century. Naturally it also was **Delta,** and so, associated with Egypt and the Nile, became **Aegyptus, Nilus, Nili Domum,** the Home of the Nile, which originally was **Nili Donum,** the Gift of the Nile, from Herodotus' ποταμοῦ δῶρον, "the river's gift."

Τρίγωνον, used by Hipparchos and Ptolemy, became **Trigonum** with Vitruvius, and **Trigonus** with Manilius, translated **Trigon** by Creech. **Tricuspis,** Three-pointed, and **Triquetrum,** the Trinal Aspect of astrology, are found for it; while Bayer had **Triplicitas** and **Orbis terrarum tripertitus** as representing the three parts of the earth, Europe, Asia, and Africa; and **Triangulus Septentrionalis,** to distinguish it from his own Southern Triangle.

Pious people of his day said that it showed the **Trinity,** its shape resembling the Greek initial letter of Δῖος; while others of the same sort likened it to the **Mitre of Saint Peter.**

Its titles **Sicilia, Trinacria,** and **Triquetra** are those of the ancients for the similarly shaped island of Sicily,— that Ceres had begged of Jove might be reproduced in the sky,— triangular from its three promontories, Lilybaeum, Pelorus, and Pachynus, and at times identified with the mythical Thrinakia of the *Odyssey*, the pasture-ground of the Oxen of the Sun, that Gower called Mela's Holy Ox-land. In modern days it has been noted as the site of the famous Palermo Observatory.

It was here that was discovered by Piazzi, on the first New Year Day of the present century, the first minor planet, which he named **Ceres Fer-**

dinandea in joint honor of the patron goddess of the island and of his king, the Bourbon Ferdinand of Naples; but the adjective has been dropped by astronomers as not conforming to their rule of mythological nomenclature for the planets,— a rule, however, much deviated from in recent times in the naming of these little bodies. Perhaps the astronomers have exhausted their classical dictionaries! It was found[1] as an 8th-magnitude star — Flammarion says as a comet — between Aries and Taurus, coincidently not far from our Triangulum, the ancient **Sicilia**; but it was little imagined at the time that 433 similar bodies would be found in the next ninety-seven years, more than 150 of them since 1892, and all but seven of these last by photography,[2] then an unknown art.

The Arabians translated our title as **Al Muthallath,** variously seen in Western usage as **Almutallath, Almutaleh, Almutlato, Mutlat, Mutlaton, Mutlathum, Mutlathun,** and **Mutlatun,** with probably still other similarly degenerated forms of the original.

The Jews are said to have known it as **Shālīsh,** from the name of an instrument of music of triangular shape, or with three cords, mentioned in the 1st *Book of Samuel*, xviii, 6. This same figure, for the three bright stars of Aries, has already been noticed at γ of that constellation.

Heis enumerates here 30 naked-eye components, but Argelander only 15.

The Chinese asterism **Tsien Ta Tseang,** Heaven's Great General, included this with λ of Andromeda and the stars of the Smaller Triangle.

α, 3.6, yellow.

Caput Trianguli was translated **Rās al Muthallath** by the Arabian astronomers.

It is a half-magnitude inferior to β, although the latter bears no name.

Together these two were the Arabs' **Al Mīzān,** the Scale-beam.

a comes to the meridian on the 6th of December.

[1] This, like many other important discoveries, was by a happy accident,— Piazzi, very differently, being in search of an extra star, the eighty-seventh of Mayer's list, wrongly laid down in Wollaston's catalogue.

Recent measurements by Barnard show that Ceres is only a little less than 500 miles in diameter, and thus the first in size of the minor planets as in order of discovery.

[2] The first of such discoveries by the camera was by Wolf on the 20th of December, 1891, of Brucia, No. 323; the first applications of the new art to the heavens having been made with the daguerreotype process by Doctor John W. Draper, of New York City, on the moon in 1840; again, by the professional Whipple of Boston, under Bond's direction, at the Harvard Observatory, on the star Wega in 1850; and at the same place on Mizar and Alcor in 1857. The first photograph of a star's spectrum was in 1872; of a nebula, in 1880; of a comet (near the sun during the latter's total eclipse), in 1882; and of a meteor, in 1891.

Triangulum Minor

was formed, and thus named, by Hevelius, from three small stars immediately to the south of the major constellation, towards Hamal of Aries; but it has been discontinued by astronomers since Flamsteed's day. Still Gore has recently revived it in the title **Triangula** on the planisphere in his translation of *l'Astronomie Populaire*, as did Proctor in his reformed list.

★

Triangulum Australe, the Southern Triangle,

much more noticeable than its northern original, first appeared in print in Bayer's *Uranometria* of 1603, although its formation is attributed to Pieter Theodor of nearly a century previous.

Caesius cited names for it drawn from the older constellation, among them **Almutabet algenubi** *Arabicē neotericis*, which would show that either the Arabians had anticipated Bayer, or were very prompt to learn of his work. But he also called it the **Three Patriarchs,** doubtless Abraham, Isaac, and Jacob, from its three prominent stars; and Julius Schiller had recourse to their descendants for his alphabetical title **Signum Tau.** Proctor catalogued it as plain **Triangulum,** the Northern Triangle being one of his **Triangula.** The French, Germans, and Italians exactly translate the Latin words. The Chinese equivalent is **San Kiō Hung.**

The constellation lies south of Ara, between the tail of Pavo and the fore feet of the Centaur, Gould assigning to it 46 components down to the 7th magnitude. The *lucida* α comes to the meridian on the 14th of July.

α, 2.2, β and γ, 3.1 each, were—perhaps are now—the seamen's **Triangle Stars.**

Ideler said that La Caille substituted for it Norma et Regula, but in maps of the present day both constellations appear side by side.

★

Tucana, the Toucan,

was published by Bayer under our English name,[1] but some one has Latinized it in ornithologists' style as we now see it. Burritt had **Toucana** and

[1] Professor Alfred Newton says that the avian word may be from the Guaranis' Tī, Nose, and Cāng, Bone; and that it first was mentioned in print by Trevét in 1558 as from that Brazilian Indian tribe. It is the *Rhamphastos toco* of the naturalists.

Touchan; the French, **Toucan**; the Italians, **Toucano**; and the Germans, **Tukan.** The Chinese translated the original word, given to them by the Jesuits, as **Neaou Chuy,** the Beak Bird, very appropriate to a creature that is almost all beak.

In the 17th century the English called it the **Brasilian Pye,** but Caesius gave it the geographically incorrect **Pica Indica**; while Kepler, Riccioli, and even later authors knew it as the **Anser Americanus,** a title that appears as late as Stieler's planisphere of 1872, in the **American Gans.**

Tucana lies immediately south of Phoenix, bordering on the south polar Octans, its tail close to the bright Achernar of Eridanus, and marks the crossing of the equinoctial colure and the antarctic circle.

Gould assigned to it 81 naked-eye stars, from 2.8 to the 7th magnitudes.

The 4th-magnitude γ is very blue, and the 5½ ν, strongly red; but its most notable object is Bode's cluster 47, N. G. C. 104. This celebrated "ball of suns" has been lettered ξ by Gould, as it shines like a hazy 4½-magnitude star. Bailey counted, within 660″ of its centre, 2235 stars, and among them six variables. The cluster seems to be completely insulated with regard to the surrounding stars.

*

Turdus Solitarius, the Solitary Thrush,

was formed by Le Monnier in 1776 from the faint stars over the tail-tip of the Hydra, where some modern seeker of fame has since substituted another avian figure, the **Noctua,** or **Night Owl.**

The title[1] is said to be that of the Solitaire, formerly peculiar to the little island Rodriguez in the Indian Ocean, 344 miles to the eastward of Mauritius; although the bird has been extinct for two centuries,— as indeed now is the constellation.

Little seems to be known of this sky figure, although Ideler wrote of it as **Einsiedler,** the German **Drossel.**

[1] The generic word *Turdus*, however, is erroneous; for the bird was not a thrush, but, as its correct name, *Pezophaps solitaria*, denotes, an extremely modified form of flightless pigeon allied to the dodos, yet larger and taller than a turkey.

'Twas noon of night, when round the pole
The sullen Bear is seen to roll.
Thomas Moore's translation of the *Odes* of Anacreon.

. . . round and round the frozen Pole
Glideth the lean white bear.
Robert Williams Buchanan's *Ballad of Judas Iscariot.*

Ursa Major, the Greater Bear,

the **Grande Ourse** of the French, the **Orsa Maggiore** of the Italians, and the **Grosse Bär** of the Germans, always has been the best known of the stellar groups, appearing in every extended reference to the heavens in the legends, parchments, tablets, and stones of remotest times. And Sir George Cornewall Lewis, quoting allusions to it by Aristotle, Strabo, and many other classical writers, thinks, from Homer's line,

Arctos, sole star that never bathes in th' ocean wave

(by reason of precession it then was much nearer the pole than it now is), that this was the only portion of the arctic sky that in the poet's time had been reduced to constellation form. This statement, however, refers solely to the Greeks; for even before Homer's day we know that earlier nations had here their own stellar groups; yet we must remember that the Ἄρκτος and Ἅμαξα of the *Iliad* and *Odyssey* consisted of but the seven stars, and that these alone bore those names till Thales formed our Ursa Minor. Later on the figure was enlarged "for the purpose of uranographic completeness," so that Heis now catalogues 227 components visible to his naked eye, although only 140 appeared to Argelander, down to the 6th magnitude.

It is almost the first object to which the attention of beginners in astronomy is called,— a fact owing partly to its circumpolar position for all points above the 41st parallel rendering it always and entirely visible above that latitude, but very largely to its great extent and to the striking conformation of its prominent stars. It is noticeable, too, that all early catalogues commenced with the two Ursine constellations.

Although the group has many titles and mythical associations, it has almost everywhere been known as a Bear, usually in the feminine, from its legendary origin. All classic writers, from Homer to those in the decline of Roman literature, thus mentioned it,— a universality of consent as to its form which, it has fancifully been said, may have arisen from Aristotle's idea that its prototype was the only creature that dared invade the frozen North.

Yet it is remarkable that the Teutonic nations did not know this stellar group under this shape, although the animal was of course familiar to them and made much of in story and worship. With them these stars were the **Wagen,** our familiar **Wain.** Aratos wrote in the *Phainomena:*

> Two Bears
> Called Wains move round it, either in her place;

Ovid, in the *Tristia,* **Magna minorque ferae;** and Propertius included both in his **Geminae Ursae;** while Horace, Vergil, and Ovid, again, called them **Gelidae Arcti.** We also meet with **Arctoi** and **Arctoe.** The Anglo-Saxon *Manual of Astronomy* of the 10th century adopted the Greek **Arctos,** although it adds "which untaught men call **Carles-wæn**"; rare old Ben Jonson, in 1609, in his *Epicoene, or the Silent Woman,* called Kallisto

> a star **Mistress Ursula** in the heavens;

and La Lande cited **Fera major, Filia Ursae,** and **Ursa cum puerulo,** referring to Arcas.

The well-known, although varied, story of Καλλιστώ,— as old as Hesiod's time,— who was changed to a bear because of Juno's jealousy and transferred to the skies by the regard of Jove, has given rise to much poetical allusion from Hesiod's day till ours, especially among the Latins. In Addison's translation of Ovid's *Metamorphoses,* where this myth is related, we read that Jove

> snatched them through the air
> In whirlwinds up to heaven and fix'd them there;
> Where the new constellations nightly rise,
> And add a lustre to the northern skies;

although the dissatisfied Juno still complained that in this location they

> proudly roll
> In their new orbs and brighten all the pole.

This version of the legend turned Kallisto's son Arcas into Ursa Minor, although he was Boötes; Matthew Arnold correctly writing of the mother and son in his *Merope:*

> The Gods had pity, made them Stars.
> Stars now they sparkle
> In the northern Heaven —
> The guard Arcturus,
> The guard-watch'd Bear.

Another version substituted her divine mistress Ἄρτεμις — also known to the Greeks as Καλλίστη, the Roman Diana — for the nymph of the celestial transformation; the last Greek word well describing the extreme beauty of this constellation. La Lande, however, referred the title to the Phoenician **Kalitsah,** or **Chalitsa,** Safety, as its observation helped to a safe voyage.

Among its names from the old story are **Kallisto** herself; **Lycaonia, Lycaonia Puella, Lycaonia Arctos,** from her father, or grandfather, king of the aboriginal race that was known as late as Saint Paul's day, with the distinct dialect alluded to in the *Acts of the Apostles*, xiv, 11; **Dianae Comes** and **Phoebes Miles** are from her companionship in arms with that goddess; and it was one of the

> arctos oceani metuentes aequore tingi,

because Tethys, at Juno's instigation, had forbidden Kallisto to enter her watery dominions. Yet Camões, from a lower latitude, wrote of **As Ursas:**

> We saw the Bears, despite of Juno, lave
> Their tardy bodies in the boreal wave.

Ovid's *arctos aequoris expertes; immunemque aequoris Arcton; liquidique immunia ponti*, and *utraque sicca*, were from the fact that, being circumpolar, neither of the Bears sets below the ocean horizon. This was a favorite conceit of the poets, and astronomically correct during millenniums before and centuries after Homer's day, although not so in recent times as to the Greater, except in high latitudes. Chaucer reproduced this in his rendering of the *De Consolatione Philosophiae* by Boëtius, whom he styles Boece:

> Ne the sterre y-cleped "the Bere," that enclyneth his ravisshinge courses abouten the soverein heighte of the worlde, ne the same sterre Ursa nis never-mo wasshen in the depe westrene see, ne coveiteth nat to deyen his flaumbe in the see of the occian, al-thogh he see other sterres y-plounged in the see;

our Bryant rendering this idea:

> The Bear that sees star setting after star
> In the blue brine, descends not to the deep.

Poetical titles induced by the legend of Arcas were **Virgo Nonacrina** and **Tegeaea Virgo,** from the Arcadian towns Nonacris and Tegea; **Erymanthis,** perhaps the Erymanthian Boar that Hercules slew, but more probably the Erymanthian Bear; **Maenalia Arctos, Maenalis,** and **Maenalis Ursa,** from those mountains; **Parrhasis, Parrhasia Virgo,** and **Parrhasides Stellae,** from the tribe, although Pluche went farther back for this to the Phoenician pilots' **Parrasis,** the Guiding Star,— the Hebrews' **Pharashah.** Sophocles wrote of it in the *Oedipus* as **Arcadium Sidus,** referring to the whole country of Arcadia, the Switzerland of Greece, famous in the classical world for its wild mountain scenery; and very early silver coins of Mantinea showed the Bear as mother of the patron god.

Such has been the myth of this constellation current for at least three millenniums; but Mueller discards it all, and says:

> The legend of Kallisto, the beloved of Zeus and mother of Arkas, has nothing to do with the original meaning of the stars. On the contrary, Kallisto was supposed to have been changed into the Arktos or Greater Bear because she was the mother of Arkas, that is to say, of the Arcadian [1] or bear race, and her name, or that of her son, reminded the Greeks of their long established name of the northern constellation.

Aratos' version of the legend, from very ancient Naxian tradition, made the two Bears the Cretan nurses of the infant Jupiter, afterwards raised to heaven for their devotion to their charge. From this came the **Cretaeae sive Arctoe** of Germanicus; but Lewis said:

> This fable is inconsistent with the natural history of the island; for the ancients testify that Crete never contained any bears or other noxious animals.

Subsequent story changed the nurses into the Cretan nymphs Helice and Melissa. Hyginus and Germanicus also used the masculine form **Ursus** as well as **Arctus.**

The Hebrew word **ʿĀsh** or **ʿAyish** in the *Book of Job*, ix, 9, and xxxviii, 32, supposed to refer to the Square in this constellation as a **Bier,** not a Bear, was translated **Arcturus** by Saint Jerome in the *Vulgate;* and this was adopted in the version of 1611 authorized by King James. Hence the popular belief that the *Bible* mentions our star *α* Boötis; but Umbreit had already corrected this to "the Bear and her young," and in the Revision of 1885 the patriarch talks to us of "the Bear with her train," these latter being represented by the three tail stars. Von Herder strangely rendered the first of these passages "Libra and the Pole Star, the Seven Stars"; but the second, more correctly, as "the Bear with her young" feeding around the pole; or, by another tradition, the nightly wanderer, a mother of the stars seeking her lost children,— those that no longer are visible. The

[1] Lucian, in *De Astrologia*, wrote that "the Arcadians were an ignorant people and despised astronomy"; and Ovid graphically described their great antiquity and primitive mode of life, well justifying their title of the Bear Race, his lines being quaintly translated by Gower:

> Therefore they naked run in sign and honour
> Of hardiness and that old bare-skinned manner.

Breeches Bible has this marginal note to its word Arcturus: "The North Star, with those that are about him.".

Hebrew observers called the constellation **Dōbh**; Phoenician, **Dub**; and Arabian, **Al Dubb al Akbar,** the Greater Bear,— **Dubhelacbar** with Bayer and **Dub Alacber** with Chilmead,— all of these perhaps adopted from Greece. Caesius cited the "Mohammedans'" **Dubbe, Dubhe,** and **Dubon**; and Robert Browning, in his *Jochanan Hakkadosh*, repeated these as **Dob.**

But whence came the same idea into the minds of our North American Indians? Was it by accident? or is it evidence of a common origin in the far antiquity of Asia? The conformation of the seven stars in no way resembles the animal,— indeed the contrary; yet they called them **Okuari** and **Paukunawa,** words for a "bear," before they were visited by the white men, as is attested by Le Clercq in 1691, by the Reverend Cotton Mather in 1712, by the Jesuit missionary La Fitau in 1724, and by the French traveler Charlevoix in 1744. And Bancroft wrote in his history of our country:

> The red men . . . did not divide the heavens, nor even a belt in the heavens, into constellations. It is a curious coincidence, that among the Algonquins of the Atlantic and of the Mississippi, alike among the Narragansetts and the Illinois, the North Star was called the Bear.

In justice, however, to their familiarity with a bear's anatomy, it should be said that the impossible tail of our Ursa was to them either **Three Hunters,** or a **Hunter with his two Dogs,** in pursuit of the creature; the star Alcor being the pot in which they would cook her. They thus avoided the incongruousness of the present astronomical ideas of Bruin's make-up, although their cooking-utensil was inadequate. The Housatonic Indians, who roamed over that valley from Pittsfield through Lenox and Stockbridge to Great Barrington, said that this chase of the stellar Bear lasted from the spring till the autumn, when the animal was wounded and its blood plainly seen in the foliage of the forest.

The long tail of the Bear, a queer appendage to a comparatively tailless animal, is thus accounted for by old Thomas Hood in his didactic style:

> *Scholar.*
>
> I marvell why (seeing she hath the forme of a beare) her tail should be so long.
>
> *Master.*
>
> Imagine that Jupiter, fearing to come too nigh unto her teeth, layde holde on her tayle, and thereby drewe her up into the heaven; so that shee of herself being very weightie, and the distance from the earth to the heavens very great, there was great likelihood that her taile must stretch. Other reason know I none.

My friend the Reverend Doctor Robert M. Luther of Newark, New Jersey, tells me that a similar story was current with the Pennsylvania Germans of forty years ago. The same "weightie" reason will apply equally well to the Smaller Bear; indeed the latter's tail is even proportionately longer, although the kink in it takes a different turn. It is probably this association of these Seven Stars with our aborigines that has given them the occasional title of the **Seven Little Indians.**

Trevisa derived the title thus: "alwey thoo sterres wyndeth and turneth rounde aboute that lyne, that is calde Axis, as a bere aboute the stake. And therefore that cercle is clepid the more bere." Boteler borrowed this for his *Hudibras*:

> And round about the pole does make
> A circle like a bear at stake.

The great epic of the Finns, the *Kalewala*, makes much of this constellation, styling it **Otawa** and **Otawainen,** in which Miss Clerke sees likeness to the names used by our aborigines for "the great Teutonic King of beasts." But that people also said that the Bear stars, and especially the pole-star, were young and beautiful maidens highly skilled in spinning and weaving,— a story originating from a fancied resemblance of their rays of light to a weaver's web.

The *Century Dictionary* has a theory as to the origin of the idea of a Bear for these seven stars, doubtless from its editor, Professor Whitney, that seems plausible,— at all events, scholarly. It is that their Sanskrit designation, **Riksha,** signifies, in two different genders, "a Bear," and "a Star," "Bright," or "to shine,"— hence a title, the **Seven Shiners,**— so that it would appear to have come, by some confusion of sound, of the two words among a people not familiar with the animal. Later on Riksha was confounded with the word **Rishi,** and so connected with the **Seven Sages,** or **Poets,** of India; afterwards with the **Seven Wise Men** of Greece, the **Seven Sleepers** of Ephesus, the **Seven Champions of Christendom,** etc.; while the **Seven Stars** of early authors, as often used for Ursa Major as for the Pleiades, certainly is much more appropriate to the Ursine figure than to the Taurine. Minsheu had "the Seven Starres called Charles Waine in the North," and three centuries earlier Chaucer wrote of "the sterres seven" with manifest reference to this constellation. The *Kalewala* had the equivalent **Seitsen tahtinen;** the Portuguese Camões, **Sete Flammas;** and the Turks, **Yidigher Yilduz.**

Hewitt says that these seven stars at first were known in India as **Seven Bears,** although also as **Seven Antelopes,** and again as **Seven Bulls,** the latter merged into one, the **Great Spotted Bull,** as the Seven Bears also were into Ursa Major, with our Arcṭurus for their keeper; and he gives their individual titles as Kratu for α, Pulaha for β, Pulastya for γ, Atri for δ, Añgiras for ϵ, and Marīci for η, the six sons of Brahma, who himself was Vashishṭha, the star ζ. The *Vishnu-Dharma*, however, claimed Atri as their ruler; indeed, there seems to be much variance in Sanskrit works as to the identity of these stars and titles.

When the figure of the Bear was extended to its present dimensions, four times as great as Homer's Arktos, we do not know, and, to quote again from Miss Clerke,

we can only conjecture; but there is evidence that it was fairly well established when Aratos wrote his description of the constellations. [He stretched it over Gemini, Cancer, and Leo.] Aratos, however, copied Eudoxus, and Eudoxus used observations made — doubtless by Accad or Chaldaean astrologers — above 2000 B. C. We infer, then, that the Babylonian Bear was no other than the modern Ursa Major. . . . Thus, circling the globe from the valley of the Ganges to the great lakes of the New World, we find ourselves confronted with the same sign in the northern skies, the relic of some primeval association of ideas, long since extinct. Extinct even in Homer's time.

And Achilles Tatios distinctly asserted that it was from Chaldaea. But Brown thinks, in regard to the identity of the archaic and modern constellations of this name in that country,

that at present there is no real evidence to connect the **Kakkabu Dabi** (or **Dabu,** the Babylonian Bear) with the Plough or Wain, still less with Ursa Major;

and identifies the latter with the Euphratean **Bel-me-Khi-ra,** the Confronter of Bel,— Bertin, with Bel himself. A group of seven stars is often shown on the cylinders from Babylonia, Lajard's *Culte de Mithra* giving many instances of this, although the reference may have been to the Pleiades; while it is Sayce's suggestion that perhaps "the god seven," so frequently mentioned in the inscriptions, is connected with Ursa Major.

Theon's attribution of the invention of the constellation to the mythical Nauplius, son of Poseidon, and a famous navigator, hardly seems worthy of mention.

Among the adjacent Syrians it was a **Wild Boar,** and in the stars of the feet of our Bear the early nomads saw the tracks of their *Ghazal.* Similarly, in the far North, it has been the **Sarw** of the Lapps, their familiar Reindeer, the **Los** of the Ostiaks, and the **Tukto** of the Greenlanders.

Smyth wrote in his *Speculum Hartwellianum:*

King Arthur, the renowned hero of the *Mabinogion,* typified the Great Bear; as his name,— Arth, bear, and Uthyr, wonderful,— implies in the Welsh language; and the constellation, visibly describing a circle in the North Polar regions of the sky, may possibly have been the true origin of the Son of Pendragon's famous Round Table, the earliest institution of a military order of knighthood.

Whatever may be the fact in this speculation, we know that the early English placed King Arthur's home here, and that the people of Great Britain long called it **Arthur's Chariot** or **Wain,** which appears in the *Lay of the Last Minstrel:*

> Arthur's slow wain his course doth roll,
> In utter darkness, round the pole.

In Ireland it has been **King David's Chariot,** from one of that island's early kings; in France, the **Great Chariot,** and it was seen on Gaulish coins. The Anglo-Norman poet De Thaun of the 12th century had it **Charere;** and La Lande cited the more modern **la Roue,** the Wheel. Occasionally it has been called the **Car of Boötes.**

And this carries us back to another of the earliest titles for our constellation, the Ἅμαξα, **Wain** or **Wagon,**— Riccioli's **Amaxa,**— of the *Iliad* and *Odyssey,* that Homer used equally with Ἄρκτος, although with the same limitation to the seven stars. Describing the shield made by Hephaistos for Achilles, the poet said, in Sir John Herschel's rendering:

> There the revolving Bear, which the Wain they call, was ensculptured,
> Circling on high, and in all its course regarding Orion;
> Sole of the starry train which refuses to bathe in the Ocean;

which I have quoted, in preference to others more rhythmical, from the interest that we all feel in the translator as an astronomer, although but little known as a poet. Homer repeated this in the 5th book of the *Odyssey,* where Ulixes, in Bryant's translation, is

> Gazing with fixed eye on the Pleiades,
> Boötes setting late and the Great Bear,
> By others called the Wain, which wheeling round,
> Looks ever toward Orion and alone
> Dips not into the waters of the deep.
> For so Calypso, glorious goddess, bade
> That, on his ocean journey, he should keep
> That constellation ever on his left;

Ithaca, whither he was bound, lying due east from Calypso's isle, Orgygia. Pope rendered the original the **Northern Team,** and the lines on Orion:

> To which, around the axle of the sky,
> The Bear, revolving, points his golden eye.

These passages clearly show the early use of the Wain stars in Greek navigation before Cynosura was known to them; as Aratos wrote:

By it on the deep
Achaians gather where to sail their ships;

Ovid imitating this in the *Fasti* and *Tristia.* Orion seems to have been often joined in this use, for Apollonius wrote:

The watchful sailor, to Orion's star
And Helice, turned heedful.

Aratos called the constellation the "Wain-like Bear"; and, alluding to the title Ἅμαξα, asserted that the word was from ἅμα, "together," the Ἅμαξαι thus circling together around the pole; but no philologist accepts this, and it might as well have come from ἄξων, "axle," referring to the axis of the heavens. In fact, Hewitt goes far back of Aratos in his statement that the Sanskrit god Akshivan, the Driver of the Axle (Aksha), was adopted in Greece as Ixion, whose well-known wheel was merely the circling course of this constellation. Anacreon mentioned it as a Chariot as well as a Bear; and Hesychios had it Ἄγαννα, an archaic word from ἄγειν, "to carry," singularly like, in orthography at least, the Akkadian title for the Wain stars, **Aganna,** or **Akanna,** the Lord of Heaven; and Aben Ezra called it **Ajala,** the Hebrew word for "wagon."

The Romans expressed the same idea in their **Currus; Plaustrum,**[1] or **Plostrum, magnum;** with the diminutive **Plaustricula,** which Capella turned into **Plaustriluca,** imitating the "Noctiluca" used by Horace for the moon. Apollinaris Sidonius, the Christian writer of the 6th century, called the constellation **Plaustra Parrhasis;** and Rycharde Eden wrote it **Plastrum,—**

al the sterres cauled Plastrum or Charles Wayne, are hydde under the Northe pole to the canibals.

In all these, of course, reference was made to the seven stars only, Bartschius plainly showing this on his chart, where he outlines them, with the title Plaustrum, included within the limits of the much larger Ursa Major.

The Italians have **Cataletto,** a Bier, and **Carro;** and the Portuguese Camões wrote it **Carreta.**

The Danes, Swedes, and Icelanders knew it as **Stori Vagn,** the Great Wagon, and as **Karls Vagn;** Karl being Thor, their greatest god, of whom the old Swedish *Rhyme Chronicle*, describing the statues in the church[2] at Upsala, says:

[1] The Latin *plaustrum*, originally a two-wheeled ox-cart, appears in the *De Re Rustica* of Cato Censorius as *plaustrum maius* for one with four wheels.

[2] It is in this church, or cathedral, that the great Linnaeus lies buried, and over its south porch is sculptured the Hebrew story of the Creation.

> The God Thor was the highest of them;
> He sat naked as a child,
> Seven stars in his hand and Charles's Wain.

The Goths similarly called the seven stars **Karl Wagen,** which has descended to modern Germans as **Wagen** and **Himmel Wagen,** the last with the story that it represents the **Chariot in which Elijah journeyed to heaven.** But in the heathen times of the northern nations it was the **Wagon of Odin, Woden,** or **Wuotan,** the father of Thor, and the **Irmines Wagen** of the Saxons. Grimm cites **Herwagen,** probably the **Horwagen** of Bayer and the **Hurwagen** of Caesius; while a common English name now is the **Waggon.** The Poles call it **Woz Niebeski,** the Heavenly Wain. In all these similes the three tail stars of our Bear were the three draught-horses in line.

The royal poet King James wrote:

> Heir shynes the charlewain, there the Harp gives light,
> And heir the Seamans Starres, and there Twinnis bright.

This old and still universally popular title, **Charles's Wain,** demands more than mere mention. It has often been derived from the Saxon *ceorl,* the *carle* of mediaeval times, our *churl*, and thus the "peasant's cart"; but this is incorrect, and the *New English Dictionary* has an exhaustive article on the words, well worthy of repetition here:

Charles's Wain. Forms: carles-wæn, Cherlemaynes-wayne, Charlmons wayn, carle wensterre, carwaynesterre, Charel-wayn, Charlewayn, Charle wane, Charles wayne or waine, Charles or Carol's wain(e), Charlemagne or Charles his wane, wain(e), Charle-waine, Charlmaigne Wain, Charles's Wain. [OE. *Carles wægn,* the wain (*ἅμαξα, plaustrum*) of Carl (Charles the Great, Charlemagne). The name appears to arise out of the verbal association of the star-name *Arcturus* with *Arturus* or Arthur, and the legendary association of Arthur and Charlemagne; so that what was originally the wain of Arcturus or Boötes ('Boötes' golden wain,' *Pope*) became at length the wain of Carl or Charlemagne. (The guess *churl's* or *carle's wain* has been made in ignorance of the history.)]

As the name *Arcturus* was formerly sometimes applied loosely to the constellation Boötes, and incorrectly to the Great Bear, the name *Carlewayne-sterre* occurs applied to the star Arcturus.

The editor cites from various authors since the year 1000, when he finds **Carleswæn** (I can make a still earlier citation of this word from one of the Anglo-Saxon *Cottonian Manuscripts* of some years previously), and quotes from Sir John Davies, the philosophical poet of the Elizabethan age:

> Those bright starres
> Which English Shepheards, Charles his waine, do name;
> But more this Ile is Charles, his waine,
> Since Charles her royall wagoner became;

and from John Taylor, "the King's water-poet," of 1630:

Charles his Cart (which we by custome call Charles his wane) is most gloriously stellifide.

The list ends with a quotation from J. F. Blake, of 1876, who even at this late day had **King Charles' Wain.**

This connection of these Seven Stars with England's kings was due to the courtiers of Charles I and II, who claimed it as in their masters' honor, and elsewhere occurs; William Bas, or Basse, about 1650, having, in *Old Tom of Bedlam:*

Bid Charles make ready his waine;

James Hogg, the Ettrick Shepherd, in the *Queen's Wake* of 1813:

Charles re-yoked his golden wain;

and Tom Hood, of fifty years ago:

looking at that Wain of Charles, the Martyr's.

This is from the *Comet*, the humorous *Astronomical Anecdote* of the great Sir William Herschel, whom the poet called the "be-knighted," and further described as

like a Tom of Coventry, sly peeping,
At Dian sleeping;
Or ogling thro' his glass
Some heavenly lass
Tripping with pails along the Milky Way.

Coverdale's Bible alludes to it and its companion as the **Waynes of Heaven,** which Edmund Becke, in his edition of 1549, transforms into **Vaynes,** and Cadmarden, in his Rouen edition of 1515, into the **Waves of Heaven.** Dutch and German versions have **Wagen am Himmel**; the Saxon versions, **Wænes Thīsl,** or Wagon-pole; and this idea of a wagon, or its parts and its driver, is seen in all the Northern tongues where the Bear is not recognized. Grimm's *Teutonic Mythology* is very full as to this branch of the stellar Wain's nomenclature.

Πλειάδα, the *Septuagint's* rendering of the Hebrew ʽĀsh, is manifestly incorrect, but may have misled the later Rabbis who applied this last word to the group in Taurus. The *Peshitta-Syriac Version* translates the **Mazzārōth** of the *Book of Job* by **ʽgaltā,** meaning our Wain.

The 15th-century German manuscript so often alluded to mentions it as the **Southern Tramontane,** a title more fully treated under Ursa Minor; and Vespucci, in his *3ª Lettera*, wrote of the two Bears:

La stella tramontana o l'orsa maggiore & minore.

Both of these have been — perhaps still are — night clocks to the English rustic, and measures of time generally, as in Poe's *Ulalume*, "star-dials that pointed to morn."

Shakespeare's Carrier at the Rochester inn-yard said:

An't be not four by the day, I'll be hang'd; Charles Wain is over the new chimney, and yet our horse not pack'd;

Tennyson, in his touching *New Year's Eve:*

We danced about the May-pole and in the hazel copse,
Till Charles's Wain came out above the tall white chimney tops;

and again, in the *Princess:*

I paced the terrace, till the Bear had wheel'd
Thro' a great arc his seven slow suns.

Spenser, in the *Faerie Queen*, thus refers to the Wain as a timepiece, and to Polaris as a guide:

By this the northern wagoner had set
His sevenfold teme behind the steadfast starre
That was in ocean waves never yet wet,
But firme is fixt, and sendith light from farre
To all that in the wide deep wandering arre.

Its well-known use by the early Greeks in navigation was paralleled in the deserts of Arabia, "through which," according to Diodorus the Sicilian, "travellers direct their course by the Bears, in the same manner as is done at sea." They serve this same purpose to the Badāwiyy of to-day, as Mrs. Sigourney describes in *The Stars*, writing of Polaris:

The weary caravan, with chiming bells,
Making strange music 'mid the desert sands,
Guides by thy pillar'd fires its nightly march.

Sophocles made a similar statement of the Bear as directing travelers generally; Falstaff, in *King Henry IV*, said:

We that take purses go by the moon and the seven stars;

and the modern Keats, in his *Robin Hood:*

the seven stars to light you,
Or the polar ray to right you.

But the astrologers of Shakespeare's time ascribed to it evil influences, which Edmund, in *King Lear*, commented upon with ridicule:

This is the excellent foppery of the world, that, when we are sick in fortune, (often the surfeit of our own behaviour), we make guilty of our disasters the sun, the moon, and the stars,—

claiming that his own

nativity was under Ursa Major, so that it follows I am rough and lecherous.

Both of the Bears have been frequently found on the old sign-boards of English inns, and, in a more important way, are emblazoned on the shields of the cities of Antwerp and Gröningen in the Netherlands.

The **Plough** has been a common title with the English down to the present time, even with so competent a scientist as Miss Clerke, one of the few astronomical writers who still continue the use of the good old names of stars and constellations. She, however, takes the three line stars as the **Handle,** not the **Team.** Minsheu mentioned it in the same way, but added *ut placet astrologis dicitur Temo, i. e.* the **Beam,** a term originating with Quintus Ennius, the Father of Roman Song, adopted by Cicero, Ovid, Statius, and Varro, and common with the astrologers. Fale, in 1593, described it as called "of countrymen the plough," the first instance in print that I have found. Thus it was, perhaps still is, the Irish **Camcheacta.** Hewitt sees this **Heavenly Plough** even in prehistoric India, and quotes from Sayce the title **Sugi,** the Wain, which later became Libra's name as the Yoke.

With the Wain and Plough naturally came the **Plough Oxen,** the **Triones** of Varro, Aulus Gellius, and the Romans generally, turned by the grammarians into **Teriones,** the Threshing-oxen, walking around the threshing-floor of the pole. Martial qualified these by *hyperborei Odrysii* and *Parrhasii*, but also called the constellation **Parrhasium Jugum**; and Claudian, *inoccidui*, "never setting." Cicero, with contemporary and later Latin writers, said **Septem-** or **Septentriones,** as did the long-haired Iopas in his *Aeneid* song of the two **Northern Cars**; and Propertius wrote of them:

Flectant Icarii sidera tarda boves;

while Claudian designated them as *pigri*; all of which remind us of similar epithets for their driver Boötes.

Septentrio seems to have been applied to either constellation; and Dante used it for the Minor, with a beautiful simile, in his *Purgatorio.* Eventually it became a term for the north pole and the north wind; then for the North generally, as the word Arctic has from the stellar *ἄρκτος*. Dante had *settentrionāle sito;* Chaucer spoke of the "Septentrioun" as a compass point; Shakespeare, in *King Henry VI:*

> as the South to the Septentrion;

Michael Drayton, the friend of Shakespeare and poet laureate in 1626, wrote in the *Poly-Olbion* of "septentrion cold"; Milton, in *Paradise Regained*, of "cold Septentrion blasts"; and, in our day, Owen Meredith in the *Wanderer* has "beyond the blue Septentrions"; while the word seems current as an adjective in nearly all modern languages. Still there is nothing new in all this, for in the *Avesta* the Seven Stars marked the North in the four quarters of the heavens.

The Persian title was **Hafturengh, Heft Averengh,** or **Heft Rengh,** qualified by **Mihin,** Greater, to distinguish it from **Kihin,** Lesser; Hewitt giving this as originally **Hapto-iringas,** the Seven Bulls, that possibly may be the origin of the Triones. Cox, however, goes far back of this classic title and says:

> They who spoke of the seven *triones* had long forgotten that their fathers spoke of the *taras* (*staras*) or strewers of light;

and Al Bīrūnī derived the word from *taraṇa*, "passage," as of the stars through the heavens. Thus from the results of modern philological research it is possible that our long received opinions as to the derivations of many star-names should be abandoned, and that we should search for them far back of Greece or Rome.

Heraclitos, the Ionic philosopher of Ephesus of about 500 B. C., asserted that this constellation marked the boundary between the East and the West, which it may be regarded as doing when on the horizon.

A coin of 74 B. C., struck by the consul Lucretius Trio, bears the Seven Stars disposed in an irregular curve around the new moon, while the word TRIO within the crescent is an evident allusion to the consul's name, albeit one hardly known in Roman history.

The Hebrew ʽĀsh, or ʽAyish, is reproduced by, or was derived from, the Arabic **Banāt Naʽash al Ḳubrā,** the Daughters of the Great Bier, *i. e.* the Mourners,— the **Benenas, Benethasch,** and **Beneth As** of Chilmead and Christmannus,— applied to the three stars in the extreme end of the group, η being Al Ḳā'id, the Chief One; from this came Bayer's **El Keid** for the whole constellation. Riccioli, quoting Kircher, said that the Arabian Christians with more definiteness termed it **Naʽash Laazar,** the Bier of Lazarus, with Mary, Martha, and Ellamath,— this last being given in Mrs.

Jameson's *Sacred and Legendary Art* as Marcella or Martilla, but by Smyth as Magdalen; Riccioli's word should be Al Amah, the Maid, the position that Marcella occupied toward the two women during their journey to Marseilles, where she was canonized. Karsten Niebuhr said that the constellation was known, even in his day, as **Na'ash** by the Arabs along the Persian Gulf; and Wetzstein tells the modern story, from that people, in which these mourners, the children of Al Na'ash, who was murdered by Al Jadī, the pole-star, are still nightly surrounding him in their thirst for vengeance, the *wālidān* among the daughters — the star Mizar — holding in her arms her new-born infant, the little Alcor, while Suhail is slowly struggling up to their help from the South. Delitzsch says that even to-day the group is known as a **Bier** in Syria; Flammarion attributing this title to the slow and solemn motion of the figure around the pole. This seems to have originated in Arabia; and from it come the titles even now occasionally heard for the quadrangle stars — the **Bier** and the **Great Coffin.** With the early Arab poets the Banāt stars were an emblem of inactivity and laziness.

It had other names also. **Cynosuris** appeared with Ovid and Germanicus for this, although it generally is applied to the Lesser Bear; Πλίνθιον, used for it or for its quarter of the sky, was from the Greek, as we see in Plutarch's *αί τῶν πλίνθιων υπογραφαί*, the "fields," or "spaces," into which the augurs divided the heavens, the *templa*, or *regiones*, *coeli* of the Latins; while Ἕλιξ, the Curved, or Spiral, One, and Ἑλίκη, apparently first used for the constellation by Aratos and Apollonius Rhodius, became common as descriptive of its twisting around the pole,— whence one of its titles now, the **Twister**; Sophocles having the same thought in Ἄρκτου στροφάδες κέλευθοι, the "circling paths of the Bear." Some, however, derived the name from the curved or twisted position of the chief stars; and others, still more probably, from the city Helice, Kallisto's birthplace in Arcadia. Ovid used this title in the *Fasti*, where he wrote of both the Bears, in navigation:

> Esse duas Arctos, quarum, Cynosura petatur
> Sidoniis, Helicen Graia carina notet;

but later on **Helice** was considered a nymph, one of the two Cretan sister nurses who nourished the infant Jupiter

> In odorous Diktē, near the Idaian hill,

whence she was transferred to the skies. Dante, in the *Paradiso*, alludes to barbarians

> coming from some region
> That every day by Helice is covered
> Revolving with her son whom she delights in.

Homer's Ἑλίκωπες has been rendered "observing Helice," and so applied to the early Grecian sailors; but there seems to be no foundation for this, as the word merely signifies "black-," "glancing-," or "rolling-eyed," and frequently was applied to various characters in the *Iliad*, with no limitation as to sex or profession.

Ancient, however, as are Ἄρκτος and Ursa, 'Āsh and the Bier, Ἅμαξα, Plaustrum, and Triones, this splendid constellation ran still further back— three or four or even more millenniums before even these titles were current — as the **Bull's Thigh,** or the **Fore Shank,** in Egypt. There it was represented on the Denderah planisphere and in the temple of Edfū by a single thigh or hind quarter of the animal, alluded to in the *Book of the Dead* as

The constellation of the Thigh in the northern sky;

and thus mentioned in inscriptions on the kings' tombs and the walls of the Ramesseum at Thebes. Sometimes the figure of the Thigh was changed to that of a cow's body with disc and horns; but, however called or represented, these stars always were prominent in the early astronomy and mythology of Egypt. **Mesχet** seems to have been their designation, and specially for some one of them, as representative of the malignant red Set,[1] Sit, or Sith, Sut or Sutech, who, with his wife Taurt or Thoueris, shown by the adjoining Hippopotamus (now a part of our Draco), represented darkness and the divinities of evil. **Set** also was a generic term applied to all circumpolar constellations, because, as always visible, they somewhat paradoxically were thought to typify darkness.

Hewitt writes of Set in his earliest form as **Kapi,** the Ape-God, stars of our Cepheus marking his head; while at one time on the Nile the Wain stars seem to have been the **Dog of Set** or **of Typhon.** This may have given rise to the title **Canis Venatica** that La Lande cited, if this be not more correctly considered as the classic Kallisto's hound; and the same idea appears in the **Catuli,** Lap-dogs, and **Canes Laconicae,** the Spartan Dogs, that Caesius cited for both of the Wains.

The myth of Horus, one of the most ancient even in ancient Egypt, deciphered from the temple walls of Edfū, 5000 B.C., as connected with the stellar Hippopotamus, was, about 3000 years afterwards, transferred to the Thigh, which then occupied the same circumpolar position that the Hippopotamus did when the original inscription was made. In view of this, Champollion alluded to the Thigh as **Horus Apollo.**

[1] Set, also Anubis, Apap, Apepi, Bes, Tebha, Temha, and Typhoeus according to Plutarch, was one of Egypt's greatest gods, who subsequently became the Greek giant Typhon, father of the fierce winds, but slain by Zeus with a thunderbolt and buried under Mount Aetna.

Towards our era, when Egypt began to be influenced by Greece, her former pupil, our Wain was regarded as the **Car of Osiris,** shown on some of that country's planispheres by an **Ark,** or **Boat,** near to the polar point, although it also seems to have been known as a Bear.

Al Bīrūnī devoted a chapter of his work on India to these seven stars, saying that they were there known as **Saptar Shayar,** the Seven Anchorites, with the pious woman Al Suhā (the star Alcor), all raised by Dharma to the sky, to a much higher elevation than the rest of the fixed stars, and all located "near Vas, the chaste woman Vumdhati"; but who was this last is not explained. And he quoted from Varāha Mihira:

The northern region is adorned with these stars, as a beautiful woman is adorned with a collar of pearls strung together, and a necklace of white lotus flowers, a handsomely arranged one. Thus adorned, they are like maidens who dance and revolve round the pole as the pole orders them.

Professor Whitney tells us that

to these stars the ancient astronomers of India, and many of the modern upon their authority, have attributed an independent motion about the pole of the heavens, at the rate of eight minutes yearly, or of a complete revolution in 2700 years;

and that this strange dogma well illustrates the character of Hindu astronomy. The matter-of-fact Al Bīrūnī, commenting on this same thing, and on the absurdly immense numbers in Hindu chronology, wrote:

The author of the theory was a man entirely devoid of scientific education, and one of the foremost in the series of fools who simply invented those years for the benefit of people who worship the Great Bear and the pole. He had to invent a vast number of years, for the more outrageous it was, the more impression it would make.

In China the **Tseih Sing,** or Seven Stars, prominent in this constellation, were known as the **Government,** although also called **Pih Tow,** the Northern Measure, which Flammarion translates the **Bushel;** while the centre of the Square was **Kwei,** an object of worship and a favorite stellar title in that country, as it occurs twice in their list of *sieu*, although there rendered the Spectre, or Striding Legs. Reeves said that the four stars of the Square were **Tien Li,** the Heavenly Reason, and Edkins, in his *Religion in China*, assigns to this spot the home of the Taouist female divinity Tow Moo. Colas gives **Ti Tche,** the Emperor's Chariot; but this was doubtless a later designation from Jesuit teaching.

Weigel of Jena figured it as the heraldic **Danish Elephant;** but Julius Schiller, as the archangel **Michael;** while Caesius said that it might represent one of the **Bears** sent by Elisha to punish his juvenile persecutors, or the **Chariot** that Pharaoh gave to Joseph.

Popular names for it have been the **Butcher's Cleaver,** somewhat similar to the Hindu figure for the other Seven Stars, the Pleiades; the **Brood Hen,** also reminding us of that cluster, as do the Gaelic **Grigirean, Crann,** and **Crannarain; Peter's Skiff,** from, or the original of, Julius Schiller's **Ship of Saint Peter;** the **Ladle;** and, what is known to every one, star-lover or not, the **Big Dipper,** the universally common title in our country. In southern France this has been changed to **Casserole,** the Saucepan.

Before the observations of the navigators of the 15th and 16th centuries the singular belief prevailed that the southern heavens contained a constellation near the pole similar to our Bear or Wain; indeed, it is said to have been represented on an early map or globe. Manilius wrote:

> The *lower Pole* resemblance bears
> To this Above, and shines with equal stars;
> With *Bears averse*, round which the *Draco* twines;

and Al Bīrūnī repeated the Sanskrit legend that at one time in the history of the Creation an attempt was made by Visvāmitra to form a southern heavenly home for the body of the dead king, the pious Somadatta; and this work was not abandoned till a southern pole and another Bear had been located in positions corresponding to the northern, this pole passing through the island Lunka, or Vadavāmukha (Ceylon). The Anglo-Saxon *Manual* made distinct mention of this duplicate constellation "which we can never see." Towards our day Eden, describing the "pole Antartike," said:

> Aloysius Cadamustus[1] wryteth in this effecte: We saw also syxe cleare bryght and great starres very lowe above the sea. And consyderynge theyr stations with our coompasse, we found them to stande ryght south, fygured in this maner, * * * * * * . We judged them to bee the chariotte or wayne of the south: But we saw not the principall starre, as we coulde not by good reason, except we shuld first lose the syght of the north pole.

And, quoting from Francisco Lopes of 1552:

> Abowt the poynt of the Southe or pole Antartike, they sawe a lyttle whyte cloude and foure starres lyke unto a crosse with three other joynynge thereunto, which resemble oure Septentrion, and are judged to bee the signes or tokens of the south exeltree of heaven.

What is referred to here is not known, for, although the figure represented is that of the Southern Cross, this constellation always is upright when on the meridian, and, as the observation was made in latitude 14° or 15°,

[1] This Alois, or Luigi, di Cada Mosto was a noted Venetian navigator in the service of Portugal, for whom is often claimed the discovery of the Cape Verd Islands in 1456; but these had been seen, at least in part, fifteen years previously, by Antonio and Bartolomeo di Nolli.

its base star was plainly visible. Still it would seem that some early knowledge of the Cross was the foundation of this idea of a southern Wain.

Pliny strangely blundered in some of his allusions to Ursa Major, asserting in one its invisibility in Egypt, and, again, describing the visit to Rome of ambassadors from Ceylon,— Milton's "utmost Indian isle Taprobane,"— wrote of them:

> Septentriones Vergiliasque apud nos veluti novo coelo mirabantur.

α, β, γ, δ, ε, ζ, and η, in this order, as one follows the line of seven stars from the north, form the familiar Dipper, of which Mr. B. F. Taylor writes in his *World on Wheels:*

> From that celestial Dipper,— or so I thought,— the dews were poured out gently upon the summer world.

All these stars, unless possibly δ, which is too faint for the Potsdam observers, are approaching our system at various rates of speed. Flammarion has a page, on this so-called star-drift, in his *l'Astronomie Populaire,* concluding that from their proper motions they will form an exaggerated Steamer Chair 50,000 years hence, as they did a magnificent Cross 50,000 years ago.

α, Binary, 2 and 11, yellow.

Dubb, more generally **Dubhe,** the Bear, is the abbreviation of the Arabians' **Ṭhahr al Dubb al Akbar,** the Back of the Greater Bear, Dubb being first found in the *Alfonsine Tables.*

Al Bīrūnī said that it was the Hindu **Kratu,** the Rishi or Sage.

Lockyer asserts that it was **Āk,** the Eye, *i. e.* the prominent one of the constellation, utilized in the alignment of the walls of the temple of Hathor at Denderah, and the orientation point of that structure perhaps before 5000 B. C.; at all events, before the Thigh became circumpolar, about 4000 B. C. This was in the times of the Hor-she-shu, the worshipers of Horus, before the reign of Mena,[1] when the star had a declination of over 64°,— now about 62° 24′. And he finds two other temples also so oriented.

As typifying a goddess of Egypt, it was **Bast Isis** and **Taurt Isis.**

The Chinese know it as **Tien Choo,** Heaven's Pivot, and as **Kow Ching.**

α is 5° from β and 10° from δ, and, being always visible, these stars afford a ready means of accurate eye measurement of others adjacent.

[1] Mena, Menes, or Min was the first historic king of Egypt, his date being variously given from 5867 B. C. to 3892 B. C., Flinders Petrie making it, from astronomical data, 4777 B. C.

The **Keepers** was Arago's name for them; while, as the **Pointers**, they indicate to beginners in astronomy the pole-star, 28¾° distant from α, and Regulus, 45° away towards the south; and they have been called the **Two Stars.**

They are circumpolar north of about 32° 45′; and, with Polaris, received much attention in the first almanac[1] that was printed in London, in 1473.

Klein surmised, in 1867, that Dubhe shows remarkable, although irregular, variations in color,— not in light,— from red to yellow, in a period of 54½ days; but this is still in doubt. Its spectrum is Solar, and it is approaching our system at the rate of twelve miles a second.

The 11th-magnitude companion, .97 of a second away, was discovered by Burnham in 1889, and is thought to be in rapid revolution around it.

β, 2.5, greenish white.

Merak, or **Mirak,** is from **Al Marāḳḳ,** the Loin (of the Bear); but Chilmead said **Miraë,** and Scaliger, **Mizar.** It may have been known by the Greeks as **Helike,** one of their names for the whole.

The Chinese called it **Tien Seuen,** an Armillary Sphere, and the Hindus, **Pulaha,** one of the Rishis.

Its spectrum is Sirian, and it is moving toward us about 18½ miles a second.

Close to it, on the west, lies the **Owl Nebula,** N. G. C. 3587, 97 M., discovered by Mechain in 1781, and so called from the two interior circular spaces, each with a central star representing the eye; although one of these stars seems to have disappeared since 1850. The angular diameter of this nebula — 2′ 40″ — indicates a magnitude sufficient to contain thousands of solar systems.

γ, 2.5, topaz yellow.

Phacd and **Phachd, Phad, Phaed, Phecda, Phekda,** and **Phegda,** are all from **Al Falidh,** the Thigh, where this star is located in the figure.

Al Bīrūnī said that it was **Pulastya,** one of the Hindu Seven Sages.

The Chinese knew it as **Ke Seuen Ke,** and as **Tien Ke,** another Armillary Sphere.

Its spectrum is similar to that of β, and the star is approaching us at the rate of 16.6 miles a second. It is 8° distant from β, and 4½° from δ.

[1] This is said to have been the second of such works; the first being variously given as published in Vienna by Purbach, or in Buda, or in Poland a few years previously.

δ, 3.6, pale yellow.

Megrez is from **Al Maghrez,** the Root of the Tail.

In China it was **Kwan,** and **Tien Kuen,** Heavenly Authority.

With the Hindus it may have been **Atri,** one of their Seven Rishis, and the *Vishṇu-Dharma* said that it ruled the other stars of the Bear.

It is 10° distant from α; 4½° from γ; 5½° from ε; and 32° from the pole, directly opposite β Cassiopeiae, and almost on the equinoctial colure. α, β, γ, and δ form the bowl of the Dipper, the body of the Bear, and the frames of the Bier, Plough, and Wain, but occupy a space of less than ¼ of the whole constellation. Within this square Heis shows eight stars.

Megrez is thought to be slightly variable, and to have decreased in lustre during the present century, on the very doubtful ground that it is much fainter than the succeeding ε. As to this Miss Clerke writes:

The immemorially observed constituents of the Plough preserve no fixed order of relative brilliancy, now one, now another of the septett having at sundry epochs assumed the primacy.

But this is uncertain, although we know that Ptolemy rated it at the 3d magnitude and Tycho at the 2d.

ε, 2.1.

Alioth, sometimes **Allioth,** seems to have originated in the first edition of the *Alfonsine Tables*, and appeared with Chaucer in the *Hous of Fame* as **Aliot**; with Bayer, as **Aliath,** from Scaliger, and as **Risalioth**; with Riccioli, as **Alabieth, Alaioth, Alhiath,** and **Alhaiath,** all somewhat improbably derived, Scaliger said, from **Alyat,**[1] the Fat Tail of the Eastern sheep. But the later Alfonsine editions adopted **Aliare** and **Aliore** — Riccioli's **Alcore** — from the *Latin Almagest* of 1515, on Al Tizini's statement that the word was **Al Ḥawar,** the White of the Eye, or the White Poplar Tree, *i. e.* Intensely Bright; Hyde transcribing the original as **Al Haur.** Ulug Beg had **Al Haun,** but Ideler, rejecting this as not being an Arabic word, substituted **Al Jaun,** the Black Courser, as if belonging to the governor, Al Ḳā'id, the star η, and its comparative faintness gives some probability to this conjecture. Assemani, however, said that on the Cufic globe it is "**Alhut,**" the Fish,— one of the many instances of blundering that Ideler attributed to him.

Bayer also assigned to it the **Micar, Mirach,** and **Mizar** that we give to

[1] The syllable Al, in this word Alyat, is not the Arabic definite article.

η, and designated it as Λαγών, the Flank, and Ὑπόζωμα, the Diaphragm, as marking those parts of the Bear's figure.

Al Bīrūnī said that it was **Añgiras** among the Hindu Seven Sages.

In China it was **Yuh Kang**, the Gemmeous Transverse, a portion of an early astronomical instrument; while other stars between it and δ were **Seang**, the Minister of State.

ϵ has a Sirian spectrum, and is in approach toward us at the rate of 19 miles a second. It is 5½° from δ, and 4½° from ζ.

In 1838 Sir John Herschel thought it the *lucida* of the seven stars, but in 1847 that η had taken its place. Franks, in 1878, considered ϵ the *lucida*, and that the sequence was ϵ, η, ζ, α, β, γ, and δ.

ζ, Double, possibly binary, 2.1 and 4.2, brilliant white and pale emerald.

Mirak was an early name for this, a repetition of that for β; but Scaliger incorrectly changed it to the present **Mizar**, from the Arabic **Mi'zar**, a Girdle or Waist-cloth, which, although inappropriate, has maintained its place in modern lists; **Mizat** and **Mirza** being other forms. There is evident confusion in the early use of this word as a stellar title, for it has also been applied to the stars β and ϵ of this constellation. The "hill Mizar" of the 42d *Psalm* sometimes is wrongly associated with this, the original Hebrew word *miṣʽar* being better rendered in the *Psalter*, from Coverdale's version, as "the little hill," *i. e.* of Hermon, of which it was a minor peak.

ζ also was the Arabic **ʽAnāḳ al Banāt**, the Necks of the Maidens, referring to the Mourners at the Bier; or perhaps this should be rendered "the Goat of the Mourners," for in some editions of Ulug Beg's *Tables* it was written **Al Inak**,—correctly **Al 'Inz.** Assemani said that it was "**Alhiac**," the Ostrich, probably another of his errors, as all these stellar birds were much farther south, in or near our River Eridanus.

With Alcor it has various combined titles noted at that star; and Wetzstein repeats an Arabic story in which Mizar is the *wālidān* of the Banāt, with Alcor as her new-born infant.

In India it may have been **Vashishṭha**, one of the Seven Sages.

ζ was the first star to be noticed as telescopically double,—by Riccioli at Bologna in 1650, and fifty years later much observed and very fully described by Gottfried Kirch and his scientific wife, Maria Margaretha Winckelmann: an association like that of the great observer Herschel and his sister, of the last century, and of Sir William and Lady Huggins in their spectroscopic work of to-day. As early as 1857 it was successfully daguerreotyped, with others surrounding, by the younger Bond of the Harvard Observatory, although Wega had been pictured by the same process at the same observatory seven years previously by the elder Bond.

The components are within 14″ of arc of each other, with a position angle of 149°.5, and may be a binary system with a long period of revolution; while Pickering has shown, by study of its spectrum photographed in 1889, that the brightest component is itself double, the two bodies, of nearly equal brightness, revolving around their common centre of gravity at a speed of 100 miles a second in 104 days, 140 millions of miles apart, and with a united mass forty times that of our sun. This spectrum is Sirian, and the star is in approach to us at the rate of 19.5 miles a second.

ζ is 4½° from ε, and 7° from η; and a straight line from it to Polaris passes through the exact pole 1° 14′ before reaching Polaris.

Mizar and Alcor are 11′ 48″ apart, and, since they have nearly identical proper motion, some think that they may also be in mutual revolution, although so distant from each other. With their attendant stars they form one of the finest objects in the sky for a small telescope, being readily resolved by a terrestrial eyepiece of 40 diameters with a 2¼-inch objective.

η, 1.9, brilliant white.

Alcaid, Alkaid, and **Benatnasch** are our present titles, from **Ḳā'id Banāt al Na'ash,** the Governor of the Daughters of the Bier, *i. e.* the Chief of the Mourners. Some of the Arabic poets wrote that these Daughters — the stars ε, ζ, and η — were

Good for nothing people whose rising and setting do not bring rain.

Bayer included **Elkeid** in his list of names for the stars as well as for the constellation, and had authority for it from Kazwini; but he added for η "**Benenaim, Bennenatz** *correctius* **Benetnasch,**" and in his text of Boötes alluded to it as **Benenacx.** The *Alfonsine Tables* of 1521 say **Bennenazc;** Riccioli, **Benat Elnanschi, Beninax, Benenath, Benenatz;** while Al Ḳā'id often has been turned into **Alchayr,** Arago's **Ackaïr,** and others' **Ackiar.** In this Al Ḳā'id we see the derivation, through the Moors, of the modern Spanish word Alcaide; and, with the same idea, Ideler translated the original as the "Stadtholder."

Assemani transcribed from the Borgian globe "**Alcatel,**" Destroying. Al Bīrūnī gave it as **Marīci,** one of the Seven Rishis of India.

In China it was known as **Yaou Kwang,** a Revolving Light.

Boteler has an amusing reference to it in *Hudibras:*

Cardan believ'd great states depend
Upon the tip o' th' Bear's tail's end;
That, as she whisk'd it t'wards the Sun,
Strew'd mighty empires up and down;
Which others say must needs be false,
Because your true bears have no tails.

η is 7° from ζ, and 26° from α; and with ζ forms another pair of Pointers—towards Arcturus. It is noted as marking the radiant of one of the richest minor meteor streams, the **Ursids** of the 10th of November.

Bradley's earliest observations for parallax were made on this star and γ Draconis, but unsuccessfully, as his instruments were inadequate; yet even in our own day Pritchard's work on η for the same purpose showed a negative result,—0''.046, and equally unsatisfactory.

Alkaid's spectrum is Sirian, and the star is approaching us at the rate of 16.1 miles a second.

Sir John Herschel thought it, in 1847, the *lucida* of the seven stars.

θ, Double, 3.4 combined, brilliant white.

This, with τ, h, v, ϕ, e, and f in the Bear's throat, breast, and fore knees, which describe somewhat of a semicircle, was the Arab star-gazers' **Sarīr Banāt al Na'ash,** the Throne of the Mourners.

This space also has been **Al Ḥauḍ,** the Pond into which the Gazelles sprang for safety at the lashing of the Lion's tail; although Hyde applied this title to the stars now our Coma Berenices, and **Ṭhufr al Ghizlān,** the Gazelles' Tracks, to the small outlying stars near the Bear's feet. But the engraver of the Borgian globe placed them at stars in the neck.

In China θ, v, and ϕ were **Wan Chang,** the Literary Illumination.

ι, Binary, 3.2 and 13, topaz yellow and purple, and κ, 3.5.

Smyth wrote that

this star has obtained the name of **Talita,** the third vertebra, the meaning of which is not quite clear. Ulug Beigh has it **Al Phikra al Thalitha,** perhaps for *Al Ḳafzah al-thālithah,* the third spring, or leap, of the ghazal;

but he was not sufficiently comprehensive, for this last title was applied by the Arabs to ι and κ together; al Ūla, the First (leap), being shown by ν and ξ, and al Thānīyah, the Second (leap), by λ and μ,— not δ and μ as that generally accurate author asserted. In popular lists ι frequently is given as **Talitha.** Hyde strangely rendered the original words of Ulug Beg as the Vertebrae of the Greater Bear,— whence probably Smyth's statement,— or the Cavity of the Heel, which, from the star's position in the figure, is a much more likely translation.

In China these two stars were **Shang Tae,** the High Dignitary.

Holden says of ι that its "companion is suspected to be a planet." It is 12″ distant from the larger, and the orbital revolution is very slow.

λ, 3.7, and μ, 3.2, red.

These are our **Tania borealis** and **Tania australis;** and together were the Arabs' **Al Ḳafzah al Thānīyah,** the Second Spring (of the Gazelle), marking the Bear's left hind foot. Baily has them in his edition of Ulug Beg's *Tables*, from Hyde's Latin translation, as **Al Phikra al Thānia,**— in the original Al Fiḳrah, the Vertebra; but this, more probably, is entirely wrong, as these three pairs of stars have always marked three of the Bear's feet.

In China they were **Chung Tae,** the Middle Dignitary.

ν, Double, 3.5 and 12, orange and cerulean blue,

ξ, Binary, 3.9 and 5.5, subdued white and grayish white,

mark the right hind foot, and are the southern of the three noted pairs.

They were the Chinese **Hea Tae,** the Lower Dignitary.

The components of ξ are but 1″ apart, with a position angle of 300°.

ν, the northern one of the two stars, is **Alula borealis,** from **Al Ḳafzah al Ūla,** the First Spring.

ξ is **Alula australis,** the southern one in the combination,— Ulug Beg's **Al Fiḳrah al Ūla.** Ideler's **Awla,** and Burritt's **Acola,** are erroneous.

This, with ζ Herculis and γ Virginis, was the most prominent of the double stars discovered to be binary systems by Sir William Herschel in his investigations for stellar parallax, when (I quote from Professor Young),

> to use his own expression, he "went out like Saul to seek his father's asses, and found a kingdom,"— the dominion of gravitation extended to the stars, unlimited by the bounds of the solar system.

ξ was the first binary of which the orbit was computed,— by Savary in 1828,— having a period of sixty-one years, and has already made more than a complete revolution since its discovery. The components are about 2″ apart, with a position angle in 1898 of 162°.7.

The foregoing three pairs, about 20° apart and the members of each pair 1½° or 2° apart, are beautifully grouped with others invisible to the naked eye. They were interesting to the Arabs, as they now are to us, and were collectively designated **Ḳafzah al Ṭibā'**, the Springs of the Gazelle, each pair marking one spring; the **Gazelle** being imagined from the unformed stars since gathered up as Leo Minor, and the springing of the animal being due to its fear of the greater Lion's tail. Ideler adopted this from Al Tizini and the Cufic globe at Dresden; while the Borgian globe shows a Gazelle and her Young in the same location. Kazwini, however, described this group as extending over the eyes, eyebrows, ears, and muzzle of the figure of our Ursa Major.

According to Williams' the Chinese knew these six stars as **San Tae**, or **Shang Tae**; but Reeves limited this title to ι and κ. Their records mention a comet seen near by in 110 B. C.

o, Double, 3.5 and 15.2.

Bayer said that "the Barbarians" called this **Muscida**, a word apparently coined in the Middle Ages for the muzzle of an animal, the feature of the Bear that the star marks.

The components are 7″ apart, at a position angle of 191°.4.

π^1, 5.6, and π^2, 4.8.

Muscida has also been applied to these, although Heis locates them nearer the eyes.

σ^1, 5.2, and σ^2, Binary, 4.8 and 9.5, flushed white and sapphire,

with o, π, ρ, A, d, and some others in the eyes, ears, and muzzle of the Bear, were the asterism that Kazwini knew as **Al Ṭibā'**, the Gazelle.

With ϕ and others they were the Chinese **San Tsze**, the Three Instructors.

The components of σ^2 are 3″ apart, with a position angle of 250°.

τ, a 5th-magnitude double, with other small stars near by, was the Chinese **Nuy Keae**, the Inner Steps.

χ, 4, red,

placed on the right foot by Burritt as **Al Kaphrah**, is wrong, for Heis puts the letter at a star on the rear of the right hind quarter, and has no letter at

Burritt's star; if entitled to a name at all, it should be **Al Ḳafzah,** as at ι and κ. Still the *Standard Dictionary* follows Burritt in its **El Kophrah.**

It was the Chinese **Tae Yang Show,** the Sun Governor, and **Shaou We,** of somewhat similar signification.

ψ, a 3½-magnitude yellow star, is **Tien Tsan,** according to Williams, but Reeves says **Ta Tsun,** Extremely Honorable.

ω, a 5th-magnitude, with near-by stars, was **Tien Laou,** Heavenly Prison.

Between ψ and ω, somewhat nearer to the former, is the 7th-magnitude Ll. 21185, one of the two or three stars that follow α Centauri in proximity to our system, and, so far as our present determinations can be trusted, 6½ light years away.

g, or 80 Fl., 4.8.

Alcor is the naked-eye companion of Mizar, and, inconspicuous though it be, has been famous in astronomical folk-lore.

This title, and that of the star ϵ, Alioth, may be from the same source, for Smyth wrote of it:

> They are wrong who pronounce the name to be an Arabian word importing sharp-sightedness: it is a supposed corruption of *al-jaún*, a courser, incorrectly written *al-jat*, whence probably the *Alioth* of the Alfonsine Tables came in, and was assigned to ϵ Ursae Majoris, the "thill-horse" of Charles's Wain. This little fellow was also familiarly termed **Suhā** [the Forgotten, Lost, or Neglected One, because noticeable only by a sharp eye], and implored to guard its viewers against scorpions and snakes, and was the theme of a world of wit in the shape of saws:

but Miss Clerke says:

> The Arabs in the desert regarded it as a test of penetrating vision; and they were accustomed to oppose "Suhel" to "Suha" (Canopus to Alcor) as occupying respectively the highest and lowest posts in the celestial hierarchy. So that *Vidit Alcor, at non lunam plenam*, came to be a proverbial description of one keenly alive to trifles, but dull of apprehension for broad facts.

Al Sahja was the rhythmical form of the usual Suhā; and it appears as **Al "Khawwar,"** the Faint One, in an interesting list of Arabic star-names, published in *Popular Astronomy* for January, 1895, by Professor Robert H. West, of the Syrian Protestant College at Beirut.

Firuzabadi called it **Our Riddle,** and **Al Ṣadāk,** the Test,—correctly **Ṣaidak,** True; while Kazwini said that "people tested their eyesight by this star." Humboldt wrote of it as being seen with difficulty, and Arago similarly alluded to it; but some now consider it brighter than formerly and no longer the difficult object that it was, even in the clear sky of the Desert; or as having increased in angular distance from Mizar.

Although the statement has been made that Alcor was not known to the Greeks, there is an old story that it was the Lost Pleiad Electra, which had wandered here from her companions and became 'Αλώπηξ, the Fox; a Latin title was **Eques Stellula,** the Little Starry Horseman; **Eques,** the Cavalier, is from Bayer; while the **Horse and his Rider,** and, popularly, in England, **Jack on the Middle Horse,** are well known, Mizar being the horse.

Al Bīrūnī mentioned its importance in the family life of the Arabs on the 18th day of the Syrian month Adar, the March equinox; and a modern story of that same people makes it the infant of the *wālidān* of the three Banāt.

In North Germany **Alkor,** as there written, has been **der Hinde,** the Hind, or Farm Hand; in Lower Germany, **Dumke**; and in Holstein, **Hans Dümken,** Hans the Thumbkin,—the legend being that Hans, a wagoner, having given the Saviour a lift when weary, was offered the kingdom of heaven for a reward; but as he said that he would rather drive from east to west through all eternity, his wish was granted, and here he sits on the highest of the horses of his heavenly team. A variant version placed Hans here for neglect in the service of his master Christ; and the Hungarians call the star **Göntzol,** with a somewhat similar tale. Another Teutonic story was that their giant Orwandil, our Orion, having frozen one of his big toes, the god Thor broke it off and threw it at the middle horse of the Wagon, where it still remains.

In China it was **Foo Sing,** a Supporting Star.

At the obtuse angle formed with Alcor and Mizar lies the **Sidus Ludovicianum,** an 8th-magnitude bluish star, just visible in a field-glass. This was first noted in 1691 by Einmart of Nuremberg, and in 1723 by another German, who, thinking that in it he had discovered a new planet, named it after his sovereign, Ludwig V, landgrave of Hesse-Darmstadt.

1830 Groombridge, or 4010 B. A. C., 6.5,

is the well-known **Flying Star,** or **Runaway Star,** that, until Kapteyn's recent discovery of a swifter one in Pictor, had shown the greatest velocity of any in the heavens, although the 7½-magnitude La Caille 9352 in Piscis Australis, and an 8½-magnitude in Sculptor, are not far behind it in this respect. According to Miss Clerke,

> Argelander discovered in 1842 its pace to be such as would carry it around the entire sphere in 185,000 years, or in 265 over as much of it as the sun's diameter covers.

Another calculator states that in 6000 years it will reach Coma Berenices. This is equivalent to a proper motion of 7″.03 of arc annually, at the rate of over 200 miles a second, and its velocity may be still greater,— a speed uncontrollable, Professor Newcomb says, by the combined attractive power of the entire sidereal universe.

The observations for its parallax do not accord in their results, but Professor Young assigns to the star a distance of 37½ light years.

It is about 16° south from γ, half-way between Coma and the stars ν and ξ on the right paw of the Bear; its exact location being 11° 46′ of right ascension and 38° 35′ of north declination, about 15° from Ll. 21258, an 8½-magnitude also much observed for its great proper motion; but 50,000 years hence the Flying Star will have separated from this by at least 100°.

From the foregoing list it will be seen that we have in the entire constellation twenty stars individually named, many of them inconspicuous, two even telescopic,— evidence enough in itself of the antiquity of, as well as the continued popular and scientific interest in, Ursa Major.

★

> **The other, less in size but valued more by sailors,**
> **Circles with all her stars in smaller orbit.**
>
> **Poste's *Aratos.***

Ursa Minor, the Lesser Bear,

the **Orsa Minore** of Italy, **Petite Ourse** of France, and **Kleine Bär** of Germany, shared with its major companion the latter's **Septentrio,** Ἄρκτος, Ἅμαξα, Ἄγαννα, and Ἑλίκη.

Similarly it was Κυνόσουρις, but solely Κυνόσουρα; this early and universal title, usually translated the "Dog's Tail," continuing as **Cynosura** down to the time of the *Rudolphine Tables;* although with us "Cynosure" is applied only to Polaris. The origin of this word is uncertain, for the star group does not answer to its name unless the dog himself be attached; still some, recalling a variant legend of Kallisto and her Dog instead of Arcas, have thought that here lay the explanation. Others have drawn this title from that of the Attican promontory east of Marathon, because sailors, on their approach to it from the sea, saw these stars shining above it and beyond; but if there be any connection at all here, the reversed derivation is more probable; while Bournouf asserted that it is in no way associated with the Greek word for "dog."

Cox identified the word with *Λυκόσουρα*, which he renders **Tail,** or **Train, of Light.** Yet this does not seem appropriate to a comparatively faint constellation, and would rather recall the city of that title in Arcadia, the country so intimately connected with the Bears. But the stellar name probably long antedated the geographical, old as this was; Pausanias considering Lycosura the most ancient city in the world, having been founded by Lycaon some time before the Deluge of Deucalion. Indeed the Arcadians asserted that they and their country antedated the creation of the moon, an assertion which gave occasion to Aristotle's term for them,—*Προσέληνοι* and the Latins' Proselenes.

Singularly coincident with the foregoing *Λυκόσουρα* was the title that the distant Gaels gave to these stars,—**Drag-blod,** the Fire Tail.

Very recently, however, Brown has suggested that the word is not Hellenic in origin, but Euphratean; and, in confirmation of this, mentions a constellation title from that valley, transcribed by Sayce as **An-ta-sur-ra,** the Upper Sphere. Brown reads this **An-nas-sur-ra,** High in Rising, certainly very appropriate to Ursa Minor; and he compares it with Κ-*υν-όσ-ου-ρα*, or, the initial consonant being omitted, **Unosoura.** This, singularly like the Euphratean original,

> might easily become Kunosoura under the influence of a popular etymology, aided by the appearance of the tail stars of the constellation. And in exact accordance with the foregoing view is the following somewhat curious passage in the *Phainomena*, 308–9:
>
> Then, too, the head of Kynosure runs very high,
> When night begins.

Ursa Minor was not mentioned by Homer or Hesiod, for, according to Strabo, it was not admitted among the constellations of the Greeks until about 600 B. C., when Thales, inspired by its use in Phoenicia, his probable birthplace, suggested it to the Greek mariners in place of its greater neighbor, which till then had been their sailing guide. Aratos, comparing the two, wrote, as in our motto, of the Minor, its Guards, β and γ, then being much nearer the pole than was α, our present pole-star. Thales is reported to have formed it by utilizing the ancient wings of Draco, perceiving that the seven chief components somewhat resembled the well-known Wain, but reversed with respect to each other. From all this come its titles *Φοινίκη*, **Phoenice,** and **Ursa Phoenicia.**

The later classical story that made sister nymphs out of the stars of our two Bears, and nurses on Mount Ida of the infant Jove, is alluded to by Manilius in his line,

The Little Bear that rock'd the mighty Jove.

Although occasionally, but wrongly, figured and described as equal in size,— Euripides wrote:

Twin Bears, with the swift-wandering rushings of their tails, guard the Atlantean pole,—

they have always occupied their present respective positions, and, as Manilius said:

stand not front to front but each doth view
The others Tayl, pursu'd as they pursue;

the scientific poet Erasmus Darwin of the last century, grandfather of Charles Robert Darwin of this, imitating this in his *Economy of Vegetation:*

Onward the kindred Bears, with footsteps rude,
Dance round the pole, pursuing and pursued.

This "dancing" of the stars generally, as well as of the planets, was a favorite simile, and in classical days specially gave name to δ and ε of this constellation, as well as in Hindu astronomy; while Dante thus applied it to all those that were circumpolar:

Like unto stars neighboring the steadfast poles,
Ladies they seemed, not from the dance released.

The Arabians knew Ursa Minor as **Al Dubb al Aṣghar,** the Lesser Bear, — Bayer's **Dhub Elezguar,** and Chilmead's **Dub Alasgar,**— although earlier it was even more familiar to them as another **Bier**; and they called the three stars in the tail of our figure **Banāt al Na'ash al Ṣughrā,** the Daughters of the Lesser Bier.

Here, and in Ursa Major, some early commentators located the **Fold,** an ancient stellar figure of the Arabs, and an appropriate title, as Firuzabadi called β and the *gammas* in Ursa Minor **Al Farḳadain,** usually rendered the Two Calves, but, better, the Two Young Ibexes; Polaris, too, was well known as a **Young He Goat,** and adjacent stars bore names of desert animals more or less associated with a fold. Perhaps Lowell had this in mind when he wrote, in *Prometheus*, of

The Bear that prowled all night about the fold
Of the North-star.

But Manilius anticipated him in writing of the Bears:

Secure from meeting they're distinctly roll'd,
Nor leave their Seats, and pass the dreadfull fold.

The Arabs also likened the constellation to a **Fish,** while with all that nation, heathen or Muḥammadan, it was **Al Fass,** the Hole in which the earth's axle found its bearing.

Others of them, as well as the Persians, figured here the **Ihlilagji,** the Myrobalanum, or Date-palm Seed or Fruit, which the grouped stars were thought to resemble; but Hyde, writing the word Myrobalanaris, said that it signified one of their geometrical figures,— described by Ideler as bounded by our α, δ, ε, ζ, η, γ, β, *a*, *b*, and the stars in the head of Camelopardalis. In Persia, where this foregoing figure was popular, Ursa Minor also was **Heft Rengh, Heft Averengh,** or **Hafturengh Kihin,** the last word designating its inferiority in size to Ursa Major.

Jensen sees here the **Leopard** of Babylonia, an emblem of darkness which this shared, there and in Egypt, with all other circumpolar constellations; while on the Nile it was the well-known **Jackal of Set** even as late as the Denderah zodiac. This Jackal also appears in the carvings on the walls of the Rameseum, but is there shown with pendent tail strikingly coinciding with the outlines of the constellation.

Plutarch said that with the Phoenicians it was **Doube** or **Dōbher** (?), similar to the Arabian title, but defined by Flammarion as the "Speaking Constellation,"— better, I think, the "Guiding One," indicating to their sailors the course to steer at sea. Jacob Bryant assigned it to Egypt, or Phoenicia, as **Cahen ourah,**— whatever that may be.

The early Danes and Icelanders knew it as the **Smaller Chariot,** or **Throne, of Thor;** and their descendants still call it **Litli Vagn,** the Little Wagon; as also, but very differently, **Fiosakonur ā lopti,** the Milkmaids of the Sky. But the Finns, apparently alone among the northern nations of Europe in this conception, have **Vähä Otawa,** the Little Bear.

Dante called the seven stars **Cornu,** doubtless then a common name, for it appeared in Vespucci's *3ª Lettera* as **Elcorno,** his editor erroneously explaining this as a typographical error for *carro*, the wain; Eden and others of his time translating this as the **Horne.** And it has been the Spanish shepherds' similarly shaped **Bocina,** a Bugle; and the Italian sailors' **Bogina,** a Boa.

Caesius mentioned **Catuli,** and **Canes Laconicae,** the Lapdogs or Puppies, and the Spartan Dogs, as titles for both of the Bears.

With the Chinese it was **Peih Sing.**

Alrucaba, or **Alruccaba,** which probably should be **Al Rukkabah,** is first found in the *Alfonsine Tables*, although the edition of 1521 applied it only to the *lucida*. While this generally is supposed to be from the Arabic **Al Rakabah,** the Riders, Grotius asserted that it is from the Chaldee **Rukub,**

a Vehicle, the Hebrew **R^{e}khūbh;** and, if so, would seem to be equivalent to the Wain and from the Hebrew editor of Alfonso. Others have thought it from **Rukbah,** the Knee, as β always has marked the forearm of the Bear, and Alrucaba, in a varied orthography, was current for that star some centuries ago, as it is now for Polaris. Riccioli gave a queerly combined name for the constellation, **Dubherukabah;** and Bayer had **Eruccabah,** ending his list of titles with **Ezra,** a blunder in some connection with the commentator Aben Ezra, whom he often cited as an authority; still Riccioli followed him in this.

The *Geneva Bible*, rendering the Hebrew 'Āsh, etc., by "Arcturus with his Sonnes," incorrectly added the marginal note, "the North Star with those that are about him."

Caesius typified the constellation as the **Chariot** sent by Joseph to bring his father down into Egypt, or that in which Elijah was carried to heaven; or as the **Bear** that David slew.

Young astronomers now know it as the **Little Dipper.**

In the old German manuscript already alluded to mention is made of

> Ursa Minor under the North Pole, which is called by another name **Tramontane** (*i. e.* because on one side of the Mons Coelius, whereon sits the Pole Star);

thus indicating another origin for this name than that found under Polaris as from the Mediterranean nations. I have seen no explanation of this, yet frequent references are met with in early records to some mountain located in the North as the seat of the gods and the habitation of life, the South being "the abode of the prince of death and of demons." Sayce writes:

> In early Sumerian days the heaven was believed to rest on the peak of "the mountain of the world" in the far northeast, where the gods had their habitations (cf. Isai. xiv, 13) [the mount of congregation, in the uttermost parts of the north], while an ocean or "deep" encircled the earth which rested upon its surface.

Von Herder referred to it as

> Albordy, the dazzling mountain, on which was held the assembly of the gods;

and identified it with "the holy mountain of God" alluded to in the *Book of the Prophet Ezekiel*, xxviii, 14; and Professor Whitney quoted from the 62d verse of the 1st chapter of the *Sūrya Siddhānta:*

> the mountain which is the seat of the gods;

and from the 34th verse of the 12th chapter:

A collection of manifold jewels, a mountain of gold, is Mēru,[1] passing through the middle of the earth-globe, and protruding on either side.

Commenting upon which, he says:

"the 'seat of the gods' is Mount Mēru, situated at the north pole."

The Norsemen had the same idea in their Himinbiorg, the Hill of Heaven, and the abode of Heĭmdallr, the guardian of the bridge Bifröst, the Rainbow, which united the earth to Āsaheimr, or Āsgard, the Yard, City, or Stronghold of the Āss, their gods, and the Olympus of Northern mythology. While far back of them the Egyptians supported their heavenly vault by four mountains, one at each of the cardinal points. Towards our day, in the report by "Christophorus Colonus, the Admyrall," recorded by Peter Martyr, we read that the great discoverer thought

that the earth is not perfectlye rounde; But that when it was created, there was a certeyne heape reysed thereon, much hygher than the other partes of the same.

Columbus called this Paria, asserting that it contained Paradise; but it would seem from his narrative that he located it somewhere in the neighborhood of his discoveries between North and South America. Even in Chilmead's *Treatise*, more than a century after Columbus, we find serious reference to this mythical mountain as

the mountaine Slotus, *which lies under the Pole, and is the highest in the world.*

May we not see in these the origin of Mons Coelius, the Heavenly Mountain, and of the name **Tramontana** from our constellation's location above that celestial elevation? And I would here call attention to the old story of the Seven Sleepers of Ephesus,[2] who, under the persecution of Decius in our 3d century, slumbered for nearly 200 years in the grotto under the similarly named Mount Coelian; these worthy successors of Epimenides the Cnosian and predecessors of our Rip Van Winkle being early associated with the seven stars of Ursa Major, and so perhaps with this, the Minor.

The latter's genethliacal influence was similar to that of its companion; the Prince, in Tennyson's *Princess*, thus accounting for his temperament:

For on my cradle shone the Northern star;

and likeness in their motions is alluded to in the same author's *In Memoriam* where

[1] Whatever geographical foundation there may be for this Mēru probably lies in the Pamir, the Roof of the World, that has lately become of strategical importance in Asia.

[2] These canonized Sleepers are still commemorated in the ritual of the Roman Catholic Church for the 27th of June.

the lesser wain
Is twisting round the polar star,—

one of the Greater Bear's titles being the Twister; and in the **Lazy Team,** a designation that it still more deserves than does Ursa Major.

In Proctor's attempt to reform constellation names he calls this simply **Minor,** the Greater Bear being **Ursa.**

Ursa Minor, as now drawn, is inclosed on three sides by the coils of Draco; formerly it was almost entirely so. Argelander here enumerates 27 stars down to the 5½ magnitude, and Heis 54.

one unchangeable upon a throne
Broods o'er the frozen heart of earth alone,
Content to reign the bright particular star
Of some who wander and of some who groan.
Christina G. Rossetti's *Later Life.*

α, Double, 2.2 and 9.5, topaz yellow and pale white.

Phoenice was the early Greek name, borrowed from its constellation, for this "lovely northern light" and the "most practically useful star in the heavens"; but for many centuries it has been **Stella Polaris,** the **Pole-star,** or simply **Polaris,**— Riccioli's **Pollaris**; this position seeming to be first recognized in literature by Dante when he wrote in the *Paradiso:*

the mouth imagine of the horn
That in the point beginneth of the axis
Round about which the primal wheel revolves.

Euclid said in his *Phainomena:*

A star is visible between the Bears, not changing its place, but always revolving upon itself;

Hipparchos, that the pole was "in a vacant spot forming a quadrangle with three other stars," both of these calling this Πόλος, the *Polus* of Lucan, Ovid, and other classical Latins; and Euphratean observers had called their pole-star **Pūl,** or **Bīl.** But, although other astronomical writers used these words for some individual star, there is no certainty as to which was intended, for it should be remembered that during many millenniums the polar point has gradually been approaching our pole-star, which 2000 years ago was far removed from it,— in Hipparchos' time 12° 24′ away according to his own statement quoted by Marinus of Tyre and cited by Ptolemy. Miss Clerke writes as to this:

The entire millennium before the Christian era may count for an interregnum as regards Pole-stars. Alpha Draconis had ceased to exercise that office; Alruccabah had not yet assumed it.

Kochab (the β of Ursa Minor), and κ of Draco, at different times in that epoch, may have been considered as this pole-star, the last a 4th-magnitude about 10° distant from the true pole; although the 5th-magnitude *b*, 4° away in Eratosthenes' day, perhaps was intended. And this is not unlikely, as this inconspicuous object, for some reason, was sufficiently noteworthy among the Chinese to bear the title How Kung, the Empress. The ἀει φανής, "ever visible," of the 5th-century Stobaeus may have referred to our Polaris, then about 7° distant from the pole.

The fact that the Polaris of his day did not exactly mark the pole was noted by Pytheas, the Greek astronomer and navigator of Massilia, the modern Marseilles, about 320 B. C.; and till this discovery the belief was prevalent that the heavenly pole was absolutely fixed.

In none of the foregoing cases does a single star seem to be mentioned as a guide in navigation; but as knowledge in this art increased, our α took the place of its constellation as **Stella Maris,** a title that Saint Jerome, in his *Onomasticon*, applied to the Virgin Mary; there, however, with no marine, or stellar, connection. But a star, being always a symbol of sanctity, was peculiarly so of the holiest of women, so that this title of the chief star of heaven was adopted as one interpretation of her Jewish name Miriam.

Bayer's **la Tramontana** was well known before his day, for Eden translated from the *First Decade*, printed in 1511, "cauled by the Italians Tramontana"; and Jehan de Mandeville ("syr Iohn Maundauile") more than a century before the discovery of our continent, in his statement of his belief in the sphericity of the earth, wrote of it as

the **Sterre Transmontane,** that is clept the **Sterre of the See,** that is unmevable, and that is toward the Northe, that we clepen the **Lode Sterre.**

One derivation of this *transmontane* is from the fact that the nations along the Mediterranean saw the star beyond their northern mountain boundary; and the word appears in the popular saying, current among the Latin races, of a man's "losing his Tramontane" when one had lost his bearings. Another earlier and much more probable origin, however, is from a title for the constellation already alluded to. Similarly the Finns know Polaris as **Taehti,** the Star at the Top of the Heavenly Mountain.

Anglo-Saxons of the 10th century said that it was the **Scip-steorra,** the Ship-star; Eden, "cauled of the Spanyardes **Nortes**"; Bayer, **Angel Stern,** the

Pivot Star, and the Latin **Navigatoria;** while it was the **Steering Star** to early English navigators, who

knew no North, but when the Pole Star shone.

Andrew Marvell, strangely the common friend of John Milton and King Charles II, said:

By night the northern star their way directs;

and Thomas Moore wrote, in his *Light of the Haram:*

that star, on starry nights
The seaman singles from the sky
To steer his bark for ever by.

Thus, as the leading star, it became the **Loadstar,** or **Lodestar,** of early English authors; Spenser saying:

The pilot can no loadstar see,

and Shakespeare's Helena, in *A Midsummer Night's Dream*, tells Hermia

Your eyes are lodestars.

Bryant beautifully alludes to its office in these verses from his *Hymn to the North Star:*

Constellations come, and climb the heavens, and go.
Star of the Pole! and thou dost see them set.
Alone in thy cold skies,
Thou keep'st thy old unmoving station yet,
Nor join'st the dances of that glittering train,
Nor dipp'st thy virgin orb in the blue western main.

On thy unaltering blaze
The half wrecked mariner, his compass lost,
Fixes his steady gaze,
And steers, undoubting, to the friendly coast;
And they who stray in perilous wastes by night,
Are glad when thou dost shine to guide their footsteps right.

A beauteous type of that unchanging good,
That bright eternal beacon, by whose ray
The voyager of time should shape his heedful way.

And Wordsworth, in the *Excursion*, thus goes back to the earliest times:

Chaldaean shepherds, ranging trackless fields,
Beneath the concave of unclouded skies
Spread like a sea, in boundless solitude,

Looked on the polar star, as on a guide
And guardian of their course, that never closed
His steadfast eye.

Milton's *Comus* had the much quoted

Our Star of Arcady,
Or Tyrian Cynosure;

and *L'Allegro:*

The Cynosure of neighb'ring eyes,—

a designation of Polaris which has everywhere become common; while **Cinosura** and **Cynosura** regularly appeared in scientific works of the 17th and 18th centuries; but this was one of the ancients' titles for the whole of Ursa Minor, and never, by them, limited to the *lucida*. The **Star of Arcady** either referred to Arcadia, the earthly home of Kallisto, or to Arcas, her son, transferred to the skies by his father Jove, when ignorantly about to slay his mother after her transformation. The poet, however, followed a common error in locating Arcas here, for he properly was identified with Boötes.

The Chinese had several names for it,— **Pih Keih**; **Ta Shin**; **Tien Hwang Ta Ti,** the Great Imperial Ruler of Heaven, the circumpolar stars circling around it in homage, the whole forming the **Purple Subtle Enclosure**; and **Ti** or **Ti Tso,** the Emperor's Seat, this last also being borne by α Herculis. And it was **Tow Kwei,** as with Ursa Major, from its square of stars, β, γ, ζ, and η. Its first use in navigation is ascribed to their emperor Hong Ti, or Hwang Ti, a grandson of Noah! However this may be, it seems certain that some polar star, or constellation, has been used in China from remote antiquity.

In earliest Northern India the star nearest the pole was known as **Grahadhāra,** the Pivot of the Planets, representing the great god Dhruva, and Al Bīrūnī said that among the Hindus of his time it was **Dhruva** himself. It was an object of their worship, as our Polaris is to-day among the Mandaeans[1] along the Tigris and lower Euphrates.

The Arabs knew Polaris as **Al Ḳiblah,** "because it is the star least distant from the pole," although then 5° away, and helped them, in any strange location distant from an established place of worship, to know the points

[1] This strange people, fast dwindling to extinction, are also known as Nasoraeans, or Saint John Christians. In their representation

the sky is an ocean of water, pure and clear, but of more than adamantine solidity, upon which the stars and planets sail. Its transparency allows us to see even to the pole-star, who is the central sun around whom all the heavenly bodies move. Wearing a jewelled crown, he stands before Abāthūr's door at the gate of the world of light; the Mandaeans accordingly invariably pray with their faces turned northward.

of the compass and thus the direction of Mecca and its Ka'bah,[1] towards which every good Muslim must turn his head in prayer. They also called it **Al Jadī, the Young He Goat,** which subsequently degenerated to **Juddah,** as Niebuhr heard it a century ago, and known in Desert story as **Giedi,** the slayer of the dead man on the Bier of Ursa Major.

Wetzstein says that in Damascus it is called **Mismār,** a Needle or Nail.

As marking the north pole it bore the latter's title, **Al Ḳuṭb al Shamāliyy,** the Northern Axle, or Spindle, from Al Ḳuṭb, the Pin fixed in the under stone of a mill around which the upper stone turns; and this same thought later appeared in English poetry, as in Marlowe's *History of Doctor Faustus*, where he says of the stars that

> All jointly move upon one axletree
> Whose terminine is term'd the world's wide pole.

The Arabian astronomers knew it as **Al Kaukab**[2] **al Shamāliyy,** the Star of the North, an appellation perhaps given by their nomad ancestors to β as nearer the pole in their time.

Kazwini mentioned the belief of the common people that a fixed contemplation of Al Kaukab would cure itching of the eyelids,— ophthalmia, then, as now, being the prevalent disease of the Desert.

The *Alfonsine Tables* of 1521 have **Alrucaba** *et est Stella polaris sive Polus;* and Bayer, **Alruccabah** *seu* **Ruccabah** *Ismaelitis*; but this was shared with the next star, as also with the constellation.

The Turks know it as **Yilduz,** the Star *par excellence;* and have a story that its light was concealed for a time after their capture of Constantinople.

Polaris is 1° 14′ distant from the exact pole, which lies on the straight line drawn from Polaris to ζ Ursae Majoris, and will continue in gradual approach to the pole till about the year 2095, when it will be only 26′ 30″

[1] This ancient Square House, probably an early Sabaean temple, was built, tradition says, first in heaven; then for Adam on earth as a tabernacle of radiant clouds let down by the angels directly under its celestial site. This, disappearing at his death, was replaced by one of stone and clay by the patriarch Seth, that in its turn was swept away by the Deluge. Lastly it was erected by Abraham and Ishmael to contain the Black Stone, Al Ḥajar al Aswad, a ruby, or jacinth, brought from heaven by Gabriel and now blackened by the pilgrims' tears, or because so often kissed by sinners; but it is generally regarded by unbelievers as a meteorite. The *Century Cyclopedia*, however, describes it as an irregular oval about seven inches in diameter, composed of about a dozen smaller stones of various shapes and sizes. The Stone is set into the northeast corner of the wall, at a convenient height for kissing.

[2] Kaukab is the same as the Assyrian and Chaldaean word Kakkab, the Hebrew Kōḥābh; this last also the fighting name of Bar Cochab, the Son of a Star, who was the leader of the second revolt of the Jews in 132–135, during the reign of Hadrian, his *shekels* bearing a star over a tetrastyle temple. The name was variously written, but correctly as Bar Coziba, from his birthplace.

away. It will then recede in favor successively of γ, π, ζ, ν, and α of Cepheus, α and δ of the Swan, and Wega of the Lyre, when, marked by this last brilliant star, 11,500 years hence the pole will be about 50° distant from its present position and within 5° of Wega, which for 3000 years will serve as the pole-star of the then existing races of mankind. The polar point will thence circle past ι and τ Herculis, θ, ι, and α Draconis, β Ursae Minoris, and κ Draconis back to our α again; the entire period being from 25,695 to 25,868 years, according to different calculations.[1] Shakespeare did not know all this when he wrote in *Julius Caesar:*

> constant as the Northern Star,
> Of whose true fixed and resting quality
> There is no fellow in the firmament.

Its distance from us has been variously estimated from 36 to 63 light years, and it is receding from our system at the rate of about 16 miles a second. The spectrum is Sirian.

The 9½-magnitude companion, 18″.6 distant, is a good test for a 2¼-inch glass with a power of 80. This was discovered by Sir William Herschel in 1779, and may be in revolution around its principal. Its present position angle is 215°. Other minute stars can be seen with a field-glass in the vicinity; and the Messrs. Henry of Paris have charted by photography 1270 stars, within 1° of the pole, where previously only about 80 were known by telescopic observation. α itself is slightly fainter than β.

While Polaris is the nearest naked-eye visible to the true pole, Smyth mentioned a nebula, now known as N. G. C. 3172, much nearer in 1843, and from its proximity called **Polarissima;** while nearer still was a 10th-magnitude star bearing the warlike title **Blücher,** then within 2′ of the exact point. Poole's *Celestial Handbook* says of some unidentified star:

> *Anonyma* — Double: magnitudes 7.5 and 9; distance 2′; it is the nearest to the pole.

β, 2, reddish.

Kochab is from the Arabic title that it shared with α; and it perhaps was this star that the Greek astronomers called Πόλος, for it was near the pole 1000 years before our era. Burritt has **Kochah.**

Alrucaba, variously written, is also common to it and Polaris, as well as to its constellation, Smyth saying that this was the *Alfonsine* **Reicchabba.**

[1] This uncertainty in the period of the cycle of precession mainly arises from the fact that the circle is not a strictly closed one, owing to the slight motion of the pole of the ecliptic due to the action of the planets upon the orbit of the earth.

Nā'ir al Farkadāin and **Anwār al Farkadāin,** the Bright One, and the Lights, of the Two Calves, were titles in the Desert for this star, from an early figure here, in the Fold, of these timid creatures keeping close to their mother. β was often designated by pre-Islamitic poets as the faithful and, from its ever visible position, the constant companion of the night traveler. Indeed the Badāwiyy claimed that they had a perpetual treaty with **Al Farkad** to this effect, and their poets made the **Two Pherkads,** β and γ, symbols of constancy. Chilmead cited **Alferkathan.**

α, β, γ^1, γ^2, δ, and ϵ constituted the group **Circitores, Saltatores, Ludentes,** or **Ludiones,** the Circlers, Leapers, or Dancers around the early pole, well known from classical times to late astronomy.

In China β was another **Ti,** the Emperor.

Its spectrum is Solar, and the star is receding from us at the rate of 8¾ miles a second.

γ^1, 3.3, and γ^2, 5.8.

These were known by the Arabs as one star, **Alifā' al Farkadain,** the Dim One of the Two Calves, but by us as **Pherkad Major** and **Pherkad Minor,** 57 minutes of arc apart.

With β and others they were the **Dancers,** and with β alone the **Guards,** or **Wardens, of the Pole,** that old Thomas Hood said were

> of the Spanish word *guardare*, which is to beholde, because they are diligently to be looked unto, in regard of the singular use which they have in navigation ;

and Recorde,

> many do call the **Shafte,** and others do name the **Guardas** after the Spanish tonge.

While Eden, in the *Arte of Navigation* which he "Englished out of the Spanyshe," in 1561, from Martin Cortes' communication to King Charles V, mentioned "two starres called the **Guardians,** or the **Mouth of the Horne**"; and still earlier, in his translation of Peter Martyr, "**the Guardens of the north pole.**" Shakespeare, in *Othello*, wrote:

> The wind-shak'd surge, with high and monstrous mane
> Seems to cast water on the burning Bear,
> And quench the guards of th' ever fixed pole.

Riccioli's title for them is **Vigiles,** to which he added

> Italicē le **guardiole,** overso **guardiane.**

These Guards, like the stars in Charles' Wain, were a timepiece to the common people, and even thought worthy of special treatises by navigators, as to their use in indicating the hours of the night.

In China γ^1 was **Ta Tsze,** the Crown Prince.

δ, 4.3, greenish.

Yildun is generally given to this, probably from the Turkish Yilduz that is better applied to α; but it has degenerated to **Vildiur,** and the *Century Dictionary* has **Gildun,** perhaps by a typographical error.

Bayer's Χορευτῆς πρώτη for δ, and Χορευτῆς δευτέρα for the adjoining ϵ, the First and the Second Dancer, were also general designations in which α, β, and the two stars γ were included.

ζ, 4.3, flushed white,

marking the junction of the handle with the bowl of the Little Dipper, is **Alifā' al Farḳadain** of some lists, η being **Anwār al Farḳadain;** but these titles certainly, and much better, belong to β and γ.

In China it was **Kow Chin.**

b, a 5th-magnitude, has been mentioned as **How Kung,** the Empress.

*

Virgin august! come in thy regal state
With soft majestic grace and brow serene;
Though the fierce Lion's reign is overpast
The summer's heat is all thine own as yet,
And all untouched thy robe of living green
By the rude fingers of the northern blast.

R. J. Philbrick's *Virgo.*

Virgo,

the Anglo-Saxon **Mæden,** the Anglo-Norman **Pulcele,** the French **Vierge,** the Italian **Virgine,** Bayer's **Junckfraw,** and the present German **Jungfrau,** — in fact a universal title,— generally has been figured with the palm branch in her right hand and the *spica*, or ear of wheat, in her left. Thus she was known in the Attic dialect as Κόρη, the Maiden, representing **Persephone,** the Roman **Proserpina,** daughter of Demeter, the Roman Ceres; while in the Ionic dialect Nonnus, of our 5th century, called her σταχυώδης Κούρη, the Wheat-bearing Maiden, **spicifera Virgo Cereris,** the **Virgo spicea munera gestans** of Manilius. When regarded as Proserpina, she was being abducted by Pluto in his Chariot, the stars of adjacent Libra; and the constellation also was **Demeter** herself, the **Ceres spicifera dea,** changed by the astrologers to **Arista,** Harvest, of which Ceres was goddess. Caesius had it **Arista Puellae,** that would seem more correct as **Aristae Puella**, the Maiden of the Harvest.

Those who claim very high antiquity for the zodiacal signs assert that the idea of these titles originated when the sun was in Virgo at the spring equinox, the time of the Egyptian harvest. This, however, carries them back nearly 15,000 years, while Aratos said that Leo first marked the harvest month; so that another signification has been given to the word σταχυώδης. We read, too, that

In Ogygian ages and among the Orientals, she was represented as a sun-burnt damsel, with an ear of corn in her hand, like a gleaner in the fields;

and, like most of that class, with a very different character from that assigned to her by the classic authors. Is it not this ancient story of the **Maiden of the Wheat-field** that is still seen in the North English and South Scottish custom of the Kern-baby, or Kernababy,—the Corn, or Kernel, Baby,—thus described by Lang in his *Custom and Myth?*

The last gleanings of the last field are bound up in a rude imitation of the human shape, and dressed in some rag-tags of finery. The usage has fallen into the conservative hands of children, but of old "the Maiden" was a regular image of the harvest-goddess, which, with a sickle and sheaves in her arms, attended by a crowd of reapers, and accompanied with music, followed the last carts home to the farm.

It is odd enough that the "Maiden" should exactly translate the old Sicilian name of the daughter of Demeter. "The Maiden" has dwindled, then, among us to the rudimentary Kernababy; but ancient Peru had her own Maiden, her Harvest Goddess.

And in Vendée the farmer's wife, as the corn-mother, is tossed in a blanket with the last sheaf to bring good luck in the subsequent threshing. Perhaps Caesius had some of this in view when he associated our sky figure with **Ruth,** the Moabitess, gleaning in the fields of Boaz.

Virgo also was **Erigone,**—perhaps from the Homeric Ἠριγένεια, the Early Born, for the constellation is very old,—a stellar title appearing in Vergil's apotheosis of his patron Augustus. This was the maiden who hung herself in grief at the death of her father Icarius, and was transported to the skies with Icarius as Boötes, and their faithful hound Maira as Procyon, or Sirius; all of which is attested by Hyginus and Ovid. It may have been this Icarian story that induced Keats' *Lines on the Mermaid Tavern:*

Sipping beverage divine,
And pledging with contented smack
The Mermaid in the Zodiac.

Sometimes she was figured with the Scales in her hands,—

Astraea's scales have weighed her minutes out,
Poised on the zodiac,—

whence she has been considered Δίκη, the divinity of Justice, the Roman **Justa** or **Justitia**; and **Astraea,** the starry daughter of Themis, the last of the celestials to leave the earth, with her modest sister Pudicitia, when the Brazen Age began. Ovid wrote of this:

Virgo caede madentes,
Ultima coelestum, terras Astraea reliquit;

when, according to Aratos, she

Soared up to heaven, selecting this abode,
Whence yet at night she shows herself to men.

Thus she is the oldest purely allegorical representation of innocence and virtue. This legend seems to be first found with Hesiod, and was given in full by Aratos, his longest constellational history in the *Phainomena.* Other authors mentioned her as Εἰρήνη, **Irene,** the sister of Astraea, and the **Pax** of the Romans, with the olive branch; as **Concordia**; as Παρθένος Δίος, the Virgin Goddess; as Σίβυλλα, the Singing Sibyl, carrying a branch into Hades; and as Τύχη, the Roman **Fortuna,** because she is a headless constellation, the stars marking the head being very faint.

Classical Latin writers occasionally called her **Ano, Atargatis,** and **Derceto,** the **Syrorum Dea** transferred here from Pisces; **Cybele** drawn by lions, for our Leo immediately precedes her; **Diana**; **Minerva**; **Panda** and **Pantica**; and even **Medusa.** Posidippus, 289 B. C., gave **Thesbia** or **Thespia,** daughter of Thespius, or of the Theban Asopus; and some said that one of the Muses, even **Urania** herself, was placed here in the sky by Apollo.

Ἄσπολια is from Kircher, who in turn took it from the Coptic Egyptians, the *Statio amoris, quem in incremento Nili dii ostendebant.* This, however, is singularly like Ἡ Πολιάς, designating Minerva as guardian of citadels and the State, already seen as a title for this constellation; and there was a Coptic Asphulia in Leo as a moon station.

In Egypt Virgo was drawn on the zodiacs of Denderah and Thebes, much disproportioned and without wings, holding an object said to be a distaff marked by the stars of Coma Berenices; while Eratosthenes and Avienus identified her with **Isis,** the thousand-named goddess, with the wheat ears in her hand that she afterwards dropped to form the Milky Way, or clasping in her arms the young Horus, the infant Southern sun-god, the last of the divine kings. This very ancient figuring reappeared in the Middle Ages as the **Virgin Mary** with the child Jesus, Shakespeare alluding to it in *Titus Andronicus* as the

Good Boy in Virgo's lap;

and Albertus Magnus, of our 13th century, asserted that the Saviour's horoscope lay here. It has been said that her initials, MV, are the symbol for the sign, ♍; although the *International Dictionary* considers this a monogram of Παρ, the first syllable of Παρθένος, one of Virgo's Greek titles; and others, a rude picturing of the wing of **Istar,** the divinity that the Semites assigned to its stars, and prominent in the *Epic of Creation.*

This **Istar,** or **Ishtar,** the Queen of the Stars, was the Ashtoreth of the 1st *Book of the Kings*, xi, 5, 33, the original of the Aphrodite of Greece and the Venus of Rome; perhaps equivalent to Athyr, Athor, or Hathor of the Nile, and the Astarte of Syria, the last philologically akin to our Esther and Star, the Greek 'Αστήρ. Astarte, too, was identified by the Venerable Bede with the Saxon goddess of spring, Eostre, at whose festival, our Easter, the stars of Virgo shine so brightly in the eastern evening sky; and the Sumerians of southern Babylonia assigned this constellation to their sixth month as the Errand, or Message, of Istar.

In Assyria Virgo represented **Baaltis, Belat, Belit,** and **Beltis,** Bēl's wife; while some thought her the Mylitta of Herodotus. But this was a very different divinity, the Babylonian Molatta, the Moon, the Mother, or Queen, of Heaven, against whose worship the Jews were warned in the *Book of the Prophet Jeremiah*, xliv, 17, 19, and should not be confounded with Ashtoreth, the goddess of the Zidonians, that our figure symbolized.

In India Virgo was **Kanya,** the Tamil **Kauni,** or Maiden,— in Hyde's transcription, **Kannae,**— mother of the great Krishna, figured as a **Goddess** sitting before a fire, or as a **Gūl;** and in the Cingalese zodiac as a **Woman in a Ship,** with a stalk of wheat in her hand. Al Bīrūnī thought this ship marked by the line of stars β, η, γ, δ, and ϵ, like a ship's keel. Varāha Mihira borrowed the Greek name, turning it into **Parthena, Partina,** or **Pathona.**

In Persia it was **Khosha,** or **Khusāk,** the Ear of Wheat, and **Secdeidos de Darzama,** this last often translated the "Virgin in Maiden Neatness"; but Ideler, doubting this, cited Beigel's conjecture that it was a Persian rendering of **Stachys,** one of the Greek titles of Virgo's star Spica. Bayer had it **Seclenidos de Darzama.**

The early Arabs made from some members of the constellation the enormous **Lion** of their sky; and of others the **Kennel Corner,** with dogs barking at the Lion. Their later astronomers, however, adopted the Greek figure, and called it **Al ʽAdhrā' al Naṭhīfah,** the Innocent Maiden, remains of which are found in the mediaeval titles **Eladari, Eleadari, Adrendesa,** and in the **Adrenedesa** of Albumasar. But as they would not draw the human form, they showed the stars as a sheaf of wheat, **Al Sunbulah,** or as some stalks with the ripened ears of the same, from the Roman Spica, its brightest star. Kazwini gave both of these Arabian names, the last degenerating into **Sunbala,** found in Bayer, and **Sumbela,** still occasionally seen. The *Almagest* of 1515 says *Virgo est Spica.*

The Turcomans knew the constellation as **Dufhiza Pakhiza,** the Pure Virgin; and the Chinese, as **She Sang Neu,** the Frigid Maiden; but before their Jesuit days it was **Shun Wei,** which Miss Clerke translates the **Serpent,** but Williams, the **Quail's Tail,** a part of the early stellar figure otherwise known as the **Red Bird, Pheasant,** or **Phoenix.**

It appears as **Ki,** the 20th in the Euphratean cycle of ecliptic constellations, and considered equivalent to Asru, a Place, *i. e.* the moon station that Spica marked; but Jensen thinks that the original should be **Siru,** or **Shiru,** perhaps meaning the "Ear of Corn"; much of this also is individually applied to Spica.

In the land of Judaea Virgo was **Bethūlah,** and, being always associated with the idea of abundance in harvest, was assigned by the Rabbis to the tribe of Asher, of whom Jacob had declared "his bread shall be fat." In Syria it was **Bethulta.**

Thus, like Isis, one of her many prototypes, Virgo always has been a much named and symbolized heavenly figure; Landseer saying of it, "so disguised, so modernized and be-Greek'd . . . that we literally don't know her when we see her."

In astrology this constellation and Gemini were the **House of Mercury,** Macrobius saying that the planet was created here; the association being plainly shown by the caduceus of that god, the herald's trumpet entwined with serpents, instead of the palm branch, often represented in her left hand. But usually, and far more appropriately, Virgo's stars have been given over to the care of Ceres, her namesake, the long-time goddess of the harvest. For her astrological colors Virgo assumed black speckled with blue; and was thought of as governing the abdomen in the human body, and as bearing rule over Crete, Greece, Mesopotamia, Turkey, Jerusalem, Lyons, and Paris, but always as an unfortunate, sterile sign. Manilius asserted that in his day it ruled the fate of Arcadia, Caria, Ionia, Rhodes, and the Doric plains. Ampelius assigned to it the charge of the wind Argestes, that blew to the Romans from the west-southwest according to Vitruvius, or from the west-northwest according to Pliny.

The latter said that the appearance of a comet within its borders implied many grievous ills to the female portion of the population.

Virgo was associated with Leo and with the star Sirius in the ancient opinion that, when with the sun, they were a source of heat; Ovid alluding to this in his *Ars Amatoria:*

> Virginis aetheriis cum caput ardet equis.

And John Skelton, the royal orator of King Henry VII, wrote:

> In autumn when the sun in **Virgine**
> By radiant heat enripened hath our corne.

A coin of Sardis, the capital of the kingdom of Lydia, bears her figure with the wheat ear in her left hand and a staff in her right; and the *stateres* of Macedonia have much the same. The *Alfonsine Tables* showed her as a very young girl with wings; the *Leyden Manuscript* and the *Hyginus* of 1488, as a young woman with branch and caduceus; and the *Albumasar* of 1489, as a woman with a fillet of wheat ears. The old German illustration also gave her wings, but dressed her in a high-necked, trailing gown; and Dürer drew her as a lovely winged angel.

Julius Schiller used her stars to represent **Saint James the Less,** and Weigel, as the **Seven Portuguese Towers.**

But all these figurings, ancient as some of them may be, are modern when compared with the still enduring Sphinx generally claimed as prehistoric, perhaps of the times of the Hor-she-shu, long anterior to the first historical Egyptian ruler, Menes; and constructed, according to Greek tradition, with Virgo's head on Leo's body, from the fact that the sun passed through these two constellations during the inundation of the Nile. Some Egyptologists, however, would upset this astronomical connection of the Virgin, Lion, and Sphinx, Mariette claiming the head to be that of the early god Harmachis, and others as of an early king.

Ptolemy extended the constellation somewhat farther to the east than we have it, the feet being carried into the modern Libra, and the stars that Hipparchos placed in the shoulder shifted to the side, to correct, as he said, the comparative distances of the stars and members of the body. Upon our maps it is about 52° in length, terminating on the east at λ and μ, and so is the longest of the zodiac figures. It is bounded on the north by Leo, Coma Berenices, and Boötes; on the east by Serpens and Libra; on the south by Hydra, Corvus and Crater; and on the west by Leo, Crater, and Corvus.

While the beautiful Spica is its most noteworthy object to the casual observer, yet the telescope shows here the densest nebular region in the heavens, in the space marked by its β, η, γ, δ, and Denebola of Leo; while other nebulae are scattered all over this region of the sky. Sir William Herschel found here no less than 323, which later search has increased to over 500,— very many more nebulae than naked-eye stars in the constellation. Argelander gives 101 of the latter, and Heis 181.

It is for these four stars in Virgo, forming with ε two sides of a right-angled triangle open towards Denebola, γ at its vertex, that Professor Young uses his mnemonic word **Begde** to recall their order. They extend along the wings through the girdle, and were the **Kennel Corner of the Barking Dogs** of the Arabs, often considered as the **Dogs** themselves.

Von Zach, of Gotha, rediscovered here on the last day of the first year of this century the minor planet Ceres, whose position had been lost some time after its discovery by Piazzi on the previous New Year's Day; Olbers repeating this, and independently, the next evening, the first anniversary of the original discovery. Here, too, Olbers found, on the 28th of March, 1802, another minor planet, Pallas, the second one discovered, and appropriately named, for the thirty-first of the *Orphic Hymns* described this goddess as "inhabiting the stars."

The sun passes through the constellation from the 14th of September to the 29th of October; and during this time

> the Virgin trails
> No more her glittering garments through the blue.

α, Spectroscopic binary, 1.3, brilliant flushed white.

Spica signifies, and marks, the Ear of Wheat shown in the Virgin's left hand—Aratos wrote "in her hands"; Vitruvius and Hyginus, "in her right hand"— when she was thought to be Ceres. All the Romans called it thus, Cicero saying **Spicum,** and their descendants, the modern Italians, **Spigha**; the French have **l'Epi.** In Old England it was the **Virgin's Spike,** and even Flamsteed thus designated it. For at least twenty-five centuries, and among all civilized peoples, the Latin word, or words of similar import, has obtained; although Smyth mentioned an attempt before his day to secure for it the illustrious name of **Newton.**

Στάχυς, perhaps of the same signification although another has been assigned to it, appeared with Aratos, Hipparchos, and Ptolemy, transcribed by the Latins as **Stachys.** Manetho had Σταχυώδης, which we have seen used for Virgo by another Graeco-Egyptian author, Nonnus. Bayer cited **Arista** for the star as for the constellation; **Aristae Puella** occurs in some Latin doggerel by Caesius; as the brightest of the figure it bore the latter's **Erigone**; while **Vindemitor** and **Vindemiator**, which better belong to ε, have been applied to it.

Other titles—**Sunbala**; **Sunbale**; **Sumbela**; Riccioli's **Sumbalet, Sombalet, Sembalet Eleandri**; and Schickard's **Sunbalon**—are from **Sunbulah** and **Al ʽAdhrā'**, Arabic words synonymous respectively with Spica and Virgo, although Hyde derived them from Σίβυλλα, the Singing Sibyl, of the constellation. Al Bīrūnī said that it was **Al Ḥulbah**, the Bristle, but his explanation of this only served to show the strange confusion in titles that existed in the Arab mind between Spica and Al Ḍafīrah in the Lion's tail. And Al Bīrūnī, again, said that it was the **Calf of the Lion**, with Arcturus as the second Calf; but Kazwini designated it as **Sāḳ al Asad**, the Shin-bone of the Lion, this Lion being the enormous figure already alluded to, of which a part of Virgo formed one of the legs.

A still more widely spread native name in the Desert was **Al Simāk al Aʽzal**, [illegible] Defenceless, or Unarmed, Simāk, *i. e.* unattended by any near-by star; [illegible] other Simāk, Arcturus, being armed with a lance, or staff, represented [illegible] adjacent stars of Boötes; and it doubtless was this isolated position of Spica that induced the Coptic title **Khoritos**, Solitary. The *Alfonsine Tables* turned Simāk al Aʽzal into **inermis Asimec**, adding **Acimon, Alaraph, Almucedie** "of [illegible] Chaldaeans," and **Alacel**; while the 1515 *Almagest* had **As**[illegible]**mis.** From all these come Bayer's **Alaazel, Alazel, Azimon,** [illegible] the Nubians," **Hazimet Alazel**, the alchemists' **Alhaiseth**, Ric[illegible]**samecti** and **Eltsamach**, and the **Azimech** still occasionally seen. Scaliger had **Hazimeth Alhacel**, and Schickard **Huzimethon.** Riccioli cited a "Nubian" title, **Eleazalet**, that some have said came from Al ʽAzalah, the Hip-bone, but it probably belongs among the derivatives from *Aʽzal*; and his **Eleadari** has been transferred to Spica from the constellation.

This star marked the 12th *manzil*, **Al Simāk**, and in early astrology was, like all of Virgo, a sign of unfruitfulness and a portent of injustice to innocence; but later on, of eminence, renown, and riches.

Chrysococca called it μικρός Κονταράτος, the Little Lance-bearer, Arcturus being Κονταράτος *par excellence.* And Hyde gave the Hebrew **Shibbōleth**, the Syrian **Shebbeltā**, the Persian **Chūshe**, and the Turkish **Salkim**, all signifying the "Ear of Wheat"; other names being the Persian **Çpur**, the **Çparegha** of the *Avesta*, the Sogdian **Shaghar** and Khorasmian **Akhshafarn**, all meaning a "Point"—*i. e.* Spica.

The Hindus knew it as **Citrā**, Bright, their 12th *nakshatra*, figured as a **Lamp**, or as a **Pearl**, with Tvashtar, the Artificer, or Shaper, as its presiding divinity; and some have thought it the **Tistar Star** that generally has been identified with Sirius.

In Babylonia, and representing the whole constellation, it personified the **wife of Bēl,** and as **Sa-Sha-Shirū,** the Virgin's Girdle, marked the 20th ecliptic asterism of that name, and the lunar asterism **Dan-nu,** the Hero of the Sky Furrow. It was also **Emuku Tin-tir-Ki,** the Might of the Abode of Life, a common title for Babylon itself.

In Chinese astronomy Spica was a great favorite as **Kió,** the Horn, or Spike, anciently **Keok** or **Guik,** the special star of springtime; and with ζ formed their 12th *sieu* under that title. Naturally it was the determinant.

It is said to have been known at one time in Egypt as the **Lute-Bearer,** and was evidently of importance, for another Egyptian name was **Repā,** the Lord; and Lockyer thinks that the great "Mena may symbolize Spica, with which star we have seen Min-worship associated." According to this same author, one of the temples at Thebes, probably dedicated to this Mena, Menat, Menes, Min, or Khem, was oriented to Spica's setting about 3200 B. C.; and the temple of the Sun at Tell al Amarna was also so oriented about 2000 B. C., or perhaps somewhat later. A similar char[illegible]acter [illegible] attached to it in Greece, for two temples have been found at Rhamn[illegible]us, "almost touching one another, both following (and with accordant d[illegible]es) the shifting places of Spica," at their erection 1092 and 747 B. C.; "and still another pair at Tegea." Temples of Herē were also so oriented at Olympia 1445 B. C., at Argos and Girgenti; and those of [illegible]eros at Athens, 1130 B. C., and of "the Great Diana of the Ephesia[illegible]

It was to the observations of this star and of Regulus about [illegible] corded by the Alexandrian Timochares, that, after comparison with [illegible] 150 years later, Hipparchos was indebted for the great discovery attributed to him of the precession of the equinoxes; although Babylonian records, and the temple orientation of Egypt and Greece, may indicate a far earlier practical knowledge of this.

According to Ptolemy, Timochares observed an occultation by the planet Venus of an unidentified star "on the tip of Virgo's wing,"— perhaps ψ or q,— on the 12th of October, 271 B. C.[1]

[1] A still earlier record of the planet, dating from 686 B. C., is on a tablet from Chaldaea now in the British Museum; while earlier still are Homer's Ἕσπερος, the Latin Hesperus,—

> the brightest star that shines in Heav'n;

and Isaiah's

> . . . day star, son of the morning,

that our *Authorized Version* rendered "Lucifer," the equivalent of the Greek alternative titles Ἑωσφόρος and Φωσφόρος, the Latin Phosphorus. The identity of this Morning Star with the Evening Star Hesperus was discovered by Pythagoras, or by Parmenides, in the 5th century before Christ.

The planet also was known as Ἀφροδίτη, as Juno's Star, and as Isis.

Spectroscopic observations by Vogel in 1890 show that Spica is in revolution with a speed of at least fifty-six miles a second in an orbit of three millions of miles' radius, around the common centre of gravity of itself and an obscure companion in a period of about four days. It is, however, never eclipsed by the latter, as is the case with the star Algol. Its spectrum is Sirian; and the system is approaching us at the rate of 9.2 miles a second. Gould thinks that it shows fluctuations in brilliancy.

It is one of the lunar stars much utilized in navigation, and lies but 2° south of the ecliptic, and 10° south of the celestial equator, coming to the meridian on the 28th of May.

With Denebola, Arcturus, and Cor Caroli it forms the **Diamond of Virgo,** 50° in extent north and south.

β, 3.9, pale yellow.

Zavijava, a universal name in modern catalogues, is first found with Piazzi, but is **Zarijan** in the *Standard Dictionary*. It is from **Al Zāwiah,** the Angle, or Corner, *i. e.* Kennel, of the Arab Dogs,— although γ exactly marks this Corner and should bear the title.

The stars β, η, γ, δ, ε, outlining this Kennel, formed the 11th *manzil*, **Al 'Awwā',** the Barker, which was considered of good omen; while Firuzabadi included it with the preceding moon station Al Ṣarfah,— β Leonis,— in the group **Al Nahrān,** the Two Rivers, as their rising was in the season of heavy rains. Other indigenous titles were **Al Bard,** the Cold, which it was said to produce; and **Warak al Asad,** the Lion's Haunches.

β marked the 18th ecliptic constellation of Babylonia, **Shēpu-arkū sha-A,** the Hind Leg of the Lion, for this country also seems to have had one of these creatures here. With η, it perhaps was **Ninsar,** the Lady of Heaven, probably a reference to Istar; and **Urra-gal,** the God of the Great City; and one of the seven pairs of stars famous in that astronomy. As a Euphratean lunar asterism it bore the same title **Ninsar,** but this included all the components of the Arabs' Kennel Corner.

These also were the Persian **Mashaha,** the Sogdian **Fastashat,** the Khorasmian **Afsasat,** and the Coptic **Abukia,** all of the Arabic signification.

In China it was **Yew Chi Fa,** the Right-hand Maintainer of Law.

β is 13° south of Denebola in Leo, culminating with it on the 3d of May.

γ, Binary and slightly variable, 3 and 3.2, white.

The Latins called this **Porrima,** or **Antevorta,** sometimes **Postvorta,** names of two ancient goddesses of prophecy, sisters and assistants of Carmenta or Carmentis, worshiped and at times invoked by their women. Porrima was known as **Prorsa** and **Prosa** by Aulus Gellius of our 2d century.

γ was specially mentioned by Kazwini as itself being **Zāwiat al 'Awwā',** the Angle, or Corner, of the Barker; and Al Tizini, with Ulug Beg, had much the same name for it; but Al Bīrūnī, quoting from Al Zajjāj, said that "these people are all wrong," and that 'Awwā' here meant "Turn," referring to the turn, or bend, in the line of stars. This interesting early figure is noticeable even to the casual observer, γ being midway between Spica and Denebola, the sides of the Kennel stretching off to the north and west, respectively marked by η and β, δ and ε.

In Babylonia it marked the 19th ecliptic constellation, **Shur-mahrū-shirū,** the Front, or West, Shur (?); while individually it was **Kakkab Dan-nu,** the Star of the Hero, and the reference point in their annals of an observation of Saturn[1] on the 1st of March, 228 B. C., the first mention of this planet that we have, and recorded by Ptolemy.

The Chinese knew γ as **Shang Seang,** the High Minister of State.

Astronomers consider the two stars alternately variable in light; and some call both yellow, so following the apparent rule of similar coloration in components of binaries when of equal brilliancy; those unequal being of contrasting colors. In 1836 they showed as a single star in the largest telescope then in use; but now are 6″ apart, moving in an orbit more eccentric than any other as yet well determined, with a period of revolution estimated at about 190 years. The position angle in 1890 was 330°. They are of special interest to astronomers, as well as a show object to all.

They culminate on the 17th of May.

δ, 3.6, golden yellow,

although individually unnamed in our lists, was one of the 'Awwā'.

On the Euphrates it was **Lu Lim,** the Gazelle, Goat, or Stag,— or perhaps King; and, with ε, probably **Mas-tab-ba,** another of the seven pairs of Twin-stars of that country. The Hindus called it **Āpa,** or **Āpas,** the Waters; and the Chinese, **Tsze Seang,** the Second Minister of State.

Secchi alluded to δ as *bellissima*, from its most beautiful banded spectrum of the 3d class of spectra, like that of α Herculis.

ε, 3.3, bright yellow,

is the **Vindemiatrix** of the *Alfonsine Tables*, whence it has descended into modern lists; but in Latin days it was **Vindemiator** with Columella, which

[1] Saturn was Χρόνος and Φαίνων, the Shiner, with the Greeks; Al Thāḳib, the Piercer, with the Arabs; and Saturnus, or Stella Solis, with the Latins.

is found as late as Flamsteed; **Vindemitor,** with Ovid and Pliny; and **Provindemiator** and **Provindemia major,** with Vitruvius; all signifying the "Grape-gatherer," from its rising in the morning just before the time of the vintage. These titles were translations of the Προτρυγετήρ, Προτρυγετής, Προτρύγετος, and Τρυγετήρ, used by Ptolemy, Plutarch, and other Greek authors, the first of these words appearing in the *Phainomena*, and rendered the "Fruit-plucking Herald"; but it is in a line of the poem considered doubtful; Riccioli had **Protrigetrix.** This profusion of titles from the earliest times indicates the singular interest with which this now inconspicuous star was regarded in classical astronomy. The *Century Cyclopedia* has the following note on it:

At the time when the zodiac seems to have been formed (2100 B. C.) this star would first be seen at Babylon before sunrise about August 20, or, since there is some evidence that it was then brighter than it is now, perhaps a week earlier. This would seem too late for the vintage, so that perhaps this tradition is older than the zodiac.

The classical name was translated by the Arabians **Muḳdim al Ḳitāf;** and another title was **Almuredin,** still seen for it, perhaps from Al Muridīn, Those Who Sent Forth. Traces of these words are found in the **Alacast, Alcalst, Alaraph,** and **Almucedie** of Bayer's *Uranometria.*

In China it was **Tsze Tseang,** the Second General.

On the Euphrates it may have been **Kakkab Mulu-izi,** the Star Man of Fire, possibly symbolizing the god **Laterak,** the Divine King of the Desert; although that title has been assigned to μ Virginis and δ Librae.

It marked the eastern boundary of the 11th *manzil*, and in astrology was a mischief-making star. It culminates on the 22d of May.

η, Variable between 3 and 4.

Zaniah is from **Al Zāwiah,** applied in German lists to this instead of to the stars β and γ, all of these being in the Kennel.

In China it was **Tso Chih Fa,** the Left-hand Maintainer of Law.

It lies on the left side of the Virgin, and just to the westward is the point of the autumnal equinox which the Chinese knew as **Yih Mun, Twan Mun,** or **Tien Mun,** Heaven's Gate. With ζ it almost exactly marks the line of the celestial equator.

θ, Triple, 4.4, 9, and 10, pale white, violet, and dusky,

is on the front of the garment, below the girdle; the components, 7″.1 and 65″ apart; the position angle of the first two stars being 345°.

Moderns have no name for it, but in the *Sūrya Siddhānta* it was **Apami-Atsa,** the Child of the Waters.

With another adjacent, but now unidentified, star, it was known in China as **Ping Taou,** the Plain and Even Way.

ι, 4.2.

Syrma is from Σύρμα, used by Ptolemy to designate this star on the Train of the Virgin's robe.

With κ and ϕ it was mentioned in the first Arabian translation of the *Syntaxis* as being in the *ḥimār*, or "skirt," of the garment; but the translator of the Latin edition of 1515, missing the point at the first letter, read the word as *ḥimār*, "an ass," so that this central one of these three stars strangely appears in that work as *in asino*. They formed the 13th *manzil*, **Al Ghafr,** the Covering, as Smyth explains,

because the beauty of the earth is hidden when they rise on the 18th Tishrīn, or 1st of November; others say on account of the shining of the stars being lessened as if covered;

but Kazwini,

because, when they rise, the earth robes herself in her splendour and finery,—her summer robes.

The Arabic word, however, is analogous to Σύρμα, and so may have been taken from Ptolemy; although Al Bīrūnī quoted from Al Zajjāj **Al Ghafar,** the Tuft in the Lion's tail, which it may have marked in the figure of the ancient Asad. Another signification of the word Ghafr is the "Young Ibex." Al Bīrūnī also said that the Arabs considered this the most fortunate of their lunar stations, as lying between the evils of the Lion's teeth and claws on one side and the tail and venom of the Scorpion on the other, and quoted from a Rajaz poet:

The best night forever
Lies between Al Zubānah and Al Asad;

adding that the horoscope of the Prophet lay here, and that the date of the birth of Moses coincided with it.

As a lunar station these stars were the Sogdian **Sarwa** and the Khorasmian **Shushak,** the Leader; the Persian **Huçru,** the Good Goer; and the Coptic **Khambalia,** Crooked-clawed, λ being substituted for ϕ; and it is said that they were the Akkadian **Lu Lim,** the He Goat, Gazelle, or Stag, the original perhaps also meaning "King," and employed for δ.

ι alone, according to Hommel, was the Death Star, **Mulu Bat.**

ι, κ, and υ constituted the 13th *sieu*, **Kang**, a Man's Neck, κ being the determining star; while, with the preceding station, the united group was **Sheu sing**, as Edkins writes it, the Star of Old Age; and, with others near, it may have been included in the Tien Mun mentioned at the star η.

μ, a 3.9-magnitude, was Al Achsasi's **Rijl al ʿAwwā'**, the Foot of the Barker. It has been included with δ Librae in the Akkadian lunar asterism **Mulu Izi**, a title also applied to ε; the Sogdian **Fasarwa**, and the Khorasmian **Sara-fsariwa**, both signifying the "One next to the Leader"—*i. e.* next to the lunar asterism ι, κ, and λ.

ν, ξ, ο, and π, forming the head of Virgo, were the Chinese **Nuy Ping**, the Inner Screen; ρ was **Kew Heang**, the Nine Officers of State, in which some smaller stars were included; σ and τ, **Tien Teen**, the Heavenly Fields; while χ and ψ, with others adjacent, were **Tsin Heen**; all of these stars being of 4th to 6th magnitudes.

*

Vulpecula cum Ansere, the Little Fox with the Goose,

is known in Italy as **Volpe colla Oca**; in Germany as **Fuchs**, or **Füchschen, mit der Gans**; and in France as **Petit Renard avec l'Oie.**

Smyth wrote that this is

> a modern constellation, crowded in by Hevelius to occupy a space between the Arrow and the Swan, where the Via Lactea divides into two branches. For this purpose he ransacked the *informes* of this bifurcation, and was so satisfied with the result, that the effigies figure in the elaborate print of his offerings to Urania. He selected it on account of the Eagle, Cerberus and Vultur Cadens. "I wished," said he, "to place a fox and a goose in the space of the sky well fitted to it; because such an animal is very cunning, voracious and fierce. Aquila and Vultur are of the same nature, rapacious and greedy."

The two members are sometimes given separately; indeed the **Anser** is often omitted. Flamsteed's *Atlas* shows both, but separates the titles; and Proctor arbitrarily combined both in his **Vulpes.** Astronomers now call the whole **Vulpecula.**

Its inventor saw 27 stars here, but Argelander catalogued 37, and Heis 62. They come to the meridian toward the end of August.

Although I have elsewhere found no named star in Vulpecula, and its general faintness would render it doubtful whether there ever has been one, yet the *Standard Dictionary* says of it under the word **Anser**:

> a small star in the constellation of the Fox and the Goose;

and the *Century Dictionary* has much the same. This may have been *a*, the *lucida*, a 4.4-magnitude just west of the Fox's head.

A meteor stream, the **Vulpeculids**, appearing from the 13th of June to the 7th of July, radiates from a point in this constellation; but the latter's most noteworthy object is the **Double-headed Shot**, or **Dumb-bell, Nebula**, N. G. C. 6853, 27 M., just visible in a 1¼-inch finder, 7° southeast from the star Albīreo.

✶

> . . . the milky way i' the sky,—
> A meeting of gentle lights without a name.
>
> Sir John Suckling.

> Torrent of light and river of the air,
> Along whose bed the glimmering stars are seen
> Like gold and silver sands in some ravine
> Where mountain streams have left their channels bare!
>
> Longfellow's *The Galaxy*.

The Galaxy, or Milky Way,

has borne arbitrary, descriptive, or fanciful titles in every age.

Anaxagoras, 550 B. C., and Aratos knew it as τo Γάλα,

> that shining wheel, men call it Milk;

Eratosthenes, as Κύκλος Γαλαξίας, the Circle of the Galaxy; other Greek authors, as Κύκλος γαλακτικός, the Galactic Circle; and Hipparchos, as ὁ Γαλάξιος, the Galaxy. Galaxurē, the Lovely One, of the *Homeric Hymn* may have been the personification of this; and Galatēa, the Milk-white, of the *Iliad*, for this nymph was a daughter of Oceanus, and the Galaxy was lŏng known as **Eridanus,** the Stream of Ocean. Indeed during all historic time it has been thought of as the **River of Heaven.**

Such, too, was the Akkadian idea of it in connection with that of a **Great Serpent**; Brown writing of this:

> No doubt the Great Serpent, in one of its mystic phases, is connected with the **Ocean-stream**—*v. g.*, the Norse **Midhgardhsormr**, the **Weltum-spanner** ("Stretcher-round-the-world"). But the Akkadian **Snake-river**, with whatever else it may be associated, cer-

tainly also in one phase, and on the three Boundary-stones referred to, represents the *Circulus Lacteus*. In *W. A. I.*, II, 51, we read:

45. Akkadian *Hid tsirra*, Assyrian *Nahru tsiri*, = "River-of-the-Snake." Thus *Hid-dagal*, "River" + "great" = *Hiddekel* (Genesis ii, 14).

46. Ak. *Hid turra An gal*, As. *Nahru markasi Ili rabi*, = "River-of-the-cord-of-the-God great."

47. Ak. *Hid zuab gal*, As. *Nahru Apshi rabi*, = "River-of-the-Abyss great."

It also was the **River-of-the-Shepherd's-hut**, dust-cloud high, and the Akkadian *Hid In-ni-na*, **River-of-the-Divine-Lady**; and, to quote again:

This Snake-river of sparkling dust, the stream of the abyss on high through which it runs, the golden cord of the heaven-god (Prof. Sayce aptly refers to Il. viii, 19), connected alike with the hill of the Sun-god and with the passage of ghosts, is the Milky Way; and it is the **River of Nana**, wife of the heaven-god, as, in Greek mythology, it is connected with Herē.

Among the Arabs it was **Al Nahr**, the River, a title that they afterwards transferred to the Greek constellation Eridanus; and those other Semites, the Hebrews, knew it as **Nehar di Nur**, the River of Light; but the Rabbi Levi recurred to the Akkadian simile in saying that it was the **Crooked Serpent** of the *Book of Job*, xxvi, 13. Usually, however, in Judaea it was **Aroch**,— in Armenia and Syria, **Arocea**,— not a lexicon word, but evidently from *Aruḥāh*, a Long Bandage, and well applied to this long band of light.

In China, as in Japan, it was **Tien Ho**, the Celestial River, and the **Silver River**, whose fish were frightened by the new moon, which they imagined to be a hook; although those countries also may have named it as we do, for in the *She King* are the lines by the emperor-poet Seuen, of the 8th century before Christ, translated by Legge:

Brightly resplendent in the sky revolved
The Milky Way;

and again:

Vast is this Milky Way,
Making a brilliant figure in the sky.

Al Bīrūnī quoted from a Sanskrit tradition that it was **Akāsh Gangā**, the Bed of the Ganges; but his other Hindu title, **Kshīra**, is not explained. In North India it was **Bhagwān ki Kachahri**, the Court of God, and **Swarga Duāri**, the Dove of Paradise.

In Rome it was often thought of as the Heavenly Girdle, **Coeli Cingulum**, and as a Circle; Pliny, calling it **Circulus lacteus**, followed Cicero, who also said **Orbis lacteus**, and made extended allusion to it in his *Vision of Scipio* as "a radiant circle of dazzling brightness amid the flaming bodies."

It is in this *Vision* that we find a graphic and beautiful description of the nine heavenly crystal circles, the foundation of the old system of astronomy, from which issued the Harmony of the Spheres universally believed in till the times of Copernicus; but Euripides already had written of it:

Thee I invoke, thou self-created Being, who gave birth to Nature, and whom light and darkness, and the whole train of globes, encircle with eternal music.

Towards our day Shakespeare, in the *Merchant of Venice*, said:

There's not the smallest orb which thou behold'st
But in his motion like an angel sings;

Milton, in *Paradise Lost:*

the fix'd stars, fix'd in their orb that flies,
And ye five other wand'ring fires that move
In mystic dance not without song;

Ben Jonson:

Spheres keep one musick, they one measure dance;

and Addison doubtless had it in mind in his beautiful astronomical hymn:

Forever singing as they shine.

Kepler assigned the various tones in music to the various planets, one issuing from each of the spheres: the bass from Saturn and Jupiter, the tenor from Mars, the contralto from Venus, and the soprano from Mercury.

The conception of the Milky Way as a pathway always and everywhere has been current. This is seen in the Romans' **Via coeli regia; Via lactis** and **Via lactea,** the **Mylke way** and **Mylke whyte way** in Eden's rendering; **Semita lactea,** the Milky Footpath; and Ovid's

High Road paved with stars to the court of Jove;

imitated, in *Paradise Lost*, by Milton's

The Way to God's eternal house,

the much quoted

Broad and ample road whose dust is gold,
And pavement stars, as stars to thee appear
Seen in the galaxy, that milky way
Which nightly as a circling zone thou seest
Powder'd with stars.

The Norsemen knew it as the Path of the Ghosts going to Valhöll (Valhalla), in the region Gladhsheimr,— the palace of their heroes slain in battle;

and our North American Indians had the same idea, as witness the "wrinkled old Nokomis," when, teaching the little Hiawatha, she

> Showed the broad white road in heaven,
> Pathway of the ghosts, the shadows,
> Running straight across the heavens,
> Crowded with the ghosts, the shadows,
> To the Kingdom of Ponemah,
> To the land of the hereafter;

the brighter stars along the Road marking their camp-fires. William Hamilton Hayne's *Indian Fancy* embodies it thus:

> Pure leagues of stars from garish light withdrawn
> Behind celestial lace-work pale as foam,—
> I think between the midnight and the dawn
> Souls pass through you to their mysterious home.

Our aborigines and the Eskimo also called it the **Ashen Path,** as did the Bushmen of Africa,—the ashes hot and glowing, instead of cold and dark, that benighted travelers might see their way home,— thus unwittingly following the classical Manilius:

> this was once the Path
> Where Phoebus drove; and in length of Years
> The heated track took Fire and burnt the Stars.
> The Colour changed, the Ashes strew'd the Way,
> And still preserve the marks of the Decay;

although he also more scientifically wrote:

> Anne magis densa stellarum turba corona.

Among the early Hindus it was the **Path of Âryamān,** leading to his throne in Elysium; in the Panjab it is **Berā dā ghās,** the Path of Noah's Ark; and in northern India, **Nagavithi,** the Path of the Snake.

The Patagonians think it the road on which their dead friends are hunting ostriches.

The Anglo-Saxons knew it as **Wætlinga Stræt,**— Hoveden's **Watlingastrete,**— the path of the Wætlings, the giant sons of King Wætla, Vate, or Ivalde; Minsheu thus defining the word:

> *howsoever* the Romans *might make it . . .* the names bee from the Saxons, *and* Roger Hoveden *saith it is so called because the sonnes of* Wethle *made it leading from the East sea to the West;*

and going into extended and very interesting details as to its course, and those of other Roman "waies" in early Britain. Old Thomas Hood similarly could see no derivation for this title,

except it be in regard of the narrowness it seemeth to have, or else in respect of that great highway that lieth between Dover and St. Albans.

This was variously known as **Werlam Street, Wadlyng Street, Vatlant Street,** and lastly **Watling Street,**[1] the ancient road still in use from Chester (the ancient Deva), through London (Londinium), to Dover (Dubris Portus); and its stellar connection appears in the *Hous of Fame:*

Lo, there, quod he, cast up thine eye.
Se yonder, lo, the Galaxyë,
Which men clepeth the Milky Wey,
For hitt is whytt, and some parfey,
Callen hit Watlinge Strete.

Another title, **Walsyngham Way,** first found in Langland's *Vision of William concerning Piers Plowman*, made it the road to the Virgin Mary in heaven, as the earthly way was to her shrine in Norfolk, where she was known as our Lady of Walsyngham; this existing till 1538, when England abolished her monasteries. The idea of this, and of other similar path-titles, may have come from the fancy that this heavenly way crowded with stars resembled the earthly roads crowded with pilgrims. Anglo-Saxon glossaries have it as **Iringes Uueg, Weg,** or **Wec,** Iringe's Way; and as **Bil-Idun's Way,** these personages being descendants of Wætla, and both Ways leading to Asgard over the bridge at which Slavonic mythology terminated this celestial way, and thus joined earth to heaven, "where four monks guard the sacred road and cut to pieces all who attempt to traverse it." Later on this **Asgard Bridge** was the title indiscriminately applied to the Milky Way and Rainbow, varied, as to the latter, by Bifröst or Asbreu.

And here I may be pardoned for repeating a quaintly beautiful passage from Minsheu's definition of the Rainbow, although not connected with the Galaxy, nor strictly astronomical:

The Bow *is the weapon of* warre *and therefore called* the Bow of the battell, ¶ Zach. 9. 10. (battle-bow) & 10. 4. (id). *The* Bow *that appeareth in the* clouds *hath no* string, *nor no deadly* arrow *prepared upon it, there is no* wrath that appeareth in it; *et dicitur* Arcus clementiae & foederis, *indicans* mundum non secundo periturum aquis. *And therefore we should love him that hath laid aside his* wrath, *and embraced us with mercie.*

It will be remembered that Minsheu's was a *polyglot* dictionary! Ves-

[1] It is only fair to say that there are other derivations for Watling Street,—one by no means improbable, Minsheu to the contrary notwithstanding, namely, that it was called after Vitellianus, the Roman director in its construction, whom the Britons knew as Guetalin.

pucci, a century before, expressed much the same sentiment where — but connecting the Bible with Science — he wrote, in Eden's rendering:

It is a pledge of peace betweene god and men, and is ever directly over ageynst the soonne.

Grimm, in *Teutonic Mythology*, cites many titles for the Galaxy. Among the Northmen it was **Wuotanes Weg,** or **Straza,** Wuotan's, or Woden's, Way, or Street; among the Midland Dutch, **Vronelden Straet,** the Women's Street, and **Hilde,** or **Hulde, Strasse,** Saint Hilda's, or Hulda's, Street; in Jutland, **Veierveien,** or **Brunel, Straet;** in Westphalia, **Wiär Strate,** the Weather Street, and **Mülen Weg,** the Milky Way; and in East Friesland, **Harmswith** and the **Melkpath.** In Hungary it was **Hada Kuttya,** the *Via Belli*, because in the journey of war and migration from Asia their ancestors followed this shining mark; and the Finns have the pretty **Linnunrata,** the Birds' Way, as the winged spirits flit thither to the free and happy land, or because the united bird-songs once were turned into a cloud of snow-white dovelets still seen overhead. This was the Lithuanian **Paukszcziu Kielis.**

In Germany the modern **Milch Strasse** is the translation of our best-known title; while it has long been, and popularly is even now, **Jakobs Strasse** and **Jakobs Weg,** Jacob's Road; as the Belt of Orion is his Staff lying alongside the road. And it has been still further associated with that patriarch as his **Ladder.**

In Sweden the Milky Way is the **Winter Street,**— so, at all events, with the peasantry,— their **Winter Gatan;** and that country's idea of it is thus beautifully given by Miss Edith M. Thomas:

Silent with star-dust, yonder it lies —
The Winter Street, so fair and so white;
Winding along through the boundless skies,
Down heavenly vale, up heavenly height.

Faintly it gleams, like a summer road
When the light in the west is sinking low,
Silent with star-dust! By whose abode
Does the Winter Street in its windings go?

And who are they, all unheard and unseen —
O, who are they, whose blessed feet
Pass over that highway smooth and sheen?
What pilgrims travel the Winter Street?

Are they not those whom here we miss
In the ways and the days that are vacant below?
As the dust of that Street their footfalls kiss
Does it not brighter and brighter grow?

Steps of the children there may stray
  Where the broad day shines though dark earth sleeps,
And there at peace in the light they play,
  While some one below still wakes and weeps.

The old Norsemen had a similar title in their **Vetrarbraut**; and the Celts knew it as **Arianrod**, the Silver Street, which also occurs for the Northern Crown, but there as the Silver Circle.

In England, for centuries, the Galaxy has been the **Way of Saint James**, sometimes the **Way to Saint James**, and thus figuratively the *Via regia ;* in Italy, the **Via lattea**; in France, the **Voie lactée.** But with the French peasantry it always has been the **Road of Saint Jacques of Compostella**, this last itself a stellar word from the Campus Stellae of Theodomir, bishop of Idria, who was guided by a star in 835 to the bones of Saint James in a field. The same title obtains in Spain, but there it is popularly known as **El Camino de Santiago**, the patron saint in battle of that country, Longfellow writing of this in his *Galaxy :*

The Spaniard sees in thee the pathway, where
  His patron saint descended in the sheen
  Of his celestial armor, on serene
And quiet nights when all the heavens were fair.

In the Basque tongue it is **Ceruco Esnibidia.**

Wherever this idea of a road was held in early times it seems to have referred to the Milky Way as traveled by the departing souls of illustrious men, who, Manilius wrote, were

loos'd from the ignoble Chain
Of Clay, and sent to their own Heaven again,—

to those stars, that were regarded not only as the homes of such, but often as the very souls themselves physically shining in the skies, as, metaphorically, they had upon the earth. Thus it was known in classical times as **Heroum Sedes.** Following out this conception, the Galaxy later became the Italian **Strada di Roma**; the Swiss **Weg uf Rom**; the Slovak **Zesta v'Rim**,—all signifying the "Way of Rome," because only through that capital of the church could access to heaven be secured.

Thomas Moore somewhat changed the figure in his *Loves of the Angels*, where he says as to the stars in general:

Rolling along like living cars
Of light, for gods to journey by!—

a thought that also is found with Pliny, and even with Saint Clement.

Romieu says that the Galaxy was **Masarati,** probably Assyrian, and identifies it with the hieroglyphic **Masrati,** the Course of the sun-god, that may be the origin of the story of Phaëthon, and we see very much the same title in the Babylonian *Creation Legend* as applied to the zodiac. This word, similar to the Hebrew **Mazzārōth** that some Rabbis positively asserted signifies the "Milky Way," appears in Stoffler's *De Sphaera* as **Maiarati,** apparently taken from Ptolemy, and supposed by Canon Cook, in the *Speaker's Commentary* on the *Book of Job*, xxxviii, 32, to be the equivalent of the Arabic **Al Majarrah,** the Milky Track.

In addition to this last,—Riccioli's **Almegiret,**—the Arabians had **Ṭarīḳ al Laban** of the same meaning, but also knew the Galaxy as **Ḍarb al Tābānīn,** the Path of the Chopped Straw Carriers, and as **Ṭarīḳ al Tibn,** the Straw Road.

Riccioli gave this as the Hebrew **Nedhībath Tebhen,** correctly **Nethībhath,** which the Syrians translated **Shᵉbhīl Tebhnā**; the Persians, **Rah Kakeshan,** or simply **Kakeshan**; the Copts, **Pimoit ende pitoh**; and the Turks, **Samān Ugh'risi.** These last also called it **Hagjiler Yūli,** the Pilgrims' Road, traversed in their annual journey to Mecca.

Riccioli also cited the "Aethiopian" **Chasara tsamangadu**; and Grimm, the same country's **Pasare Zamanegade,** the Straw Stalks lying in the Road; —both probably from one original differently transcribed. And a singular legend, from some unknown source, tells us that these Stalks, or Chopped Straw, marking the Pilgrims' Road, were dropped by Saint Venus (!) after her theft from Saint Peter; hence her Armenian title Hartacol, or Hartacogh, the Straw-thief. In China it shared the zodiac's name of the **Yellow Road,** from the color of this scattered straw.

In classic folk-lore the Milky Way was marked out by the corn ears dropped by Isis in her flight from Typhon; or was the result of some of Juno's nursery troubles with the infant Hercules. Alluding to these, Manilius wrote that it

> justly draws
> Its name, the Milky Circle, from its cause.

From this doubtless came the Roman **Circulus Junonius.** Early India accounted for it in somewhat the same way in connection with Saramā; and a similar thought is expressed by the Arabic **Umm al Samā',** the Mother of the Sky.

Caer Gwydyon, the Castle of Gwydyon, the enchanter son of Don, the King of the Fairies, is one of its Celtic titles in more modern times, others of the family appearing in Cassiopeia and Corona Borealis. But the Celts also thought it the road along which Gwydyon pursued his erring wife.

The Incas of Peru said that it was the dust of stars, and gave titles to its various parts; the Ottawa Indians, that it was the muddy water stirred up by a turtle swimming along the bottom of the sky; while the Polynesian islanders know it as the **Long, Blue, Cloud-eating Shark.**

In poetry, too, the Milky Way has ever been a favorite — indeed, a hackneyed — subject. Miss Myra Reynolds tells us in her *Treatment of Nature in English Poetry:*

> From Waller on, the Milky Way typifies virtues so numerous that they shine in one undistinguished blaze;

and that Swift's *Apollo's Edict* of 1720, among its prohibitions to authors of the use of some of the more wearisomely frequent similitudes, specifically forbids their even naming the Milky Way,— a rule that would have been equally applicable to the classical authors as to those of our day. Among the former, Manilius wrote of it:

> as a beaten Path that spreads between
> A troden Meadow, and divides the Green.
> Or as when Seas are plow'd behind the Ship,
> Foam curls on the green surface of the Deep.
> In Heaven's dark surface such this Circle lies,
> And parts with various Light the Azure skies.
> Or as when Iris draws her radiant Bow
> Such seems this Circle to the World below.

Among recent poetical similes we find Edward Young's

> this midnight pomp,
> This gorgeous arch with golden worlds inlaid;

Joseph Rodman Drake's

> The milky baldric of the skies,

and in the *Culprit Fay:*

> the bank of the milky way;

Tennyson's

> marvelous round of milky light
> Below Orion;

while in the *Lady of Shalott* he likens the "gemmy bridle" of Sir Lancelot to

> some branch of stars we see
> Hung in the golden Galaxy.

The Finnish Topelius made it the

> starry bridge of light,
> Which now smiles down upon the earth from heaven's placid face,
> And firmly binds together still the shores of boundless space.

This was built by the lovers Zulamith and Salami that they might be united in heaven as they had been on earth.

> They toiled and built a thousand years
> In love's all powerful might:
> And so the Milky Way was made—
> A starry bridge of light;

and when the task was successfully accomplished they were merged together in the single star Sirius.

Homer strangely did not allude to it, unless he may have personified it in the *Iliad.* Nor did Ptolemy express any opinion as to its nature, although he called it the **Band,— Fascia** in one Latin translation,— and fully described it in the 8th book of the *Syntaxis;* his account of it being considered "certainly superior to all the rather fantastic representations given in the maps published before the last quarter of our century."

Dante gave much attention to it in his *Convito*, repeating various of the opinions of the ancient philosophers. He said that Anaxagoras considered it reflected light from the sun, an opinion shared by Aristotle, Democritus, and even by the later Avicenna (Ibn Sina of Bokhara) of about A. D. 1000; and he attributed to Aristotle another theory — that it was the gathering of vapors under the stars of that region. His own lines in the *Paradiso*—

> distinct with less and greater lights
> Glimmers between the two poles of the world—

accurately describe it, as does his

> Galassia si, che fa dubbiar ben saggi;

for speculation concerning it was almost as varied as its observers.

Aristotle expressed still a third opinion, that it was the gases from the earth set on fire in the sky; Oinopides and Metrodorus considered it the early course of the sun abandoned after the bloody banquet of Thyestes; the Pythagoreans and others, that it marked the blazing path of the disastrous runaway when, as in the *Inferno*,

> Phaeton abandoned the reins,
> Whereby the heavens, as still appears, were scorched;

or, as in Longfellow's *The Galaxy:*

> Phaeton's wild course that scorched the skies
> Where'er the hoofs of his hot coursers trod.

Some thought it the sunbeams left behind in the track of the sun's chariot, — the **Vestigium Solis,** that Macrobius termed **Zona perusta,** the Girdle Burned; and others, **Via perusta.** Plutarch said that it was the shadow of the earth as the sun passed beneath us. Diodorus the Sicilian, of the 1st century before Christ, and the philosopher-naturalist Theophrastus, of the 3d, asserted that it marked the junction of the two starry hemispheres, — a statement thus versified by Manilius:

> Whether the Skies grown old here shrink their frame,
> And through the chinks admit an upper Flame,
> Or whether here the Heaven's two Halves are joyn'd,
> But odly clos'd, still leave a Seam behind.
> Or here the parts in Wedges closely prest,
> To fix the Frame, are thicker than the Rest.
> Like Clouds condens'd appear, and bound the Sight,
> The Azure being thickened into White.

Even as late as 1603 Bayer wrote:

> Constat hic circulus ex tenui nebulosa substantia;

and such probably was the general scientific conception of the Galaxy until seven years later Galileo's "glazed optic tube" revealed its larger constituent stars, and, as he wrote in the *Nuncius Sidereus*,

> got rid of disputes about the Galaxy . . . for it is nothing else but a mass of innumerable stars planted together in clusters.

A few, however, even in antiquity seem to have known, or at least suspected, its true character; for Democritus, the master of Epicurus, about 460 B. C., and Pythagoras before him, said that it was a vast assemblage of very distant stars, in which belief Aristotle seems to have coincided; although several other, and absurd, opinions are attributed to this eminent man, as well as to Democritus. Manilius thus expressed this belief:

> Or is the spatious Bend serenely bright
> From little Stars, which there their Beams unite,
> And make one solid and continued Light?

Arabian poets wrote similarly, as Ta'abbata Sharran, whose verse is quoted in the *Ḥamasah*,—

> The Mother of clustered stars.

Our knowledge of it may thus briefly be summed up: It covers more than one tenth of the visible heavens, containing nine tenths of the visible stars, and seems a vast zone-shaped nebula, nearly a great circle of the sphere, the poles being in Coma and Cetus. In a measure it can be resolved by slight optical aid into innumerable stars, although even the largest telescopes will not resolve the faintest parts. Many of these stars are small, "not at all comparable with our sun in dimensions." It is inclined about 63° to the celestial equator, and, Sir John Herschel wrote,

is to sidereal what the invariable ecliptic is to planetary astronomy — a plane of ultimate reference, the ground-plane of the sidereal system.

Our position close to its central plane is not favorable to a correct survey; but, as we see it, it is marked by strange cavities and excrescences, with branches in all directions, and is interrupted in its course, especially at Ophiuchus and Argo, apparently by the operation of some force still at work,— these interruptions being in its width as well as in its course. Its apparent structure is not uniform, but curdled or flaky,— bright patches alternating with faint or with almost absolute vacancies.

While it contains a large number of star-clusters, it has but few true nebulae, although among these are the important Horseshoe Nebula below Scutum, the Dumb-bell in Vulpecula, and the Trifid in Sagittarius; yet large diffused masses of nebulosity are found in several portions of it.

Pickering's spectroscopic work seems to indicate that the Milky Way forms a system separate from the rest of the sidereal universe; but Gould inclined to the opinion that it is "the resultant of two or more superposed galaxies," which will perhaps account for the brighter portions in Cassiopeia and Crux as representing "the intersection of the two crossed rings visibly diverging in Ophiuchus." And Miss Clerke thus concludes the chapter on the Milky Way in her *System of the Stars:*

What is unmistakable is that the entire formation, whether single or compound, is no isolated phenomenon. All the contents of the firmament are arranged with reference to it. It is a large part of a larger scheme exceeding the compass of finite minds to grasp in its entirety.

L'ENVOI

Unto those Three Things which the Ancients held impossible, there should be added this Fourth, to find a Book Printed without erratas.

Alfonso de Cartagena

That this book has its faults, no one can doubt,
Although the Author could not find them out.
The faults you find, good Reader, please to mend,
Your comments to the Author kindly send.

Kitchiner's *The Economy of the Eyes.— Part II.*

GENERAL INDEX

Not including Authors' Names, Arabic Titles, Biblical references, nor Greek matter.
Separate indices for these follow this General Index.

ARABIC INDEX

Accentuation of the originals of the corrupted words has been followed for the latter as far as practicable, but in many cases necessarily is arbitrary. The Arabic alphabet, with its English equivalents, follows these pages.

THE ARABIC ALPHABET

'	ا	Alif	glottal catch.
b	ب	Bā	
t	ت	Tā	
th	ث	Thā	
j	ج	Jīm	like *j* in *Jack*, or *g* in *gem*.
ḥ	ح	Ḥā	smooth guttural aspirate.
ḣ	خ	Ḣā	like *ch* in the Scotch word *loch;* in the German *rache*. Velar spirant.
d	د	Dāl	
dh	ذ	Dhāl	like *th* in *the, that*.
r	ر	Rā	
z	ز	Zāy	
s	س	Sīn	
sh	ش	Shīn	
ṣ	ص	Ṣād	like *ts;* or, as in modern Arabic, a sharp palatal *s*.
ḍ	ض	Ḍād	*d* with a glottal catch.
ṭ	ط	Ṭā	emphatic palatal *t*.
th	ظ	Ṭhā	emphatic *z*.
'	ع	'Ain	strong glottal catch.
gh	غ	Ghain	post-palatal guttural.
f	ف	Fā	
ḳ	ق	Ḳāf	pronounced by the tongue and the velum palati.
k	ك	Kāf	
l	ل	Lām	
m	م	Mīm	
n	ن	Nūn	
h	ه	Hā	
w	و	Wāw	
y	ى	Yā	

At the beginning of words and syllables the Alif (') is not represented. The termination of feminine nouns (*at*) is represented by *ah*, except where a genitive follows. The case terminations (nom. *u;* gen. *i;* acc. *a*) and their nasalized forms (*un; in; an*) are not represented. The article is invariably transcribed *al;* no account is taken of the assimilation of the *l* to a following consonant. The vowels are used in their so-called Continental pronunciation.

GREEK INDEX

INDEX TO ASTRONOMICAL REFERENCES

AS FOUND IN THE REVISED VERSION OF THE BIBLE.

THE OLD TESTAMENT.

THE NEW TESTAMENT.

PARTIAL LIST

OF AUTHORS, AUTHORITIES, AND BOOKS OF REFERENCE CITED IN THIS WORK.

The page number refers to the first, or to some important mention.

A CATALOGUE OF SELECTED DOVER BOOKS IN ALL FIELDS OF INTEREST

THE SENSE OF BEAUTY, George Santayana. Masterfully written discussion of nature of beauty, materials of beauty, form, expression; art, literature, social sciences all involved. 168pp. 5⅜ x 8½. 20238-0 Pa. $3.50

ON THE IMPROVEMENT OF THE UNDERSTANDING, Benedict Spinoza. Also contains *Ethics, Correspondence,* all in excellent R. Elwes translation. Basic works on entry to philosophy, pantheism, exchange of ideas with great contemporaries. 402pp. 5⅜ x 8½. 20250-X Pa. $5.95

THE TRAGIC SENSE OF LIFE, Miguel de Unamuno. Acknowledged masterpiece of existential literature, one of most important books of 20th century. Introduction by Madariaga. 367pp. 5⅜ x 8½.
20257-7 Pa. $6.00

THE GUIDE FOR THE PERPLEXED, Moses Maimonides. Great classic of medieval Judaism attempts to reconcile revealed religion (Pentateuch, commentaries) with Aristotelian philosophy. Important historically, still relevant in problems. Unabridged Friedlander translation. Total of 473pp. 5⅜ x 8½. 20351-4 Pa. $6.95

THE I CHING (THE BOOK OF CHANGES), translated by James Legge. Complete translation of basic text plus appendices by Confucius, and Chinese commentary of most penetrating divination manual ever prepared. Indispensable to study of early Oriental civilizations, to modern inquiring reader. 448pp. 5⅜ x 8½. 21062-6 Pa. $6.00

THE EGYPTIAN BOOK OF THE DEAD, E. A. Wallis Budge. Complete reproduction of Ani's papyrus, finest ever found. Full hieroglyphic text, interlinear transliteration, word for word translation, smooth translation. Basic work, for Egyptology, for modern study of psychic matters. Total of 533pp. 6½ x 9¼. (USCO) 21866-X Pa. $8.50

THE GODS OF THE EGYPTIANS, E. A. Wallis Budge. Never excelled for richness, fullness: all gods, goddesses, demons, mythical figures of Ancient Egypt; their legends, rites, incarnations, variations, powers, etc. Many hieroglyphic texts cited. Over 225 illustrations, plus 6 color plates. Total of 988pp. 6⅛ x 9¼. (EBE)
22055-9, 22056-7 Pa., Two-vol. set $20.00

THE STANDARD BOOK OF QUILT MAKING AND COLLECTING, Marguerite Ickis. Full information, full-sized patterns for making 46 traditional quilts, also 150 other patterns. Quilted cloths, lame, satin quilts, etc. 483 illustrations. 273pp. 6⅞ x 9⅝. 20582-7 Pa. $5.95

CORAL GARDENS AND THEIR MAGIC, Bronsilaw Malinowski. Classic study of the methods of tilling the soil and of agricultural rites in the Trobriand Islands of Melanesia. Author is one of the most important figures in the field of modern social anthropology. 143 illustrations. Indexes. Total of 911pp. of text. 5⅝ x 8¼. (Available in U.S. only)
23597-1 Pa. $12.95

THE PHILOSOPHY OF HISTORY, Georg W. Hegel. Great classic of Western thought develops concept that history is not chance but a rational process, the evolution of freedom. 457pp. 5⅜ x 8½. 20112-0 Pa. $6.00

LANGUAGE, TRUTH AND LOGIC, Alfred J. Ayer. Famous, clear introduction to Vienna, Cambridge schools of Logical Positivism. Role of philosophy, elimination of metaphysics, nature of analysis, etc. 160pp. 5⅜ x 8½. (USCO) 20010-8 Pa. $2.50

A PREFACE TO LOGIC, Morris R. Cohen. Great City College teacher in renowned, easily followed exposition of formal logic, probability, values, logic and world order and similar topics; no previous background needed. 209pp. 5⅜ x 8½. 23517-3 Pa. $4.95

REASON AND NATURE, Morris R. Cohen. Brilliant analysis of reason and its multitudinous ramifications by charismatic teacher. Interdisciplinary, synthesizing work widely praised when it first appeared in 1931. Second (1953) edition. Indexes. 496pp. 5⅜ x 8½. 23633-1 Pa. $7.50

AN ESSAY CONCERNING HUMAN UNDERSTANDING, John Locke. The only complete edition of enormously important classic, with authoritative editorial material by A. C. Fraser. Total of 1176pp. 5⅜ x 8½. 20530-4, 20531-2 Pa., Two-vol. set $16.00

HANDBOOK OF MATHEMATICAL FUNCTIONS WITH FORMULAS, GRAPHS, AND MATHEMATICAL TABLES, edited by Milton Abramowitz and Irene A. Stegun. Vast compendium: 29 sets of tables, some to as high as 20 places. 1,046pp. 8 x 10½. 61272-4 Pa. $17.95

MATHEMATICS FOR THE PHYSICAL SCIENCES, Herbert S. Wilf. Highly acclaimed work offers clear presentations of vector spaces and matrices, orthogonal functions, roots of polynomial equations, conformal mapping, calculus of variations, etc. Knowledge of theory of functions of real and complex variables is assumed. Exercises and solutions. Index. 284pp. 5⅝ x 8¼. 63635-6 Pa. $5.00

THE PRINCIPLE OF RELATIVITY, Albert Einstein et al. Eleven most important original papers on special and general theories. Seven by Einstein, two by Lorentz, one each by Minkowski and Weyl. All translated, unabridged. 216pp. 5⅜ x 8½. 60081-5 Pa. $3.50

THERMODYNAMICS, Enrico Fermi. A classic of modern science. Clear, organized treatment of systems, first and second laws, entropy, thermodynamic potentials, gaseous reactions, dilute solutions, entropy constant. No math beyond calculus required. Problems. 160pp. 5⅜ x 8½. 60361-X Pa. $4.00

ELEMENTARY MECHANICS OF FLUIDS, Hunter Rouse. Classic undergraduate text widely considered to be far better than many later books. Ranges from fluid velocity and acceleration to role of compressibility in fluid motion. Numerous examples, questions, problems. 224 illustrations. 376pp. 5⅝ x 8¼. 63699-2 Pa. $7.00

THE AMERICAN SENATOR, Anthony Trollope. Little known, long unavailable Trollope novel on a grand scale. Here are humorous comment on American vs. English culture, and stunning portrayal of a heroine/villainess. Superb evocation of Victorian village life. 561pp. 5⅜ x 8½.
23801-6 Pa. $7.95

WAS IT MURDER? James Hilton. The author of *Lost Horizon* and *Goodbye, Mr. Chips* wrote one detective novel (under a pen-name) which was quickly forgotten and virtually lost, even at the height of Hilton's fame. This edition brings it back—a finely crafted public school puzzle resplendent with Hilton's stylish atmosphere. A thoroughly English thriller by the creator of Shangri-la. 252pp. 5⅜ x 8. (Available in U.S. only)
23774-5 Pa. $3.00

CENTRAL PARK: A PHOTOGRAPHIC GUIDE, Victor Laredo and Henry Hope Reed. 121 superb photographs show dramatic views of Central Park: Bethesda Fountain, Cleopatra's Needle, Sheep Meadow, the Blockhouse, plus people engaged in many park activities: ice skating, bike riding, etc. Captions by former Curator of Central Park, Henry Hope Reed, provide historical view, changes, etc. Also photos of N.Y. landmarks on park's periphery. 96pp. 8½ x 11. 23750-8 Pa. $4.50

NANTUCKET IN THE NINETEENTH CENTURY, Clay Lancaster. 180 rare photographs, stereographs, maps, drawings and floor plans recreate unique American island society. Authentic scenes of shipwreck, lighthouses, streets, homes are arranged in geographic sequence to provide walking-tour guide to old Nantucket existing today. Introduction, captions. 160pp. 8⅞ x 11¾. 23747-8 Pa. $7.95

STONE AND MAN: A PHOTOGRAPHIC EXPLORATION, Andreas Feininger. 106 photographs by *Life* photographer Feininger portray man's deep passion for stone through the ages. Stonehenge-like megaliths, fortified towns, sculpted marble and crumbling tenements show textures, beauties, fascination. 128pp. 9¼ x 10¾. 23756-7 Pa. $5.95

CIRCLES, A MATHEMATICAL VIEW, D. Pedoe. Fundamental aspects of college geometry, non-Euclidean geometry, and other branches of mathematics: representing circle by point. Poincare model, isoperimetric property, etc. Stimulating recreational reading. 66 figures. 96pp. 5⅝ x 8¼.
63698-4 Pa. $3.50

THE DISCOVERY OF NEPTUNE, Morton Grosser. Dramatic scientific history of the investigations leading up to the actual discovery of the eighth planet of our solar system. Lucid, well-researched book by well-known historian of science. 172pp. 5⅜ x 8½. 23726-5 Pa. $3.50

THE DEVIL'S DICTIONARY. Ambrose Bierce. Barbed, bitter, brilliant witticisms in the form of a dictionary. Best, most ferocious satire America has produced. 145pp. 5⅜ x 8½. 20487-1 Pa. $2.50

THE COMPLETE BOOK OF DOLL MAKING AND COLLECTING, Catherine Christopher. Instructions, patterns for dozens of dolls, from rag doll on up to elaborate, historically accurate figures. Mould faces, sew clothing, make doll houses, etc. Also collecting information. Many illustrations. 288pp. 6 x 9. 22066-4 Pa. $4.95

THE DAGUERREOTYPE IN AMERICA, Beaumont Newhall. Wonderful portraits, 1850's townscapes, landscapes; full text plus 104 photographs. The basic book. Enlarged 1976 edition. 272pp. 8¼ x 11¼. 23322-7 Pa. $7.95

CRAFTSMAN HOMES, Gustav Stickley. 296 architectural drawings, floor plans, and photographs illustrate 40 different kinds of "Mission-style" homes from *The Craftsman* (1901-16), voice of American style of simplicity and organic harmony. Thorough coverage of Craftsman idea in text and picture, now collector's item. 224pp. 8⅛ x 11. 23791-5 Pa. $6.50

PEWTER-WORKING: INSTRUCTIONS AND PROJECTS, Burl N. Osborn. & Gordon O. Wilber. Introduction to pewter-working for amateur craftsman. History and characteristics of pewter; tools, materials, step-by-step instructions. Photos, line drawings, diagrams. Total of 160pp. 7⅞ x 10¾. 23786-9 Pa. $3.50

THE GREAT CHICAGO FIRE, edited by David Lowe. 10 dramatic, eyewitness accounts of the 1871 disaster, including one of the aftermath and rebuilding, plus 70 contemporary photographs and illustrations of the ruins—courthouse, Palmer House, Great Central Depot, etc. Introduction by David Lowe. 87pp. 8¼ x 11. 23771-0 Pa. $4.00

SILHOUETTES: A PICTORIAL ARCHIVE OF VARIED ILLUSTRATIONS, edited by Carol Belanger Grafton. Over 600 silhouettes from the 18th to 20th centuries include profiles and full figures of men and women, children, birds and animals, groups and scenes, nature, ships, an alphabet. Dozens of uses for commercial artists and craftspeople. 144pp. 8⅜ x 11¼. 23781-8 Pa. $4.50

ANIMALS: 1,419 COPYRIGHT-FREE ILLUSTRATIONS OF MAMMALS, BIRDS, FISH, INSECTS, ETC., edited by Jim Harter. Clear wood engravings present, in extremely lifelike poses, over 1,000 species of animals. One of the most extensive copyright-free pictorial sourcebooks of its kind. Captions. Index. 284pp. 9 x 12. 23766-4 Pa. $8.95

INDIAN DESIGNS FROM ANCIENT ECUADOR, Frederick W. Shaffer. 282 original designs by pre-Columbian Indians of Ecuador (500-1500 A.D.). Designs include people, mammals, birds, reptiles, fish, plants, heads, geometric designs. Use as is or alter for advertising, textiles, leathercraft, etc. Introduction. 95pp. 8¾ x 11¼. 23764-8 Pa. $4.50

SZIGETI ON THE VIOLIN, Joseph Szigeti. Genial, loosely structured tour by premier violinist, featuring a pleasant mixture of reminiscenes, insights into great music and musicians, innumerable tips for practicing violinists. 385 musical passages. 256pp. 5⅝ x 8¼. 23763-X Pa. $4.00

TONE POEMS, SERIES II: TILL EULENSPIEGELS LUSTIGE STREICHE, ALSO SPRACH ZARATHUSTRA, AND EIN HELDENLEBEN, Richard Strauss. Three important orchestral works, including very popular *Till Eulenspiegel's Marry Pranks,* reproduced in full score from original editions. Study score. 315pp. 9⅜ x 12¼. (Available in U.S. only)
23755-9 Pa. $8.95

TONE POEMS, SERIES I: DON JUAN, TOD UND VERKLARUNG AND DON QUIXOTE, Richard Strauss. Three of the most often performed and recorded works in entire orchestral repertoire, reproduced in full score from original editions. Study score. 286pp. 9⅜ x 12¼. (Available in U.S. only) 23754-0 Pa. $8.95

11 LATE STRING QUARTETS, Franz Joseph Haydn. The form which Haydn defined and "brought to perfection." *(Grove's).* 11 string quartets in complete score, his last and his best. The first in a projected series of the complete Haydn string quartets. Reliable modern Eulenberg edition, otherwise difficult to obtain. 320pp. 8⅜ x 11¼. (Available in U.S. only)
23753-2 Pa. $8.95

FOURTH, FIFTH AND SIXTH SYMPHONIES IN FULL SCORE, Peter Ilyitch Tchaikovsky. Complete orchestral scores of Symphony No. 4 in F Minor, Op. 36; Symphony No. 5 in E Minor, Op. 64; Symphony No. 6 in B Minor, "Pathetique," Op. 74. Bretikopf & Hartel eds. Study score. 480pp. 9⅜ x 12¼. 23861-X Pa. $10.95

THE MARRIAGE OF FIGARO: COMPLETE SCORE, Wolfgang A. Mozart. Finest comic opera ever written. Full score, not to be confused with piano renderings. Peters edition. Study score. 448pp. 9⅜ x 12¼. (Available in U.S. only) 23751-6 Pa. $12.95

"IMAGE" ON THE ART AND EVOLUTION OF THE FILM, edited by Marshall Deutelbaum. Pioneering book brings together for first time 38 groundbreaking articles on early silent films from *Image* and 263 illustrations newly shot from rare prints in the collection of the International Museum of Photography. A landmark work. Index. 256pp. 8¼ x 11.
23777-X Pa. $8.95

AROUND-THE-WORLD COOKY BOOK, Lois Lintner Sumption and Marguerite Lintner Ashbrook. 373 cooky and frosting recipes from 28 countries (America, Austria, China, Russia, Italy, etc.) include Viennese kisses, rice wafers, London strips, lady fingers, hony, sugar spice, maple cookies, etc. Clear instructions. All tested. 38 drawings. 182pp. 5⅜ x 8.
23802-4 Pa. $2.75

THE ART NOUVEAU STYLE, edited by Roberta Waddell. 579 rare photographs, not available elsewhere, of works in jewelry, metalwork, glass, ceramics, textiles, architecture and furniture by 175 artists—Mucha, Seguy, Lalique, Tiffany, Gaudin, Hohlwein, Saarinen, and many others. 288pp. 8⅜ x 11¼. 23515-7 Pa. $8.95

SECOND PIATIGORSKY CUP, edited by Isaac Kashdan. One of the greatest tournament books ever produced in the English language. All 90 games of the 1966 tournament, annotated by players, most annotated by both players. Features Petrosian, Spassky, Fischer, Larsen, six others. 228pp. 5⅜ x 8½. 23572-6 Pa. $3.50

ENCYCLOPEDIA OF CARD TRICKS, revised and edited by Jean Hugard. How to perform over 600 card tricks, devised by the world's greatest magicians: impromptus, spelling tricks, key cards, using special packs, much, much more. Additional chapter on card technique. 66 illustrations. 402pp. 5⅜ x 8½. (Available in U.S. only) 21252-1 Pa. $5.95

MAGIC: STAGE ILLUSIONS, SPECIAL EFFECTS AND TRICK PHOTOGRAPHY, Albert A. Hopkins, Henry R. Evans. One of the great classics; fullest, most authorative explanation of vanishing lady, levitations, scores of other great stage effects. Also small magic, automata, stunts. 446 illustrations. 556pp. 5⅜ x 8½. 23344-8 Pa. $6.95

THE SECRETS OF HOUDINI, J. C. Cannell. Classic study of Houdini's incredible magic, exposing closely-kept professional secrets and revealing, in general terms, the whole art of stage magic. 67 illustrations. 279pp. 5⅜ x 8½. 22913-0 Pa. $4.00

HOFFMANN'S MODERN MAGIC, Professor Hoffmann. One of the best, and best-known, magicians' manuals of the past century. Hundreds of tricks from card tricks and simple sleight of hand to elaborate illusions involving construction of complicated machinery. 332 illustrations. 563pp. 5⅜ x 8½. 23623-4 Pa. $6.95

THOMAS NAST'S CHRISTMAS DRAWINGS, Thomas Nast. Almost all Christmas drawings by creator of image of Santa Claus as we know it, and one of America's foremost illustrators and political cartoonists. 66 illustrations. 3 illustrations in color on covers. 96pp. 8⅜ x 11¼. 23660-9 Pa. $3.50

FRENCH COUNTRY COOKING FOR AMERICANS, Louis Diat. 500 easy-to-make, authentic provincial recipes compiled by former head chef at New York's Fitz-Carlton Hotel: onion soup, lamb stew, potato pie, more. 309pp. 5⅜ x 8½. 23665-X Pa. $3.95

SAUCES, FRENCH AND FAMOUS, Louis Diat. Complete book gives over 200 specific recipes: bechamel, Bordelaise, hollandaise, Cumberland, apricot, etc. Author was one of this century's finest chefs, originator of vichyssoise and many other dishes. Index. 156pp. 5⅜ x 8. 23663-3 Pa. $2.75

TOLL HOUSE TRIED AND TRUE RECIPES, Ruth Graves Wakefield. Authentic recipes from the famous Mass. restaurant: popovers, veal and ham loaf, Toll House baked beans, chocolate cake crumb pudding, much more. Many helpful hints. Nearly 700 recipes. Index. 376pp. 5⅜ x 8½. 23560-2 Pa. $4.95

ILLUSTRATED GUIDE TO SHAKER FURNITURE, Robert Meader. Director, Shaker Museum, Old Chatham, presents up-to-date coverage of all furniture and appurtenances, with much on local styles not available elsewhere. 235 photos. 146pp. 9 x 12. 22819-3 Pa. $6.95

COOKING WITH BEER, Carole Fahy. Beer has as superb an effect on food as wine, and at fraction of cost. Over 250 recipes for appetizers, soups, main dishes, desserts, breads, etc. Index. 144pp. 5⅜ x 8½. (Available in U.S. only) 23661-7 Pa. $3.00

STEWS AND RAGOUTS, Kay Shaw Nelson. This international cookbook offers wide range of 108 recipes perfect for everyday, special occasions, meals-in-themselves, main dishes. Economical, nutritious, easy-to-prepare: goulash, Irish stew, boeuf bourguignon, etc. Index. 134pp. 5⅜ x 8½. 23662-5 Pa. $3.95

DELICIOUS MAIN COURSE DISHES, Marian Tracy. Main courses are the most important part of any meal. These 200 nutritious, economical recipes from around the world make every meal a delight. "I . . . have found it so useful in my own household,"—*N.Y. Times*. Index. 219pp. 5⅜ x 8½. 23664-1 Pa. $3.95

FIVE ACRES AND INDEPENDENCE, Maurice G. Kains. Great back-to-the-land classic explains basics of self-sufficient farming: economics, plants, crops, animals, orchards, soils, land selection, host of other necessary things. Do not confuse with skimpy faddist literature; Kains was one of America's greatest agriculturalists. 95 illustrations. 397pp. 5⅜ x 8½. 20974-1 Pa. $4.95

A PRACTICAL GUIDE FOR THE BEGINNING FARMER, Herbert Jacobs. Basic, extremely useful first book for anyone thinking about moving to the country and starting a farm. Simpler than Kains, with greater emphasis on country living in general. 246pp. 5⅜ x 8½. 23675-7 Pa. $3.95

PAPERMAKING, Dard Hunter. Definitive book on the subject by the foremost authority in the field. Chapters dealing with every aspect of history of craft in every part of the world. Over 320 illustrations. 2nd, revised and enlarged (1947) edition. 672pp. 5⅜ x 8½. 23619-6 Pa. $8.95

THE ART DECO STYLE, edited by Theodore Menten. Furniture, jewelry, metalwork, ceramics, fabrics, lighting fixtures, interior decors, exteriors, graphics from pure French sources. Best sampling around. Over 400 photographs. 183pp. 8⅜ x 11¼. 22824-X Pa. $6.95

ACKERMANN'S COSTUME PLATES, Rudolph Ackermann. Selection of 96 plates from the *Repository of Arts*, best published source of costume for English fashion during the early 19th century. 12 plates also in color. Captions, glossary and introduction by editor Stella Blum. Total of 120pp. 8⅜ x 11¼. 23690-0 Pa. $5.00

THE ANATOMY OF THE HORSE, George Stubbs. Often considered the great masterpiece of animal anatomy. Full reproduction of 1766 edition, plus prospectus; original text and modernized text. 36 plates. Introduction by Eleanor Garvey. 121pp. 11 x 14¾. 23402-9 Pa. **$8.95**

BRIDGMAN'S LIFE DRAWING, George B. Bridgman. More than 500 illustrative drawings and text teach you to abstract the body into its major masses, use light and shade, proportion; as well as specific areas of anatomy, of which Bridgman is master. 192pp. 6½ x 9¼. (Available in U.S. only) 22710-3 Pa. **$4.50**

ART NOUVEAU DESIGNS IN COLOR, Alphonse Mucha, Maurice Verneuil, Georges Auriol. Full-color reproduction of *Combinaisons ornementales* (c. 1900) by Art Nouveau masters. Floral, animal, geometric, interlacings, swashes—borders, frames, spots—all incredibly beautiful. 60 plates, hundreds of designs. 9⅜ x 8-1/16. 22885-1 Pa. **$4.50**

FULL-COLOR FLORAL DESIGNS IN THE ART NOUVEAU STYLE, E. A. Seguy. 166 motifs, on 40 plates, from *Les fleurs et leurs applications decoratives* (1902): borders, circular designs, repeats, allovers, "spots." All in authentic Art Nouveau colors. 48pp. 9⅜ x 12¼. 23439-8 Pa. **$6.00**

A DIDEROT PICTORIAL ENCYCLOPEDIA OF TRADES AND INDUSTRY, edited by Charles C. Gillispie. 485 most interesting plates from the great French Encyclopedia of the 18th century show hundreds of working figures, artifacts, process, land and cityscapes; glassmaking, papermaking, metal extraction, construction, weaving, making furniture, clothing, wigs, dozens of other activities. Plates fully explained. 920pp. 9 x 12. 22284-5, 22285-3 Clothbd., Two-vol. set **$50.00**

HANDBOOK OF EARLY ADVERTISING ART, Clarence P. Hornung. Largest collection of copyright-free early and antique advertising art ever compiled. Over 6,000 illustrations, from Franklin's time to the 1890's for special effects, novelty. Valuable source, almost inexhaustible.
Pictorial Volume. Agriculture, the zodiac, animals, autos, birds, Christmas, fire engines, flowers, trees, musical instruments, ships, games and sports, much more. Arranged by subject matter and use. 237 plates. 288pp. 9 x 12. 20122-8 Clothbd. **$15.00**

Typographical Volume. Roman and Gothic faces ranging from 10 point to 300 point, "Barnum," German and Old English faces, script, logotypes, scrolls and flourishes, 1115 ornamental initials, 67 complete alphabets, more. 310 plates. 320pp. 9 x 12. 20123-6 Clothbd. $15.00

CALLIGRAPHY (CALLIGRAPHIA LATINA), J. G. Schwandner. High point of 18th-century ornamental calligraphy. Very ornate initials, scrolls, borders, cherubs, birds, lettered examples. 172pp. 9 x 13. 20475-8 Pa. **$7.95**